5.1 THE SIX TRIGONOMETRIC FUNCTIONS OF θ

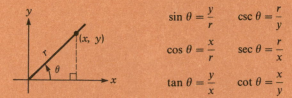

$$\sin \theta = \frac{y}{r} \qquad \csc \theta = \frac{r}{y}$$

$$\cos \theta = \frac{x}{r} \qquad \sec \theta = \frac{r}{x}$$

$$\tan \theta = \frac{y}{x} \qquad \cot \theta = \frac{x}{y}$$

5.1 FUNDAMENTAL IDENTITIES

$$\csc \theta = \frac{1}{\sin \theta} \qquad \cot \theta = \frac{\cos \theta}{\sin \theta}$$

$$\sec \theta = \frac{1}{\cos \theta} \qquad \sin^2 \theta + \cos^2 \theta = 1$$

$$\cot \theta = \frac{1}{\tan \theta} \qquad \tan^2 \theta + 1 = \sec^2 \theta$$

$$\tan \theta = \frac{\sin \theta}{\cos \theta} \qquad \cot^2 \theta + 1 = \csc^2 \theta$$

$$\sin(-\theta) = -\sin \theta \qquad \csc(-\theta) = -\csc \theta$$
$$\cos(-\theta) = \cos \theta \qquad \sec(-\theta) = \sec \theta$$
$$\tan(-\theta) = -\tan \theta \qquad \cot(-\theta) = -\cot \theta$$

5.2 FUNCTIONS OF IMPORTANT ANGLES AND COFUNCTION RELATIONSHIPS

θ	$\sin \theta$	$\cos \theta$	$\tan \theta$
0°	0	1	0
30°	$\frac{1}{2}$	$\frac{\sqrt{3}}{2}$	$\frac{\sqrt{3}}{3}$
45°	$\frac{\sqrt{2}}{2}$	$\frac{\sqrt{2}}{2}$	1
60°	$\frac{\sqrt{3}}{2}$	$\frac{1}{2}$	$\sqrt{3}$
90°	1	0	undefined
180°	0	-1	0
270°	-1	0	undefined

$$\sin \theta = \cos(90° - \theta) \qquad \cos \theta = \sin(90° - \theta)$$
$$\tan \theta = \cot(90° - \theta) \qquad \cot \theta = \tan(90° - \theta)$$
$$\csc \theta = \sec(90° - \theta) \qquad \sec \theta = \csc(90° - \theta)$$

5.3 RADIAN MEASURE

π radians = 180 degrees

$$1 \text{ degree} = \frac{\pi}{180} \text{ radians} \qquad 1 \text{ radian} = \frac{180}{\pi} \text{ degrees}$$

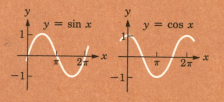

$s \quad s = r\theta \quad (\theta \text{ in radians})$ — Formula for arc length

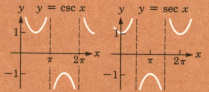

$A = \frac{1}{2}r^2\theta \quad (\theta \text{ in radians})$ — Formula for the area of a sector

$v = r\omega \qquad$ Formula for linear velocity

5.4–5.5 GRAPHS OF THE TRIGONOMETRIC FUNCTIONS

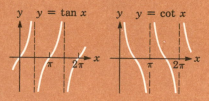

5.4–5.5 AMPLITUDE, PERIOD, AND PHASE SHIFT

Function	Amplitude	Period	Phase Shift
$y = a \sin b(x + c)$	$|a|$	$\frac{2\pi}{b}$	$|c|$
$y = a \cos b(x + c)$	$|a|$	$\frac{2\pi}{b}$	$|c|$
$y = \tan b(x + c)$	—	$\frac{\pi}{b}$	$|c|$

FUNCTIONS AND GRAPHS

Books in the Gustafson and Frisk Series

BEGINNING ALGEBRA, SECOND EDITION
INTERMEDIATE ALGEBRA, SECOND EDITION
ALGEBRA FOR COLLEGE STUDENTS, SECOND EDITION
COLLEGE ALGEBRA, THIRD EDITION
COLLEGE ALGEBRA AND TRIGONOMETRY, SECOND EDITION
PLANE TRIGONOMETRY, SECOND EDITION

FUNCTIONS AND GRAPHS

R. DAVID GUSTAFSON
PETER D. FRISK

Rock Valley College

Brooks/Cole Publishing Company
Monterey, California

Brooks/Cole Publishing Company
A Division of Wadsworth, Inc.

Printed in the United States of America

10 9 8 7 6 5 4 3 2 1

Library of Congress Cataloging-in-Publication Data
Gustafson, R. David (Roy David), [date]
   Functions and graphs.

   Includes index.
   1. Mathematics—1961– .   I. Frisk, Peter D.,
[date].   II. Title.
QA39.2.G86 1987   510   86-26429
ISBN 0-534-07230-5

Sponsoring Editor: *Jeremy Hayhurst*
Marketing Representative: *Jay Honeck*
Editorial Assistant: *Amy Mayfield*
Production Editor: *Candyce L. Cameron*
Production Assistant: *Dorothy Bell*
Manuscript Editor: *Trisha Cain*
Permissions Editor: *Carline Haga*
Interior and Cover Design: *Sharon L. Kinghan*
Cover Illustration: *David Aguero*
Art Coordinator: *Lisa Torri*
Interior Illustration: *Lori Heckelman*
Typesetting: *Syntax International, Singapore*
Cover Printing: *Phoenix Color Corp., Long Island City, NY*
Printing and Binding: *R. R. Donnelley & Sons Co., Crawfordsville, Indiana*

Roy and Lois
Donald and Evelyn

PREFACE

TO OUR COLLEAGUES

Our conversations with mathematics faculty at both two- and four-year institutions have convinced us that we share a common concern. It is not primarily with students who are poorly prepared. Rather, it is with those who understand a concept and are either *unable to adapt* it to the situation at hand or *adapt it inappropriately*. For example, each of us has encountered the student who "completely understands" factoring the difference of two squares, but *cannot adapt* the concept to factor $x^2 + 2xy + y^2 - 9$. Or, the student who would never dream of adding $\frac{1}{2}$ and $\frac{1}{3}$ to get $\frac{1}{5}$, but *adapts* the concept of linearity *inappropriately* to produce the "formula" $x^{-1} + y^{-1} = (x + y)^{-1}$.

How are we to teach students to use the tools of mathematics creatively and appropriately? We have found one classroom technique to be effective: Through copious explanation and example, we demonstrate the use of a mathematical idea while guiding the student in proper technique. We have implemented this philosophy in *Functions and Graphs*. Through discussion and example, we provide new material as well as extensions of previous material applied to new situations. At every new concept, we remind students of the tools already a part of their toolbox and we suggest new ways the tools might be used. In the early stages of the course, when students are most vulnerable to frustration and discouragement, the instruction and examples are supportive and prescriptive: we provide mathematical *intensive care*. Through this care, our students learn concepts, gain confidence, and are less likely to withdraw from the course. As the course progresses, however, the text's language grows more mature, its logic more rigorous, its examples less tightly structured, and its exercises more varied in scope. Our students, no longer in need of intensive care, are then able to use their new mathematical tools with skill, confidence, and imagination.

As teacher of the course, you will appreciate the many features of *Functions and Graphs*:

Solid Mathematics   The development of pre-calculus mathematics—the concepts and skills necessary for successful entry into calculus—is direct and straightforward. The level of mathematical rigor is always in step with the students' growing mathematical sophistication. Every effort has been made to ensure the accuracy of the mathematics and of the answers to the exercises. The book has been critiqued by many reviewers, and both authors and several problem checkers have worked each of the 3900 exercises. Although the exercise sets provide primarily practice and drill, they also contain problems that will challenge even the best of your students.

Calculators   Appropriate use of calculators is assumed throughout the text. We believe that students should learn calculator skills in the mathematics classroom. They will then be prepared to use a calculator in science and business classes and for nonacademic purposes. The directions within the exercise sets clearly indicate which exercises require the use of a calculator.

Accessibility to Students   The text is written for students to read and understand. On the Frey Readability Test, the writing is at the tenth-grade level. The numerous problems in each exercise set parallel the more than 500 worked examples in which author's notes explain many of the steps in the solution process. Students will like the review exercises with *all* answers provided in Appendix I, the functional use of a second color, and the endpapers that list the important facts and formulas developed in the text. They will also appreciate the *Student Solutions Manual*, which contains solutions to the even-numbered exercises. Answers to the odd-numbered problems appear in Appendix I.

Teacher Support   For compiling individualized examinations, a computer-based bank of test items on a full-featured, test-generating, EXP-TEST® system is available to those who adopt the text.

ORGANIZATION AND COVERAGE

The text can be used in a variety of ways. Several chapters are sufficiently independent to allow you to pick and choose topics that are relevant to student and classroom needs. The diagram on the following page indicates how the chapters are interrelated.

Chapter One reviews basic algebra—the real number system and its properties, exponents and radicals, polynomial arithmetic, solutions of equations and inequalities, and absolute value.

The concept of function is introduced in Chapter Two and developed further in Chapter Three. Emphasis is placed on functional notation and graphing. Inverse functions also receive a thorough treatment.

Chapter Four includes exponential functions, logarithms, and many of their applications. Base-*e* and base-10 logarithms are covered in detail. Use of the calculator is encouraged throughout.

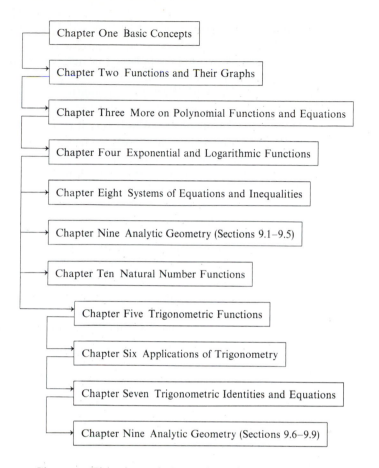

Chapter One Basic Concepts

Chapter Two Functions and Their Graphs

Chapter Three More on Polynomial Functions and Equations

Chapter Four Exponential and Logarithmic Functions

Chapter Eight Systems of Equations and Inequalities

Chapter Nine Analytic Geometry (Sections 9.1–9.5)

Chapter Ten Natural Number Functions

Chapter Five Trigonometric Functions

Chapter Six Applications of Trigonometry

Chapter Seven Trigonometric Identities and Equations

Chapter Nine Analytic Geometry (Sections 9.6–9.9)

Chapters Five through Seven introduce the trigonometric functions as functions of trigonometric angles in standard position. This is how trigonometry developed historically and we are convinced that this approach makes sense to students. However, we make the transition from angle domains to real number domains very early in the discussion. Again, calculators are emphasized throughout.

Chapter Eight develops techniques for solving systems of linear equations. Matrix methods are developed and some matrix algebra is presented. Linear programming in two variables is developed and the topic of partial fractions is introduced as an application of systems of linear equations.

Chapter Nine develops the basic forms of the equations of the conic sections and provides the opportunity for graphing these equations. Solutions of simultaneous second-degree equations are obtained both graphically and algebraically. Polar forms of the conics, polar equations, and parametric equations are also covered.

Chapter Ten introduces proof by mathematical induction, discusses infinite series, and concludes with the binomial theorem.

TO THE STUDENT

We have tried to write a text that you can read and understand. We have provided an extensive number of worked examples for you to study and parallel exercises for you to work. Read the explanations carefully because they provide insight into not only *how*, but also *why* the solution proceeds as it does. The *Student Solutions Manual*, available through your bookstore, provides complete solutions to all of the even-numbered exercises.

The material presented here will be of value to you in later years. Therefore, we suggest that you keep your text after completing this course. It is a durable and dependable reference that will keep at your fingertips the material you have learned. It may also serve you well in a calculus course, for review or reference.

We wish you well.

ACKNOWLEDGMENTS

We are grateful to the following people who reviewed the manuscript in its various stages. Each of them had valuable suggestions that have been incorporated into this text. They are Nancy Angle, Colorado School of Mines; Sam Buchanan, University of Central Arkansas; Paul Campbell, Beloit College; Donald Devine, Western Illinois University; Gerald Flynn, SUNY-Farmingdale; Jon Scott, Montgomery College; Lynn Tooley, Bellevue Community College; and Gloria White, Oberlin College.

We wish to thank Jeanne Wyatt for her help in preparing the manuscript. We also thank the staff at Brooks/Cole, especially Jeremy Hayhurst, Craig Barth, Candy Cameron, Dorothy Bell, Trisha Cain, Sharon Kinghan, Lisa Torri, and Lori Heckelman for all their help.

R. David Gustafson
Peter D. Frisk

CONTENTS

FUNCTIONS AND GRAPHS

1

BASIC CONCEPTS

In a technical society, mathematics is essential. It is basic to engineering and all of the sciences. Its many applications include business and economics, medicine, and even politics. As you study the material in this course, you will see and appreciate the power and diversity of mathematical methods.

Because the concept of *number* is basic to all of mathematics, we begin by discussing various subsets of the set of real numbers.

1.1 THE REAL NUMBER SYSTEM

A **set** is a collection of objects. To denote a set, we often use braces to enclose a list of its **members** or **elements**. For example, the notation

$$\{a, b, c\}$$

denotes the set with elements a, b, and c. To indicate that b is an element of this set, we write

$$b \in \{a, b, c\} \qquad \text{Read } \in \text{ as "is an element of."}$$

A familiar set of numbers, called the **set of real numbers** and denoted by $\mathscr{R}$, includes elements such as

$$12, \quad 0, \quad -75.2, \quad \frac{3}{5}, \quad -\sqrt{5}, \quad \text{and} \quad \pi \text{ (pi, approximately 3.14)}$$

Certain subsets of $\mathscr{R}$ have special names. For example,

$$\{1, 2, 3, \ldots\}$$

is the **set of positive integers** or the **set of natural numbers**, and the set

$$\{-1, -2, -3, \ldots\}$$

is the **set of negative integers**. The positive and negative integers together with 0 form the set of **integers**, often denoted as **J**.

$$\mathbf{J} = \{\ldots, -3, -2, -1, 0, 1, 2, 3, \ldots\}$$

Integers such as -10 and 6 that can be divided by 2 without a remainder are called **even integers**. The other integers, such as -15 and 11, are called **odd integers**.

1

A positive integer other than 1 that can be divided only by 1 and itself is called a **prime number**. The prime numbers form the set

$$\{2, 3, 5, 7, 11, 13, 17, 19, 23, \ldots\}$$

Positive integers other than 1 that are not prime are called **composite numbers**. The composite numbers form the set

$$\{4, 6, 8, 9, 10, 12, 14, 15, 16, \ldots\}$$

Note that 1 is neither a prime number nor a composite number.

A real number is called a **rational number** if it can be written as a **fraction** $\frac{a}{b}$, where a and b are integers and $b \neq 0$. For example, the real number 3.25 is a rational number because it can be written as the fraction $\frac{13}{4}$. The set of rational numbers is denoted by the letter **Q**.

$$\mathbf{Q} = \{x : x = \frac{a}{b}, \text{ where } a \text{ and } b \text{ are integers and } b \neq 0\}$$

The previous notation, called **set-builder notation**, indicates the set of all numbers x such that x is the quotient of two integers and the denominator is not 0.

It can be shown that any rational number can be written as a decimal that either repeats or terminates. A real number such as π or $\sqrt{2}$ that cannot be written as a repeating or terminating decimal is called an **irrational number**. The set of all irrational numbers is denoted by the letter **H**. Because every real number is either a rational number or an irrational number, the **union** of the sets **Q** and **H** is the set of all real numbers. In symbols we write

$$\mathbf{Q} \cup \mathbf{H} = \mathscr{R}$$

Also, no real number is both a rational and an irrational number. Thus, the **intersection** of sets **Q** and **H** is the **empty set**, denoted by $\varnothing$. We say that the sets **Q** and **H** are **disjoint**, and we write

$$\mathbf{Q} \cap \mathbf{H} = \varnothing$$

Real numbers such as -98, $-\frac{7}{2}$, and $-\pi$ are called **negative numbers**, and we say that their sign is negative. Real numbers such as 35, $\frac{7}{8}$, and $+\sqrt{2}$ are called **positive numbers**, and their sign is positive. *The number 0 is neither positive nor negative.*

Example 1 **a.** 6 is a positive number, a natural number, an even integer, a composite number, a rational number, and a real number.

b. $\frac{14}{13}$ is a rational number, a real number, and a positive number.

c. 0 is an integer, an even integer, a rational number, and a real number. ∎

For every real number a there exists a single real number $-a$ such that

$$a + (-a) = -a + a = 0$$

The numbers a and $-a$ are called **negatives** or **additive inverses** of each other. Because the negative of $-a$, denoted as $-(-a)$, must be a, we have

$$-(-a) = a$$

If a represents a positive number, then $-a$ represents a negative number. If, however, a is a *negative* number, then $-a$ must be a *positive* number. The **absolute value of a real number a** is either a itself or $-a$, *whichever is nonnegative.*

Definition.   For any real number a, the **absolute value of a**, denoted by $|a|$, is

$$|a| = \begin{cases} a & \text{if } a \text{ is positive or } 0 \\ -a & \text{if } a \text{ is negative} \end{cases}$$

Example 2   **a.** $|5| = 5$ **b.** $|-7| = -(-7)$ **c.** $|0| = 0$ **d.** $-|-13| = -(13)$
$\qquad\qquad\qquad\qquad\qquad\qquad\quad = 7 \qquad\qquad\qquad\qquad\qquad\qquad = -13$ ∎

If x and y represent real numbers, their **sum** is the real number denoted by $x + y$. If x and y have the same sign, we find their sum by adding their absolute values and using their common sign. If x and y have different signs, we find their sum by subtracting their absolute values (the smaller from the larger) and using the sign of the number with the greater absolute value. Note that, for any real number x,

$$x + 0 = 0 + x = x$$

Because adding 0 to x gives x, 0 is called the **additive identity**.

The **difference** obtained when y is subtracted from x is the real number denoted by $x - y$. To subtract y from x, we add the negative of y to x. Thus,

$$x - y = x + (-y)$$

Example 3   **a.** $-3 + (-5) = -[|-3| + |-5|] = -[3 + 5] = -8$

b. $4 + (-10) = -[|-10| - |4|] = -[10 - 4] = -6$

c. $-5 + 6 = +[|6| - |-5|] = +[6 - 5] = 1$

d. $11 - (-12) = 11 + [-(-12)] = 11 + 12 = 23$ ∎

The **product** of two real numbers x and y is the real number denoted by $x \cdot y$, $(x)(y)$, or just xy. The numbers x and y are called **factors** of the product. If a, b, and c are integers and $a = bc$, then b and c are called **divisors** of a, and a is said to be **divisible** by both b and c. Note that, for any real number x,

$$x \cdot 1 = 1 \cdot x = x$$

Because multiplying x by 1 gives x, 1 is called the **multiplicative identity**.

For any nonzero real number x, there is a unique real number $\frac{1}{x}$ such that

$$x\left(\frac{1}{x}\right) = \left(\frac{1}{x}\right)x = 1$$

The numbers x and $\frac{1}{x}$ are called **multiplicative inverses** or **reciprocals** of each other.

If $y \neq 0$, the **quotient** obtained when x is divided by y is the real number denoted by $\frac{x}{y}$ or $x \div y$. To divide x by y, we multiply x by the multiplicative inverse of y:

$$\frac{x}{y} = x\left(\frac{1}{y}\right)$$

The product (or quotient) of two real numbers with the same sign is the product (or quotient) of their absolute values. The product (or quotient) of two real numbers with different signs is the negative of the product (or quotient) of their absolute values. If one of the factors of a product is zero, the product is zero.

Example 4

a. $-3(4) = -[|-3||4|] = -[(3)(4)] = -12$

b. $-5(-7) = |-5||-7| = (5)(7) = 35$

c. $\dfrac{14}{-7} = -\dfrac{|14|}{|-7|} = -\dfrac{14}{7} = -2$

d. $\dfrac{-36}{-9} = \dfrac{|-36|}{|-9|} = \dfrac{36}{9} = 4$ ■

Many properties of negatives are summarized as follows:

If x and y denote real numbers, then

1. $(-1)x = -x$
2. $-(-x) = x$
3. $(-x) + (-y) = -(x + y)$
4. $(-x)y = x(-y) = -xy$
5. $(-x)(-y) = xy$

To avoid ambiguity, arithmetic operations are performed in the following order: Unless parentheses indicate otherwise, we perform multiplications and/or divisions in order from left to right. We then perform additions and/or subtractions in order from left to right. In a fraction, we simplify the numerator and the denominator separately and then perform the division, if possible.

Example 5   If $a = 2$, $b = -3$, and $c = -5$, evaluate $\dfrac{ab + 3c}{b(c - a)}$.

Solution Substitute values for each variable and simplify.

$$\frac{ab + 3c}{b(c - a)} = \frac{2(-3) + 3(-5)}{-3(-5 - 2)}$$

$$= \frac{-6 + (-15)}{-3(-7)}$$

$$= \frac{-21}{21}$$

$$= -1$$

The following properties govern the arithmetic of real numbers:

Let x, y, and z be real numbers. If there are no divisions by 0, then

The Closure Properties: $x + y$, $x - y$, xy, and $\frac{x}{y}$ are real numbers.

The Commutative Properties: $x + y = y + x$ $\qquad$ $xy = yx$

The Associative Properties: $x + (y + z) = (x + y) + z$ $\quad$ $x(yz) = (xy)z$

The Distributive Property: $x(y + z) = xy + xz$

The Identity Properties: $x + 0 = 0 + x = x$ $\qquad$ $x \cdot 1 = 1 \cdot x = x$

The Inverse Properties: $x + (-x) = -x + x = 0$ $\qquad$ $x\left(\frac{1}{x}\right) = \frac{1}{x}(x) = 1$

Properties of Equality: If $x = y$, then

$$x + z = y + z \qquad\qquad x - z = y - z$$

$$xz = yz \qquad\qquad \frac{x}{z} = \frac{y}{z}$$

Another useful property of real numbers is called the **zero-factor theorem**.

The Zero-Factor Theorem. If x and y represent real numbers, then

$$xy = 0 \qquad \text{if and only if} \qquad x = 0 \quad \text{or} \quad y = 0$$

Because of the words *if and only if*, the zero-factor theorem is equivalent to both a statement and its converse. Thus, the zero-factor theorem implies each of the following statements:

1. If $xy = 0$, then $x = 0$ or $y = 0$.
2. If $x = 0$ or $y = 0$, then $xy = 0$.

If a and b are real numbers and the difference $a - b$ is positive, we say that **a is greater than b** and write $a > b$. If $a > b$, we can also say that **b is less than a** and write $b < a$. The expression $a \geq b$ indicates that **a is greater than or equal to b**, and the expression $a \leq b$ indicates that **a is less than or equal to b**. If a and b are real numbers, one and only one of the following statements is true:

$$a < b \quad \text{or} \quad a > b \quad \text{or} \quad a = b$$

Graphing Subsets of Real Numbers

Real numbers can be represented as points on the **number line**, often called the **real line**. To do so, we choose some point on the line as the **origin**, mark off equal distances to the right and to the left of the origin, and label the points as in Figure 1–1. The point on the number line corresponding to the real number 1.5, for example, is midway between the points labeled 1 and 2. To every real number there corresponds exactly one point, called its **graph**, on the number line, and to each point there corresponds exactly one real number, called its **coordinate**.

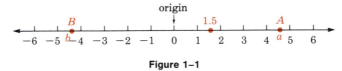

Figure 1–1

Suppose that point A is the graph of a on the number line and B is the graph of b, as in Figure 1–1. Then $a > b$ if and only if A lies to the right of B. If $a = b$, then A and B must be the same point. The *positive* numbers are coordinates of points to the *right* of the origin, and the *negative* numbers are coordinates of points to the *left*.

Graphs of sets of real numbers are often portions of the number line called **intervals**. Figure 1–2 shows the graph of the real numbers between -2 and 4. This interval, which does not include its endpoints, is called an **open interval** and is denoted by the **inequality** $-2 < x < 4$, or more briefly as $(-2, 4)$. The parentheses indicate that the endpoints are not included.

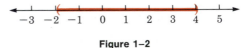

Figure 1–2

The interval shown in Figure 1–3 includes its endpoints and is called a **closed interval**. It is denoted by the inequality $-1 \leq x \leq 2$, or by $[-1, 2]$. The brackets indicate that both endpoints are included.

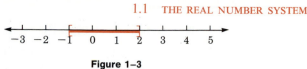

Figure 1–3

The **half-open interval** shown in Figure 1–4 is described by the inequality $1 \le x < 4$, or by $[1, 4)$.

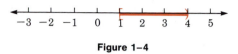

Figure 1–4

Example 6 If $A = (2, 4)$ and $B = [3, 5)$, find the graph of **a.** $A \cup B$ and **b.** $A \cap B$.

Solution **a.** The union of intervals A and B is the set of all real numbers that are elements of *either A or B or both*. Thus, $A \cup B = (2, 4) \cup [3, 5) = (2, 5)$. The graph of this interval is shown in Figure 1–5a.

b. The intersection of intervals A and B is the set of all real numbers that are elements of *both A and B*. Thus, $A \cap B = (2, 4) \cap [3, 5) = [3, 4)$. The graph of this interval is shown in Figure 1–5b.

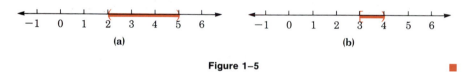

(a)   (b)

Figure 1–5

On the number line $|a|$ represents the distance between the point with coordinate a and the origin. In Figure 1–6 the coordinate of A is 3 and the coordinate of B is -3. Note that both A and B are 3 units from the origin and that $|3| = |-3| = 3$.

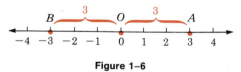

Figure 1–6

In Figure 1–7 the distance $d(CD)$ between the point C with coordinate -2 and point D with coordinate 4 is 6 units. This distance is the difference obtained by subtracting the coordinate of C from the coordinate of D. If we use absolute value notation, the order of the subtraction is not important. The distance is

$$d(CD) = |4 - (-2)| = |-2 - 4| = 6$$

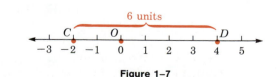

Figure 1–7

This discussion suggests the following definition.

> **Definition.** If a is the coordinate of point A on the number line and b the coordinate of point B, then the **distance between A and B** is
> $$d(AB) = |b - a| = |a - b|$$

Example 7   On the number line the coordinate of point A is 3 and the coordinate of point B is -2. The origin is point O. Find **a.** $d(BO)$ and **b.** $d(AB)$.

Solution   **a.** $d(BO) = |0 - (-2)| = |2| = 2$

   b. $d(AB) = |-2 - 3| = |-5| = 5$ ■

■ Exercise 1.1

In Exercises 1–10 classify each number as a positive or a negative integer, a prime number or a composite number, an even or an odd integer, and a rational or an irrational number, if possible.

1. 3   **2.** -7 **3.** 0   **4.** 2

5. $\dfrac{2}{3}$ **6.** $\dfrac{6}{3}$ **7.** -0.25 **8.** π

9. 111   **10.** 113

In Exercises 11–16 evaluate each expression.

11. $|17|$ **12.** $|-19|$ **13.** $|-17 - (-12)|$

14. $|-15 - (-19)|$ **15.** $|2 - \pi|$ **16.** $|4 - \pi|$

In Exercises 17–22 let $a = 3$, $b = -2$, $c = -1$, and $d = 2$. Evaluate each expression.

17. $ab + cd$ **18.** $d(b + a)$ **19.** $\dfrac{ac - bd}{cd - ad}$

20. $\dfrac{bc - ad}{bd + ac}$ **21.** $\dfrac{3a - 2b + c}{3(a + 2b)}$ **22.** $\dfrac{5d(b - 2c)}{3c(d - 5)}$

In Exercises 23–32 indicate the property of the real numbers that justifies each statement.

23. $3 + 7$ is a real number.   **24.** $5 + 6 = 6 + 5$

25. $9 \cdot 8$ is a real number.   **26.** $a(bc) = (bc)a$

27. $a + (b + c) = (a + b) + c$ **28.** $7(2 + 5) = 7 \cdot 7$

29. $7(2 + 5) = 7 \cdot 2 + 7 \cdot 5$ **30.** If $19x = 0$, then $x = 0$.

31. $23.5 + (-23.5) = 0$ **32.** $2(0.5) = 1$

In Exercises 33–36 graph each interval on the number line.

33. $-3 < x < 5$ **34.** $2 \le x < \dfrac{7}{2}$ **35.** $(-4, 1]$ **36.** $\left[0, \dfrac{2}{3}\right]$

In Exercises 37–40 let A = (−2, 3), B = [0, 2), and C = (0, 4), where A, B, and C are intervals. Graph each indicated set.

37. $A \cup C$ **38.** $B \cap C$ **39.** $A \cap B$ **40.** $A \cup B$

In Exercises 41–44 the coordinate of point A on the number line is −3, point B is 2, point C is 5, and O is the origin. Find the indicated distance.

41. $d(OA)$ **42.** $d(AB)$ **43.** $d(AC)$ **44.** $d(OC)$

The properties of real numbers discussed in this section can be used to prove other properties of real numbers. For example, consider the proof of the following theorem.

Theorem.   If $a + b = 0$, then $a = -b$.

Proof

$a + b = 0$	Given.
$(a + b) + (-b) = 0 + (-b)$	Addition property of equality: Add $-b$ to both sides.
$(a + b) + (-b) = -b$	0 is the additive identity.
$a + [b + (-b)] = -b$	Addition is associative.
$a + 0 = -b$	The sum of additive inverses is 0.
$a = -b$	0 is the additive identity. $\square$

In Exercises 45–50 prove each theorem.

45. If $a \neq 0$ and $a \cdot a = a$, then $a = 1$.   **46.** If $a \cdot a = 0$, then $a = 0$.

47. If $a - b = 0$, then $a = b$.   **48.** If $a \neq 0$ and $ab = a$, then $b = 1$.

49. If $a = -a$, then $a = 0$.   **50.** $a(x - y) = ax - ay$

Prove the zero-factor theorem by proving the statements in Exercises 51–52.

51. If $ab = 0$ and $a \neq 0$, then $b = 0$.   **52.** If $a = 0$, then $ab = 0$.

1.2 EXPONENTS AND RADICALS

Positive integer exponents indicate repeated multiplication.

Definition.   If x is a real number and n is a positive integer, then

$$x^n = \underbrace{x \cdot x \cdot x \cdots x}_{n \text{ factors of } x}$$

In the exponential expression x^n, the number x is called the **base** and n is called an **exponent**. The expression x^n is called the **nth power of x**. If $x \neq 0$, we can extend the definition of exponents to nonpositive exponents.

Definition.   If $x \neq 0$, then

$$x^0 = 1 \quad \text{and} \quad x^{-n} = \frac{1}{x^n}$$

Example 1   If $x \neq 0$, then

a. $3x^0 + (3x)^0 = 3(1) + 1 = 4 \qquad (x \neq 0)$

b. $3^{-2} + 3x^{-2} = \frac{1}{3^2} + 3\left(\frac{1}{x^2}\right) = \frac{1}{9} + \frac{3}{x^2} \qquad (x \neq 0)$ ∎

The following properties of exponents can be proved by using the definitions of exponents and the properties of real numbers.

If x and y are real numbers, m and n are integers, and there are no divisions by 0, then

$$x^m x^n = x^{m+n} \qquad \frac{x^m}{x^n} = x^{m-n} \qquad (xy)^n = x^n y^n$$

$$\left(\frac{x}{y}\right)^n = \frac{x^n}{y^n} \qquad (x^m)^n = x^{mn} \qquad \left(\frac{x}{y}\right)^{-n} = \left(\frac{y}{x}\right)^n$$

Example 2   **a.** $x^5 x^7 = x^{5+7} = x^{12}$

b. $x^3 x^5 x^7 = (x^3 x^5)x^7 = x^8 x^7 = x^{15}$

c. $\frac{y^5}{y^7} = y^{5-7} = y^{-2} = \frac{1}{y^2} \qquad (y \neq 0)$

d. $(y^3)^4 = y^{(3)(4)} = y^{12}$

e. $\left(\frac{2}{5}\right)^{-3} = \left(\frac{5}{2}\right)^3 = \frac{5^3}{2^3} = \frac{125}{8}$

f. $-(x^2 y)^4 = -(x^2)^4 y^4 = -x^8 y^4$

g. $\left(-\frac{5x^2 y}{z^3}\right)^2 = (-1)^2 \frac{5^2 (x^2)^2 y^2}{(z^3)^2}$

$$= \frac{25x^4 y^2}{z^6} \qquad (z \neq 0)$$ ∎

When simplifying expressions involving exponents, find the power of each base first. Then follow the priority rules for the order of operations. Thus, $5 \cdot 2^3$ means $5 \cdot 8$, or 40. However, $(5 \cdot 2)^3$ means 10^3, or 1000.

Example 3   If $a = 2$, $b = -3$, and $c = -1$, evaluate $\dfrac{4b^2 + 2c}{ac + (2c)^2}$.

Solution

$$\frac{4b^2 + 2c}{ac + (2c)^2} = \frac{4(-3)^2 + 2(-1)}{2(-1) + [2(-1)]^2}$$

$$= \frac{4(9) + 2(-1)}{2(-1) + (-2)^2}$$

$$= \frac{36 - 2}{-2 + 4}$$

$$= \frac{34}{2}$$

$$= 17$$

∎

Radical Expressions

If x and y are real numbers and $y^2 = x$, then y is called a **square root of x**. Because $8^2 = 64$ and $(-8)^2 = 64$, for example, both 8 and -8 are square roots of 64. The positive square root, 8, is called the **principal square root** of 64. In general,

---

Definition.   If x and y are real numbers, then y is a **square root** of x if and only if $y^2 = x$.

If x is a positive real number, the **principal square root of x**, denoted as $\sqrt{x}$, is the positive square root of x.

The principal square root of 0 is 0: $\sqrt{0} = 0$.

The expression $\sqrt{x}$ is called a **radical expression**, the symbol $\sqrt{}$ is called a **radical sign**, and the number under the radical sign is called a **radicand**.

Note that square roots of negative numbers are not real numbers, because no real number has a negative square. If x is nonnegative, however, then $\sqrt{x}$ is a real number, and $\sqrt{x} \geq 0$.

The negative square root of a positive number x is denoted by the symbol $-\sqrt{x}$. Thus,

$$-\sqrt{81} = -9 \qquad \text{and} \qquad -\sqrt{225} = -15$$

If x is any real number, the expression x^2 represents a nonnegative number and $\sqrt{x^2}$ represents its principal square root. If $x \geq 0$, then $\sqrt{x^2}$ is x because x is nonnegative. However, if $x < 0$, then $\sqrt{x^2}$ is $-x$ because $-x$ is positive. Since x can be any real number in this discussion, we must use absolute value symbols to guarantee that the principal square root of x^2 is a nonnegative number. Thus, if x is any real number, then

$$\sqrt{x^2} = |x|$$

Note that $\sqrt{x^4} = x^2$. Absolute value symbols are not needed here because x^2 is nonnegative.

Example 4   If x can be any real number, then

a. $\sqrt{49x^2} = |7x| = 7|x|$

b. $\sqrt{81x^4y^2} = |9x^2y| = 9x^2|y|$

c. $\sqrt{36x^4y^8} = 6x^2y^4$ ∎

Numbers such as $1, 4, 9, \frac{4}{9}$, and $\frac{49}{36}$ are called **perfect squares** because their square roots are rational numbers. The square root of a positive number that is not a perfect square is an irrational number.

The concept of *square root* can be extended to *n*th roots, where n is a positive integer.

> **Definition.**   If x and y are real numbers and n is a positive integer, then y is an **nth root of x** if and only if $y^n = x$.

The **principal *n*th root of *x*** is denoted as $\sqrt[n]{x}$. In the notation $\sqrt[n]{x}$, the number n is called the **index** of the radical, and the radical is said to be of **order *n***. If $n = 3$, the radical is called a **cube root**. If the index is not written, it is understood to be 2.

The possibilities that can occur when considering the radical $\sqrt[n]{x}$ are summarized as follows.

> If x is a positive real number and n is a positive integer, then $\sqrt[n]{x}$ is positive.
> If x is negative and n is a positive *odd* integer, then $\sqrt[n]{x}$ is negative.
> If x is negative and n is a positive *even* integer, then $\sqrt[n]{x}$ is not a real number.
> The *n*th root of 0 is 0: $\sqrt[n]{0} = 0$.

Example 5   a. $\sqrt[5]{32} = 2$ Note that $2^5 = 32$ and both 32 and 2 are positive.

b. $\sqrt[3]{-64} = -4$ Note that $(-4)^3 = -64$, both -64 and -4 are negative, and the index is odd.

c. $\sqrt[4]{-64}$ is not a real number. Note that -64 is negative and the index is even.   ∎

The following facts about radicals can be proved.

If x is a real number and n is a positive integer, then

$$(\sqrt[n]{x})^n = x$$

If $x \geq 0$ and n is a positive integer, or if $x < 0$ and n is a positive *odd* integer, then

$$\sqrt[n]{x^n} = x$$

If x is any real number and n is a positive *even* integer, then

$$\sqrt[n]{x^n} = |x|$$

If the indicated roots are real numbers, then

$$\sqrt[n]{xy} = \sqrt[n]{x}\,\sqrt[n]{y}$$

$$\sqrt[n]{\frac{x}{y}} = \frac{\sqrt[n]{x}}{\sqrt[n]{y}} \qquad (y \neq 0)$$

$$\sqrt[m]{\sqrt[n]{x}} = \sqrt[n]{\sqrt[m]{x}} = \sqrt[mn]{x}$$

Example 6   **a.** $(\sqrt[3]{5})^3 = 5$ **b.** $(\sqrt[3]{-5})^3 = -5$

c. $(\sqrt[7]{-4})^7 = -4$ **d.** $\sqrt[6]{(-4)^6} = |-4| = 4$ ■

The properties of radicals can be used to simplify many expressions.

Example 7   **a.** $\sqrt[3]{8x^3} = \sqrt[3]{8}\,\sqrt[3]{x^3} = 2x$

b. $\sqrt[4]{81x^4y^{-8}} = \sqrt[4]{81}\,\sqrt[4]{x^4}\,\sqrt[4]{y^{-8}} = 3|x|y^{-2} = \dfrac{3|x|}{y^2}$

c. $\sqrt[6]{8} = \sqrt[2]{\sqrt[3]{8}} = \sqrt{2}$ ■

Example 8   **a.** $\sqrt{900x^6y^4} = \sqrt{900}\,\sqrt{x^6}\,\sqrt{y^4}$

$\qquad\qquad = 30|x^3|y^2$

b. $\sqrt[3]{216x^9y} = \sqrt[3]{216}\,\sqrt[3]{x^9}\,\sqrt[3]{y}$

$\qquad\qquad = 6x^3\sqrt[3]{y}$

c. $\sqrt{50} + \sqrt{200} = \sqrt{25 \cdot 2} + \sqrt{100 \cdot 2}$

$\qquad\qquad\quad = \sqrt{25}\sqrt{2} + \sqrt{100}\sqrt{2}$

$\qquad\qquad\quad = 5\sqrt{2} + 10\sqrt{2}$

$\qquad\qquad\quad = (5 + 10)\sqrt{2}$

$\qquad\qquad\quad = 15\sqrt{2}$

d. $3\sqrt[4]{x^9 y} + x^2 \sqrt[4]{xy} = 3\sqrt[4]{x^8 xy} + x^2\sqrt[4]{xy}$

$\qquad\qquad\qquad = 3\sqrt[4]{x^8}\sqrt[4]{xy} + x^2\sqrt[4]{xy}$

$\qquad\qquad\qquad = 3x^2\sqrt[4]{xy} + 1x^2\sqrt[4]{xy}$

$\qquad\qquad\qquad = (3+1)x^2\sqrt[4]{xy}$

$\qquad\qquad\qquad = 4x^2\sqrt[4]{xy}$ ∎

A fraction with a radical in its denominator can be rewritten as an equivalent fraction with no radical in its denominator. The process to use is called **rationalizing the denominator**.

Example 9 Each of the following expressions is simplified by multiplying both the numerator and denominator by an expression that will free the denominator from radicals.

a. $\dfrac{1}{\sqrt{3}} = \dfrac{1\sqrt{3}}{\sqrt{3}\sqrt{3}}$

$\qquad = \dfrac{\sqrt{3}}{3}$ By definition $\sqrt{3}\sqrt{3} = (\sqrt{3})^2 = 3$.

b. $\sqrt[5]{\dfrac{3}{8x}} = \dfrac{\sqrt[5]{3}}{\sqrt[5]{8x}}$

$\qquad = \dfrac{\sqrt[5]{3}\sqrt[5]{4x^4}}{\sqrt[5]{8x}\sqrt[5]{4x^4}}$

$\qquad = \dfrac{\sqrt[5]{3\cdot 4x^4}}{\sqrt[5]{32x^5}}$

$\qquad = \dfrac{\sqrt[5]{12x^4}}{2x}$ $(x \neq 0)$ ∎

Fractional Exponents

Fractional exponents are related to radicals. Consider the symbol $3^{1/2}$. Because we would like fractional exponents to obey the rules of integral exponents, $(3^{1/2})^2$ should be found by multiplying exponents:

$$(3^{1/2})^2 = 3^{(1/2)2} = 3^1 = 3$$

Because $(\sqrt{3})^2$ and $(3^{1/2})^2$ are both equal to 3, we define $3^{1/2}$ to be $\sqrt{3}$. In general,

Definition. If n is any positive integer and $x \geq 0$ when n is even, then

$$x^{1/n} = \sqrt[n]{x}$$

and, if $x \neq 0$, then

$$x^{-1/n} = \dfrac{1}{\sqrt[n]{x}}$$

We extend the above definition to cover exponential expressions such as $3^{5/4}$ that have fractional exponents with numerators different from 1. To be consistent with the power rule of exponents, $3^{5/4}$ should mean either

$$(3^5)^{1/4}, \text{ which is } \sqrt[4]{3^5}$$

or

$$(3^{1/4})^5, \text{ which is } (\sqrt[4]{3})^5$$

Thus, we make the following definition.

Definition.   If m and n are positive integers and $\sqrt[n]{x}$ is a real number, then

$$x^{m/n} = \sqrt[n]{x^m} = (\sqrt[n]{x})^m$$

and, if $x \neq 0$,

$$x^{-m/n} = \frac{1}{x^{m/n}}$$

Example 10   In the left column, radicals are used to simplify expressions containing fractional exponents. In the right column, the same expressions are simplified by using the rules of exponents.

a. $\begin{aligned}32^{1/5} &= \sqrt[5]{32} \\ &= 2\end{aligned}$ $\begin{aligned}32^{1/5} &= (2^5)^{1/5} \\ &= 2^{5(1/5)} \\ &= 2^1 \\ &= 2\end{aligned}$

b. $\begin{aligned}(-8)^{1/3} &= \sqrt[3]{-8} \\ &= -2\end{aligned}$ $\begin{aligned}(-8)^{1/3} &= [(-2)^3]^{1/3} \\ &= -2^1 \\ &= -2\end{aligned}$

c. $\begin{aligned}(0.01)^{-1/2} &= \frac{1}{\sqrt{0.01}} \\ &= \frac{1}{0.1} \\ &= 10\end{aligned}$ $\begin{aligned}(0.01)^{-1/2} &= (10^{-2})^{-1/2} \\ &= 10^1 \\ &= 10\end{aligned}$

d. $\begin{aligned}25^{3/2} &= (\sqrt{25})^3 \\ &= 5^3 \\ &= 125\end{aligned}$ $\begin{aligned}25^{3/2} &= (5^2)^{3/2} \\ &= 5^{2(3/2)} \\ &= 5^3 \\ &= 125\end{aligned}$

Example 11   Simplify the expression $\left(\dfrac{16y^{1/3}}{x^{4/3}}\right)^{3/2}$.

Solution
$$\left(\frac{16y^{1/3}}{x^{4/3}}\right)^{3/2} = \frac{(16)^{3/2}(y^{1/3})^{3/2}}{(x^{4/3})^{3/2}}$$

$$= \frac{4^3 y^{3/6}}{x^{12/6}}$$

$$= \frac{64y^{1/2}}{x^2}$$

∎

Fractional exponents can be used to simplify many radical expressions. For example,

$$\sqrt[6]{4} = 4^{1/6} = (2^2)^{1/6} = 2^{2/6} = 2^{1/3} = \sqrt[3]{2}$$

Example 12   Simplify **a.** $\sqrt{2}\sqrt[3]{4}$ and **b.** $\dfrac{\sqrt[3]{3}}{\sqrt{5}}$.

Solution   **a.** $\sqrt{2}\sqrt[3]{4} = 2^{1/2}4^{1/3} = 2^{1/2}(2^2)^{1/3} = 2^{1/2}2^{2/3} = 2^{1/2+2/3} = 2^{7/6} = 2 \cdot 2^{1/6} = 2\sqrt[6]{2}$

b. $\dfrac{\sqrt[3]{3}}{\sqrt{5}} = \dfrac{\sqrt[3]{3}\sqrt{5}}{\sqrt{5}\sqrt{5}} = \dfrac{3^{1/3}5^{1/2}}{5} = \dfrac{3^{2/6}5^{3/6}}{5} = \dfrac{(3^2 5^3)^{1/6}}{5} = \dfrac{\sqrt[6]{9 \cdot 125}}{5} = \dfrac{\sqrt[6]{1125}}{5}$

∎

Scientific Notation and Significant Digits

Scientists and engineers often work with numbers that are either very large or very small. These numbers can be written compactly in **scientific notation**.

Definition.   A number is written in **scientific notation** if it is written in the form

$$x \times 10^n$$

where n is an integer and $1 \le x < 10$.

Example 13   Light travels 29,980,000,000 centimeters per second. To write this number in scientific notation, you must express it as a product of a number between 1 and 10 and some power of 10. The number 2.998 lies between 1 and 10. To get the number 29,980,000,000, the decimal point in 2.998 must be moved 10 places to the right. To accomplish this, multiply 2.998 by 10^{10}. Thus,

$$29,980,000,000 = 2.998 \times 10^{10}$$

∎

Example 14   One meter is approximately 0.00062 mile. To write this number in scientific notation, you must express it as a product of a number between 1 and 10 and some power of 10.

The number 0.00062 may be obtained by moving the decimal point in 6.2 four places to the left. To accomplish this, multiply 6.2 by $\frac{1}{10^4}$, or 10^{-4}. Thus,

$$0.00062 = 6.2 \times 10^{-4}$$ ■

Example 15   Express 5.73×10^{-3} in standard notation.

Solution   The factor of 10^{-3} indicates that 5.73 is to be divided by three factors of 10. Because each division by 10 moves the decimal point one place to the left, you must move the decimal point in 5.73 three places to the left.

$$5.73 \times 10^{-3} = 0.00573$$ ■

Numbers in applied problems are often obtained by measurement and are only approximations to exact values. *The results of a calculation involving approximate values cannot be more accurate than the least accurate of the data* and must be **rounded off** to an appropriate number of decimal places.

Definition.   If a number y is written in scientific notation as $y = x \times 10^n$, and x is rounded to k decimal places, then y is accurate to $k + 1$ **significant digits**.

It is not always easy to decide how many significant digits a number has. If the number 24,000 is a measurement rounded to the nearest thousand, it would be written in scientific notation as

$$24{,}000 = 2.4 \times 10^4$$

and have 2 significant digits.

If 24,000 is accurate to the nearest unit, it would be written in scientific notation as

$$24{,}000 = 2.4000 \times 10^4$$

and have 5 significant digits.

When approximate numbers are multiplied or divided, the result should be rounded to the same number of *significant digits* as the factor with the least number of significant digits.

When approximate numbers are added or subtracted, the result should be rounded to the same number of *decimal places* as the factor with the least number of decimal places.

Example 16   If 2.3 and 42.576 are approximate numbers, then their product is not 97.9248, as a calculator would suggest. This result must be rounded to 2 significant digits. The answer is 98.

The sum of 2.3 and 42.576 is not 44.876, as a calculator would suggest. This result must be rounded to 1 decimal place. The answer is 44.9   ■

━━ Exercise 1.2 ━━━━━━━━━━━━━━━━━━━━━━━━━━━━━━━━━━━━━━

In Exercises 1–20 simplify each expression. Express all answers with positive exponents only. Assume that all variables are restricted to those numbers for which the expression is defined.

1. $(5x)^0$

2. $5x^0$

3. $(-x)^3$

4. $(-x)^4$

5. x^2x^3

6. $(x^2x)^3$

7. $(a^3a^2)^{-2}$

8. $(b^5b^{-6})^{-1}$

9. $(-3x^2y)^5$

10. $(-2x^{-2}y)^{-5}$

11. $-[(3x^{-2})^{-1}y^2]^{-2}$

12. $[(-x^2)^{-2}y^{-3}]^{-3}$

13. $\left(\dfrac{x^2}{y^3}\right)^3$

14. $\left(\dfrac{-x^2}{y^2}\right)^5$

15. $\left(\dfrac{3x^2z^{-1}}{z^3}\right)^2$

16. $\left(\dfrac{2x^2y^3}{xy^{-2}z^{-1}}\right)^2$

17. $\left(\dfrac{5x^{-3}y^{-2}}{3x^2y^{-2}}\right)^{-2}$

18. $\left(\dfrac{3x^{-2}y^{-5}}{2x^2y^5}\right)^{-3}$

19. $\left[\dfrac{3(x^2)^3(3y^3)^{-2}}{9x^{10}(3y)^{-2}}\right]^{-2}$

20. $\left[\dfrac{(2y^3)^{-1}(3x^3)^{-2}}{4x^{-5}(3y)^{-2}}\right]^2$

In Exercises 21–26 let $a = 3$, $b = -2$, and $c = -1$. Evaluate each expression.

21. $3a^2 + (2b)^2$

22. $9a^b - b^a$

23. $2^c + b^c$

24. $3(c^b + b^a)$

25. $\dfrac{b^2 - 2(b + c)^a}{a^a - b}$

26. $\dfrac{5(a + c)^{a+b}}{3b^{c-b} - c}$

In Exercises 27–50 evaluate each expression.

27. $\sqrt{49}$

28. $-\sqrt{64}$

29. $\sqrt[3]{-27}$

30. $-\sqrt[4]{81}$

31. $-\sqrt[3]{-8}$

32. $\sqrt[5]{-32}$

33. $-(\sqrt{7})^2$

34. $(-\sqrt{3})^2$

35. $-\sqrt[3]{7^3}$

36. $\sqrt[3]{(-7)^3}$

37. $-(\sqrt[5]{17})^5$

38. $\sqrt[4]{(-13)^4}$

39. $9^{1/2}$

40. $-16^{1/2}$

41. $(-8)^{2/3}$

42. $-1000^{2/3}$

43. $100^{3/2}$

44. $64^{-1/4}$

45. $25^{-1/2}$

46. $64^{-3/2}$

47. $32^{-3/5}$

48. $-9^{-3/2}$

49. $-(27^{-2/3})$

50. $-(-8^{-1/3})$

In Exercises 51–68 assume that all variables are unrestricted. Simplify each expression. Use absolute value symbols as necessary.

51. $\sqrt{x^2y^2}$

52. $\sqrt{y^4z^2}$

53. $\sqrt{4x^2y^4}$

54. $\sqrt{25x^4y^6}$

55. $\sqrt[4]{x^4y^8}$

56. $\sqrt[5]{x^5y^{10}}$

57. $\sqrt[6]{x^6y^{12}}$

58. $\sqrt[3]{-27x^6y^9}$

59. $\sqrt{25x^3}$

60. $\sqrt{8x^4}$

61. $\sqrt[3]{16x^4y^3}$

62. $\sqrt[5]{64x^6y^5}$

63. $-\sqrt[4]{32x^3y^5}$

64. $-\sqrt[7]{-x^6y^7z^8}$

65. $\dfrac{\sqrt[3]{x^6y^3}}{\sqrt[3]{x^3}}$

66. $\dfrac{\sqrt[6]{128x^6y^8}}{\sqrt[6]{x^{12}y^2}}$

67. $\dfrac{\sqrt[4]{81x^{-6}y^8}}{\sqrt[4]{x^{-2}y^4}}$

68. $\dfrac{\sqrt[4]{625x^8y^4}}{\sqrt{16x^4y^2}}$

In Exercises 69–76 assume that all variables represent positive real numbers. Simplify each expression.

69. $2\sqrt[3]{81} + 3\sqrt[3]{24}$

70. $3\sqrt[4]{32} - 2\sqrt[4]{162}$

71. $\sqrt{200x^2} + \sqrt{98x^2}$

72. $\sqrt{128a^3} - a\sqrt{162a}$

73. $2\sqrt{48y^5} - 3y\sqrt{12y^3}$

74. $y\sqrt{112y} + 4\sqrt{175y^3}$

75. $\sqrt[4]{768z^5} + \sqrt[4]{48z^5}$

76. $-2\sqrt[5]{64y^2} + 3\sqrt[5]{-2y^2}$

In Exercises 77–86 assume that all variables represent positive real numbers. Simplify each expression and write all answers without using negative exponents.

77. $(x^{10}y^5)^{3/5}$ **78.** $(a^6b^{12})^{5/6}$ **79.** $(x^9y^{12})^{-2/3}$ **80.** $(r^8s^{16})^{-3/4}$

81. $(r^{-2}s^4)^{1/2}$ **82.** $(u^2v^{-2})^{-1/2}$ **83.** $(32y^{-10}z^5)^{-2/5}$ **84.** $(625a^4b^{-8})^{-3/4}$

85. $\left(\dfrac{125}{64}x^9y^{-6}\right)^{-4/3}$ **86.** $\left(\dfrac{32}{243}x^{-15}y^{10}z^{-5}\right)^{-3/5}$

In Exercises 87–96 assume that all variables are positive real numbers. Rationalize each denominator and simplify.

87. $\dfrac{5}{\sqrt{5}}$ **88.** $\dfrac{10}{\sqrt{2}}$ **89.** $\dfrac{2}{\sqrt{x}}$ **90.** $\dfrac{3}{\sqrt[3]{y}}$

91. $\dfrac{2}{\sqrt[5]{2}}$ **92.** $\dfrac{3}{\sqrt[3]{9}}$ **93.** $\dfrac{5a}{\sqrt[3]{25a}}$ **94.** $\dfrac{16}{\sqrt[3]{16b^2}}$

95. $\sqrt[5]{\dfrac{3x^3}{8y^2}}$ **96.** $\sqrt[4]{\dfrac{y}{9x^3}}$

In Exercises 97–100 write each expression as a single radical.

97. $\sqrt{2}\,\sqrt[3]{2}$ **98.** $\sqrt{3}\,\sqrt[5]{3}$ **99.** $\dfrac{\sqrt{3}}{\sqrt[4]{3}}$ **100.** $\dfrac{\sqrt[3]{5}}{\sqrt[4]{5}}$

101. For what values of x does $\sqrt{x^2} = x$?   **102.** For what values of x does $\sqrt[4]{x^4} = -x$?

In Exercises 103–106 express each number in scientific notation.

103. 23,000   **104.** 0.0000468   **105.** 24.7×10^3 **106.** 0.0037×10^5

In Exercises 107–110 express each number in standard notation.

107. 2.57×10^{-5} **108.** -5.582×10^7 **109.** 0.003×10^5 **110.** $34{,}000 \times 10^{-8}$

In Exercises 111–116 the numbers given are approximations. Perform the indicated operations and round the answers as appropriate.

111. $(7.2)(93.271)$ **112.** $(0.00014)(376.12)$ **113.** $9.237 + 21.1 + 32.32$

114. $37.2 - 16.3791 - 12.25$ **115.** $(4.38)(7.91) - 2.6834$ **116.** $5.371 + (2.1)(5.3)$

1.3 ARITHMETIC OF ALGEBRAIC EXPRESSIONS AND FACTORING

A **monomial** is either a number or the product of a number and one or more variables with positive integer exponents. The number is called the **numerical coefficient**, or simply the **coefficient** of the monomial. For example, x, $-5x^2y$, and -12 are monomials with coefficients of 1, -5, and -12, respectively.

The **degree** of a monomial is the sum of the exponents of its variables. The degree of x is 1, the degree of $-5x^2y$ is 3, and the degree of -12 is 0 (note that $-12 = -12x^0$). All nonzero constants have a degree of 0. The constant 0 does not have a defined degree.

A single monomial or a finite sum of monomials is called a **polynomial**. Each of the monomials in that sum is a **term** of the polynomial. A polynomial with two terms is called a **binomial**, and one with three terms is called a **trinomial**. The **degree of a polynomial** is the degree of the term with highest degree in that polynomial. The only polynomial with no defined degree is the **zero polynomial**: 0.

Example 1   **a.** $3x^2y^3 + 5xy^2 + 7$ is a trinomial of degree 5 because its term of highest degree (its first term) has degree 5.

b. $3xy + 5x^2y$ is a binomial of degree 3.

c. $5x + 3y + \sqrt[3]{3}z - \sqrt{7}$ is a polynomial because its variables have positive integral exponents. It is a first-degree polynomial.

d. $5x^{1/2} + 3\sqrt[5]{y}$ is not a polynomial because its variables do not have positive integral exponents. ■

If two terms have the same variables with the same exponents, they are called **similar** or **like terms**. Because the variables represent real numbers, the distributive property enables us to combine similar terms. For example, the terms in the binomial $3x^2y + 5x^2y$ are similar terms and

$$3x^2y + 5x^2y = (3 + 5)x^2y$$
$$= 8x^2y$$

Polynomials can be added or subtracted by removing parentheses and combining similar terms.

Example 2   Perform the subtraction $3(5x^2 + 2y) - 5(x^2 - 3y)$ and simplify.

Solution   Use the distributive property to remove each set of parentheses. Then use the commutative and associative properties to rearrange and regroup the terms within the polynomial, and simplify the result by combining terms:

$$3(5x^2 + 2y) - 5(x^2 - 3y) = 3 \cdot 5x^2 + 3 \cdot 2y - 5x^2 - 5(-3)y$$
$$= 15x^2 + 6y - 5x^2 + 15y$$
$$= 15x^2 - 5x^2 + 6y + 15y$$
$$= 10x^2 + 21y$$ ■

Polynomials can be multiplied, as the following examples illustrate.

Example 3   Find the product of $3x^2y^3z$ and $5xyz^2$.

Solution   Use the commutative and associative properties of multiplication to regroup and rearrange the factors, and use the properties of exponents to simplify the resulting expression.

$$(3x^2y^3z)(5xyz^2) = 3 \cdot 5 \cdot x^2 \cdot x \cdot y^3 \cdot y \cdot z \cdot z^2$$
$$= 15x^3y^4z^3$$ ■

Example 4   Find the product of $3xy^2$ and $2xy + x^2$.

Solution   Use the distributive property to multiply each term of the binomial by the monomial, and simplify.

$$3xy^2(2xy + x^2) = 3xy^2 \cdot 2xy + 3xy^2 \cdot x^2$$
$$= 6x^2y^3 + 3x^3y^2$$

∎

Example 5   Find the product of $2x + 3$ and $3y - 5$.

Solution   Use the distributive property to multiply each term of $3y - 5$ by the quantity $2x + 3$. Then use the distributive property again to remove parentheses.

$$(2x + 3)(3y - 5) = (2x + 3)3y - (2x + 3)5$$
$$= 6xy + 9y - 10x - 15$$

Observe that the product is simply the result of multiplying each term of the second binomial by each term of the first.

∎

Example 6   Expand $(x + 3)^2$.

Solution   The expression $(x + 3)^2$ indicates the product of two binomials: $(x + 3)(x + 3)$. Multiply each term of the second binomial by each term of the first:

$$(x + 3)^2 = (x + 3)(x + 3)$$
$$= x \cdot x + 3x + 3x + 3 \cdot 3$$
$$= x^2 + 6x + 9$$

Students often make the mistake of interpreting $(x + 3)^2$ as $x^2 + 9$. This is not correct. When squaring binomials, always remember the middle term:

$$(x + 3)^2 = x^2 + 6x + 9$$

∎

Example 7   Find the product of $x + \sqrt{7}$ and $x - \sqrt{7}$.

Solution   Multiply each term of the second binomial by each term of the first, and simplify.

$$(x + \sqrt{7})(x - \sqrt{7}) = x^2 + \sqrt{7}x - \sqrt{7}x - (\sqrt{7})^2$$
$$= x^2 - 7$$

Note that the middle term of this product—the term that would involve x—subtracts out.

∎

The discussion in Examples 6 and 7 suggests the following product formulas:

$$(x + y)^2 = x^2 + 2xy + y^2$$
$$(x - y)^2 = x^2 - 2xy + y^2$$
$$(x + y)(x - y) = x^2 - y^2$$

Example 8   Simplify the fraction $\dfrac{6}{\sqrt{7} + 2}$ by rationalizing the denominator.

Solution   Multiply both the numerator and the denominator of the fraction by $\sqrt{7} - 2$. Because $(\sqrt{7} + 2)(\sqrt{7} - 2) = 7 - 4 = 3$, this multiplication will clear the fraction's denominator of radicals.

$$\frac{6}{\sqrt{7} + 2} = \frac{6(\sqrt{7} - 2)}{(\sqrt{7} + 2)(\sqrt{7} - 2)}$$

$$= \frac{6(\sqrt{7} - 2)}{3}$$

$$= 2(\sqrt{7} - 2) \qquad \blacksquare$$

In calculus it is sometimes necessary to rationalize a numerator.

Example 9   Rationalize the numerator of the fraction $\dfrac{\sqrt{x + h} - \sqrt{x}}{h}$.

Solution $\dfrac{\sqrt{x + h} - \sqrt{x}}{h} = \dfrac{(\sqrt{x + h} - \sqrt{x})(\sqrt{x + h} + \sqrt{x})}{h(\sqrt{x + h} + \sqrt{x})}$

$$= \frac{x + h - x}{h(\sqrt{x + h} + \sqrt{x})}$$

$$= \frac{h}{h(\sqrt{x + h} + \sqrt{x})}$$

$$= \frac{1}{\sqrt{x + h} + \sqrt{x}} \qquad \text{Divide numerator and} \atop \text{denominator by } h. \qquad \blacksquare$$

To divide a polynomial by a monomial, we proceed as in the next example.

Example 10 $\dfrac{4x^3y^2 + 3xy^5 - 12xy}{3x^2y^3} = \dfrac{4x^3y^2}{3x^2y^3} + \dfrac{3xy^5}{3x^2y^3} + \dfrac{-12xy}{3x^2y^3}$

$$= \frac{4x}{3y} + \frac{y^2}{x} - \frac{4}{xy^2} \qquad \blacksquare$$

The division of polynomials is accomplished by using a process similar to that used for long division. We consider the division

$$x + 7 \,\overline{)\, 2x^2 + 11x - 30}$$

The binomial $x + 7$ is called the **divisor**, and the trinomial $2x^2 + 11x - 30$ is called the **dividend**. The answer, called the **quotient**, will appear above the division symbol.

We begin the division process, called the **division algorithm**, by asking "What expression when multiplied by x gives $2x^2$?" The answer is $2x$ because $x \cdot 2x =$

$2x^2$. We place **$2x$** as the first term in the quotient, multiply each term of the divisor by $2x$, subtract, and bring down the -30:

$$\begin{array}{r} \color{red}{2x} \\ x+7{\overline{\smash{\big)}\,2x^2+11x-30}} \\ \underline{2x^2+14x} \\ -3x-30 \end{array}$$

We continue the division by asking the question "What expression when multiplied by x gives $-3x$?" We place the answer, **-3**, as the second term in the quotient, multiply each term of the divisor by -3, and subtract:

$$\begin{array}{r} 2x\color{red}{-\;3} \\ x+7{\overline{\smash{\big)}\,2x^2+11x-30}} \\ \underline{2x^2+14x} \\ -3x-30 \\ \underline{-3x-21} \\ -9 \end{array}$$

Because the degree of the remainder, -9, is less than the degree of the divisor, the division process is completed. You can express the result in *quotient* $+$ $\dfrac{remainder}{divisor}$ form as follows:

$$2x-3+\frac{-9}{x+7}$$

Note that you have subtracted from $2x^2+11x-30$ first $2x$ times $x+7$, then -3 times $x+7$, leaving a remainder of -9.

The division algorithm works most effectively when the terms of both the divisor and the dividend are written with their exponents in descending order.

Example 11 Divide $-3x^3-3+x^5-4x^2$ by x^2-x.

Solution Write the terms of both the dividend and divisor in descending order and leave space for the missing terms of x^4 and x.

$$\begin{array}{r} x^3+x^2-2x\;-6 \\ x^2-x{\overline{\smash{\big)}\,x^5-3x^3-4x^2-3}} \\ \underline{x^5-x^4} \\ x^4-3x^3 \\ \underline{x^4-\;x^3} \\ -2x^3-4x^2 \\ \underline{-2x^3+2x^2} \\ -6x^2 \\ \underline{-6x^2+6x} \\ -6x-3 \end{array}$$

Thus, the result is $x^3+x^2-2x-6+\dfrac{-6x-3}{x^2-x}$.

Factoring Polynomials

To **factor** a polynomial is to write the polynomial as the product of several factors. In this section we will limit the discussion to factoring over the integers. If a polynomial cannot be factored using integers only, that polynomial is called **irreducible** or **prime** over the set of integers.

Example 12   Factor the trinomial $3xy^2 - 6x^2y + 3xy$.

Solution   Note that each term contains a factor of $3xy$. Use the distributive property to factor out that common factor of $3xy$:

$$3xy^2 - 6x^2y + 3xy = \mathbf{3xy} \cdot y - \mathbf{3xy} \cdot 2x + \mathbf{3xy} \cdot 1$$
$$= \mathbf{3xy}(y - 2x + 1)$$

Hence, $3xy^2 - 6x^2y + 3xy$ factors as $3xy(y - 2x + 1)$.

This type of factoring is called **factoring out a common** (monomial) **factor**.   ■

Example 13   Factor the expression $ax + bx - a - b$.

Solution   There is no factor common to all four terms, but the factor x is common to the first two terms, and the factor -1 is common to the last two. Factor out the x and -1, as follows:

$$ax + bx - a - b = x(a + b) - 1(a + b)$$

The common factor $(a + b)$ is present in each term and can be factored out:

$$ax + bx - a - b = x\mathbf{(a + b)} - 1\mathbf{(a + b)}$$
$$= \mathbf{(a + b)}(x - 1)$$

The technique used in this example is called **factoring by grouping**.   ■

Example 14   Factor the expression $49x^2 - 9y^2$.

Solution   Observe that each term is a perfect square and that the given expression is the difference of those squares. The difference of the squares of two quantities factors as the product of two binomials: one binomial is the *sum* of those quantities and the other is their *difference*.

$$49x^2 - 9y^2 = (\mathbf{7x})^2 - (\mathbf{3y})^2$$
$$= (\mathbf{7x + 3y})(\mathbf{7x - 3y})$$

This type of factoring, called **factoring the difference of two squares**, is based on the special product

$$(x + y)(x - y) = x^2 - y^2$$ ■

Example 15   Factor the quadratic trinomial $2x^2 - 5x - 12$.

Solution   If the given trinomial is factorable, it will factor as the product of two binomials of the form $ax + b$ and $cx + d$. Because $2x \cdot x$ gives the first term of the given trinomial, begin by considering this partial factorization:

$$2x^2 - 5x - 12 = (2x + ?)(x + ?)$$

Because the last term of the trinomial is -12, the product of the second terms of the binomial factors must be -12. There are several possibilities, but only one gives the required middle term of $-5x$:

$$2x^2 - 5x - 12 = (2x + 3)(x - 4)$$

This type of problem is called **factoring the general quadratic trinomial**.   ■

We can also factor the **sum of two cubes** and the **difference of two cubes**. Like the difference of two squares, they follow a pattern:

$$x^3 + y^3 = (x + y)(x^2 - xy + y^2)$$
$$x^3 - y^3 = (x - y)(x^2 + xy + y^2)$$

Example 16   Factor the binomial $27x^6 + 64y^3$.

Solution   This binomial is the sum of two cubes and can be factored as follows:

$$27x^6 + 64y^3 = (3x^2)^3 + (4y)^3$$
$$= (3x^2 + 4y)[(3x^2)^2 - (3x^2)(4y) + (4y)^2]$$
$$= (3x^2 + 4y)(9x^4 - 12x^2y + 16y^2)$$ ■

The following examples illustrate some miscellaneous factoring methods.

Example 17   Factor $x^2 - xy - 2y^2 - 2x + 4y$.

Solution   The polynomial factors by grouping. Group and factor the first three terms, factor -2 from the last two terms, and proceed as follows:

$$x^2 - xy - 2y^2 - 2x + 4y = \underbrace{x^2 - xy - 2y^2}_{} \underbrace{- 2x + 4y}_{}$$
$$= (x + y)(x - 2y) - 2(x - 2y)$$
$$= (x + y - 2)(x - 2y)$$ ■

Example 18   Factor the trinomial $z^4 - 3z^2 + 1$.

Solution   An attempt to factor $z^4 - 3z^2 + 1$ as the product of two binomials will fail. One reasonable attempt is $(z^2 - 1)(z^2 - 1)$, but this product is $z^4 - 2z^2 + 1$. However, if you add and subtract z^2, you obtain

$$z^4 - 3z^2 + 1 = z^4 - 3z^2 + z^2 + 1 - z^2$$
$$= (z^4 - 2z^2 + 1) - z^2$$
$$= (z^2 - 1)^2 - z^2$$

The expression $(z^2 - 1)^2 - z^2$ is the difference of two squares and can be factored as

$$= (z^2 - 1 + z)(z^2 - 1 - z)$$

or

$$= (z^2 + z - 1)(z^2 - z - 1)$$ ■

Although the expressions in the next two examples are not polynomials, they can be factored by using the previous techniques.

Example 19   Factor $x^{3/2}y^{1/2} + x^{1/2}y^{3/2}$.

Solution   Factor out the common factor $x^{1/2}y^{1/2}$.

$$x^{3/2}y^{1/2} + x^{1/2}y^{3/2} = x^{1/2}y^{1/2}(x) + x^{1/2}y^{1/2}(y)$$
$$= x^{1/2}y^{1/2}(x + y)$$ ■

Example 20   Factor $y^{-3} - x^{-2}y^{-1}$.

Solution   The factor y^{-3} is the smallest power of y that appears in the two terms of the given expression. Similarly, x^{-2} is the smallest power of x that appears. Factor out $y^{-3}x^{-2}$ and proceed as follows:

$$y^{-3} - x^{-2}y^{-1} = y^{-3}x^{-2}x^2 - y^{-3}x^{-2}y^2$$
$$= x^{-2}y^{-3}(x^2 - y^2)$$
$$= x^{-2}y^{-3}(x + y)(x - y)   \text{Factor the difference of two squares.}$$ ■

Exercise 1.3

In Exercises 1–6 indicate whether the given expression is a polynomial. If it is a polynomial, give its degree and tell if it is a monomial, binomial, or trinomial.

1. $x^2 + 3x + 4$

2. $\sqrt{5}x^3 + 4x^2$

3. $4x^{-3} - 3x^{-2} + 2x + 1$

4. $\sqrt{x}$

5. 0

6. $\dfrac{5}{x} + \dfrac{x}{5} + 5$

In Exercises 7–44 perform the indicated operations and simplify.

7. $(x^5 + 2x^3 + 7) - (x^5 - 2x^3 - 7)$

8. $(3x^7 - 7x^3 + 3) - (7x^7 - 3x^3 + 7)$

9. $2(x^2 + 3x - 1) - 3(x^2 + 2x - 4) + 4$

10. $5(x^3 - 8x + 3) + 2(3x^2 + 5x) - 7$

11. $8(x^2 - 2x + 5) + 4(x^2 - 3x + 2) - 6(2x^2 - 8)$

12. $3x(x - 5) - x(2 + 3x) + 3(x + 2)$

13. $x(x^2 - 1) - x^2(x + 2) - x(2x - 2)$

14. $x(x - 4) - (x^2 + 3) + x(2x + 3)$

15. $(2x^2y^3)(4xy^4)$

16. $(-15a^3b)\left(\dfrac{1}{3}ab^4\right)$

17. $-4rs(r^2 + s^2)$

18. $6u^2v(2uv^3 - y)$

19. $(x - 5)(x + 5)$

20. $(x - 6)(x + 5)$

21. $(3x - 2)(x + 2)$

22. $(2x - 3)(4x + 1)$

23. $(2x + 3)(x - 4)$

24. $(3 - x)(2 + 3x)$

25. $(5x - 1)(2x + 3)$

26. $(-x + 1)(2x - 3)$

27. $(2 - 4x)(3 - 2x)$

28. $(3x - 4)(4x + 3)$

29. $(-2x + 3)(3x + 1)$

30. $(4x + 5)(3x - 4)$

31. $(9x - 1)(x^2 - 3)$

32. $(8x^2 + 1)(x + 2)$

33. $(5x + 2)(x^2 - 1)$

34. $(1 - 2x^2)(x^2 + 3)$

35. $(3x + 9)^2$

36. $(5x^2 - 1)^2$

37. $(3x - 1)^3$

38. $(2x - 3)^3$

39. $(xy^2 - 1)(x^2y + 2)$

40. $(xy - z^2)(xy + z^2)$

41. $(3x + 1)(2x^2 + 4x - 3)$

42. $(2x - 5)(x^2 - 3x + 2)$

43. $(x^2 + x + 1)(x^2 + x - 1)$

44. $(x^2 - x + 1)(x^2 + x + 1)$

In Exercises 45–48 multiply the expressions as you would multiply polynomials.

45. $x^{1/2}(x^{1/2}y + xy^{1/2})$

46. $ab^{1/2}(a^{1/2}b^{1/2} + b^{1/2})$

47. $(a^{1/2} + b^{1/2})(a^{1/2} - b^{1/2})$

48. $(x^{3/2} + y^{1/2})^2$

In Exercises 49–56 rationalize each denominator.

49. $\dfrac{2}{\sqrt{3} - 1}$

50. $\dfrac{1}{\sqrt{5} + 2}$

51. $\dfrac{6}{\sqrt{5} - \sqrt{2}}$

52. $\dfrac{15}{\sqrt{7} + \sqrt{2}}$

53. $\dfrac{x}{x - \sqrt{3}}$

54. $\dfrac{y}{2y + \sqrt{7}}$

55. $\dfrac{y + \sqrt{2}}{y - \sqrt{2}}$

56. $\dfrac{\sqrt{3} - \sqrt{2}}{1 + \sqrt{2}}$

In Exercises 57–58 rationalize each numerator.

57. $\dfrac{\sqrt{x + 3} - \sqrt{x}}{3}$

58. $\dfrac{\sqrt{2 + h} - \sqrt{2}}{h}$

In Exercises 59–66 simplify each fraction by performing a division.

59. $\dfrac{6x^2 + 12x + 9}{3x}$

60. $\dfrac{8x^2y - 12xy^3 + x^4y^4}{4xy^2}$

61. $\dfrac{2x^2 - 19x + 35}{2x - 5}$

62. $\dfrac{2x^2 - 19x + 35}{x - 7}$

63. $\dfrac{x^3 - 2x^2 + 3 - 4x}{x^2 + x - 1}$

64. $\dfrac{x^3 - x - 3x^2 + 8}{-x - 3 + x^2}$

65. $\dfrac{x^5 - 2x^3 - 3x^2 + 9}{x^2 - 2}$

66. $\dfrac{x^5 - 2x^3 - 3x^2 + 9}{x^3 - 3}$

In Exercises 67–112 completely factor each algebraic expression over the set of integers, if possible.

67. $3x - 6$

68. $5y - 15$

69. $8x^2 + 4x^3$

70. $9y^3 + 6y^2$

71. $7xy^2 + 14x^2y$

72. $25y^2z - 15yz^2$

73. $3a^2bc + 6ab^2c + 9abc^2$

74. $5x^3y^3z^3 + 25x^2y^2z^2 - 125xyz$

75. $b(x + y) - a(x + y)$

76. $b(x - y) + a(x - y)$

77. $4a + b - 12a^2 - 3ab$

78. $x^2 + 4x + xy + 4y$

79. $3x^3 + 3x^2 - x - 1$

80. $4x + 6xy - 9y - 6$

81. $4x^2 - 9$

82. $4 - 9r^2$

83. $(x + z)^2 - 25$

84. $x^2 - (y - z)^2$

85. $x^4 - y^4$

86. $1 - y^8$

87. $3x^2 - 12$

88. $18xy^2 - 8x$

89. $x^2 - 2x - 15$

90. $24y^2 + 15 - 38y$

91. $-15 + 2a + 24a^2$

92. $6x^2 + 29x + 35$

93. $35 - x + 6x^2$

94. $12y^2 - 58y - 70$

95. $6x^3 - 23x^2 - 35x$

96. $6x^2 - 11x - 35$

97. $35 - 47x + 6x^2$

98. $x^4 + 2x^2 - 15$

99. $8z^3 - 27$

100. $(x + y)^3 - 64$

101. $64a^6 - y^6$

102. $4x^2 + 6x^2y + 6x + 9xy$

103. $xy - y + x^2 + x - 2$

104. $x^2 - 6x + 9 - 144y^2$

105. $(a + b)^2 - 3(a + b) - 10$

106. $6(u + v)^2 + 11u + 4 + 11v$

107. $x^6 + 7x^3 - 8$

108. $x^2 + ax + 6x + 3a + 9$

109. $x^4 + 3x^2 + 4$

110. $x^4 + 7x^2 + 16$

111. $4a^4 + 1 + 3a^2$

112. $2x^4 + 8$

In Exercises 113–120 factor the indicated monomial from the given expression.

113. $3x + 2$; 2

114. $a + b$; a

115. $x + x^{1/2}$; $x^{1/2}$

116. $2x + \sqrt{2}y$; $\sqrt{2}$

117. $ab^{3/2} - a^{3/2}b$; ab

118. $ab^2 + b$; b^{-1}

119. $x^{1/2}y + x^{-1/2}y^2$; $x^{-1/2}y$

120. $x^{2/3}y^{-2/3} - x^{-1/3}y^{1/3}$; $x^{-1/3}y^{-2/3}$

1.4 ALGEBRAIC FRACTIONS

If $y \neq 0$, the indicated division $\frac{x}{y}$ is called a **fraction**. The number x is called the **numerator** of the fraction, and y is called the **denominator**. Remember that, because division by zero is not defined, *the denominator of a fraction can never be 0.*

Quotients of algebraic expressions are called **algebraic fractions**. If the algebraic expressions in a fraction are polynomials, the fraction is called a **rational expression**. Each of the following fractions is an algebraic fraction, and the third is a rational expression.

$$\frac{2\sqrt{x} + 3xy^2}{3\sqrt{x} + 2} \qquad \frac{5 + \dfrac{x}{y}}{3y - 2} \qquad \frac{5x^2 - 3x}{2x^2 - x + 1}$$

As with all fractions, denominators cannot be 0.

The properties of fractions are summarized as follows.

If a, b, c, and d represent algebraic expressions and no denominators are 0, then

Equality of Fractions: $\dfrac{a}{b} = \dfrac{c}{d}$ if and only if $ad = bc$

The Fundamental Property of Fractions: $\dfrac{ax}{bx} = \dfrac{a}{b}$

Multiplication of Fractions: $\dfrac{a}{b} \cdot \dfrac{c}{d} = \dfrac{ac}{bd}$

Division of Fractions: $\dfrac{a}{b} \div \dfrac{c}{d} = \dfrac{a}{b} \cdot \dfrac{d}{c} = \dfrac{ad}{bc}$

Addition of Fractions with Like Denominators: $\dfrac{a}{b} + \dfrac{c}{b} = \dfrac{a + c}{b}$

A fraction has been **reduced to lowest terms** if all factors common to the numerator and the denominator of that fraction have been removed, or **canceled**. To **simplify a fraction** means to reduce it to lowest terms.

We use the fundamental property of fractions to simplify a fraction. Remember that only factors of the entire numerator may cancel factors of the entire denominator.

Example 1 Simplify the fraction $\dfrac{x^2 - 9}{x^2 - 3x}$.

Solution Factor both the numerator and the denominator and use the fundamental property of fractions to cancel the common factor of $x - 3$.

$$\frac{x^2 - 9}{x^2 - 3x} = \frac{(x + 3)(x - 3)}{x(x - 3)}$$

$$= \frac{(x + 3)(x - 3)}{x(x - 3)}$$

$$= \frac{x + 3}{x}$$

∎

Properties of negatives and quotients are related by the following theorem.

Theorem.   If $b \neq 0$, then

$$\frac{-a}{b} = \frac{a}{-b} = -\frac{a}{b} \qquad \text{and} \qquad \frac{-a}{-b} = \frac{a}{b}$$

Example 2 Simplify the fraction $\dfrac{x^2 - y^2}{y - x}$.

Solution Factor the difference of two squares in the numerator and -1 from the denominator. The fraction can then be simplified.

$$\frac{x^2 - y^2}{y - x} = \frac{(x + y)(x - y)}{-(x - y)} = \frac{x + y}{-1} = -(x + y)$$

∎

Example 3 Find the product of $\dfrac{x^2 - x - 2}{x^2 - 1}$ and $\dfrac{x^2 + 2x - 3}{x^2 - 4x + 4}$.

Solution Multiply the fractions, factor the numerator and the denominator, and cancel all common factors:

$$\frac{x^2 - x - 2}{x^2 - 1} \cdot \frac{x^2 + 2x - 3}{x^2 - 4x + 4} = \frac{(x^2 - x - 2)(x^2 + 2x - 3)}{(x^2 - 1)(x^2 - 4x + 4)}$$

$$= \frac{(x - 2)(x + 1)(x + 3)(x - 1)}{(x + 1)(x - 1)(x - 2)(x - 2)}$$

$$= \frac{x + 3}{x - 2}$$

∎

Example 4   Simplify $\dfrac{2x^2 - 5x - 3}{3x - 1} \div \dfrac{2x^2 + x}{3x} \cdot \dfrac{3x - 1}{x - 3}$.

Solution   Perform the division and multiplication in order from left to right. Then factor and simplify.

$$\frac{2x^2 - 5x - 3}{3x - 1} \div \frac{2x^2 + x}{3x} \cdot \frac{3x - 1}{x - 3}$$

$$= \frac{2x^2 - 5x - 3}{3x - 1} \cdot \frac{3x}{2x^2 + x} \cdot \frac{3x - 1}{x - 3} \qquad \text{Change the division to a multiplication.}$$

$$= \frac{(2x^2 - 5x - 3)(3x)(3x - 1)}{(3x - 1)(2x^2 + x)(x - 3)}$$

$$= \frac{(2x + 1)(x - 3)\,3x(3x - 1)}{(3x - 1)x(2x + 1)(x - 3)}$$

$$= 3 \qquad\qquad\blacksquare$$

Example 5   Combine and simplify:

$$\frac{x - 2}{x^2 - 1} - \frac{x + 3}{x^2 + 3x + 2} + \frac{3}{x^2 + x - 2}$$

Solution   Begin by finding a common denominator. To do so, factor each denominator. Then multiply the numerator and the denominator of each fraction by whatever expression is required to produce the same denominator in all the fractions.

$$\frac{x - 2}{x^2 - 1} - \frac{x + 3}{x^2 + 3x + 2} + \frac{3}{x^2 + x - 2}$$

$$= \frac{x - 2}{(x + 1)(x - 1)} - \frac{x + 3}{(x + 1)(x + 2)} + \frac{3}{(x - 1)(x + 2)}$$

$$= \frac{(x - 2)(x + 2)}{(x + 1)(x - 1)(x + 2)} - \frac{(x + 3)(x - 1)}{(x + 1)(x + 2)(x - 1)} + \frac{3(x + 1)}{(x - 1)(x + 2)(x + 1)}$$

Then add the fractions, combine similar terms in the numerator, and simplify.

$$= \frac{(x^2 - 4) - (x^2 + 2x - 3) + (3x + 3)}{(x + 1)(x - 1)(x + 2)}$$

$$= \frac{x^2 - 4 - x^2 - 2x + 3 + 3x + 3}{(x + 1)(x - 1)(x + 2)}$$

$$= \frac{x + 2}{(x + 1)(x - 1)(x + 2)}$$

$$= \frac{1}{(x + 1)(x - 1)} \qquad\qquad\blacksquare$$

A fraction that has a fraction in its numerator or its denominator is called a **complex fraction**.

Example 6 Simplify the complex fraction $\dfrac{\dfrac{1}{x}+\dfrac{1}{y}}{\dfrac{x}{y}}$.

Solution Determine that a common denominator of the three fractions in the complex fraction is xy. Multiply both the numerator and the denominator of the complex fraction by xy, and simplify.

$$\frac{\dfrac{1}{x}+\dfrac{1}{y}}{\dfrac{x}{y}}=\frac{\left(\dfrac{1}{x}+\dfrac{1}{y}\right)xy}{\left(\dfrac{x}{y}\right)xy}$$

$$=\frac{\dfrac{xy}{x}+\dfrac{xy}{y}}{\dfrac{xxy}{y}}$$

$$=\frac{y+x}{x^2}$$ ∎

Exercise 1.4

In Exercises 1–4 simplify each fraction.

1. $\dfrac{x^2-16}{x^2-8x+16}$

2. $\dfrac{4-x^2}{x^2-5x+6}$

3. $\dfrac{6x^2+x-12}{4x^2+4x-3}$

4. $\dfrac{x^3-8}{x^2+ax-2x-2a}$

In Exercises 5–38 perform the indicated operations and simplify.

5. $\dfrac{x^2-1}{x}\cdot\dfrac{x^2}{x^2+2x+1}$

6. $\dfrac{y^2-2y+1}{y}\cdot\dfrac{y+2}{y^2+y-2}$

7. $\dfrac{2x^2+32}{8}\div\dfrac{x^2+16}{2}$

8. $\dfrac{x^2+x-6}{x^2-6x+9}\div\dfrac{x^2-4}{x^2-9}$

9. $\dfrac{z^2+z-20}{z^2-4}\div\dfrac{z^2-25}{z-5}$

10. $\dfrac{ax+bx+a+b}{a^2+2ab+b^2}\div\dfrac{x^2-1}{x^2-2x+1}$

11. $\dfrac{3x^2+7x+2}{x^2+2x}\cdot\dfrac{x^2-x}{3x^2+x}$

12. $\dfrac{x^2+x}{2x^2+3x}\cdot\dfrac{2x^2+x-3}{x^2-1}$

13. $\dfrac{x^2+x}{x-1}\cdot\dfrac{x^2-1}{x+2}$

14. $\dfrac{x^2+5x+6}{x^2+6x+9}\cdot\dfrac{x+2}{x^2-4}$

15. $\dfrac{3x^2+5x-2}{x^2+2x}\cdot\dfrac{2x^2+5x}{6x^2+13x-5}$

16. $\dfrac{x^2+7x+12}{x^2-x-6}\cdot\dfrac{x^2-3x-10}{x^2+2x-3}\cdot\dfrac{x^2-4x+3}{x^2-x-20}$

17. $\dfrac{x^2+13x+12}{8x^2-6x-5}\div\dfrac{2x^2-x-3}{8x^2-14x+5}$

18. $\dfrac{x^2-2x-3}{21x^2-50x-16}\cdot\dfrac{3x-8}{x-3}\div\dfrac{x^2+6x+5}{7x^2-33x-10}$

19. $\dfrac{x^3+27}{x^2-4}\div\left(\dfrac{x^2+4x+3}{x^2+2x}\div\dfrac{x^2+x-6}{x^2-3x+9}\right)$

20. $\dfrac{x(x-2)-3}{x(x+7)-3(x-1)}\cdot\dfrac{x(x+1)-2}{x(x-7)+3(x+1)}$

21. $\dfrac{3}{x + 3} + \dfrac{x + 2}{x + 3}$

22. $\dfrac{3}{x + 1} + \dfrac{x + 2}{x + 1}$

23. $\dfrac{4}{x - 1} - \dfrac{3x}{x + 1}$

24. $\dfrac{2}{5 - x} + \dfrac{1}{x - 5}$

25. $\dfrac{a + 3}{a^2 + 7a + 12} + \dfrac{a}{a^2 - 16}$

26. $\dfrac{x}{x^2 - 4} - \dfrac{1}{x + 2}$

27. $\dfrac{1}{3a + 4} - \dfrac{a - 7}{3a^2 + 13a + 12} - \dfrac{4}{a^2 + 4a + 3}$

28. $\dfrac{x + 1}{6x^2 + x - 1} + \dfrac{x}{4x^2 - 1} - \dfrac{x - 1}{6x^2 - 5x + 1}$

29. $\dfrac{1}{x - 2} + \dfrac{3}{x + 2} - \dfrac{3x - 2}{x^2 - 4}$

30. $\dfrac{x}{x - 3} - \dfrac{5}{x + 3} + \dfrac{3(3x - 1)}{x^2 - 9}$

31. $\left(\dfrac{1}{x - 2} + \dfrac{1}{x - 3} \right) \cdot \dfrac{x - 3}{2x}$

32. $\left(\dfrac{1}{x + 1} - \dfrac{1}{x - 2} \right) \div \dfrac{1}{x - 2}$

33. $\dfrac{3x}{x - 4} - \dfrac{x}{x + 4} + \dfrac{3x + 1}{x^2 - 16}$

34. $\dfrac{7x}{x - 5} - \dfrac{3x}{x - 5} + \dfrac{3x - 1}{x^2 - 25}$

35. $\dfrac{1}{x^2 + 3x + 2} - \dfrac{2}{x^2 + 4x + 3} + \dfrac{1}{x^2 + 5x + 6}$

36. $\dfrac{-2}{x - y} + \dfrac{2}{x - z} - \dfrac{2z - 2y}{(y - x)(z - x)}$

37. $\dfrac{3x - 2}{x^2 + x - 20} - \dfrac{4x^2 + 2}{x^2 - 25} + \dfrac{3x^2}{x^2 - 16}$

38. $\dfrac{3x + 2}{8x^2 - 10x - 3} + \dfrac{x + 4}{6x^2 - 11x + 3} - \dfrac{1}{4x + 1}$

In Exercises 39–64 simplify each complex fraction.

39. $\dfrac{\dfrac{3a}{b}}{\dfrac{6ac}{b^2}}$

40. $\dfrac{\dfrac{3t^2}{9x}}{\dfrac{t}{18x}}$

41. $\dfrac{\dfrac{3a^2b}{ab}}{27}$

42. $\dfrac{\dfrac{3u^2v}{4t}}{3uv}$

43. $\dfrac{\dfrac{x - y}{ab}}{\dfrac{y - x}{ab}}$

44. $\dfrac{\dfrac{x^2 - 5x + 6}{2x^2y}}{\dfrac{x^2 - 9}{2x^2y}}$

45. $\dfrac{\dfrac{1}{x} + \dfrac{1}{y}}{xy}$

46. $\dfrac{xy}{\dfrac{11}{x} - \dfrac{11}{y}}$

47. $\dfrac{\dfrac{1}{x} + \dfrac{1}{y}}{\dfrac{1}{x} - \dfrac{1}{y}}$

48. $\dfrac{\dfrac{1}{x} - \dfrac{1}{y}}{\dfrac{1}{x} + \dfrac{1}{y}}$

49. $\dfrac{\dfrac{3a}{b} - \dfrac{4a^2}{x}}{\dfrac{1}{b} + \dfrac{1}{ax}}$

50. $\dfrac{1 - \dfrac{x}{y}}{\dfrac{x^2}{y^2} - 1}$

51. $\dfrac{x + 1 - \dfrac{6}{x}}{x + 5 + \dfrac{6}{x}}$

52. $\dfrac{2z}{1 - \dfrac{3}{z}}$

53. $\dfrac{3xy}{1 - \dfrac{1}{xy}}$

54. $\dfrac{x - 3 + \dfrac{1}{x}}{\dfrac{1}{x} - x + 3}$

55. $\dfrac{3x}{x + \dfrac{1}{x}}$

56. $\dfrac{2x^2 + 4}{2 + \dfrac{4x}{5}}$

57. $\dfrac{\dfrac{x}{x + 2} - \dfrac{2}{x - 1}}{\dfrac{3}{x + 2} + \dfrac{x}{x - 1}}$

58. $\dfrac{\dfrac{2x}{x - 3} + \dfrac{1}{x - 2}}{\dfrac{3}{x - 3} - \dfrac{x}{x - 2}}$

59. $\dfrac{1}{1 + x^{-1}}$

60. $\dfrac{y^{-1}}{x^{-1} + y^{-1}}$

61. $\dfrac{3(x + 2)^{-1} + 2(x - 1)^{-1}}{(x + 2)^{-1}}$

62. $\dfrac{2x(x - 3)^{-1} - 3(x + 2)^{-1}}{(x - 3)^{-1}(x + 2)^{-1}}$

63. $\dfrac{4}{1 + \dfrac{1}{2x^{-1}}}$

64. $\dfrac{ab}{2 + \dfrac{3}{2a^{-1}}}$

In Exercises 65–67 each equation represents an unlikely result. Verify each one.

65. $\dfrac{a^2}{(a - 1)^2 - 1} = \dfrac{a}{(a - 1) - 1}$

66. $\dfrac{(a + b)^3 + a^3}{(a + b)^3 + b^3} = \dfrac{(a + b) + a}{(a + b) + b}$

67. $\sqrt{a + \dfrac{a}{a^2 - 1}} = |a|\sqrt{\dfrac{a}{a^2 - 1}}$

68. Prove that $\dfrac{a}{b} + \dfrac{c}{d} = \dfrac{ad + bc}{bd}$.

1.5   EQUATIONS

Francois Vieta (Viête) (1540–1603) Vieta simplified the subject of algebra by developing the symbolic notation that we use today.

An **equation** is a statement indicating that two mathematical expressions are equal. An equation such as $3x - 2 = 10$ could be true or false, depending on the value of the variable x. If $x = \mathbf{4}$, the equation is true and the number 4 is said to **satisfy** the equation:

$$3x - 2 = 10$$
$$3 \cdot 4 - 2 \overset{?}{=} 10$$
$$12 - 2 \overset{?}{=} 10$$
$$10 = 10$$

However, the equation is false for all other numbers x. Because 4 satisfies the equation, it is called a **solution** or **root**. We also say that the equation has the solution $x = 4$ and that the **solution set** of the equation is $\{4\}$. To **solve an equation** is to find all of its solutions.

The set of all permissible replacements for a variable is called the **domain of the variable**. An equation such as $3x - 2 = 10$ that is satisfied by some but not all values in the domain of the variable is called a **conditional equation**. For some equations every value in the domain of the variable satisfies the equation. Such equations are called **identities**. For example, the equation

$$x^2 - 9 = (x + 3)(x - 3)$$

is an identity because it is true for every real number x.

Two conditional equations that have the same solution set are said to be **equivalent equations**. Any equation equivalent to an equation of the form $ax + b = 0$, where a and b are real numbers, is called a **linear** or a **first-degree equation**. We will solve such an equation by transforming it into simpler, but equivalent, equations, continuing until we obtain one whose solution is obvious. To do this, we use various properties of equality. For example, we can substitute a quantity for its equal in any equation to produce an equivalent equation. We can add any algebraic expression to both sides of an equation (or subtract it from both sides) to form an equivalent equation. Similarly, we can multiply or divide both sides of an equation by any expression that represents a nonzero number, and not change the equation's solutions.

Example 1   Solve the equation $2x(5 - x) + 3x^2 = x(x + 7) + 3$.

Solution
$$2x(5 - x) + 3x^2 = x(x + 7) + 3$$

$10x - 2x^2 + 3x^2 = x^2 + 7x + 3$ Remove parentheses.

$10x + x^2 = x^2 + 7x + 3$ Combine terms.

$10x = 7x + 3$ Subtract x^2 from both sides.

$3x = 3$ Subtract $7x$ from both sides.

$x = 1$ Divide both sides by 3.

All of the above equations are equivalent, and the final equation has exactly one solution: the number 1. Verify that 1 is a solution of the original equation by showing that it satisfies the equation. ◼

Example 2   Solve the equation $\dfrac{x}{x + 5} + \dfrac{x - 1}{x} = 2$.

Solution   Clear the equation of fractions by multiplying both sides by $x(x + 5)$, the **least common denominator** of the fractions. Then use the distributive property to remove parentheses and proceed as follows:

$$\frac{x}{x + 5} + \frac{x - 1}{x} = 2$$

$$x(x + 5)\left(\frac{x}{x + 5} + \frac{x - 1}{x}\right) = x(x + 5)2$$

$$x(x + 5)\,\frac{x}{x + 5} + x(x + 5)\,\frac{x - 1}{x} = 2x(x + 5)$$

$$x^2 + (x + 5)(x - 1) = 2x^2 + 10x$$

$$x^2 + x^2 + 4x - 5 = 2x^2 + 10x$$

$$2x^2 + 4x - 5 = 2x^2 + 10x$$

$$4x - 5 = 10x$$

$$-5 = 6x$$

$$-\frac{5}{6} = x$$

Check the solution $x = -\dfrac{5}{6}$. ◼

We can multiply both sides of an equation by a *nonzero quantity*. However, if we multiply by an expression that is equal to zero for some numbers x, we might produce an equation that is not equivalent to the original. This situation is illustrated in the next example.

Example 3   Solve the equation $\dfrac{x + 1}{x - 3} = \dfrac{4}{x - 3}$.

Solution   If $x \neq 3$, you can clear the equation of fractions by multiplying both sides by $x - 3$. Then proceed as follows:

$$\frac{x+1}{x-3} = \frac{4}{x-3}$$

$$(x-3)\frac{x+1}{x-3} = \frac{4}{x-3}(x-3) \qquad \text{Multiply both sides by } x-3.$$

$$x + 1 = 4 \qquad \text{Simplify.}$$

$$x = 3 \qquad \text{Subtract 1 from both sides.}$$

Because x cannot be 3, this false solution, called an **extraneous solution**, must be discarded. Thus, the original equation has no solution. ■

Many equations, called **literal equations**, contain several variables. Often these equations are **formulas** such as

$$\frac{1}{R} = \frac{1}{R_1} + \frac{1}{R_2}$$

which is used in electronics to calculate the total resistance R of two parallel resistances R_1 and R_2. It is often necessary to solve such equations for one of the variables.

Example 4   Solve $\dfrac{1}{R} = \dfrac{1}{R_1} + \dfrac{1}{R_2}$ for R_1.

Solution   Clear the equation of fractions by multiplying both sides by RR_1R_2, and proceed as follows:

$$\frac{1}{R} = \frac{1}{R_1} + \frac{1}{R_2}$$

$$RR_1R_2\,\frac{1}{R} = RR_1R_2\left(\frac{1}{R_1} + \frac{1}{R_2}\right)$$

$$R_1R_2 = RR_2 + RR_1 \qquad \text{Remove parentheses.}$$

$$R_1R_2 - RR_1 = RR_2 \qquad \text{Subtract } RR_1 \text{ from both sides.}$$

$$R_1(R_2 - R) = RR_2 \qquad \text{Factor } R_1 \text{ from the left side.}$$

$$R_1 = \frac{RR_2}{R_2 - R} \qquad \text{Divide both sides by } R_2 - R.$$

■

Definition.   An equation of the form $ax^2 + bx + c = 0$, where $a \neq 0$, is called a **quadratic** or **second-degree polynomial equation**.

If the trinomial $ax^2 + bx + c$ is factorable, the quadratic equation can be solved by using the zero-factor theorem.

Example 5   Solve the equation $3x^2 - 5x - 2 = 0$.

Solution   Factor the left side of the equation

$$3x^2 - 5x - 2 = 0$$

and write it in the form

$$(3x + 1)(x - 2) = 0$$

By the zero-factor theorem, either

$$3x + 1 = 0 \quad \text{or} \quad x - 2 = 0$$

The solutions of these equations, *and the solutions of the given equation,* are

$$x = -\frac{1}{3} \quad \text{and} \quad x = 2$$

Because no other numbers will make the individual factors equal to zero, no other numbers will satisfy the original equation. ∎

Example 6   Solve the equation $15x^3 - 7x^2 = 2x$.

Solution   To use the zero-factor theorem to solve this equation, its right side must be 0. Add $-2x$ to both sides of the equation and factor the left side.

$$15x^3 - 7x^2 = 2x$$
$$15x^3 - 7x^2 - 2x = 0$$
$$x(15x^2 - 7x - 2) = 0 \qquad \text{Factor out the common factor of } x.$$
$$x(3x - 2)(5x + 1) = 0 \qquad \text{Factor the trinomial.}$$

$$x = 0 \quad \text{or} \quad 3x - 2 = 0 \quad \text{or} \quad 5x + 1 = 0 \qquad \text{Set each factor equal to 0.}$$
$$x = \tfrac{2}{3} \qquad\qquad x = -\tfrac{1}{5}$$

The given equation has three solutions: $x = 0$, $x = \frac{2}{3}$, and $x = -\frac{1}{5}$. ∎

Example 7   Solve the equation $x^4 - 13x^2 + 36 = 0$.

Solution   Solve the equation by factoring:

$$x^4 - 13x^2 + 36 = 0$$
$$(x^2 - 4)(x^2 - 9) = 0$$
$$(x + 2)(x - 2)(x + 3)(x - 3) = 0$$

$$x + 2 = 0 \quad \text{or} \quad x - 2 = 0 \quad \text{or} \quad x + 3 = 0 \quad \text{or} \quad x - 3 = 0$$
$$x = -2 \qquad\qquad x = 2 \qquad\qquad x = -3 \qquad\qquad x = 3$$

The given equation has four solutions: 2, -2, 3, and -3. ∎

Equations are used to solve many word problems.

Example 8   A woman invests \$10,000, part at 9% annual interest and the rest at 14%. In each case the interest is compounded annually. The total annual interest income is \$1,275. How much is invested at each rate?

Solution   Let x represent the amount invested at 9% interest. Then $10,000 - x$ (the rest of \$10,000) represents the amount invested at 14%.

   The annual income from each investment is the product of the interest rate and the amount invested. The total income from the woman's two investments can be expressed in two ways: as $1,275 (which is given) and as the sum of the incomes of the two investments:

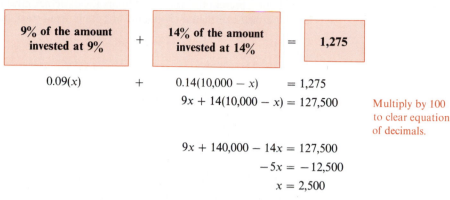

$$0.09(x) + 0.14(10,000 - x) = 1,275$$
$$9x + 14(10,000 - x) = 127,500$$

Multiply by 100 to clear equation of decimals.

$$9x + 140,000 - 14x = 127,500$$
$$-5x = -12,500$$
$$x = 2,500$$

The investment at 9% was $2,500, and $10,000 − $2,500, or $7,500, was invested at 14%. These amounts are correct because 9% of $2,500 is $225, and 14% of $7,500 is $1,050. The total income from both investments is $225 + $1,050, or $1,275. ■

Example 9 If a bottle contains 3 liters of whole milk that is $3\frac{1}{2}\%$ butterfat, how much skimmed milk must be added to produce a mixture that is 2% butterfat?

Solution Let L represent the number of liters of skimmed milk that must be added to the 3 liters of whole milk. Then $3 + L$ represents the number of liters in the final mixture. The butterfat in the final mixture is the sum of the butterfat in the original 3 liters and the butterfat in the skimmed milk. This leads to the following equation:

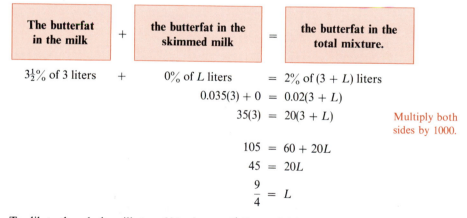

$$3\frac{1}{2}\% \text{ of 3 liters} + 0\% \text{ of } L \text{ liters} = 2\% \text{ of } (3 + L) \text{ liters}$$
$$0.035(3) + 0 = 0.02(3 + L)$$
$$35(3) = 20(3 + L)$$

Multiply both sides by 1000.

$$105 = 60 + 20L$$
$$45 = 20L$$
$$\frac{9}{4} = L$$

To dilute the whole milk to a 2% mixture, $2\frac{1}{4}$ liters of skimmed milk must be added. To check this answer, note that the final mixture contains $0.02(3 + 2.25) = 0.02(5.25) = 0.105$ liter of pure butterfat and that this is equal to the amount of pure butterfat, $0.035(3) = 0.105$ liter, in the whole milk. ■

Example 10 A man leaves home driving at the rate of 50 miles per hour. After discovering that he has forgotten his wallet, his daughter drives at the rate of 55 miles per hour to catch up with him. How long will it take her to overtake him and return the wallet if the man had a 15-minute head start?

Solution   Uniform-motion problems are based on the formula $d = rt$, where d is distance traveled at the rate r for a time t. You can organize the information of this problem as in Figure 1–8. Let t represent the number of hours the daughter must drive to overtake her father. Because the father had a 15-minute, or $\frac{1}{4}$-hour, head start, he will have been on the road for $t + \frac{1}{4}$ hours.

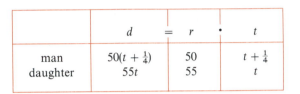

	d	$=$	r	$\cdot$	t
man	$50(t + \frac{1}{4})$		50		$t + \frac{1}{4}$
daughter	$55t$		55		t

Figure 1–8

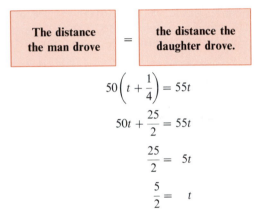

$$50\left(t + \frac{1}{4}\right) = 55t$$

$$50t + \frac{25}{2} = 55t$$

$$\frac{25}{2} = 5t$$

$$\frac{5}{2} = t$$

It will take the daughter 2.5 hours to overtake her father.   ■

Example 11   If a man could paint a house in six days, and his daughter could paint the same house in eight days, how long would it take them to paint the house working together?

Solution   Let n represent the number of days it would take them to paint the house if they work together. In one day the two working together could paint $\frac{1}{n}$ of the house. In one day the father working alone could paint $\frac{1}{6}$ of the house. In one day the daughter working alone could paint $\frac{1}{8}$ of the house. The work they can do together in one day is the sum of what each could do in one day. This gives the equation

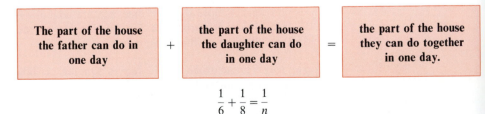

$$\frac{1}{6} + \frac{1}{8} = \frac{1}{n}$$

Multiply both sides of this equation by $24n$ to clear the fractions, and solve for n.

$$24n\left(\frac{1}{6}+\frac{1}{8}\right)=24n\left(\frac{1}{n}\right)$$

$$4n+3n=24$$

$$7n=24$$

$$n=\frac{24}{7}$$

The paint job will take $\frac{24}{7}$, or $3\frac{3}{7}$, days. To check this answer verify that if the father paints $\frac{1}{6}$ of the house each day and the daughter paints $\frac{1}{8}$ of the house each day and if they both work for $\frac{24}{7}$ days, then one complete house is painted.

$$\frac{1}{6}\cdot\frac{24}{7}+\frac{1}{8}\cdot\frac{24}{7}\overset{?}{=}1$$

$$\frac{4}{7}+\frac{3}{7}\overset{?}{=}1$$

$$1=1$$

■

Example 12   A man drives 600 miles to a convention. On the return trip, he increases his speed by 10 miles per hour and saves 2 hours of driving time. How fast did he drive each way?

Solution   Let s represent the speed in miles per hour driving to the convention. On the return trip the speed is $s+10$ miles per hour. Recall that the distance d traveled by an object moving at a constant rate r for a certain time t is given by the formula $d=rt$. Thus, $t=\frac{d}{r}$.

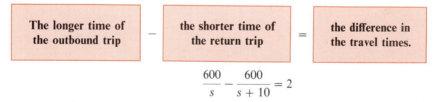

$$\frac{600}{s}-\frac{600}{s+10}=2$$

Multiply both sides of this equation by $s(s+10)$ to clear the equation of fractions, and solve for s.

$$s(s+10)\left(\frac{600}{s}-\frac{600}{s+10}\right)=2s(s+10)$$

$$600s+6000-600s=2s^2+20s$$

$$6000=2s^2+20s$$

$$0=s^2+10s-3000$$

$$0=(s-50)(s+60)$$

$$s-50=0 \quad\text{or}\quad s+60=0$$

$$s=50 \qquad\qquad s=-60$$

The solution $s = -60$ must be discarded because it is negative. The man drove 50 miles per hour to the convention, and $50 + 10$, or 60, miles per hour on the return trip. Verify that these answers are correct. ∎

Exercise 1.5

In Exercises 1–24 solve each equation, if possible.

1. $2x + 5 = 15$

2. $3x + 2 = x + 8$

3. $2(x + 2) = 4(x - 3)$

4. $3(x + 2) - x = 2(x + 3)$

5. $2(a + 1) = 3(a - 2) - a$

6. $x^2 = (x + 4)(x - 4) + 16$

7. $\dfrac{3x + 5}{2} = 7$

8. $\dfrac{x}{2} - 7 = 14$

9. $\dfrac{4x}{3} - \dfrac{x}{2} = 5$

10. $\dfrac{x}{3} + \dfrac{2x}{5} = 11$

11. $\dfrac{2}{y} + \dfrac{4}{5} = \dfrac{2(5 + 2y)}{5y}$

12. $\dfrac{4}{5} + \dfrac{1}{b} = \dfrac{9}{b}$

13. $\dfrac{2}{x + 1} - 2 = \dfrac{-4}{x + 1}$

14. $\dfrac{3}{x + 1} - 1 = \dfrac{-2}{x + 1}$

15. $\dfrac{2}{x - 2} + \dfrac{1}{x + 1} = \dfrac{-3(x - 2)}{(x - 2)(x + 1)}$

16. $\dfrac{3}{x - 2} + \dfrac{1}{x} = \dfrac{3}{x - 2}$

17. $\dfrac{x - 1}{x + 3} - \dfrac{1 - 2x}{3 - x} = \dfrac{2 - x}{x - 3}$

18. $\dfrac{2}{a - 2} + \dfrac{1}{a + 1} = \dfrac{-1}{(2 - a)(a + 1)}$

19. $\dfrac{9x + 6}{x^2 + 3x} = \dfrac{7}{x + 3}$

20. $\dfrac{5}{x - 1} + \dfrac{3}{x - 3} = \dfrac{8}{x - 2}$

21. $\dfrac{1}{11 - n} - \dfrac{2(3n - 1)}{-7n^2 + 74n + 33} = \dfrac{1}{7n + 3}$

22. $\dfrac{1}{n + 8} - \dfrac{3n - 4}{5n^2 + 42n + 16} = \dfrac{1}{5n + 2}$

23. $\dfrac{3y}{3y - 6} - \dfrac{2y}{2y + 4} = \dfrac{8}{y^2 - 4}$

24. $\dfrac{2a + 3}{a^2 + 5a + 6} + \dfrac{3a - 2}{a^2 + a - 6} = \dfrac{5a - 2}{a^2 - 4}$

In Exercises 25–36 solve each equation for the indicated variable.

25. $p = 2l + 2w$; for w

26. $A = \dfrac{1}{2} h(b_1 + b_2)$; for b_2

27. $z = \dfrac{x - \mu}{\sigma}$; for μ

28. $P_n = L + \left(\dfrac{s}{f}\right)i$; for f

29. $V = \pi h^2\left(r - \dfrac{h}{3}\right)$; for r

30. $V = \dfrac{1}{3}(B_1 + B_2 + \sqrt{B_1 B_2})h$; for h

31. $V = \dfrac{1}{6} h(B + B' + 4M)$; for h

32. $A = \dfrac{1}{2} r^2(\theta - \phi)$; for θ

33. $S = \dfrac{a - lr}{1 - r}$; for r

34. $a = (n - 2)\dfrac{180}{n}$, for n

35. $r = \dfrac{x + y}{1 - xy}$; for x

36. $r = \dfrac{x + y}{1 - xy}$; for y

In Exercises 37–60 solve each equation by factoring.

37. $x^2 - x = 0$

38. $x^2 + 8x = 0$

39. $x^2 - 9 = 0$

40. $x^2 - 25 = 0$

41. $x^2 - 9x + 14 = 0$

42. $x^2 + x - 12 = 0$

43. $5x^2 - 13x + 6 = 0$

44. $2x^2 + 5x - 12 = 0$

45. $15x^2 + 16x = 15$

46. $6x^2 - 25x = -25$

47. $12x^2 + 9 = 24x$

48. $24x^2 + 6 = 24x$

49. $x^3 - 8x^2 + 15x = 0$

50. $x^3 + 5x^2 - 14x = 0$

51. $3x^3 - x^2 = 2x$

52. $10x^3 = 7x^2 - x$

53. $x^3 = x$

54. $x^3 = 9x$

55. $x^4 - 10x^2 + 9 = 0$

56. $x^4 - 29x^2 + 100 = 0$

57. $2x^4 + 72 = 26x^2$

58. $4x^4 = 5x^2 + 9$

59. $(x^2 - 2x)^2 - 11(x^2 - 2x) + 24 = 0$

60. $2(y + 2)^2 - 9(y + 2) + 9 = 0$

In Exercises 61–78 use a linear equation to solve each word problem.

61. If a certain number is added to both the numerator and the denominator of the fraction $\frac{3}{5}$, the resulting fraction is $\frac{5}{6}$. Find the number.

62. The denominator of a fraction exceeds the numerator by 1. If 4 is added to the numerator of the fraction, and 3 is subtracted from the denominator, the result is 7. Find the original fraction.

63. An executive invests some money at 7% and some money at 6% annual interest. If a total of $22,000 is invested and the annual return is $1,420, how much is invested at 7%?

64. A student invests some money at 8% and twice as much at 9% annual interest. After one year, the total income from these two investments is $2080. How much was invested at each rate?

65. An adult ticket for a college basketball game costs $2.50, and a student ticket costs $1.75. The total receipts from the game were $1217.25 and 585 tickets were sold. How many of these were student tickets?

66. Of the 800 tickets sold to a movie, 480 were adult tickets. The gate receipts totaled $2080. What was the cost of a student ticket if an adult ticket cost $3?

67. A child has equal numbers of nickels, dimes, and quarters. The coins are worth $3.20. How many of each type are there?

68. A small car radiator has a 6-liter capacity. If the liquid in the radiator is 40% antifreeze, how much liquid must be replaced with pure antifreeze to bring the mixture up to a 50% solution?

69. A nurse has 1 liter of a solution that is 20% alcohol. How much pure alcohol must she add to bring it to a solution that contains 25% alcohol?

70. If there are 400 milliliters of a chemical in 1 liter of solution, how many milliliters of water must be added to dilute it to a 25% solution? (There are 1000 milliliters in 1 liter.)

71. How many liters of water must evaporate to make 12 liters of a 24% salt solution into a 36% salt solution?

72. John scored 5 points higher on his midterm, and 13 points higher on his final, than on his first exam. What did he score on that first exam if his mean (average) score was 90?

73. Sally took 4 tests in science class. On each successive test, her score improved by 3 points. If her mean score is 69.5%, what did she get on the first test?

74. A motorboat goes 5 miles upstream in the same time that it requires to go 7 miles downstream. The river flows at 2 miles per hour. What is the speed of the boat in still water?

75. John drove to his uncle's house in a distant city in 5 hours. When he returned home, there was less traffic and the trip took only 3 hours. If John drove 26 miles per hour faster on the return trip, how fast did he drive each way?

76. Suzi drove home for spring vacation at 55 miles per hour, but her brother Jim, who left at the same time, could drive only 48 miles per hour. When Suzi got home, Jim still had 35 miles to go. How far did Suzi drive?

77. If a woman can mow a yard with a lawn tractor in 2 hours, and another woman can mow the lawn with a push mower in 4 hours, how long will it take them if they work together?

78. A garden hose can fill a swimming pool in 3 days, and a larger hose can fill the pool in 2 days. How long will it take to fill the pool if both hoses are used?

In Exercises 79–96 use a quadratic equation to solve each word problem.

79. The length of a rectangle is 4 feet longer than its width. Its area is 32 square feet. What are its dimensions?

80. The length of a rectangle is 5 times longer than its width. Its area is 125 square feet. What is its perimeter?

81. The side of a square is 4 centimeters shorter than the side of a second square. The sum of the areas of the squares is 106 square centimeters. Find the length of the side of the larger square.

82. The base of a triangle is one-third as long as its height. How long is the base if the triangle's area is 24 square meters?

83. A boy rides a bicycle a distance of 40 miles. His return trip takes 2 hours longer because his speed decreases by 10 miles per hour. How fast does he ride each way?

84. A farmer drives a tractor from one town to another, a distance of 120 kilometers. He drives 10 kilometers per hour faster on the return trip, cutting 1 hour off the time. How fast does he drive each way?

85. If the speed were increased by 10 miles per hour, a 420-mile trip would take 1 hour less time. How long does the trip take at the slower speed?

86. By increasing her usual speed by 25 kilometers per hour, a bus driver decreases the time on a 25-kilometer trip by 10 minutes. What is the usual speed?

87. Two pipes are used to fill a water storage tank. The first pipe can fill the tank in 4 hours. The two pipes together can fill the tank in 2 hours less time than the second pipe alone. How long would it take for the second pipe alone to fill the tank?

88. A hose can fill a swimming pool in 6 hours. Another hose requires 3 more hours to fill the pool than the two hoses together. How long would it take the second hose alone to fill the pool?

89. Working together, Sarah and Heidi can milk the cows in 1 hour. If they work alone, it takes Heidi 50 minutes longer than it takes Sarah. How long does it take Sarah to milk the cows alone?

90. If two opposite sides of a square are increased by 10 meters and the other sides are decreased by 8 meters, the area of the rectangle that is formed is 63 square meters. Find the area of the original square.

91. Maude and Matilda have each invested some money for their retirement. Maude invested $1000 more than Matilda but at an interest rate that was 1% less. Last year Maude received interest of $280 on her investment and Matilda received $240. At what rates were their investments made?

92. Scott and Laura have both invested some money. Scott invested $3000 more than Laura and at a 2% higher rate. After one year, Scott received $800 interest and Laura received $400. How much did Scott invest?

93. Some mathematics professors at a college would like to purchase a $150 microwave oven for their department workroom. If four of the professors do not contribute, everyone else's share will increase by $10. How many mathematics professors teach at the college?

94. A farmer intends to construct a windscreen by planting pine trees in a quarter-mile row. His daughter points out that 44 fewer trees will be needed if they are planted 5 feet farther apart. If the farmer takes her advice, how many trees will he plant? (*Hint:* 1 mile = 5280 feet.)

95. The height of a projectile fired upward with an initial velocity of 400 feet per second is given by the formula $h = -16t^2 + 400t$, where h is the height (in feet) and t is the time in seconds. Find the time required for the projectile to return to earth.

96. The height of an object tossed upward with an initial velocity of 104 feet per second is given by the formula $h = 104t - 16t^2$, where h is the height (in feet) and t is the time in seconds. In how many seconds will the object return to its point of departure?

1.6 COMPLEX NUMBERS AND MORE ON EQUATIONS

Many equations do not have real solutions. For example, the equation $x^2 = -1$ has no real solutions, because no real number is a square root of -1. At one time mathematicians believed that square roots of negative numbers, such as $\sqrt{-1}$, were nonsense. In the seventeenth century these numbers were named **imaginary numbers** by René Descartes (1596–1650). It was not until the nineteenth century that mathematicians began to think of imaginary numbers as an extension of the real number system.

The imaginary number $\sqrt{-1}$, often called the **imaginary unit**, is denoted as i. Thus,

$$i^2 = -1$$

Definition.   An **imaginary number** is a number that can be written in the form bi, where b is a nonzero real number and $i = \sqrt{-1}$.

If a is a positive real number, then the **principal square root** of $-a$, denoted by $\sqrt{-a}$, is the imaginary number $\sqrt{a}\,i$.

The powers of i form a repeating pattern. Assuming that $i^0 = 1$, we have

$$i^0 = 1$$
$$i^1 = i$$
$$i^2 = -1$$
$$i^3 = i^2 i = -1i = -i$$
$$i^4 = (i^2)^2 = (-1)^2 = 1$$
$$i^5 = i^4 i = 1i = i$$

The pattern continues: $i, -1, -i, 1, i, -1, -i, 1, \ldots$. Thus,

$$i^{21} = i^{20}i = (i^4)^5 i = 1^5 i = i$$

It is also possible to raise i to negative powers. For example,

$$i^{-7} = \frac{1}{i^7} = \frac{1i}{i^7 i} = \frac{i}{i^8} = \frac{i}{1} = i$$

Example 1   Find the product of $\sqrt{-4}$ and $\sqrt{-9}$.

Solution   First express each radical in *bi* form. Then multiply, remembering that $i^2 = -1$:

$$\sqrt{-4}\sqrt{-9} = \sqrt{4}i\sqrt{9}i$$
$$= 2i \cdot 3i$$
$$= 2 \cdot 3i^2$$
$$= 6(-1)$$
$$= -6$$

In Example 1 it was first necessary to write each number in *bi* form, because the equation

$$\sqrt{x}\sqrt{y} = \sqrt{xy}$$

is *not* true if x and y are both negative. Note that

$$\sqrt{(-4)(-9)} = \sqrt{36} = 6$$

but

$$\sqrt{-4}\sqrt{-9} = 2i \cdot 3i = 6i^2 = -6$$

To avoid errors, always express numbers in *bi* form before attempting any algebraic manipulations.

If we use imaginary numbers, it is possible to factor the sum of two squares. For example,

$$x^2 + 4 = x^2 + 2^2$$
$$= x^2 - 2^2i^2$$
$$= (x + 2i)(x - 2i)$$

Expressions such as $3 + 2i$ and $-\sqrt{5} - \sqrt{7}i$ that represent the sum or difference of a real number and an imaginary number are called **complex numbers**.

Definition.   A **complex number** is a number of the form $a + bi$, where a and b are real numbers and $i = \sqrt{-1}$. The number a is called the **real part** of the complex number. The number b is called the **imaginary part**.

Two complex numbers are equal if and only if their real parts are equal and their imaginary parts are equal:

$$a + bi = c + di \qquad \text{if and only if} \qquad a = c \text{ and } b = d$$

If $b = 0$, the complex number $a + bi$ is the real number a. If $a = 0$ and $b \neq 0$, the complex number $a + bi$ is the imaginary number bi. Thus, *both the set of real numbers and the set of imaginary numbers are subsets of the set of complex numbers*. Figure 1–9 indicates the relationships among the various number sets discussed thus far.

Complex numbers can be added and subtracted as if they were binomials. They can also be multiplied as binomials if we remember that $i^2 = -1$.

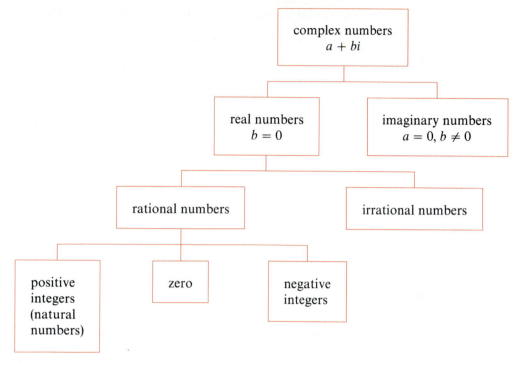

Figure 1-9

Example 2   **a.** $(2 + 3i) + (4 - 5i) = 2 + 3i + 4 - 5i$

$$= 2 + 4 + 3i - 5i$$

$$= 6 - 2i$$

b. $(2 + 3i)(4 - 5i) = 8 - 10i + 12i - 15i^2$

$$= 8 - 10i + 12i + 15 \qquad i^2 = -1$$

$$= 23 + 2i \qquad \qquad \qquad \blacksquare$$

We generalize the results of the previous example and define addition and multiplication of complex numbers as follows:

Definition.   If a, b, c, and d are real numbers and $i = \sqrt{-1}$, then

$$(a + bi) + (c + di) = (a + c) + (b + d)i$$

$$(a + bi)(c + di) = (ac - bd) + (ad + bc)i$$

Because the sum and the product of two complex numbers are also complex numbers, we say that the set of complex numbers is closed under the operations of addition and multiplication.

The complex numbers $a + bi$ and $a - bi$ are called **conjugates** of each other. Note that *the product of a complex number and its conjugate is a real number*:

$$(a + bi)(a - bi) = a^2 - abi + abi - b^2i^2$$

$$= a^2 + b^2$$

This fact enables us to divide one complex number by another, as the next example illustrates.

Example 3   Perform the division $\dfrac{3-2i}{2+i}$ by writing the quotient in $a + bi$ form.

Solution   The given expression is a fraction that is not in simplified form because it has a square root (the imaginary unit i) in its denominator. To perform the indicated division, rationalize the denominator of the fraction by multiplying both the numerator and the denominator of the fraction by $2 - i$, the conjugate of the denominator.

$$\frac{3-2i}{2+i} = \frac{(3-2i)(2-i)}{(2+i)(2-i)}$$

$$= \frac{6 - 3i - 4i + 2i^2}{4 - 2i + 2i - i^2}$$

$$= \frac{6 - 7i - 2}{4 + 1} \qquad -i^2 = -(-1) = 1$$

$$= \frac{4 - 7i}{5}$$

$$= \frac{4}{5} + \left(-\frac{7}{5}\right)i$$

$$= \frac{4}{5} - \frac{7}{5}i$$

It is acceptable to write the complex number $a + (-b)i$ as $a - bi$.   ■

Note that, if there is no division by 0, the quotient of two complex numbers is another complex number.

Completing the Square

If the solutions of a quadratic equation are irrational or complex numbers, the equation cannot be solved easily by factoring. We now develop a method known as **completing the square**, which can be used to solve *all* quadratic equations.

Consider the equation $x^2 = c$. Its roots can be found by subtracting c from both sides, factoring, setting each factor equal to zero, and solving for x:

$$x^2 = c$$
$$x^2 - c = 0$$
$$(x + \sqrt{c})(x - \sqrt{c}) = 0$$
$$x + \sqrt{c} = 0 \qquad \text{or} \qquad x - \sqrt{c} = 0$$
$$x = -\sqrt{c} \qquad \qquad \qquad x = \sqrt{c}$$

Thus, the roots of the equation $x^2 = c$ are $x = -\sqrt{c}$ and $x = \sqrt{c}$. If $c \geq 0$, the roots are real numbers. If $c < 0$, the roots are imaginary numbers.

This fact, often called the **square-root property of equality**, can be used to solve quadratic equations. To solve $x^2 + 8x + 7 = 0$, for example, we begin by adding -7 to both sides of the equation:

$$x^2 + 8x + 7 = 0$$
$$x^2 + 8x \quad = -7$$

We then add 16 to both sides, combine terms, and factor the trinomial on the left. *Note that adding* 16 *makes the left side a perfect square.* Because of this, the process is called *completing the square.*

$$x^2 + 8x \ \mathbf{+\ 16} = -7 \ \mathbf{+\ 16}$$
$$x^2 + 8x + 16 = \quad 9$$
$$(x + 4)^2 = \quad 9$$

$$x + 4 = -\sqrt{9} \qquad \text{or} \qquad x + 4 = \sqrt{9} \qquad \color{red}{\text{Use the square-root property.}}$$
$$x + 4 = -3 \qquad\qquad\qquad\ x + 4 = 3$$
$$x = -7 \qquad\qquad\qquad\quad x = -1$$

To complete the square, we must know what number must be added to $x^2 \pm bx$ to obtain a perfect trinomial square. This question is answered by observing that

$$\left(x + \frac{b}{2}\right)^2 = x^2 + 2\left(\frac{b}{2}\right)x + \left(\frac{b}{2}\right)^2$$
$$= x^2 + bx + \left(\frac{b}{2}\right)^2$$

and

$$\left(x - \frac{b}{2}\right)^2 = x^2 - 2\left(\frac{b}{2}\right)x + \left(-\frac{b}{2}\right)^2$$
$$= x^2 - bx + \left(-\frac{b}{2}\right)^2$$

In each case the constant term that must be added is the square of one-half of the coefficient of x.

Example 4 Solve the equation $9x^2 - 6x + 5 = 0$.

Solution To complete the square, the coefficient of x^2 must be 1. To make this so, begin by dividing both sides of the equation by 9. Then proceed as follows:

$$9x^2 - 6x + 5 = \quad 0$$
$$x^2 - \frac{6}{9}x + \frac{5}{9} = \frac{0}{9}$$
$$x^2 - \frac{2}{3}x + \frac{5}{9} = \quad 0$$
$$x^2 - \frac{2}{3}x \quad = -\frac{5}{9}$$

To complete the square on x, add $[\frac{1}{2}(-\frac{2}{3})]^2$, or $\frac{1}{9}$, to both sides.

$$x^2 - \frac{2}{3}x + \frac{1}{9} = -\frac{5}{9} + \frac{1}{9}$$

$$x^2 - \frac{2}{3}x + \frac{1}{9} = -\frac{4}{9}$$

$$\left(x - \frac{1}{3}\right)^2 = -\frac{4}{9}$$

$$x - \frac{1}{3} = -\sqrt{-\frac{4}{9}} \quad \text{or} \quad x - \frac{1}{3} = \sqrt{-\frac{4}{9}}$$

$$x - \frac{1}{3} = -\frac{2}{3}i \qquad\qquad x - \frac{1}{3} = \frac{2}{3}i$$

$$x = \frac{1}{3} - \frac{2}{3}i \qquad\qquad x = \frac{1}{3} + \frac{2}{3}i$$

Note that the two roots are conjugate complex numbers.   ■

The Quadratic Formula

The method of completing the square can be used to solve the **general quadratic equation** $ax^2 + bx + c = 0$, where $a \neq 0$. To do so, we add $-c$ to both sides and divide by a to get

$$x^2 + \frac{b}{a}x = \frac{-c}{a}$$

To complete the square on x, we add the square of $\frac{b}{2a}$ to both sides. We then factor, combine terms, and proceed as follows:

$$x^2 + \frac{b}{a}x + \frac{b^2}{4a^2} = \frac{b^2}{4a^2} - \frac{c}{a}$$

$$\left(x + \frac{b}{2a}\right)^2 = \frac{b^2 - 4ac}{4a^2}$$

$$x + \frac{b}{2a} = \sqrt{\frac{b^2 - 4ac}{4a^2}} \quad \text{or} \quad x + \frac{b}{2a} = -\sqrt{\frac{b^2 - 4ac}{4a^2}}$$

$$x = -\frac{b}{2a} + \frac{\sqrt{b^2 - 4ac}}{2a} \qquad\qquad x = -\frac{b}{2a} - \frac{\sqrt{b^2 - 4ac}}{2a}$$

$$x = \frac{-b + \sqrt{b^2 - 4ac}}{2a} \qquad\qquad x = \frac{-b - \sqrt{b^2 - 4ac}}{2a}$$

The previous two solutions are often written as a single expression called the **quadratic formula**.

The Quadratic Formula. The roots of the general quadratic equation $ax^2 + bx + c = 0$, where $a \neq 0$, are

$$x = \frac{-b \pm \sqrt{b^2 - 4ac}}{2a}$$

Example 5   Use the quadratic formula to solve $2x^2 - 5x + 3 = 0$.

Solution   In this equation $a = 2$, $b = -5$, and $c = 3$. Substitute these values into the quadratic formula and simplify:

$$x = \frac{-b \pm \sqrt{b^2 - 4ac}}{2a}$$

$$= \frac{-(-5) \pm \sqrt{(-5)^2 - 4(2)(3)}}{2(2)}$$

$$= \frac{5 \pm \sqrt{25 - 24}}{4}$$

$$= \frac{5 \pm \sqrt{1}}{4}$$

$$= \frac{5 \pm 1}{4}$$

$$x = \frac{3}{2} \quad \text{or} \quad x = 1$$

Check both of these solutions.   ■

Example 6   Find three cube roots of 1.

Solution   The cube roots of 1 are roots of the equation $x^3 = 1$, or $x^3 - 1 = 0$. Factor the difference of two cubes on the left side of the equation and apply the zero-factor theorem.

$$x^3 - 1 = 0$$
$$(x - 1)(x^2 + x + 1) = 0$$
$$x - 1 = 0 \quad \text{or} \quad x^2 + x + 1 = 0$$
$$x = 1$$

Thus, 1 is one of the three cube roots of 1. To find the remaining two, solve the equation $x^2 + x + 1 = 0$ by letting $a = 1$, $b = 1$, and $c = 1$ in the quadratic formula:

$$x = \frac{-b \pm \sqrt{b^2 - 4ac}}{2a} = \frac{-1 \pm \sqrt{1^2 - 4(1)(1)}}{2(1)} = \frac{-1 \pm \sqrt{-3}}{2} = \frac{-1 \pm \sqrt{3}i}{2}$$

Thus, the three cube roots of 1 are 1, $-\frac{1}{2} + \frac{\sqrt{3}}{2}i$, and $-\frac{1}{2} - \frac{\sqrt{3}}{2}i$.   ■

An interesting number known as the **golden ratio** and denoted as ϕ is related to aesthetically pleasing proportions in art and architecture. It is the *positive* number that is 1 less than its reciprocal.

Example 7   Determine the golden ratio, ϕ.

Solution   The equation to solve is

$$\phi = \frac{1}{\phi} - 1 \qquad \text{where } \phi > 0$$

Multiply both sides by ϕ to obtain the quadratic equation

$$\phi^2 + \phi - 1 = 0$$

Solve it by letting $a = 1$, $b = 1$, and $c = -1$ in the quadratic formula:

$$\phi = \frac{-b \pm \sqrt{b^2 - 4ac}}{2a} = \frac{-1 \pm \sqrt{1^2 - 4(1)(-1)}}{2(1)} = \frac{-1 \pm \sqrt{5}}{2}$$

The solutions are $\phi = -\frac{1}{2} + \frac{\sqrt{5}}{2}$ and $\phi = -\frac{1}{2} - \frac{\sqrt{5}}{2}$. Because $\phi > 0$, the second solution (which is negative) must be discarded. Thus,

$$\phi = -\frac{1}{2} + \frac{\sqrt{5}}{2}$$

Use a calculator to determine that

$$\phi \approx 0.618034 \qquad\blacksquare$$

Example 8   Solve the equation $xy^2 + x^2y + 1 = 0$ for y.

Solution   This is a quadratic equation in the variable y. Solve it for y by letting $a = x$, $b = x^2$, and $c = 1$ in the quadratic formula.

$$y = \frac{-b \pm \sqrt{b^2 - 4ac}}{2a}$$

$$y = \frac{-x^2 \pm \sqrt{(x^2)^2 - 4(x)(1)}}{2x}$$

$$y = \frac{-x^2 \pm \sqrt{x^4 - 4x}}{2x} \qquad (x \neq 0) \qquad \blacksquare$$

Many equations containing radicals become quadratic equations after using this fact:

If a and b are real numbers, n is an integer, and $a = b$, then

$$a^n = b^n$$

Example 9   Solve the equation $\sqrt{x + 3} = 3x - 1$.

Solution   Rid this equation of the radical by squaring both sides of the equation and simplifying.

$$\sqrt{x + 3} = 3x - 1$$
$$(\sqrt{x + 3})^2 = (3x - 1)^2$$
$$x + 3 = 9x^2 - 6x + 1$$
$$0 = 9x^2 - 7x - 2$$
$$0 = (9x + 2)(x - 1)$$

$$9x + 2 = 0 \qquad \text{or} \qquad x - 1 = 0$$
$$x = -\frac{2}{9} \qquad\qquad\qquad x = 1$$

Because squaring both sides of an equation does not guarantee a result that is equivalent to the original equation, you *must check each proposed root.*

For $x = -\dfrac{2}{9}$:

$$\sqrt{x + 3} = 3x - 1$$
$$\sqrt{-\frac{2}{9} + 3} = 3\left(-\frac{2}{9}\right) - 1$$
$$\sqrt{\frac{25}{9}} = -\frac{2}{3} - 1$$
$$\frac{5}{3} \neq -\frac{5}{3}$$

Thus, $-\frac{2}{9}$ is not a solution.

For $x = 1$:

$$\sqrt{x + 3} = 3x - 1$$
$$\sqrt{1 + 3} = 3(1) - 1$$
$$\sqrt{4} = 3 - 1$$
$$2 = 2$$

Thus, 1 is a solution.

The only solution of $\sqrt{x + 3} = 3x - 1$ is the number 1.   ■

Example 10   Solve the equation $\sqrt{2x + 3} + \sqrt{x - 2} = 4$.

Solution   To simplify the work, rewrite the equation in the form

$$\sqrt{2x + 3} = 4 - \sqrt{x - 2}$$

so that its left side contains only one radical. Square both sides to get

$$(\sqrt{2x + 3})^2 = (4 - \sqrt{x - 2})^2$$
$$2x + 3 = 16 - 8\sqrt{x - 2} + (x - 2)$$

Combine terms and add $-x - 14$ to both sides, leaving $-8\sqrt{x - 2}$ by itself on the right side of the equation.

$$2x + 3 = 14 - 8\sqrt{x - 2} + x$$
$$x - 11 = -8\sqrt{x - 2}$$

Square both sides again, simplify, and solve the resulting equation for x.

$$x^2 - 22x + 121 = 64(x - 2)$$
$$x^2 - 86x + 249 = 0$$
$$(x - 3)(x - 83) = 0$$
$$x = 3 \quad \text{or} \quad x = 83$$

Substituting these proposed roots into the given equation shows that $x = 83$ does not work; 83 is an extraneous solution and must be discarded. The value $x = 3$ does satisfy the given equation. Hence, 3 is a root.   ■

Example 11   Solve the equation $2x^{2/3} + 5x^{1/3} - 3 = 0$.

Solution   The given equation is quadratic in the variable $x^{1/3}$ and can be written as

$$2(x^{1/3})^2 + 5x^{1/3} - 3 = 0$$

and solved by factoring:

$$(2x^{1/3} - 1)(x^{1/3} + 3) = 0$$

$$
\begin{array}{ll}
2x^{1/3} - 1 = 0 \quad\quad \text{or} & x^{1/3} + 3 = 0 \\
2x^{1/3} = 1 & x^{1/3} = -3 \\
x^{1/3} = \dfrac{1}{2} & (x^{1/3})^3 = (-3)^3 \\
& x = -27 \\
(x^{1/3})^3 = \left(\dfrac{1}{2}\right)^3 & \\
x = \dfrac{1}{8} &
\end{array}
$$

Verify that both $\frac{1}{8}$ and -27 are solutions.   ■

Example 12   Solve the equation $x^{3/2} - 3x^{1/2} + 2x^{-1/2} = 0$.

Solution   Proceed as follows:

$$x^{3/2} - 3x^{1/2} + 2x^{-1/2} = 0$$
$$x^{-1/2}(x^2 - 3x + 2) = 0 \quad\quad \text{Factor out } x^{-1/2}.$$
$$x^{-1/2}(x - 1)(x - 2) = 0 \quad\quad \text{Factor } x^2 - 3x + 2.$$

$$
\begin{array}{lllll}
x^{-1/2} = 0 & \text{or} & x - 1 = 0 & \text{or} & x - 2 = 0 \\
\text{no solution} & & x = 1 & & x = 2
\end{array}
$$

Use the zero-factor theorem.

Note that the first of these equations has no solution. The given equation has the two solutions $x = 1$ and $x = 2$.   ■

The Discriminant

The nature of the roots of a quadratic equation can be determined without solving the equation. If the coefficients of the quadratic equation $ax^2 + bx +$

$c = 0$ are real numbers and the radicand $b^2 - 4ac$, called the **discriminant**, in the quadratic formula is a negative number, then the roots of the equation are nonreal complex numbers. If $b^2 - 4ac \geq 0$, the roots are real numbers. The possibilities are summarized in the following table.

If a, b, and c are real numbers and $b^2 - 4ac$ is ...	then the solutions are ...
positive	real numbers and unequal
zero	real numbers and equal
negative	nonreal complex numbers and complex conjugates

If the coefficients of the quadratic equation are rational numbers and the discriminant is a perfect square, then the roots of the equation are also rational numbers. If the discriminant is equal to zero, then the two roots of the equation are equal rational numbers. The possibilities are summarized in the following table.

If a, b, and c are rational numbers and $b^2 - 4ac$ is ...	then the solutions are ...
a nonzero perfect square	rational numbers and unequal
positive and not a perfect square	irrational numbers and unequal
zero	rational numbers and equal

Example 13   Determine the nature of the roots of the quadratic equation $3x^2 + 4x + 1 = 0$.

Solution   Calculate the discriminant, $b^2 - 4ac$, as follows:

$$b^2 - 4ac = 4^2 - 4(3)(1)$$
$$= 4$$

Because a, b, and c are rational numbers and the discriminant is a perfect square, the two roots of the given quadratic equation are rational and unequal.   ■

Example 14   If k is an arbitrary constant, many quadratic equations are represented by the equation

$$(k - 2)x^2 + (k + 1)x + 4 = 0$$

What values of k will give an equation with roots that are real and equal?

Solution   Compute the discriminant and demand that it be zero.

$$b^2 - 4ac = (k + 1)^2 - 4(k - 2)4$$
$$0 = k^2 + 2k + 1 - 16k + 32$$
$$0 = k^2 - 14k + 33$$
$$0 = (k - 3)(k - 11)$$
$$k - 3 = 0 \quad \text{or} \quad k - 11 = 0$$
$$k = 3 \qquad\qquad k = 11$$

If $k = 3$, then the equation $(k - 2)x^2 + (k + 1)x + 4 = 0$ becomes

$$(3 - 2)x^2 + (3 + 1)x + 4 = 0$$

or

$$x^2 + 4x + 4 = 0$$

Solve this equation to show that its roots are the real and equal numbers -2 and -2. Similarly, $k = 11$ produces an equation with roots that are real and equal. ■

Exercise 1.6

In Exercises 1–4 simplify each expression.

1. i^9 **2.** i^{27} **3.** i^{-38} **4.** i^{-104}

In Exercises 5–6 use imaginary numbers to factor each expression.

5. $x^2 + 16$ **6.** $9x^2 + 4$

In Exercises 7–24 perform any indicated operations, and express the final answer in $a + bi$ form.

7. $(2 - 7i) + (3 + i)$ **8.** $(-7 + 2i) + (2 - 8i)$ **9.** $(5 - 6i) - (7 + 4i)$

10. $(11 + 2i) - (13 - 5i)$ **11.** $(14i + 2) + (2 - 4i)$ **12.** $(5 + 8i) - (23i - 32)$

13. $(2 + 3i)(-3 - 4i)$ **14.** $(-2 - 4i)(4 + 5i)$ **15.** $(-11 + \sqrt{-25})(-2 - \sqrt{-36})$

16. $(6 + \sqrt{-49})(6 - \sqrt{-49})$ **17.** $(\sqrt{-16} + 3)(2 + \sqrt{-9})$ **18.** $(12 - \sqrt{-4})(\sqrt{-25} + 7)$

19. $(2 + 3i)^2$ **20.** $(2 - i)^3$ **21.** $\dfrac{1}{2 + i}$

22. $\dfrac{3}{4i^2}$ **23.** $\dfrac{2i}{7 + i}$ **24.** $\dfrac{2 + i}{3 - i}$

25. Verify that $-\dfrac{1}{2} + \dfrac{\sqrt{3}}{2}i$ and $-\dfrac{1}{2} - \dfrac{\sqrt{3}}{2}i$ are roots of $x^2 + x + 1 = 0$.

26. Verify that $\dfrac{1}{2} + \dfrac{\sqrt{5}}{2}i$ and $\dfrac{1}{2} - \dfrac{\sqrt{5}}{2}i$ are not roots of $x^2 - x - 1 = 0$.

In Exercises 27–34 use the square-root property of equality to solve each equation. You may need to factor an expression.

27. $x^2 = 9$ **28.** $x^2 = 20$ **29.** $y^2 - 50 = 0$

30. $x^2 - 75 = 0$ **31.** $(x - 1)^2 = 4$ **32.** $(y + 2)^2 - 49 = 0$

33. $a^2 + 2a + 1 = 9$ **34.** $x^2 - 6x + 9 = 4$

In Exercises 35–44 solve each equation by completing the square.

35. $x^2 - 8x + 15 = 0$

36. $x^2 + 10x + 21 = 0$

37. $x^2 + x - 6 = 0$

38. $x^2 - 9x + 20 = 0$

39. $x^2 - 25x = 0$

40. $x^2 - 4 = 0$

41. $3x^2 - 4 = -4x$

42. $6x^2 = x + 2$

43. $x^2 + 5 = -5x$

44. $2x^2 = 3x + 1$

In Exercises 45–56 solve each equation by using the quadratic formula.

45. $x^2 - 12 = 0$

46. $x^2 + 5x = 0$

47. $2x^2 - x - 15 = 0$

48. $6x^2 + x - 2 = 0$

49. $2x(x + 3) + 1 = 0$

50. $5x(3 - x) = 3x + 4$

51. $5x\left(x + \dfrac{1}{5}\right) = 3$

52. $3x\left(x - \dfrac{2}{3}\right) = -x$

53. $x^2 + 2x + 2 = 0$

54. $x^2 + 4x + 8 = 0$

55. $x^2 + 4x = -5$

56. $x^2 = -(2x + 5)$

57. Find the three cube roots of 8.

58. Find the three cube roots of -1.

In Exercises 59–64 solve each equation.

59. $x + 1 = \dfrac{12}{x}$

60. $x - 2 = \dfrac{15}{x}$

61. $8x - \dfrac{3}{x} = 10$

62. $15x - \dfrac{4}{x} = 4$

63. $x\left(30 - \dfrac{13}{x}\right) = \dfrac{10}{x}$

64. $x\left(20 - \dfrac{17}{x}\right) = \dfrac{10}{x}$

In Exercises 65–68 solve for the indicated variable.

65. $y = ax^2 + ax + a$; for x

66. $h = -16t^2 + 400t$; for t

67. $xy = ax - y^2$; for y

68. $x(x + y) = y^2$; for x

In Exercises 69–90 solve each equation.

69. $\sqrt{x^2 + 21} = x + 3$

70. $\sqrt{3 + x^2} = -(x + 1)$

71. $\sqrt{3x + 1} = x - 1$

72. $\sqrt{6x^2 + 6} = \sqrt{11 - 7x}$

73. $3 + \sqrt{x + 7} = \sqrt{x + 34}$

74. $x + 4 = \sqrt{\dfrac{6x + 6}{5}} + 3$

75. $\sqrt{\dfrac{x^2 - 1}{x - 2}} = 2\sqrt{2}$

76. $\dfrac{\sqrt{x^2 - 1}}{\sqrt{3x - 5}} = \sqrt{2}$

77. $x - 13x^{1/2} + 12 = 0$

78. $x + x^{1/2} - 20 = 0$

79. $2x^{1/3} + 3x^{1/6} = 2$

80. $6x^{2/3} - x^{1/3} = 2$

81. $x^{1/2} - x^{-3/2} = 0$

82. $2x^{3/2} - x^{1/2} - x^{-1/2} = 0$

83. $(x^2 - 2x)^2 - 11(x^2 - 2x) + 24 = 0$

84. $6(y + 2)^2 - 27(y + 2) + 27 = 0$

85. $\dfrac{1}{x} + \dfrac{3}{x + 2} = 2$

86. $\dfrac{1}{x - 1} + \dfrac{1}{x - 4} = \dfrac{5}{4}$

87. $\dfrac{1}{x + 1} + \dfrac{5}{2x - 4} = 1$

88. $x + 1 + \dfrac{x + 2}{x - 1} = \dfrac{3}{x - 1}$

89. $x^{-4} - 13x^{-2} + 36 = 0$

90. $2x^{-4} - 11x^{-2} = 14$

In Exercises 91–96 use the discriminant to determine the nature of the roots of each equation. **Do not solve the equations.**

91. $x^2 + 6x + 9 = 0$ **92.** $x^2 - 5x + 2 = 0$ **93.** $3x^2 - 2x + 5 = 0$

94. $9x^2 + 42x + 49 = 0$ **95.** $10x^2 + 29x = 21$ **96.** $10x^2 + x = 21$

97. Find two values of k such that the equation $x^2 + kx + 3k - 5 = 0$ will have a double root; that is, the roots will be equal.

98. For what value of b will the solutions of the equation $x^2 - 2bx + b^2 = 0$ be equal?

99. Does the equation $1492x^2 + 1984x - 1776 = 0$ have any roots that are real numbers?

100. Does the equation $2004x^2 + 10x + 1985 = 0$ have any roots that are real numbers?

101. If r_1 and r_2 are the roots of the quadratic equation $ax^2 + bx + c = 0$, find the value of $r_1 + r_2$ and $r_1 r_2$ in terms of a, b, and c. These results can be used to check the solution of a quadratic equation.

102. Find five consecutive integers a, b, c, d, and e such that $a^2 + b^2 + c^2 = d^2 + e^2$.

In Exercises 103–104 let $z = a + bi$. Asssume that $\bar{z}$ represents the conjugate of z: $\bar{z} = a - bi$. These exercises will be referred to later in the book.

103. Prove that the conjugate of the sum of two complex numbers is equal to the sum of their conjugates:

$$\overline{(a + bi) + (c + di)} = \overline{a + bi} + \overline{c + di}$$

104. Prove that the conjugate of the product of two complex numbers is equal to the product of their conjugates:

$$\overline{(a + bi)(c + di)} = \overline{(a + bi)}\,\overline{(c + di)}$$

105. Show that the addition of two complex numbers is commutative. Do this by adding the complex numbers $a + bi$ and $c + di$ in both orders, and observing that the sums are equal.

106. Show that the multiplication of two complex numbers is commutative. Do this by multiplying the complex numbers $a + bi$ and $c + di$ in both orders, and observing that the products are equal.

107. Show that the addition of complex numbers is associative.

108. Show that the multiplication of complex numbers is associative.

1.7   INEQUALITIES AND ABSOLUTE VALUE

Recall that, if a and b are real numbers, then $a < b$ if and only if there is a positive number p such that $b - a = p$. Recall also that, if $a < b$, then $b > a$. If a is less than or equal to b, we write $a \leq b$, and if a is greater than or equal to b, we write $a \geq b$.

Inequalities such as $3x - 7 < 2$ are called **linear inequalities**. Values of x that satisfy the inequality are called its **solutions**. If the numbers **0** and **1**, for example, are substituted for x in the inequality $3x - 7 < 2$, the true statements $-7 < 2$ and $-4 < 2$ result. Thus, **0** and **1** are two of the inequality's infinitely many solutions. To **solve** an inequality is to find all of its solutions. Two inequalities with the same solution set are called **equivalent inequalities**.

Several facts enable us to solve inequalities as if they were equations: Any real number can be added to (or subtracted from) both sides of an inequality to obtain another inequality with the same order (direction). Similarly, an inequality can be multiplied (or divided) by any *positive* real number to obtain another inequality with the *same* order. However, if both sides of an inequality are multiplied (or divided) by a *negative* number, the order of the inequality is reversed. If a, b, and c are real numbers, the following theorem holds.

Theorem.

1. If $a < b$, then $a + c < b + c$.
2. If $a < b$, then $a - c < b - c$.
3. If $a < b$ and $c > 0$, then $ca < cb$ and $\dfrac{a}{c} < \dfrac{b}{c}$.
4. If $a < b$ and $c < 0$, then $ca > cb$ and $\dfrac{a}{c} > \dfrac{b}{c}$.
5. If $a < b$ and $b < c$, then $a < c$.

We prove the first part of this theorem and leave the rest as exercises.

Proof *Part 1.* Suppose that a, b, and c are real numbers and that $a < b$. Then by definition there is a *positive* number p such that

$$b - a = p$$

We add zero in the form of $c - c$ to the left side of the equation and complete the proof as follows:

$$b - a + c - c = p$$
$$b + c - a - c = p \qquad \text{Use the commutative property of addition.}$$
$$(b + c) - (a + c) = p \qquad \text{Use the associative property of addition and factor out } -1.$$

By the definition of $<$ and the fact that p is positive, we have

$$a + c < b + c \qquad\qquad\qquad\qquad\qquad\qquad\qquad \square$$

Example 1 Solve $-5(x - 2) \le 20 + x$ and graph the solution set.

Solution Proceed as with equations, removing parentheses and solving for x:

$$-5(x - 2) \le 20 + x$$
$$-5x + 10 \le 20 + x$$
$$-6x + 10 \le 20 \qquad \text{Add } -x \text{ to both sides.}$$
$$-6x \le 10 \qquad \text{Subtract 10 from both sides.}$$
$$x \ge \frac{10}{-6} \qquad \text{Divide both sides by } -6 \text{ and change the order of the inequality.}$$

or

$$x \ge -\frac{5}{3}$$

The graph of the solution set of the given inequality is shown in Figure 1–10. The solution is the **infinite interval** from $-\frac{5}{3}$ to infinity, denoted as $[-\frac{5}{3}, \infty)$.

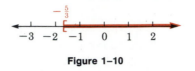

Figure 1–10 ▪

Example 2   Solve $5 < 3x - 7 \le 8$ and graph the solution set.

Solution   Two inequalities are involved in this expression. Because a constant can be added to both sides of each inequality, add 7 to each part of the **double inequality** and proceed as follows:

$$5 < 3x - 7 \le 8$$
$$5 + 7 < 3x - 7 + 7 \le 8 + 7$$
$$12 < 3x \le 15$$
$$4 < x \le 5 \qquad \text{Divide each part by 3.}$$

This solution is the interval $(4, 5]$ shown in Figure 1–11.

Figure 1–11 ▪

Example 3   Solve $3 + x \le 3x + 1 < 7x - 2$.

Solution   Each part of this double inequality must be solved separately. Note that $3 + x \le 3x + 1 < 7x - 2$ is equivalent to

$$3 + x \le 3x + 1 \qquad \text{and} \qquad 3x + 1 < 7x - 2$$

Thus,

$$
\begin{array}{ccl}
3 + x \le 3x + 1 & \text{and} & 3x + 1 < 7x - 2 \\
3 \le 2x + 1 & & 1 < 4x - 2 \\
2 \le 2x & & 3 < 4x \\
1 \le x & & \dfrac{3}{4} < x
\end{array}
$$

The solutions of the given inequality are those numbers x that are greater than or equal to 1 *and also* greater than $\frac{3}{4}$. Because any numbers greater than or equal to 1 will also be greater than $\frac{3}{4}$, the solution is $x \ge 1$. In interval notation, the solution is $[1, \infty) \cap (\frac{3}{4}, \infty)$, or $[1, \infty)$. Its graph is shown in Figure 1–12.

Figure 1–12 ▪

The following theorem, which we accept without proof, provides a method for solving inequalities involving polynomials.

Theorem. Suppose that $P(x)$ is a polynomial and the real numbers a and b are successive solutions of the equation $P(x) = 0$. Then either $P(x)$ is positive for all values of x on the interval (a, b), or $P(x)$ is negative for those values.

To solve a polynomial inequality such as $P(x) < 0$, we would first solve the equation $P(x) = 0$ and list the real roots in increasing order. For each pair a and b of successive roots in this list, we would then choose a **test value** t on the interval (a, b). If $P(t) < 0$, then t is a solution of the given inequality and, by the previous theorem, so are *all* numbers on the interval (a, b). Thus, the interval (a, b) is part of the solution set.

Example 4 Solve $x^3 + 5x^2 - 6x < 0$.

Solution Form the equation $x^3 + 5x^2 - 6x = 0$ and find its roots:

$$x^3 + 5x^2 - 6x = 0$$
$$x(x^2 + 5x - 6) = 0$$
$$x(x + 6)(x - 1) = 0$$

$x = 0$ or $x + 6 = 0$ or $x - 1 = 0$
$x = -6$ $x = 1$

The roots listed in increasing order are -6, 0, and 1. They determine four intervals: $(-\infty, -6)$, $(-6, 0)$, $(0, 1)$, and $(1, \infty)$. In each interval choose a test value t such as those indicated in Figure 1–13. Because $-10 \in (-\infty, -6)$, for example, choose $t = -10$ and substitute -10 for x in the inequality:

$$x^3 + 5x^2 - 6x < 0$$
$$(-10)^3 + 5(-10)^2 - 6(-10) < 0$$
$$-1000 + 500 + 60 < 0$$
$$-440 < 0$$

Because the inequality is satisfied, *all* numbers on the interval $(-\infty, -6)$ are part of the solution set. Now try a number such as -1, which lies in the interval $(-6, 0)$. Substituting $t = -1$ for x gives

$$x^3 + 5x^2 - 6x < 0$$
$$(-1)^3 + 5(-1)^2 - 6(-1) < 0$$
$$-1 + 5 + 6 < 0$$
$$10 < 0$$

Because the inequality is *not* satisfied, *no* numbers on the interval $(-6, 0)$ are part of the solution set.

Similarly, try test values t such as $\frac{1}{2}$ and 2 to determine that the interval $(0, 1)$ is part of the solution set and that the interval $(1, \infty)$ is not. The graph of the final solution appears in Figure 1–13.

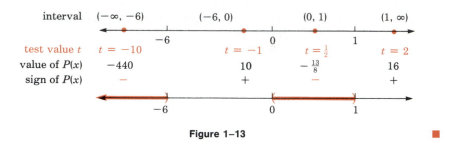

Figure 1–13

Example 5   Solve the inequality $\dfrac{x^2 - 9}{x^2} \geq 0$.

Solution   Although the left side of the inequality is not a polynomial, the technique of Example 4 still applies. However, the intervals are determined by the solutions of the two equations $x^2 - 9 = 0$ and $x^2 = 0$. These solutions, arranged in increasing order, are -3, 0, and 3. They divide the number line into four intervals, as in Figure 1–14. Use test values t in each of these intervals to determine which intervals are part of the solution set.

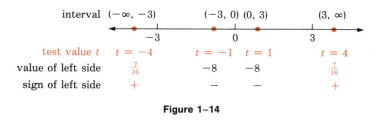

Figure 1–14

Now determine which endpoints of the intervals are solutions of the inequality. Verify that 3 and -3 are solutions and that 0 is not. Thus, the solution set of the given inequality is the set

$$(-\infty, -3] \cup [3, \infty)$$

The graph appears in Figure 1–15.

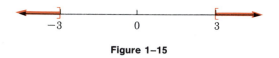

Figure 1–15

Absolute Value Equations and Inequalities

Solving many equations and inequalities that involve absolute values requires the use of the following theorem.

> **Theorem.**   If $a > 0$, then
> 1. $|x| = a$ is equivalent to $x = a$ or $x = -a$.
> 2. $|x| < a$ is equivalent to $-a < x < a$.
> 3. $|x| > a$ is equivalent to $x > a$ or $x < -a$.
>
> The symbols $>$ and $<$ can be replaced by $\geq$ and $\leq$, respectively.

We will prove Parts 2 and 3 of the preceding theorem and leave the proof of Part 1 as an exercise.

Proof *Part 2.*   Suppose that $a > 0$ and that $|x| < a$. We consider two cases.

Case 1

Let $x \geq 0$. Then, by the definition of absolute value, $|x| = x$, and the inequality $|x| < a$ is equivalent to $0 \leq x < a$. Because $-a$ is negative, we have

$$-a < x < a$$

Case 2

Let $x < 0$. Then, by the definition of absolute value, $|x| = -x$, and the inequality $|x| < a$ is equivalent to $0 < -x < a$, or $0 > x > -a$. Because a is positive, we have $a > x > -a$, or

$$-a < x < a$$

Hence, for all x

$$\text{if} \quad |x| < a, \qquad \text{then} \qquad -a < x < a$$

Part 2 of the theorem is proved.   □

Amalie Noether (1882–1935) Emmy Noether worked in abstract algebra. Einstein called her the most creative woman mathematical genius since higher education of women began.

Proof *Part 3.*   Suppose that $a > 0$ and $|x| > a$. Again we consider two cases.

Case 1

Let $x \geq 0$. Then $|x| = x$ and the inequality $|x| > a$ becomes $x > a$.

Case 2

Let $x < 0$. Then $|x| = -x$ and the inequality $|x| > a$ becomes $-x > a$, or $x < -a$.

Combining the results of Cases 1 and 2, it is apparent that positive values of x must be greater than a, whereas negative values of x must be less than $-a$.
Hence,

$$\text{if } |x| > a, \qquad \text{then either} \quad x > a \quad \text{or} \quad x < -a$$

Part 3 of the theorem is proved.   □

The graph of the inequality $|x| < a$ (for $a > 0$) includes all points lying *less* than a units from the origin. See Figure 1–16a. The graph of the inequality $|x| > a$ (for $a > 0$) includes all points lying *more* than a units from the origin,

for $a > 0$

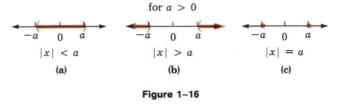

$\|x\| < a$	$\|x\| > a$	$\|x\| = a$
(a)	(b)	(c)

Figure 1–16

as in Figure 1–16b. The graph of the equation $|x| = a$ (for $a > 0$) includes the two points that lie *exactly* a units from the origin, as in Figure 1–16c.

Example 6 Solve the equation $|3x - 5| = 7$.

Solution The equation $|3x - 5| = 7$ is equivalent to the two equations

$$3x - 5 = 7 \quad \text{or} \quad 3x - 5 = -7$$

which can be solved separately:

$$
\begin{array}{lcl}
3x - 5 = 7 & \text{or} & 3x - 5 = -7 \\
3x = 12 & & 3x = -2 \\
x = 4 & & x = -\dfrac{2}{3}
\end{array}
$$

Figure 1–17

The graph of the solution set consists of the two points shown in Figure 1–17. ∎

Example 7 Solve the equation $|2x| = |x - 3|$.

Solution The equation $|2x| = |x - 3|$ will be satisfied if the quantities $2x$ and $x - 3$ are either equal to each other or are the negatives of each other. That idea determines two equations, which can be solved separately:

$$
\begin{array}{lcl}
2x = x - 3 & \text{or} & 2x = -(x - 3) \\
x = -3 & & 2x = -x + 3 \\
& & 3x = 3 \\
& & x = 1
\end{array}
$$

Both $x = -3$ and $x = 1$ satisfy the given equation. ∎

Example 8 Find the solution of the inequality $|x - 2| \le 7$ and graph it on the number line.

Solution The inequality $|x - 2| \le 7$ is equivalent to the double inequality

$$-7 \le x - 2 \le 7$$

Add 2 to each part to obtain

Figure 1–18

$$-5 \le x \le 9$$

The graph of this solution set is shown in Figure 1–18. ∎

Example 9   Solve the inequality $\left| \dfrac{2x + 3}{2} \right| > 5$ and graph its solution set.

Solution   The given inequality is equivalent to the following two inequalities, which can be solved separately:

$$\dfrac{2x + 3}{2} > 5 \qquad \text{or} \qquad \dfrac{2x + 3}{2} < -5$$
$$2x + 3 > 10 \qquad\qquad 2x + 3 < -10$$
$$2x > 7 \qquad\qquad\qquad 2x < -13$$
$$x > \dfrac{7}{2} \qquad\qquad\qquad x < -\dfrac{13}{2}$$

The graph of the solution set appears in Figure 1–19.

Figure 1–19

Example 10   Express the double inequality $-3 \le x \le 7$ as an absolute value inequality.

Solution   Midway between -3 and 7 is their average,

$$\dfrac{-3 + 7}{2} = 2$$

Subtracting **2** from each part of the inequality gives

$$-3 \le x \le 7$$
$$-3 - \mathbf{2} \le x - \mathbf{2} \le 7 - \mathbf{2}$$
$$-5 \le x - 2 \le 5$$

The double inequality is equivalent to the absolute value inequality

$$\left| x - 2 \right| \le 5$$

If a is the coordinate of a fixed point A on the number line and if δ (delta) is a positive real number, the graph of all numbers x that satisfy the inequality

$$\left| x - a \right| < \delta$$

is the interval $(a - \delta, a + \delta)$ shown in Figure 1–20. This interval, called a **δ-neighborhood of a**, consists of all points lying less than δ units from A.

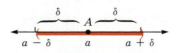

Figure 1–20

Example 11   If $\delta = 0.02$, graph a δ-neighborhood of 5.

Solution   Solve the inequality

$$|x - a| < \delta$$

where $\delta = 0.02$ and $a = 5$, as follows. The graph of the solution set appears in Figure 1–21.

$$|x - 5| < 0.02$$
$$-0.02 < x - 5 < 0.02$$
$$-0.02 + 5 < x < 0.02 + 5$$
$$4.98 < x < 5.02$$

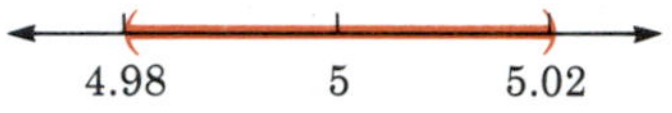

Figure 1–21   ■

Recall that, if a is any real number, we must use absolute value notation to guarantee that the principal square root of a^2 is nonnegative. Thus,

$$|a| = \sqrt{a^2}$$

This fact can be used to solve certain inequalities.

Example 12   Solve the inequality $|x + 2| > |x + 1|$.

Solution
$$|x + 2| > |x + 1|$$
$$\sqrt{(x + 2)^2} > \sqrt{(x + 1)^2} \qquad \text{Use } |a| = \sqrt{a^2}.$$
$$(x + 2)^2 > (x + 1)^2 \qquad \text{Square both sides.}$$
$$x^2 + 4x + 4 > x^2 + 2x + 1$$
$$2x > -3$$
$$x > -\frac{3}{2}$$

In interval notation the solution is $(-\frac{3}{2}, \infty)$.   ■

▬▬ Exercise 1.7 ▬▬▬▬▬▬▬▬▬▬▬▬▬▬▬▬▬▬▬

In Exercises 1–26 solve each inequality and graph the solution set.

1. $3x + 2 < 5$

2. $-2x + 4 < 6$

3. $3x + 2 \geq 5$

4. $-2x + 4 \geq 6$

5. $-5x + 3 > -2$

6. $4x - 3 > -4$

7. $\frac{3}{5}x + 4 > 2$

8. $\frac{1}{4}x - 3 \geq 5$

9. $2(x - 3) \leq -2(x - 3)$

10. $3(x + 2) \geq 2(x + 5)$

11. $\frac{3 + x}{4} \leq \frac{2x - 4}{3}$

12. $\frac{2 + x}{5} > \frac{x - 1}{2}$

13. $\frac{6(x - 4)}{5} \geq \frac{3(x + 2)}{4}$

14. $\frac{3(x + 3)}{2} < \frac{2(x + 7)}{3}$

15. $4 < 2x - 8 \leq 10$

16. $3 \le 2x + 2 < 6$

17. $9 \ge \dfrac{x - 4}{2} > 2$

18. $5 < \dfrac{x - 2}{6} < 20$

19. $0 \le \dfrac{4 - x}{3} \le 5$

20. $0 \ge \dfrac{5 - x}{6} \ge -10$

21. $-3x > -2x > -x$

22. $-3x < -2x < -x$

23. $2x + 1 < 3x - 2 < 12$

24. $2 - x < 3x + 5 < 18$

25. $2 + x \le 3x - 2 \le 5x + 2$

26. $x > 2x + 3 > 4x - 7$

In Exercises 27–40 solve each inequality and graph the solution set.

27. $x(x - 2) \ge 0$

28. $(x + 2)(x - 3) \le 0$

29. $x(x^2 + 1) \le 0$

30. $x^2(x + 3) > 0$

31. $x^2 + 3x > 4$

32. $x^2 + 2x < -1$

33. $2x^3 - 5x^2 - 3x < 0$

34. $x^3 - 3x^2 - 10x > 0$

35. $\dfrac{x + 3}{x - 2} \ge 0$

36. $\dfrac{x - 1}{x^2 - 4} \le 0$

37. $\dfrac{x^2 - 9}{x^2 - 4} \le 0$

38. $\dfrac{x^2 + 3x + 2}{x^2 - x} < 0$

39. $\dfrac{3}{x} > 2$ (*Hint:* Subtract 2 from both sides and add terms.)

40. $\dfrac{4}{x + 2} \le 5$

In Exercises 41–48 solve each equation.

41. $|3x - 1| = 5$

42. $|7x - 5| = 3$

43. $\left|\dfrac{3x - 4}{2}\right| = 5$

44. $\left|5x + \dfrac{1}{2}\right| = \dfrac{9}{2}$

45. $|x + 3| = |x|$

46. $|x + 5| = |5 - x|$

47. $\left|\dfrac{x + 3}{2}\right| = |2x - 3|$

48. $\left|\dfrac{x - 2}{3}\right| = |6 - x|$

In Exercises 49–62 solve each inequality and graph the solution set.

49. $|x - 3| < 6$

50. $|x - 2| \ge 4$

51. $|2x + 4| \ge 10$

52. $|5x - 2| \le 7$

53. $|x + 3| > 0$

54. $|x - 3| \ge 0$

55. $\left|\dfrac{5x + 2}{3}\right| < 1$

56. $\left|\dfrac{3x + 2}{4}\right| > 2$

57. $\left|\dfrac{x + 3}{2}\right| < \dfrac{2 - x}{4}$

58. $\left|\dfrac{3 - x}{2}\right| > \dfrac{x + 1}{6}$

59. $0 < |2x + 1| < 3$

60. $0 < |2x - 3| \le 1$

61. $2 < \left|\dfrac{x - 5}{3}\right| \le 4$

62. $8 > \left|\dfrac{3x - 1}{3}\right| > 3$

In Exercises 63–70 express each inequality as an absolute value inequality.

63. $3 < x < 7$

64. $-3 < x < 9$

65. $-\dfrac{1}{2} \le x \le \dfrac{9}{2}$

66. $\dfrac{3}{2} \le x \le \dfrac{11}{2}$

67. $x \ne 2$ and $0 < x < 4$

68. $x \ne 0$ and $-5 < x < 5$

69. $x \le 0$ or $x \ge 4$

70. $x \le -3$ or $x \ge 7$

In Exercises 71–72 graph a δ-neighborhood of a for the given values of a and δ.

71. $a = 3.2,\ \delta = 0.1$ **72.** $a = -2.1,\ \delta = 0.01$

In Exercises 73–78 solve each inequality and express each solution set in interval notation.

73. $|x + 1| \geq |x|$ **74.** $|x + 1| < |x + 2|$

75. $|2x + 1| < |2x - 1|$ **76.** $|3x - 2| \geq |3x + 1|$

77. $|2x - 1| > |x + 2|$ **78.** $|x - 3| < |3x - 11|$

79. An airline pilot flies at altitudes from 27,720 feet to 33,000 feet. Express this interval in miles. (*Hint:* One mile = 5280 feet.)

80. The perimeter of a rectangle is to be between 180 inches and 200 inches. What is the range of values for its length if the width is to be 40 inches?

81. The perimeter of an equilateral triangle is to be between 50 centimeters and 60 centimeters. What is the range of values for the length of one side?

82. The perimeter of a square is to be between 25 meters and 60 meters. What is the range of values for its area?

83. Express the relationship $20 < l \leq 30$ in terms of P, where $P = 2l + 2w$.

84. Express the relationship $10 < C < 20$ in terms of F, where $F = \frac{9}{5}C + 32$.

85. A department store manager is paid $7.50 per hour, which is from $2.00 to $3.50 per hour more than that earned by the sales clerks in her department. What is the range of hourly rates for the clerks? What is the range of hours these clerks must work to earn as much as the manager earns in 40 hours?

86. One acute angle of a right triangle is not less than 30° more than twice the other acute angle. What are the possible measures of the larger angle?

In Exercises 87–94 assume that a, b, and c are real numbers. Prove each statement.

87. If $a < b$, then $a - c < b - c$.   **88.** If $a < b$ and $c > 0$, then $ca < cb$.

89. If $a < b$ and $c > 0$, then $a/c < b/c$.   **90.** If $a < b$ and $c < 0$, then $ca > cb$.

91. If $a < b$ and $c < 0$, then $a/c > b/c$.   **92.** If $a < b$ and $b < c$, then $a < c$.

93. $|x| = a$ is equivalent to $x = a$ and $x = -a$.   **94.** If $a < b$ and $c > 0$, then $a < b + c$.

95. Assume that a, b, c, and d are positive numbers. Prove that, if $\dfrac{a}{b} < \dfrac{c}{d}$, then $\dfrac{b}{a} > \dfrac{d}{c}$.

REVIEW EXERCISES

In Review Exercises 1–2 classify each number as a positive or a negative integer, a prime number or a composite number, an even or an odd integer, and a rational or an irrational number, if possible.

1. -87 **2.** 3π

In Review Exercises 3–4 let x = -2, y = 3, and z = 1. Find each value.

3. $|xyz|$ **4.** $\dfrac{-3y + xz}{4x - yz}$

In Review Exercises 5–6 indicate which property of real numbers justifies each statement.

5. $(a + b) + 2 = a + (b + 2)$ **6.** $2(a + b) = 2a + 2b$

In Review Exercises 7–8 graph each interval on the number line.

7. $-3 < x \le \dfrac{1}{3}$

8. $[-2, 4] \cap (0, 5]$

In Review Exercises 9–10 the coordinate of point A on the number line is -6, the coordinate of point B is 3, and the coordinate of point C is -1. Find each distance.

9. $d(AB)$

10. $d(AC)$

In Review Exercises 11–14 simplify each expression. Express all answers with positive exponents only. Assume that all variables are positive numbers.

11. $(x^2y^3)^2$

12. $\left(\dfrac{x^{-2}y^2}{3x^2}\right)^3$

13. $\left(\dfrac{3x^2y}{x^2y^2}\right)^{-2}$

14. $\left(\dfrac{-3x^3y}{xy^3}\right)^{-3}$

In Review Exercises 15–16 evaluate each expression if $x = 0$, $y = -1$, and $z = 2$.

15. $3^x y^2$

16. $(y^{zy}z^y)^y$

In Review Exercises 17–20 simplify each expression.

17. $-\sqrt{36}$

18. $-\sqrt[5]{-32}$

19. $36^{-1/2}$

20. $-16^{3/2}$

In Review Exercises 21–24 assume that all variables are unrestricted. Simplify each expression and use absolute value symbols as needed.

21. $\sqrt{x^2y^4}$

22. $\dfrac{\sqrt[3]{x^9y^3}}{\sqrt[3]{x^3}}$

23. $\sqrt{50} + \sqrt{8}$

24. $y\sqrt[3]{54x^4} - \sqrt[3]{250x^4y^3}$

In Review Exercises 25–28 assume that all variables represent positive real numbers. Simplify each expression and write all answers without using negative exponents.

25. $(x^{12}y^2)^{1/2}$

26. $\left(\dfrac{x^{14}}{y^4}\right)^{-1/2}$

27. $\left(\dfrac{64y^6}{27x^3}\right)^{-2/3}$

28. $(256a^8b^{-4})^{3/4}$

In Review Exercises 29–30 assume that all variables represent positive real numbers. Rationalize each denominator and simplify.

29. $\dfrac{8}{\sqrt{6}}$

30. $\sqrt[4]{\dfrac{y^4}{8x}}$

In Review Exercises 31–32 write each expression as a single radical.

31. $\sqrt{3}\sqrt[3]{3}$

32. $\dfrac{\sqrt{2}}{\sqrt[3]{2}}$

33. Write the number 0.000345 in scientific notation.

34. Write the number 6.21×10^5 in standard notation.

In Review Exercises 35–38 multiply and combine terms, if possible.

35. $2(3x^3 + 4) + 4(2x^2 - 3x)$

36. $3(2x^2 - 3x + 4) - 3(4x^2 - 5x + 1)$

37. $(3x - 2)(4x + 3)$

38. $(x^2 + 3)(2x^3 + 4x^2 - 1)$

39. Rationalize the denominator of $\dfrac{2}{\sqrt{x} - \sqrt{2}}$.

40. Rationalize the numerator of $\dfrac{3\sqrt{x} - h}{h}$.

In Review Exercises 41–42 simplify each fraction by performing a division.

41. $\dfrac{6x^3y^3 + 12x^2y^4 + 18xy^5}{12x^2y^2}$

42. $\dfrac{5 + x^3 - x}{x^2 + 3x}$

In Review Exercises 43–54 factor each polynomial, if possible.

43. $3x^3 - 3x$

44. $5x^4 - 5y^4$

45. $6x^2 + 7x - 24$

46. $3a^2 + ax - 3a - x$

47. $8x^3 + 125$

48. $6x^2 - 20x - 16$

49. $x^2 + 6x + 9 - 4t^2$

50. $1 + 14b + 4b^2$

51. $121z^2 + 4 - 44z$

52. $128y^3 - 2000$

53. $2xy - 4zx - wy + 2zw$

54. $x^4 + x^2 + 1$

In Review Exercises 55–60 perform the indicated operations and simplify.

55. $\dfrac{x^2 - 4x + 4}{x + 2} \cdot \dfrac{x^2 + 5x + 6}{x - 2}$

56. $\dfrac{2x^2 + x - 3}{3x^2 - 7x + 4} \div \dfrac{10x + 15}{3x^2 - x - 4}$

57. $\dfrac{2}{x - 4} + \dfrac{3x}{x + 5}$

58. $\dfrac{5x}{x - 2} - \dfrac{3x + 1}{x + 3}$

59. $\dfrac{x}{x + 1} - \dfrac{3x + 7}{x + 2} + \dfrac{2x + 1}{x + 2}$

60. $\dfrac{3x}{x + 1} + \dfrac{x^2 + 4x + 3}{x^2 + 3x + 2} - \dfrac{x^2 + x - 6}{x^2 - 4}$

In Review Exercises 61–62 simplify each complex fraction.

61. $\dfrac{\dfrac{5x}{2}}{\dfrac{3x^2}{8}}$

62. $\dfrac{\dfrac{1}{x} + \dfrac{1}{y}}{\dfrac{1}{y} - \dfrac{1}{x}}$

In Review Exercises 63–66 solve each equation, if possible.

63. $8(3x - 5) - 4(2x + 3) = 12$

64. $\dfrac{x + 3}{x + 4} + \dfrac{x + 3}{x + 2} = 2$

65. $\dfrac{3x}{x - 1} - \dfrac{5}{x + 3} = 3$

66. $x + \dfrac{1}{2x - 3} = \dfrac{2x^2}{2x - 3}$

67. Solve $\dfrac{1}{f} = \dfrac{1}{f_1} + \dfrac{1}{f_2}$ for f_1.

68. Solve $\dfrac{y}{x} = \dfrac{x + y}{2}$ for x.

In Review Exercises 69–70 solve each equation by factoring.

69. $5x^2 - 8x = 0$

70. $27x^3 = 30x^2 - 8x$

In Review Exercises 71–74 perform all indicated operations and express all answers in $a + bi$ form.

71. $(2 + 3i) + (4 - 2i)$

72. $(3 - \sqrt{-9})(2 + \sqrt{-25})$

73. $\dfrac{3}{1 + i}$

74. $\dfrac{3 + i}{1 - i}$

75. Use the method of completing the square to solve $x^2 + 8x + 15 = 0$.

76. Use the quadratic formula to solve $a^2 + 2a = -5$.

In Review Exercises 77–84 solve each equation.

77. $\dfrac{3x}{2} - \dfrac{2x}{x-1} = x - 3$

78. $\dfrac{12}{x} - \dfrac{x}{2} = x - 3$

79. $x^4 - 2x^2 + 1 = 0$

80. $x^4 + 36 = 37x^2$

81. $\sqrt{x-1} + x = 7$

82. $\sqrt{5-x} + \sqrt{5+x} = 4$

83. $\sqrt{a+9} - \sqrt{a} = 3$

84. $\sqrt{y+5} + \sqrt{y} = 1$

85. Find the value of k that will cause the roots of $kx^2 + 4x + 12 = 0$ to be equal.

86. Find the values of k that will cause the roots of $4y^2 + (k+2)y = 1 - k$ to be equal.

In Review Exercises 87–92 solve each word problem.

87. A liter of fluid is 50% alcohol. How much water must be added to dilute it to a 20% solution?

88. Scott can wash 37 windows in 3 hours, and Bill can wash 27 windows in 2 hours. How long will it take the two of them to wash 100 windows?

89. A tank can be filled in 9 hours by one pipe, and in 12 hours by another. How long will it take both pipes to fill the empty tank?

90. How many ounces of pure zinc must be alloyed with 20 ounces of brass that is 30% zinc and 70% copper to produce brass that is 40% zinc?

91. A bank loans $10,000, part of it at 11% annual interest, and the rest at 14%. If the annual income is $1265, how much was loaned at each rate?

92. A farmer wishes to enclose a rectangular garden with 300 yards of fencing. A river runs along one side of the garden, so no fencing is needed there. What will be the dimensions of the rectangle if the area is 10,450 square yards?

In Review Exercises 93–104 solve each equation or inequality and graph the solution set.

93. $4x - 2 \le 10$

94. $-3(x+4) > -2(x-2)$

95. $\dfrac{3-x}{5} \ge \dfrac{2x+4}{2}$

96. $5 < \dfrac{x+2}{3} < 20$

97. $x^2 - x - 6 \ge 0$

98. $\dfrac{x^2 + 2x + 1}{x-1} < 0$

99. $\left| \dfrac{5x+3}{2} \right| = 14$

100. $|2x+3| = |3x-6|$

101. $|x+3| \le 4$

102. $|x-2| > 5$

103. $2 < \dfrac{|x+5|}{2} < 4$

104. $|x+1| > |x+2|$

2 FUNCTIONS AND THEIR GRAPHS

Mathematical expressions often indicate relationships between several quantities. For example, the formula $d = rt$ indicates how distance traveled depends on the rate of speed and on time. Because d depends on the values of r and t, we say that d is a **function** of r and t. The concept of *function*, an idea used in all of mathematics, is the topic of this chapter.

2.1 THE CARTESIAN COORDINATE SYSTEM

Any point on a number line is associated with a single real number, called its **coordinate**. René Descartes is credited with the idea of associating each point in the *plane* with a *pair* of real numbers.

Descartes' idea is based on two perpendicular number lines, called the **coordinate axes**, that divide the plane into four **quadrants**, numbered as shown in Figure 2–1a. One axis, usually called the **x-axis**, is drawn horizontally. Its positive direction is to the right. The other axis, usually called the **y-axis**, is drawn vertically. Its positive direction is upward. These axes have the same unit distance and intersect at a point called the **origin**, which is the zero point on each

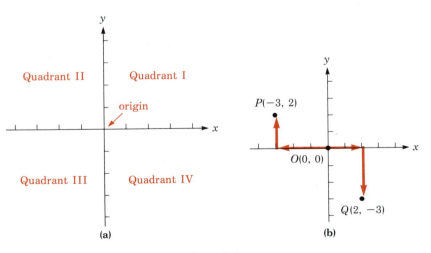

Figure 2–1

number line. These axes determine a **rectangular coordinate system**, or a **Cartesian coordinate system**, so named in honor of its inventor.

To **plot** the point associated with the pair of real numbers $(-3, 2)$, for example, we start at the origin, count 3 units to the left, and then 2 units up. (See Figure 2-1b.) Point P, called the **graph** of the point with **coordinates** $(-3, 2)$, lies in the second quadrant. Similarly, point Q with coordinates $(2, -3)$ lies in the fourth quadrant. The coordinates of the origin are $(0, 0)$.

Note that the pairs $(-3, 2)$ and $(2, -3)$ represent different points. Because the order of the numbers in the pair (a, b) is important, such pairs are called **ordered pairs**. The first coordinate, a, is called the **x-coordinate**, or the **abscissa**. The second coordinate, b, is called the **y-coordinate**, or the **ordinate**. It is proper to say "the point P with coordinates (a, b)," but it is also acceptable to say simply "the point $P(a, b)$." The set of all ordered pairs (x, y) is called the **xy-plane**, or the **Cartesian plane**.

Example 1   Graph the points $P(2, 3)$, $Q(-4, -2)$, $R(-5, 0)$, $S(0, 4)$, and $T(3, -2)$, and indicate in which quadrant each point lies or on which axis it lies.

Solution

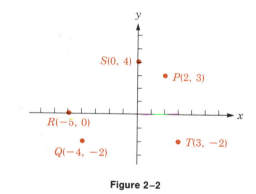

Figure 2-2

Coordinate systems enable us to draw "pictures" or graphs of equations in two variables. The **graph of an equation** in the two variables x and y is the set of all points on the xy-plane with coordinates (x, y) that satisfy the equation.

Example 2   Graph the equation $x + 2y = 5$.

Solution   To determine ordered pairs (x, y) that satisfy the equation, pick some arbitrary values for either x or y, substitute those values in the equation, and solve for the other variable. For example, if $y = -1$, you can find x as follows.

$$x + 2y = 5$$
$$x + 2(-1) = 5 \qquad \text{Substitute } -1 \text{ for } y.$$
$$x - 2 = 5 \qquad \text{Simplify.}$$
$$x = 7 \qquad \text{Add 2 to both sides.}$$

Thus, one ordered pair that satisfies the equation is $(7, -1)$. This ordered pair and others that satisfy the equation $x + 2y = 5$ are shown in the table of values in Figure 2–3. Plot each of these ordered pairs on a rectangular coordinate system as in the figure. Later, we will show that all points (x, y) that satisfy an equation such as $x + 2y = 5$ lie on a straight line. The line that joins the five points is the graph of the equation.

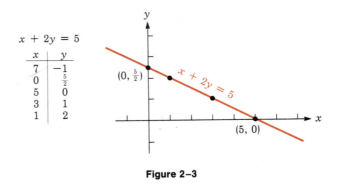

$x + 2y = 5$

x	y
7	-1
0	$\frac{5}{2}$
5	0
3	1
1	2

Figure 2–3

Because the line in Example 2 intersects the y-axis at the point $(0, \frac{5}{2})$, the number $\frac{5}{2}$ is called the **y-intercept** of the line. Similarly, 5 is the **x-intercept** of the line.

Example 3   Use the x- and y-intercepts to graph the line represented by the equation $4x - 3y = 12$.

Solution   Begin by finding the x- and y-intercepts. To find the x-intercept, substitute **0** for y and solve for x.

$$4x - 3y = 12$$
$$4x - 3(0) = 12 \qquad \text{Substitute 0 for } y.$$
$$4x = 12 \qquad \text{Simplify.}$$
$$x = 3 \qquad \text{Divide both sides by 4.}$$

The x-intercept is 3. The line intersects the x-axis at the point $(3, 0)$.
To find the y-intercept, substitute **0** for x and solve for y.

$$4x - 3y = 12$$
$$4(0) - 3y = 12 \qquad \text{Substitute 0 for } x.$$
$$-3y = 12 \qquad \text{Simplify.}$$
$$y = -4 \qquad \text{Divide both sides by } -3.$$

The y-intercept is -4. The line intersects the y-axis at the point $(0, -4)$.
Although these two points, if calculated correctly, are sufficient to draw the line, you should find and plot a third point as a check. If $x = 6$, for example, then $y = 4$. Thus, the line also passes through $(6, 4)$. The graph appears in Figure 2–4.

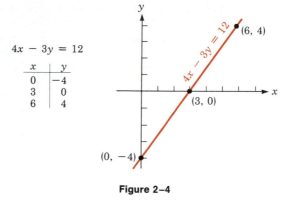

$4x - 3y = 12$

x	y
0	-4
3	0
6	4

Figure 2–4

Graphing an equation is often easier if the equation is first solved for one of its variables.

Example 4   Graph the equation $y + x - x^2 = 1$.

Solution   To determine several points whose coordinates satisfy the equation $y + x - x^2 = 1$, first solve the equation for y to obtain

$$y = x^2 - x + 1$$

Substitute several numbers x into the equation and calculate the corresponding values of y. Plot these points and join them with a smooth curve, as in Figure 2–5.

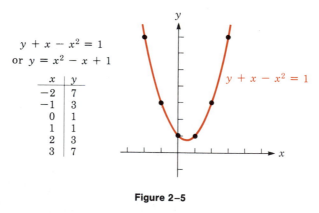

$y + x - x^2 = 1$
or $y = x^2 - x + 1$

x	y
-2	7
-1	3
0	1
1	1
2	3
3	7

Figure 2–5

The Distance Formula

To distinguish between two points on the Cartesian plane, subscript notation is often used. One point might be denoted as $P(x_1, y_1)$, read as "point P with coordinates x sub 1, y sub 1," and the second as $Q(x_2, y_2)$. We can derive a **distance formula** to calculate the distance $d(PQ)$ between these two points.

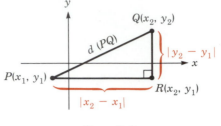

Figure 2–6

If the points $P(x_1, y_1)$ and $Q(x_2, y_2)$ do not lie on either a vertical or a horizontal line, PQ can be drawn as the hypotenuse of the right triangle PQR, with R having coordinates (x_2, y_1). (See Figure 2–6.) Because the line segment RQ is vertical, the square of its length is $(y_2 - y_1)^2$. Because PR is horizontal, the square of its length is $(x_2 - x_1)^2$. By the Pythagorean theorem we know that the square of the hypotenuse of right triangle PQR is equal to the sum of the squares of the two legs. Thus,

$$d(PQ)^2 = (x_2 - x_1)^2 + (y_2 - y_1)^2$$

Because $d(PQ)$ is nonnegative, taking the principal square root of both sides of this equation gives the **distance formula**.

The Distance Formula.   The distance $d(PQ)$ between points $P(x_1, y_1)$ and $Q(x_2, y_2)$ is given by

$$d(PQ) = \sqrt{(x_2 - x_1)^2 + (y_2 - y_1)^2}$$

The distance formula is valid even if points P and Q lie on a vertical line or a horizontal line.

Example 5   Show that the points $P(1, 7)$, $Q(-4, 2)$, and the origin O are the vertices of an isosceles triangle.

Solution   After plotting the points as in Figure 2–7, it appears that sides PQ and OP are of equal length. To verify this, find $d(OP)$ and $d(PQ)$. To do so, let $P(x_1, y_1) = P(1, 7)$, let $Q(x_2, y_2) = Q(-4, 2)$, and let $O(x_o, y_o) = O(0, 0)$ be the origin. Then,

$$d(OP) = \sqrt{(x_1 - x_o)^2 + (y_1 - y_o)^2} \qquad \text{and} \qquad d(PQ) = \sqrt{(x_2 - x_1)^2 + (y_2 - y_1)^2}$$
$$= \sqrt{(1 - 0)^2 + (7 - 0)^2} \qquad\qquad\qquad = \sqrt{(-4 - 1)^2 + (2 - 7)^2}$$
$$= \sqrt{1 + 49} \qquad\qquad\qquad\qquad\qquad = \sqrt{25 + 25}$$
$$= \sqrt{50} \qquad\qquad\qquad\qquad\qquad\qquad = \sqrt{50}$$
$$= 5\sqrt{2} \qquad\qquad\qquad\qquad\qquad\qquad = 5\sqrt{2}$$

Because the lengths of sides OP and PQ are equal, the triangle is isosceles.

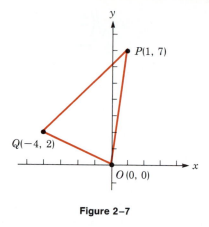

Figure 2–7

The Midpoint Formula

The **midpoint** of a line segment is the point on the segment that is midway between the endpoints. The x-coordinate of the midpoint is the average of the x-coordinates of the segment's endpoints. Similarly, the y-coordinate of the midpoint is the average of the y-coordinates of the segment's endpoints. For example, the midpoint of the segment joining $P(3, 4)$ and $Q(5, 7)$ is

$$M\left(\frac{3 + 5}{2}, \frac{4 + 7}{2}\right) \quad \text{or} \quad M\left(4, \frac{11}{2}\right)$$

Theorem (Midpoint Formula).   The midpoint $M(x_M, y_M)$ of the line segment joining $P(x_1, y_1)$ and $Q(x_2, y_2)$ is given by the formulas

$$x_M = \frac{x_1 + x_2}{2} \quad \text{and} \quad y_M = \frac{y_1 + y_2}{2}$$

Proof   To prove that M is the midpoint of segment PQ, we must show that $d(PM) = d(MQ)$ and that points P, Q, and M all lie on the same line. (See Figure 2–8.)

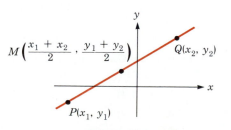

Figure 2–8

$$d(PM) = \sqrt{\left(\frac{x_1 + x_2}{2} - x_1\right)^2 + \left(\frac{y_1 + y_2}{2} - y_1\right)^2}$$

$$= \sqrt{\left(\frac{x_2 - x_1}{2}\right)^2 + \left(\frac{y_2 - y_1}{2}\right)^2}$$

$$= \sqrt{\frac{(x_2 - x_1)^2}{4} + \frac{(y_2 - y_1)^2}{4}}$$

$$= \frac{1}{2}\sqrt{(x_2 - x_1)^2 + (y_2 - y_1)^2}$$

$$d(MQ) = \sqrt{\left(x_2 - \frac{x_1 + x_2}{2}\right)^2 + \left(y_2 - \frac{y_1 + y_2}{2}\right)^2}$$

$$= \sqrt{\left(\frac{x_2 - x_1}{2}\right)^2 + \left(\frac{y_2 - y_1}{2}\right)^2}$$

$$= \sqrt{\frac{(x_2 - x_1)^2}{4} + \frac{(y_2 - y_1)^2}{4}}$$

$$= \frac{1}{2}\sqrt{(x_2 - x_1)^2 + (y_2 - y_1)^2}$$

Thus, $d(PM) = d(MQ)$. The distance between P and Q is

$$d(PQ) = \sqrt{(x_2 - x_1)^2 + (y_2 - y_1)^2}$$

Because $d(PM) + d(MQ) = d(PQ)$, points P, Q, and M must lie on the same line. The theorem is proved.   □

Example 6   The midpoint of the line segment joining $P(-3, 2)$ and $Q(x_2, y_2)$ is $M(1, 4)$. Find the coordinates of Q.

Solution   Let $P(x_1, y_1) = P(-3, 2)$, and let $M(x_M, y_M) = M(1, 4)$. Find $Q(x_2, y_2)$ by using the definition of midpoint and solving for x_2 and y_2.

$$x_M = \frac{x_1 + x_2}{2} \qquad \text{and} \qquad y_M = \frac{y_1 + y_2}{2}$$

$$1 = \frac{-3 + x_2}{2} \qquad\qquad 4 = \frac{2 + y_2}{2}$$

$$2 = -3 + x_2 \qquad\qquad 8 = 2 + y_2 \qquad \text{Multiply both sides by 2.}$$

$$5 = x_2 \qquad\qquad\qquad 6 = y_2$$

The coordinates of the endpoint Q are $(5, 6)$.   ■

▬ Exercise 2.1 ▬▬▬▬▬▬▬▬▬▬▬▬▬▬▬▬▬▬▬▬▬▬▬▬▬▬▬▬▬▬▬▬▬▬

In Exercises 1–12 graph each point on the Cartesian plane, and indicate the quadrant in which it lies or the axis on which it lies.

1. $(2, 5)$	**2.** $(-3, 4)$	**3.** $(-4, -5)$	**4.** $(6, -2)$
5. $(5, 2)$	**6.** $(3, -4)$	**7.** $(4, 0)$	**8.** $(-2, 2)$
9. $(0, 2)$	**10.** $(3, 4)$	**11.** $(-7, 0)$	**12.** $(2, -2)$

In Exercises 13–18 graph each equation.

13. $x + y = 5$ **14.** $x - y = 3$ **15.** $2x - y = 4$

16. $3x + y = 9$ **17.** $3x + 2y = 6$ **18.** $4x - 5y = 20$

In Exercises 19–24 use the x- and y-intercepts to graph each equation.

19. $2x + 7y = 14$ **20.** $3x - 5y = 15$ **21.** $y - 2x = 7$

22. $y + 3 = -4x$ **23.** $6x - 3y = 10$ **24.** $4x + 8y - 1 = 0$

In Exercises 25–30 first solve each equation for y, and then graph the equation.

25. $2(x - y) = 3x + 2$ **26.** $5(x + 2) = 3y - x$ **27.** $3x + y = 3(x - 1)$

28. $2(y - x) = 3(x + 2)$ **29.** $y + 2 = x(x + 1)$ **30.** $x(x + 3) = y - 2$

In Exercises 31–36 find the distance between P and the origin.

31. $P(4, -3)$ **32.** $P(-5, 12)$ **33.** $P(-3, 2)$

34. $P(-5, 0)$ **35.** $P(a, b)$ **36.** $P(b, -a)$

In Exercises 37–46 find the distance between P and Q.

37. $P(0, 5); Q(6, -3)$ **38.** $P(-2, 15); Q(-12, -10)$

39. $P(3, 3); Q(5, 5)$ **40.** $P(6, -3); Q(-3, 2)$ **41.** $P(3, 7); Q(5, 7)$ **42.** $P(4, -6); Q(4, -8)$

43. $P(\pi, 2); Q(\pi, 5)$ **44.** $P(\sqrt{3}, \pi); Q(\pi, \sqrt{3})$ **45.** $P(x, 0); Q(0, x)$ **46.** $P(a, b); Q(b, a)$

In Exercises 47–54 find the midpoint of the line segment PQ.

47. $P(2, 4); Q(6, 8)$ **48.** $P(3, -6); Q(-1, -6)$ **49.** $P(2, -5); Q(-2, 7)$

50. $P(0, 3); Q(-10, -13)$ **51.** $P(0, 0); Q(\sqrt{5}, \sqrt{5})$ **52.** $P(\sqrt{3}, 0); Q(0, \sqrt{3})$

53. $P(a, 2); Q(2, a)$ **54.** $P(a, b); Q(0, 0)$

In Exercises 55–58 one endpoint P and the midpoint M of line segment PQ are given. Find the coordinates of the other endpoint Q.

55. $P(1, 4); M(3, 5)$ **56.** $P(2, -7); M(-5, 6)$ **57.** $P(5, -5); M(5, 5)$ **58.** $P(-7, 3); M(0, 0)$

59. Show that a triangle with vertices at $(13, -2)$, $(9, -8)$, and $(5, -2)$ is isosceles (has two equal sides).

60. Show that a triangle with vertices at $(-1, 2)$, $(3, 1)$, and $(4, 5)$ is isosceles.

61. Show that the points $(-3, -1)$, $(3, 1)$, $(1, 7)$, and $(-5, 5)$ are the vertices of a rhombus by showing that its four sides are equal.

62. Find the coordinates of the three points that divide the line joining $(-3, 5)$ and $(8, -7)$ into four equal segments.

63. Use the distance formula to find the radius of a circle with center at $(4, -2)$ and passing through $(6, -9)$.

64. Find the center of a circle if the endpoints of a diameter are $(-17, 6)$ and $(9, -12)$.

65. Find the coordinates of the two points on the y-axis that are $\sqrt{74}$ units from the point $(5, -3)$.

66. The diagonals of a square with an area of 50 square units lie on the x- and y-axes. Find the coordinates of the vertices.

67. Show that the distance between the points (a, b) and (c, d) is equal to the distance between the point $(a - c, b - d)$ and the origin.

68. Use the distance formula to show that the midpoint of a line segment is the same distance from each of the endpoints.

69. A square has an area of one square unit. Find the area of the smaller square formed by joining the midpoints of the sides of the larger square. See Illustration 1.

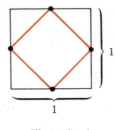

Illustration 1

70. In Illustration 2, M is the midpoint of the hypotenuse of right triangle AOB. Show that the area of rectangle $OLMN$ is one-half of the area of triangle AOB.

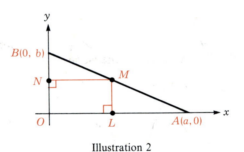

Illustration 2

71. The rectangle in Illustration 3 is twice as long as it is wide, and its sides are parallel to the coordinate axes. If its perimeter is 66 units, find the coordinates of vertex C.

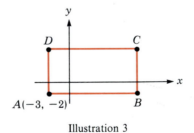

Illustration 3

72. The base AB of the isosceles triangle ABC in Illustration 4 is 5 units greater than side AC. If the perimeter is 77 units and its base is parallel to the x-axis, find the coordinates of vertex B.

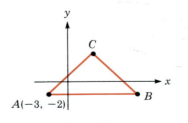

Illustration 4

2.2 SLOPE AND EQUATIONS OF LINES

The slope of a line drawn in the Cartesian plane is a measure of its inclination. We consider the line l in Figure 2–9a, which passes through $P(x_1, y_1)$ and $Q(x_2, y_2)$. If line QR is perpendicular to the x-axis and PR is perpendicular to the y-axis, then triangle PRQ is a right triangle and point R has coordinates of (x_2, y_1). The distance from R to Q, called the **rise** of segment PQ, is denoted as Δy (delta y). Thus, Δy is the change in the y-coordinates of P and Q: $\Delta y = y_2 - y_1$. The horizontal distance from P to R, called the **run** and denoted as Δx, is the change in the x-coordinates of points P and Q: $\Delta x = x_2 - x_1$. If $x_1 \neq x_2$, then the slope of the line, denoted as m, in Figure 2–9a is the *rise* divided by the *run*:

$$m = \frac{\Delta y}{\Delta x}$$

Definition. If $P(x_1, y_1)$ and $Q(x_2, y_2)$ are two points on a nonvertical line l in the Cartesian plane, the **slope of line l** is given by

$$m = \frac{\Delta y}{\Delta x} = \frac{y_2 - y_1}{x_2 - x_1}$$

If $x_1 = x_2$, then line l is a vertical line and has no defined slope.

In Figure 2–9b point S represents an arbitrary point on line l. Because triangles PRQ and PTS are similar, their corresponding sides are in proportion. Thus, the ratios of the rise to the run in the two triangles are equal, and we have

$$m = \frac{y_2 - y_1}{x_2 - x_1} = \frac{y_3 - y_1}{x_3 - x_1}$$

This fact implies that the slope of a nonvertical line is a constant that can be calculated by using *any* two points on the line. Furthermore, if point P is on a line with slope m and the ratio of rise to run of a segment PQ is also m, then point Q is also on the line.

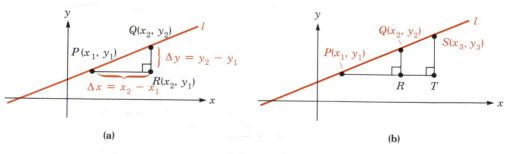

(a)   **(b)**

Figure 2–9

Example 1   Find the slope of the line passing through the points $P(-1, -2)$ and $Q(7, 8)$.

Solution   Let $x_1 = -1$, $y_1 = -2$, $x_2 = 7$, and $y_2 = 8$ in the formula that defines the slope, and simplify.

$$m = \frac{y_2 - y_1}{x_2 - x_1}$$

$$= \frac{8 - (-2)}{7 - (-1)}$$

$$= \frac{10}{8}$$

$$= \frac{5}{4}$$

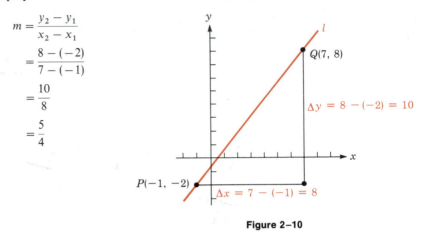

Figure 2–10

Thus, the slope of the line is $\frac{5}{4}$. (See Figure 2–10.)   ∎

If a line *rises* as it moves to the right—that is, if increasing x values result in increasing y values—the slope of the line is *positive* (see Figure 2–11a). If the line *drops* as it moves to the right—that is, if increasing x values result in decreasing y values—the slope of the line is *negative* (see Figure 2–11b). If a line is parallel to the x-axis, its slope is 0 (see Figure 2–11c). If a line is parallel to the y-axis, its slope is undefined (see Figure 2–11d).

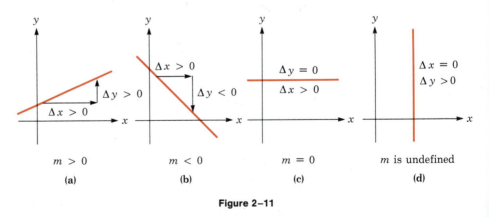

Figure 2–11

A theorem relates parallel lines to their slopes.

Theorem.   Two nonvertical lines are parallel if and only if they have the same slope.

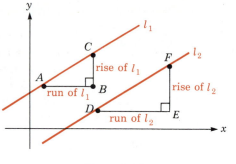

Figure 2–12

Proof   Suppose that the nonvertical lines l_1 and l_2 of Figure 2–12 have slopes of m_1 and m_2, respectively, and are parallel. Then the right triangles ABC and DEF are similar, and it follows that

$$m_1 = \frac{\text{rise of } l_1}{\text{run of } l_1}$$

$$= \frac{\text{rise of } l_2}{\text{run of } l_2}$$

$$= m_2$$

Thus, two nonvertical parallel lines have the same slope.

In the exercises you will be asked to prove that if two lines have the same slope, they are parallel.   □

It is also true that two lines with no defined slope are vertical lines, and vertical lines are parallel.

Example 2   The line passing through $P(-2, 3)$ and $Q(3, -1)$ is parallel to the line passing through $R(4, -5)$ and $S(x, 7)$. Find x.

Solution   Because the lines PQ and RS are parallel, the slopes of PQ and RS are equal. Find the slope of each line, set them equal to each other, and solve the resulting equation.

$$\text{slope of } PQ = \text{slope of } RS$$

$$\frac{-1 - 3}{3 - (-2)} = \frac{7 - (-5)}{x - 4}$$

$$\frac{-4}{5} = \frac{12}{x - 4}$$

$$-4(x - 4) = 5 \cdot 12 \qquad \text{Multiply both sides by } 5(x - 4).$$

$$-4x + 16 = 60$$

$$-4x = 44$$

$$x = -11$$

Thus, $x = -11$. The line passing through $P(-2, 3)$ and $Q(3, -1)$ is parallel to the line passing through $R(4, -5)$ and $S(-11, 7)$.   ■

Two numbers with a product of -1 are called **negative reciprocals** of each other. The following theorem describes the relationship between the slopes of perpendicular lines.

Theorem.   Two nonvertical lines are perpendicular if and only if their slopes are negative reciprocals.
  A line with a slope of 0 is perpendicular to a line with no defined slope.

Proof   Suppose that l_1 and l_2 are lines with slopes m_1 and m_2 intersecting at the origin. Let $P(a, b)$ be a point on l_1, let $Q(c, d)$ be a point on l_2, and let neither point P nor point Q be the origin. (See Figure 2–13.)

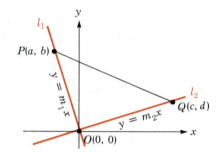

Figure 2–13

First, suppose lines l_1 and l_2 are perpendicular. Then triangle POQ is a right triangle with right angle at O, and by the Pythagorean theorem $d(OP)^2 + d(OQ)^2 = d(PQ)^2$. Thus,

$$d(OP)^2 + d(OQ)^2 = d(PQ)^2$$
$$(a - 0)^2 + (b - 0)^2 + (c - 0)^2 + (d - 0)^2 = (a - c)^2 + (b - d)^2$$
$$a^2 + b^2 + c^2 + d^2 = a^2 - 2ac + c^2 + b^2 - 2bd + d^2$$
$$0 = -2ac - 2bd$$
$$bd = -ac$$

1.
$$\frac{b}{a} = -\frac{c}{d}$$

The coordinates of P are (a, b) and the coordinates of O are $(0, 0)$. Using the definition of slope, we have

$$m_1 = \frac{b - 0}{a - 0} = \frac{b}{a}$$

Similarly, we have

$$m_2 = \frac{d}{c}$$

or

$$\frac{1}{m_2} = \frac{c}{d}$$

Substitute m_1 for $\dfrac{b}{a}$ and $\dfrac{1}{m_2}$ for $\dfrac{c}{d}$ in Equation 1 to get

$$m_1 = -\frac{1}{m_2}$$

Hence, if lines l_1 and l_2 are perpendicular, they have slopes that are negative reciprocals.

Conversely, suppose that the slopes of lines l_1 and l_2 are negative reciprocals of each other. Because the steps of the proof are reversible, $d(OP)^2 + d(OQ)^2 = d(PQ)^2$. By the Pythagorean theorem, triangle POQ is a right triangle. Thus, l_1 and l_2 are perpendicular.

□

It is also true that a line with a slope of 0 is horizontal and thus perpendicular to a vertical line, which has no defined slope.

Example 3   If two lines intersect at the point $P(-5, 3)$, one passes through the point $Q(-1, -3)$, and the other passes through $R(1, 7)$, are the lines perpendicular?

Solution   First find the slopes of lines PQ and PR.

$$\text{slope of } PQ = \frac{\Delta y}{\Delta x} = \frac{-3 - 3}{-1 - (-5)} = \frac{-6}{4} = -\frac{3}{2}$$

$$\text{slope of } PR = \frac{\Delta y}{\Delta x} = \frac{7 - 3}{1 - (-5)} = \frac{4}{6} = \frac{2}{3}$$

Because the slopes of these lines are negative reciprocals, the lines are perpendicular.

■

Point–Slope Form of the Equation of a Line

If two points on a line are known, its equation can be written by using the *point–slope* form of the equation of the line. Suppose that the nonvertical line l of Figure 2–14 has a slope of m and passes through the fixed point $P(x_1, y_1)$.

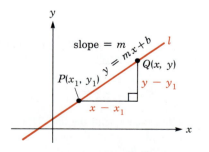

Figure 2–14

If $Q(x, y)$ is another point on that line, then by the definition of *slope* we have

$$m = \frac{y - y_1}{x - x_1}$$

or

$$y - y_1 = m(x - x_1)$$

Because the equation $y - y_1 = m(x - x_1)$ displays the coordinates of a fixed point on the line and the line's slope, it is called the **point–slope form** of the equation of a line.

> **Point–Slope Form of the Equation of a Line.**   The equation of the line passing through the point $P(x_1, y_1)$ and having a slope of m is
>
> $$y - y_1 = m(x - x_1)$$

Proof   The proof that the equation $y - y_1 = m(x - x_1)$ is the equation of the line l has two parts. We must show that

1. the coordinates of every point on the line satisfy the equation, and
2. every point whose coordinates satisfy the equation lies on the line.

Part 1.   Let $Q(a, b)$ be any point on the line except point $P(x_1, y_1)$. Substitute **a** for x and **b** for y in the equation $y - y_1 = m(x - x_1)$ and solve for m.

$$y - y_1 = m(\mathbf{x} - x_1)$$
$$\mathbf{b} - y_1 = m(\mathbf{a} - x_1)$$
$$m = \frac{b - y_1}{a - x_1}$$

Because the right side of this equation is the slope m of the line, we have the identity

$$m = m$$

Thus, the coordinates of the point Q satisfy the equation of the line. The coordinates of point $P(x_1, y_1)$ itself satisfy the equation $y - y_1 = m(x - x_1)$, because $y_1 - y_1 = m(x_1 - x_1)$ reduces to $0 = 0$. Thus, *any point on line l has coordinates that satisfy the equation.*

Part 2.   Suppose that the coordinates of some point $R(a, b)$ satisfy the equation $y - y_1 = m(x - x_1)$ and that point R is not point P. Then $b - y_1 = m(a - x_1)$, and

$$m = \frac{b - y_1}{a - x_1}$$

Thus, the slope of the line RP is m. Because line l is the only line with slope m passing through P, point R must lie on line l. Thus, *any point with coordinates that satisfy the equation lies on the line l.*

The theorem is proved.   □

Example 4   Find the equation of the line passing through the point $P(3, -1)$ with a slope of $-\frac{5}{3}$. Then solve that equation for y.

Solution   Substitute 3 for x_1, -1 for y_1, and $-\frac{5}{3}$ for m into the point–slope form of the equation of a line.

$$y - y_1 = m(x - x_1)$$

$$y - (-1) = -\frac{5}{3}(x - 3)$$

$$y + 1 = -\frac{5}{3}x + 5$$

$$y = -\frac{5}{3}x + 4$$

■

We can use the concept of *slope* to help graph the equation obtained in Example 4. We first plot the point $P(3, -1)$ as in Figure 2–15. Because the slope is $-\frac{5}{3}$, the line drops 5 units in a run of 3 units. Thus, we can locate another point Q by moving 3 units to the right of P and 5 units down. The graph of the equation is the line PQ.

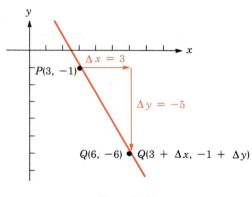

Figure 2–15

Slope–Intercept Form of the Equation of a Line

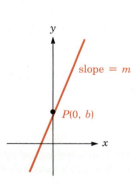

Figure 2–16

If the y-intercept of the line with slope m shown in Figure 2–16 is b, then the line intersects the y-axis at $P(0, b)$.

You can derive the equation of a line with slope m and y-intercept b from the point–slope form of the line:

$$y - y_1 = m(x - x_1)$$

$$y - b = m(x - 0) \qquad \text{Let } x_1 = 0 \text{ and } y_1 = b.$$

$$y = mx + b$$

Because the equation $y = mx + b$ displays both the slope and the y-intercept of a line, it is called the **slope–intercept form** of the equation of a line.

> **Slope–Intercept Form of the Equation of a Line.**   The equation of the line with slope m and y-intercept b is
>
> $$y = mx + b$$

Example 5   Use the slope–intercept form to find the equation of the line with slope $\frac{7}{3}$ and y-intercept -6.

Solution   Substitute $\frac{7}{3}$ for m and -6 for b into the slope–intercept form of the equation of the line and simplify.

$$y = mx + b$$

$$y = \frac{7}{3}x + (-6)$$

$$y = \frac{7}{3}x - 6$$

■

Example 6   Find the slope and the y-intercept of the line $3(y + 2) = 6x - 1$.

Solution   Write the equation in the form $y = mx + b$ to determine the slope m and the y-intercept b.

$$3(y + 2) = 6x - 1$$

$$3y + 6 = 6x - 1$$

$$3y = 6x - 7$$

$$y = 2x - \frac{7}{3}$$

The slope of the line is 2, and the y-intercept is $-\frac{7}{3}$.

■

Example 7   Use the slope–intercept form to find the equation of the line perpendicular to the line

$$y = \frac{1}{3}x + 7$$

and passing through the point $(2, 5)$.

Solution   The slope of the given line is $\frac{1}{3}$. The slope m of the required line must be -3, which is the negative reciprocal of $\frac{1}{3}$. Because the required line must pass through the point $(2, 5)$, the values $x = 2$ and $y = 5$ must satisfy the equation. Thus,

$$y = mx + b$$

$$5 = -3(2) + b$$

$$11 = b$$

The equation of the required line is

$$y = -3x + 11$$

■

If a line is parallel to the x-axis, its slope is **0**, and its equation is

$$y = \mathbf{0}x + b$$

or

$$y = b$$

If a line is parallel to the y-axis, it has no defined slope and cannot be the graph of any equation of the form $y = mx + b$. A vertical line does have an equation, however. If a vertical line passes through a point with an x-coordinate of a, then *every* point with an x-coordinate of a lies on the line and its equation is

$$x = a$$

Equations of Horizontal and Vertical Lines.   The equation of the horizontal line passing through the point $P(a, b)$ is

$$y = b$$

The equation of the vertical line passing through the point $P(a, b)$ is

$$x = a$$

Example 8   Find the equation of the line passing through the points $P(-3, 4)$ and $Q(-3, -2)$.

Solution   Because the x-coordinates of points P and Q are equal, the line has no defined slope. It is a vertical line with equation $x = -3$ (see Figure 2–17).

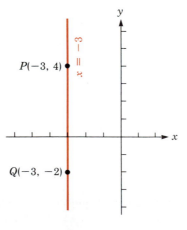

Figure 2–17

General Form of the Equation of a Line

We have shown that the graph of the equation

$$y - y_1 = m(x - x_1)$$

is a line. We can rewrite this equation as

$$-mx + y = y_1 - mx_1$$

which has the form $Ax + By = C$, where A, B, and C are constants: $A = -m$, $B = 1$, and $C = y_1 - mx_1$. In general, for any real numbers A, B, and C (where A and B are not *both* 0), the equation $Ax + By = C$ represents a line and is called the **general form** of the equation of a line. Any equation that can be written in general form is called a **linear equation in x and y.**

The General Form of the Equation of a Line.   If A, B, and C are real numbers and $B \neq 0$, the graph of the equation

$$Ax + By = C$$

is a nonvertical line with slope $-\dfrac{A}{B}$ and y-intercept $\dfrac{C}{B}$.

If $B = 0$ and $A \neq 0$, then the equation $Ax + By = C$ represents a vertical line with x-intercept $\frac{C}{A}$.

Example 9   Show that the two lines $3x - 2y = 5$ and $-6x + 4y = 7$ are parallel.

Solution   To show that two lines are parallel, show that they have the same slope. The first equation, $3x - 2y = 5$, is in general form, with $A = 3$, $B = -2$, and $C = 5$. By the previous theorem, the slope of the line is

$$m_1 = -\frac{A}{B} = -\frac{3}{-2} = \frac{3}{2}$$

Similarly, the second equation, $-6x + 4y = 7$ is in general form, with $A = -6$, $B = 4$, and $C = 7$. The slope of this line is

$$m_2 = -\frac{A}{B} = -\frac{-6}{4} = \frac{3}{2}$$

Because the slopes of the two lines are equal, the lines are parallel.   ■

Exercise 2.2

In Exercises 1–8 find the slope of the line passing through each pair of points, if possible.

1. $P(2, 5)$; $Q(3, 10)$ **2.** $P(3, -1)$; $Q(5, 3)$ **3.** $P(8, -7)$; $Q(4, 1)$

4. $P(5, 17)$; $Q(17, 17)$ **5.** $P(-4, 3)$; $Q(-4, -3)$ **6.** $P(2, \sqrt{2})$; $Q(\sqrt{7}, \sqrt{2})$

7. $P(a + b, c)$; $Q(b + c, a)$; $a \neq c$ **8.** $P(b, 0)$; $Q(a + b, 0)$

In Exercises 9–14 find two points on the line and determine the slope of the line.

9. $5x - 10y = 3$ **10.** $8y + 2x = 5$ **11.** $3(y + 2) = 2x - 3$

12. $4(x - 2) = 3y + 2$ **13.** $3(y + x) = 3(x - 2)$ **14.** $2x + 5 = 2(y + x)$

In Exercises 15–20 determine whether the lines with the given slopes are parallel, perpendicular, or neither.

15. $m_1 = 3, m_2 = -\dfrac{1}{3}$

16. $m_1 = \dfrac{2}{3}, m_2 = \dfrac{3}{2}$

17. $m_1 = \sqrt{8}, m_2 = 2\sqrt{2}$

18. $m_1 = 1, m_2 = -1$

19. $m_1 = -\sqrt{2}, m_2 = \dfrac{\sqrt{2}}{2}$

20. $m_1 = 2\sqrt{7}, m_2 = \sqrt{28}$

In Exercises 21–24 determine whether the line through the given points and the line through $R(-3, 5)$ and $S(2, 7)$ are parallel, perpendicular, or neither.

21. $P(2, 4); Q(7, 6)$

22. $P(-3, 8); Q(-13, 4)$

23. $P(-4, 6); Q(-2, 1)$

24. $P(0, -9); Q(4, 1)$

In Exercises 25–28 find the slopes of lines PQ and PR, and determine whether points P, Q, and R lie on the same line.

25. $P(-2, 8); Q(-6, 9); R(2, 5)$

26. $P(1, -1); Q(3, -2); R(-3, 0)$

27. $P(-3, 9); Q(-5, 5); R(-6, 3)$

28. $P(-1, 1); Q(13, -1); R(-8, 2)$

In Exercises 29–32 determine which, if any, of the three lines PQ, PR, and QR are perpendicular.

29. $P(5, 4), Q(2, -5), R(8, -3)$

30. $P(8, -2), Q(4, 6), R(6, 7)$

31. $P(0, 0), Q(a, b), R(-b, a)$

32. $P(1, 3), Q(1, 9), R(7, 3)$

In Exercises 33–36 use the point–slope form to write the equation of the line passing through the given point and having the given slope. Express the answer in slope–intercept form.

33. $P(2, 4); m = 2$

34. $P(3, 5); m = -3$

35. $P\left(-\dfrac{3}{2}, \dfrac{1}{2}\right); m = 2$

36. $P\left(\dfrac{1}{4}, \dfrac{4}{3}\right); m = 0$

In Exercises 37–40 use the slope–intercept form to write the equation of the line with the given slope and y-intercept. Give the answer in general form.

37. $m = 3, b = -2$

38. $m = -\dfrac{1}{3}, b = \dfrac{2}{3}$

39. $m = a, b = \dfrac{1}{a}$

40. $m = a, b = 2a$

In Exercises 41–44 find the slope and the y-intercept of the line defined by the given equation, if possible.

41. $3(14x + 12) = 5(y + 3)$

42. $-2(3x + 6) = -2y + 15$

43. $2(y - 3) + x = 2y - 7$

44. $3(y + 2) + 2x = 2(y + x)$

In Exercises 45–48 find the equation of the line parallel to the given line and passing through the given point. Give the answer in slope–intercept form.

45. $3x = 2y + 1; (-1, 1)$

46. $5y - 7x = 2; (5, -3)$

47. $3(x - y) = 5(y - 2x); (2, 3)$

48. $3x - 2(y + 2x) = x - 3; (1, -5)$

In Exercises 49–52 find the equation of the line perpendicular to the given line and passing through the given point. Give the answer in general form.

49. $3x = 2y + 1; (-1, 1)$

50. $5y - 7x = 2; (5, -3)$

51. $3(x - y) = 5(y - 2x); (2, 3)$

52. $3x - 2(y + 2x) = x - 3; (1, -5)$

In Exercises 53–56 write the equation of the line passing through the given point and parallel to the given axis.

53. $P(1, 0); y$-axis

54. $P(-3, -7); x$-axis

55. $P(-3, 5); x$-axis

56. $P(-5, 9); y$-axis

In Exercises 57–62 find the slope and the y-intercept of the line that has the given properties.

57. The line passes through $P(-3, 5)$ and $Q(7, -5)$.

58. The line passes through $P(15, 3)$ and $Q(-3, -6)$.

59. The line is parallel to the line $y = 3(x - 7)$ and passes through $P(-2, 5)$.

60. The line is parallel to the line $3y + 1 = 6(x - 2)$ and passes through $P(5, 0)$.

61. The line is perpendicular to the line $y + 3x = 8$ and passes through $P(2, 3)$.

62. The line is perpendicular to the line $x = 2y - 7$ and passes through $P(8, 0)$.

63. Show that the three points $A(-1, -1)$, $B(-3, 4)$, and $C(4, 1)$ are the vertices of a right triangle.

64. Show that the four points $A(1, -1)$, $B(3, 0)$, $C(2, 2)$, and $D(0, 1)$ are the vertices of a square.

65. Show that the four points $A(-2, -2)$, $B(3, 3)$, $C(2, 6)$, and $D(-3, 1)$ are the vertices of a parallelogram. (Show that both pairs of opposite sides are parallel.)

66. Show that the four points $E(1, -2)$, $F(5, 1)$, $G(3, 4)$, and $H(-3, 4)$ are the vertices of a trapezoid. (Show that only one pair of opposite sides are parallel.)

67. Prove that the equation of the line with x-intercept a and y-intercept b may be written in the form

$$\frac{x}{a} + \frac{y}{b} = 1$$

68. Prove that if $B \neq 0$ the graph of $Ax + By = C$ has a slope of $-\dfrac{A}{B}$ and a y-intercept of $\dfrac{C}{B}$.

69. Prove that, if $B = 0$ and $A \neq 0$, the graph of $Ax + By = C$ is a vertical line with x-intercept $\dfrac{C}{A}$.

70. Prove that the lines $Ax + By = C$ and $Bx - Ay = C$ are perpendicular.

71. Prove that the equation of the line passing through the points (x_1, y_1) and (x_2, y_2) can be written in the form

$$y - y_1 = \frac{y_2 - y_1}{x_2 - x_1}(x - x_1)$$

This equation is called the **two-point form** of the equation of a line.

72. Use the distance formula to find the equation of the perpendicular bisector of the line segment joining $A(3, 5)$ and $B(5, -3)$. [*Hint:* Let $P(x, y)$ be a point on the perpendicular bisector. Then $d(PA) = d(PB)$.]

2.3 FUNCTIONS AND FUNCTION NOTATION; VARIATION

Many equations describe a correspondence between two variables. The equation $y = 3x + 2$, for example, describes a correspondence in which each number x determines a single value of y. Such correspondences are called **functions**.

> **Definition.**   A **function** is a correspondence that assigns to *each* element x of some set X a *single* value y of some set Y.
>
> The set X is called the **domain** of the function. The value y that corresponds to a particular x in the domain is called the **image** of x under the function.
>
> The collection of all images of elements in the domain is called the **range** of the function.

Because the value of y depends on a number x, y is called the **dependent variable** and x is called the **independent variable**. If a function is defined by some mathematical expression and the domain is not given explicitly, we will assume the domain of the function to be its **implied domain**—the set of all real numbers for which the expression is meaningful.

Example 1 Does the equation $y = 5 - 7x$ define y to be a function of x? If so, find its domain and range.

Solution To determine y in the equation $y = 5 - 7x$, the product $7x$ is subtracted from 5. Because the multiplication and subtraction gives a single result, only one value of y is produced. Thus, the equation does determine a function.

Because x can represent any real number, the implied domain of the function is the set of real numbers.

Because y can be any real number, the range of the function is the set of real numbers.

∎

Example 2 Does the equation $y = \dfrac{3}{x - 2}$ define y to be a function of x? If so, find its domain and range.

Solution Because division by zero is not defined, x cannot be 2. Because each number x other than 2 determines exactly one value of y, the equation does define a function.

The domain of the function is the set of all real numbers except 2: the domain is $\{x : x \neq 2\}$.

Because the fraction has a nonzero numerator, y cannot be 0. To see that all values other than 0 are in the range of the function, solve the equation $y = \dfrac{3}{x - 2}$ for x to get

$$x = \frac{3 + 2y}{y} \qquad \text{provided } y \neq 0$$

Because any value of y except 0 is the image of some number x, where $x = \dfrac{3 + 2y}{y}$, the range of the function is $\{y : y \neq 0\}$.

∎

Example 3 Find the domain and range of the function $f(x) = \sqrt{1 - x^2}$.

Solution The implied domain of a function consists of those real numbers whose images under the function are also real numbers. Because $f(x)$ is to be a real value, the radicand $1 - x^2$ must be nonnegative. Solve the inequality $1 - x^2 \geq 0$ to determine that x is in the interval $[-1, 1]$. Thus, the domain of f is

$$\{x : -1 \leq x \leq 1\}$$

If x is in the domain of f, the nonnegative number $1 - x^2$ is also less than or equal to 1. Thus, $\sqrt{1 - x^2}$ lies in $[0, 1]$. The range of f is

$$\{y : 0 \leq y \leq 1\}$$

∎

To indicate that y depends on x, we often use **function notation** and write

$$y = f(x) \qquad \text{Read as "}y\text{ equals } f \text{ of } x.\text{"}$$

Function notation provides a convenient way of indicating the image of a particular number x. If $y = f(x) = 3x + 2$ defines a function f, for example, the symbol $f(2)$ indicates the image of **2** under f.

$$f(x) = 3x + 2$$
$$f(2) = 3(2) + 2 \qquad \text{Substitute 2 for } x.$$
$$= 8$$

Thus, when $x = 2$, $y = f(2) = 8$. Similarly, the image of $-\frac{2}{3}$ is $f(-\frac{2}{3})$, or 0, because

$$f(x) = 3x + 2$$
$$f(-\tfrac{2}{3}) = 3(-\tfrac{2}{3}) + 2 \qquad \text{Substitute } -\tfrac{2}{3} \text{ for } x.$$
$$= 0$$

Because $f(-\frac{2}{3}) = 0$, we say that $-\frac{2}{3}$ is a **zero** of the function f.

We also say that $f(x)$ is the **value of the function f at x**. If x is in the domain of f, we say that f **is defined at x**. If A is a subset of the domain of f, we say that f **is defined on A**. Strictly speaking, the function f is the correspondence from set X to set Y, and the equation $y = f(x)$ defines that function. Yet it is common to read $y = f(x)$ as "y is a function of x." Other notations also indicate that f is a function from X to Y:

$$X \xrightarrow{\ f\ } Y \qquad \text{and} \qquad f : X \longrightarrow Y$$

Functions are often denoted by letters other than f.

Example 4   Let $g(x) = 3x^2 + x - 4$. Find **a.** $g(1)$, **b.** $g(k)$, **c.** $g(k + 1)$, and **d.** $\dfrac{g(k+1) - g(k)}{k}$.

Solution   **a.** $g(x) = 3x^2 + x - 4$

$$g(1) = 3(1)^2 + (1) - 4 \qquad \text{Substitute 1 for } x.$$
$$= 3(1) + 1 - 4$$
$$= 0$$

b. $g(x) = 3x^2 + x - 4$

$$g(k) = 3k^2 + k - 4 \qquad \text{Substitute } k \text{ for } x.$$

c. $g(x) = 3x^2 + x - 4$

$$g(k + 1) = 3(k + 1)^2 + (k + 1) - 4 \qquad \text{Substitute } k + 1 \text{ for } x.$$
$$= 3(k^2 + 2k + 1) + k + 1 - 4$$
$$= 3k^2 + 6k + 3 + k + 1 - 4$$
$$= 3k^2 + 7k$$

d. $\dfrac{g(k+1) - g(k)}{k} = \dfrac{3k^2 + 7k - (3k^2 + k - 4)}{k} \qquad \text{See parts \textbf{b} and \textbf{c}.}$

$$= \dfrac{6k + 4}{k}$$

■

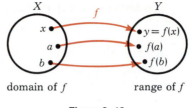

Figure 2–18

By definition a function f is a correspondence, often called a **mapping**, from the elements of one set X to the elements of another set Y. We can visualize this mapping with Figure 2–18. A function f that maps the element $x \in X$ to the element $y = f(x) \in Y$ can be represented by an arrow leaving x and pointing to y. The set of all those elements in X from which arrows originate is the *domain* of the function. The set of all the images in Y to which arrows point is the *range* of the function.

A correspondence is still a function if arrows point from several different elements of X to the same element y in the range, as in Figure 2–19. However, if arrows leave some element x in the domain and point to several elements in the range, as in Figure 2–20, the correspondence is *not* a function because a single y is not assigned to each x.

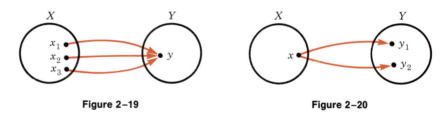

Figure 2–19   **Figure 2–20**

Any correspondence between the variables x and y is called a **relation**.

Definition.   A **relation** is a correspondence that assigns to each element x in a set X one or more values of y in a set Y.

The **domain** of the relation is set X, and the **range** is the set of all values y corresponding to numbers x in the domain.

Note that a function is a relation but a relation is not necessarily a function.

Example 5   Does the equation $|y| = x$ determine y as a function of x? If not, give the domain and the range of the relation.

Solution   If some value of x determines more than one value of y, then the equation does not define a function. If $x = 2$, for example, then $|y| = 2$, and y could be either 2 or -2. Thus, $x = 2$ determines two values of y, and the equation $|y| = x$ does not define a function. The equation does determine a relation, however.

The domain of the relation is the set of nonnegative real numbers, and the range is the set of real numbers. ■

A relation determines a set of ordered pairs (x, y), where x is an element of the domain of the relation and y is a corresponding value in the range. The **graph of the relation** is the graph of all these ordered pairs. If the relation is a function, the **graph of the function** consists of those points and only those points with coordinates $(x, f(x))$.

A test called the **vertical line test** can be used to determine whether the graph of a relation in the xy-plane represents a function. If each vertical line that intersects the graph does so exactly once, then each number x determines exactly one value y and the graph represents a function (see Figure 2–21a). If any vertical line intersects the graph more than once, then to some number x there corresponds more than one value y, and the graph does *not* represent a function (see Figure 2–21b).

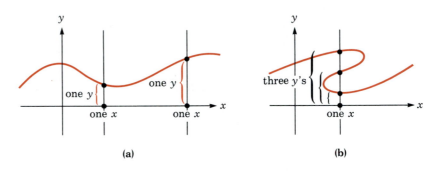

(a)   (b)

Figure 2–21

Example 6 Graph the points (x, y) for which $x = y^2$ and determine whether this relation is a function.

Solution Several ordered pairs (x, y) that satisfy the equation are listed in the table of values in Figure 2–22. Plot those points and draw the graph as shown in the figure. Because to some numbers x there correspond two y values, the graph does not pass the vertical line test. Thus, the relation is *not* a function.

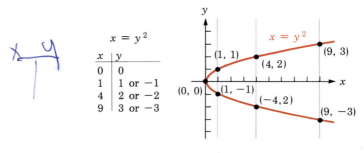

$x = y^2$

x	y
0	0
1	1 or −1
4	2 or −2
9	3 or −3

Figure 2–22   ■

Example 7 Graph the points (x, y) for which $y = x^2$. Is this relation a function?

Solution Plot several ordered pairs (x, y) that satisfy the equation and draw the graph as shown in Figure 2–23. Because *every* vertical line intersects the graph only once, the graph *does* pass the vertical line test, and the relation is a function.

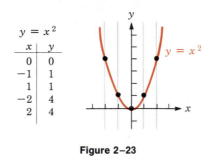

$y = x^2$

x	y
0	0
−1	1
1	1
−2	4
2	4

Figure 2–23

The slope–intercept form of the equation of a line determines a function called a **linear function**.

Definition. A **linear function** is a function defined by an equation of the form

$$f(x) = mx + b$$

The graph of a linear function is a line.

Example 8 The cost C of electricity is a linear function of x, the number of kilowatt-hours used. If the cost of 100 kilowatt-hours is \$17 and the cost of 500 kilowatt-hours is \$57, what formula expresses the function?

Solution Because C is a linear function of x, there are constants m and b such that

$$C = mx + b$$

Because $C = 17$ when $x = 100$, the point $(x_1, C_1) = (100, 17)$ lies on the straight-line graph of $C = mx + b$. Because $C = 57$ when $x = 500$, the point $(x_2, C_2) = (500, 57)$ also lies on the line. The slope of the line is

$$m = \frac{C_2 - C_1}{x_2 - x_1}$$

$$= \frac{57 - 17}{500 - 100}$$

$$= \frac{40}{400}$$

$$= 0.10$$

Thus, $m = 0.10$. To find b, substitute $C = $ **17**, $x = $ **100**, and $m = $ **0.10** into the equation $C = mx + b$ and solve for b.

$$C = mx + b$$
$$17 = 0.10(100) + b$$
$$17 = 10 + b$$
$$7 = b$$

Thus, $C = 0.10x + 7$. The electric company charges a flat rate of $7, plus 10¢ per kilowatt-hour used.   ∎

Example 9   Express the area of a square as a function of its perimeter.

Solution   The area A of a square is a function of the length s of a side, expressed by

    1. $A = f(s) = s^2$

The perimeter P is also a function of the length s of a side:

    2. $P = g(s) = 4s$

To determine a function h so that $A = h(P)$, solve Equation 2 for s and substitute the result into Equation 1.

$$P = 4s$$
$$s = \frac{P}{4} \qquad \text{Divide both sides by 4.}$$
$$A = \left(\frac{P}{4}\right)^2 \qquad \text{Substitute } \frac{P}{4} \text{ for } s \text{ in Equation 1.}$$
$$= \frac{P^2}{16}$$

Thus, $A = h(P) = \dfrac{P^2}{16}$.

In the context of this example, the domain of h is not the implied domain of the set of real numbers. Because the perimeter of a square is a positive number, the domain of h is the set of positive real numbers.   ∎

Example 10   Four squares with edge of length x centimeters are cut from the corners of a 12- by 19-centimeter rectangular sheet of cardboard. The edges are then folded up to form a

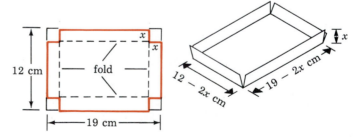

Figure 2–24

shallow tray (see Figure 2–24). Express the volume V of the tray formed as a function of x and give the domain of the function.

Solution   When the cardboard is folded, the length of the base of the tray is $19 - 2x$ centimeters, and its width is $12 - 2x$ centimeters. Because the volume of the tray is the product of length, width, and depth, we have

$$V = f(x) = (19 - 2x)(12 - 2x)x$$
$$= (228 - 62x + 4x^2)x$$
$$= 4x^3 - 62x^2 + 228x$$

In the context of this example, the domain of f is not the implied domain of the set of real numbers. Because lengths of edges of rectangles must be positive numbers, x must be positive. Because the width of the original rectangle is 12 centimeters, the edge of the cut-out square must be less than 6 centimeters. Thus, the domain of f is $\{x : 0 < x < 6\}$. ■

Variation

Scientists often use the language of **variation** to express certain functional relationships among several variables. There are three major types of variation.

Types of Variation.   Suppose that k is a constant.
If $y = kx$, then y **varies directly** as x, or y is **directly proportional** to x.
If $y = \dfrac{k}{x}$, then y **varies inversely** as x, or y is **inversely proportional** to x.
If $y = kxz$, then y **varies jointly** as x and z.
The number k is called the **constant of proportionality**.

Example 11   The energy an object has because of its motion is called **kinetic energy**. The kinetic energy of an object varies jointly as its mass and the square of its velocity. A 25-gram mass moving at the rate of 30 centimeters per second has a kinetic energy of 11,250 dyne-centimeters. What will be the kinetic energy of a 10-gram mass moving at 40 centimeters per second?

Solution   Let E represent the kinetic energy of the object and m and v its mass and velocity, respectively. The phrase "energy varies jointly as its mass and the square of its velocity" means that there is a constant k such that

$$E = kmv^2$$

The constant k can be evaluated by using the given information: $E = 11{,}250$, $m = 25$, and $v = 30$. Substitute these values into the formula $E = kmv^2$ and solve for k.

$$E = k\,mv^2$$
$$11{,}250 = k(25)(30)^2$$
$$11{,}250 = k(22{,}500)$$
$$\frac{1}{2} = k$$

Now substitute $\frac{1}{2}$ for k in the formula to evaluate E when $m = $ **10** and $v = $ **40**.

$$E = \mathbf{k}mv^2$$

$$E = \frac{1}{2}mv^2$$

$$= \frac{1}{2}\mathbf{(10)(40)}^2$$

$$= 8000$$

A 10-gram mass that is moving at 40 centimeters per second has a kinetic energy of 8000 dyne-centimeters. ∎

Exercise 2.3

In Exercises 1–8 indicate whether the equation determines y to be a function of x. Assume that all variables represent real numbers.

1. $y = x$ **2.** $y - 2x = 0$ **3.** $y^2 = x$ **4.** $y = |x|$

5. $y = x^2 + 4$ **6.** $y - 7 = 7$ **7.** $|x - 2| = y$ **8.** $|x| = |y|$

In Exercises 9–20 find the domain and range of each function. Then find $f(0)$, $f(3)$, and $f(-4)$, if possible.

9. $f(x) = 3x + 5$ **10.** $f(x) = -5x + 2$ **11.** $f(x) = x^2$

12. $f(x) = x^2 + 3$ **13.** $f(x) = \dfrac{3}{x + 1}$ **14.** $f(x) = \dfrac{-7}{x + 3}$

15. $f(x) = |x - 3|$ **16.** $f(x) = \sqrt{16 - x^2}$ **17.** $f(x) = \sqrt{x^2 - 16}$

18. $f(x) = \dfrac{2}{x^2 + 2x}$ **19.** $f(x) = \dfrac{x}{x - 3}$ **20.** $f(x) = \dfrac{x - 3}{x}$

In Exercises 21–26 some correspondences are illustrated. If the correspondence represents y as a function of x, so indicate. If the illustration does not represent such a function, explain why.

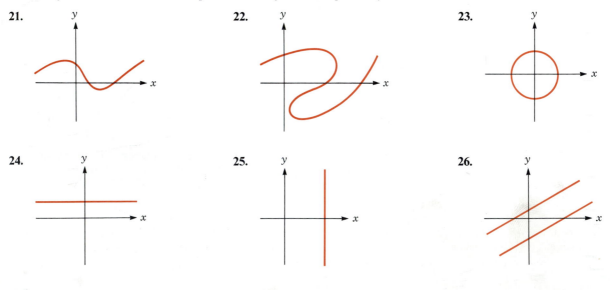

In Exercises 27–44 graph each correspondence. Use the vertical line test to decide whether the correspondence defines y to be a function of x.

27. $y = 2x + 3$

28. $y = 3x + 2$

29. $2x = 3y - 3$

30. $3x = 2(y + 1)$

31. $y = -x^2$

32. $y = 1 - x^2$

33. $y = \sqrt{x}$

34. $y = -|x|$

35. The number y is the greatest integer less than or equal to x.

36. The number y is the least integer greater than or equal to x.

37. $x = y^3$

38. $x = y^2 + 3$

39. $y = |x| + x$

40. $|y| = 1 + x$

41. $y = \dfrac{1}{x}$

42. $y = \dfrac{1}{x^2}$

43. $|x| = y^2$

44. $|y| = x^2$

45. The Fahrenheit temperature reading F is a linear function of the Celsius reading C. If $C = 0$ when $F = 32$, and $C = 100$ when $F = 212$, express F as a function of C.

46. The velocity v of a falling object is a linear function of the time t it has been falling. If $v = 15$ when $t = 0$, and $v = 79$ when $t = 2$, express v as a function of t.

47. The cost C of water is a linear function of n, the number of gallons used. If 1000 gallons cost \$4.70, and 9000 gallons cost \$14.30, what equation describes this function?

48. The amount A of money on deposit for t years in an account earning simple interest is a linear function of t. Express that function as an equation if $A = \$272$ when $t = 3$ and $A = \$320$ when $t = 5$.

49. Steel beams expand when heated, and their length is a linear function of temperature. When the temperature is $40°$ F, a certain steel beam is 50 feet long. At $95°$ F, that beam has expanded by 0.12 feet. Express the beam's length L as a linear function of its temperature T.

50. Express the radius of a circle as a function of its circumference.

51. Express the radius of a circle as a function of its area.

52. Express the area of a circle as a function of the circumference.

53. Express the volume of a cube as a function of the area of one face.

54. Express the length of the diagonal of a square as a function of the length of an edge.

55. Express the volume of a cube as a function of the total surface area.

56. Express the volume of a sphere as a function of its surface area. (*Hint:* $V = \frac{4}{3}\pi r^3$ and $A = 4\pi r^2$.)

57. Express the surface area of a cube as a function of the length of a diagonal of one face.

58. Express the volume of a cube as a function of the area of one face.

59. If the cylindrical can of Illustration 1 has a constant volume of V cubic units, express the height h as a function of the can's radius r. (*Hint:* $V = \pi r^2 h$.)

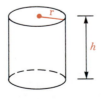

Illustration 1

60. The U.S. Post Office will accept a package for mailing if the sum of its length and girth does not exceed 108 inches. The rectangular box of Illustration 2 has a square end with edge x and just passes the Post Office's restrictions. Express the volume of the box as a function of x.

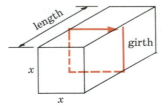

Illustration 2

61. A right triangle with a given constant hypotenuse s generates the cone of Illustration 3 when it is rotated about one leg. If x is the length of the other leg, express the cone's volume V as a function of x. (*Hint:* $V = \frac{1}{3}\pi r^2 h$.)

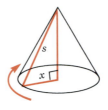

Illustration 3

62. A rectangular sheet of tin with area A is rolled into a cylinder, as in Illustration 4. Express the volume of the cylinder as a function of dimension x.

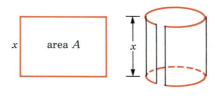

Illustration 4

63. A rectangular beam of thickness x is cut from a circular log of diameter d (see Illustration 5). Express the cross-sectional area of the beam as a function of x. (*Hint:* Use the Pythagorean theorem to find the width of the beam.)

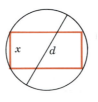

Illustration 5

64. A window is in the form of a rectangle surmounted by a semicircle, with width x as indicated in Illustration 6. If the rectangular portion is three times as tall as it is wide, express the total window area as a function of x.

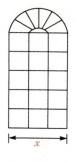

Illustration 6

65. A baseball player runs from first base to second. Express the player's distance d from home plate as a function of x, his distance from first base (see Illustration 7).

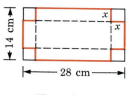

Illustration 7

66. Four squares with edge x are cut from the corners of a 14- by 28-centimeter rectangular sheet of cardboard. What is the volume of the shallow tray formed when the flaps are folded up (see Illustration 8)?

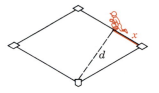

Illustration 8

67. A rectangular sheet of metal with a perimeter of 10 meters is bent into a cylinder with height h. Express the volume of the cylinder as a function of h.

68. A piece of length x is cut from a wire 1 meter long and is bent into a circle. The remaining section is formed into a square. Express the total area enclosed as a function of x.

69. y is directly proportional to x. If $y = 15$ when $x = 4$, find y when $x = \frac{7}{5}$.

70. P varies jointly with r and s. If $P = 16$ when $r = 5$ and $s = -8$, find P when $r = 4$ and $s = 10$.

71. The volume of a gas varies directly as the temperature and inversely as the pressure. If the temperature of a certain gas is 350 K (degrees Kelvin), the pressure is 40 dynes per square centimeter, and the volume is 20 cubic centimeters, find the volume if the pressure is 50 dynes per square centimeter and the temperature decreases to 250 K.

72. The power, in watts, dissipated by a resistor in the form of heat varies directly as the square of the voltage and inversely as the resistance. A 20-ohm resistor dissipates 20 watts when connected to a 20-volt power source. What voltage across a 10-ohm resistor will cause it to dissipate 40 watts?

73. The area of a circle varies directly as the square of its radius. What is the constant of proportionality?

74. The diagonal of a cube varies directly as an edge. What is the constant of proportionality?

In Exercises 75–76 suppose that a is a constant and that f is a function for which $f(x + a) = f(x)$ for all real numbers x.

75. Prove that $f(2a) = f(a)$.

76. Prove that $f(2a) = f(0)$.

In Exercises 77–78 suppose that g is a function for which $g(x + 5) = g(x) + g(5)$ for all real numbers x.

77. Prove that $g(0) = 0$.

78. Prove that $g(-5) = -g(5)$.

2.4 POLYNOMIAL FUNCTIONS

A linear function defined by the equation $y = mx + b$ is a polynomial function because its right side is a polynomial in the variable x. In this section we will discuss functions involving polynomials of higher degree.

> **Definition.**   A **polynomial function** in one independent variable, say x, is defined by an equation of the form $y = P(x)$, where $P(x)$ is a polynomial in the variable x.
>   The **degree of the polynomial function $y = P(x)$** is the degree of $P(x)$.

A polynomial function of second degree is called a **quadratic function**.

> **Definition.**   A **quadratic function** is a second-degree polynomial function in one variable. It is defined by an equation of the form $y = ax^2 + bx + c$, where a, b, and c are constants and $a \neq 0$.

Example 1   Graph the quadratic function $y = x^2 - 2x - 3$.

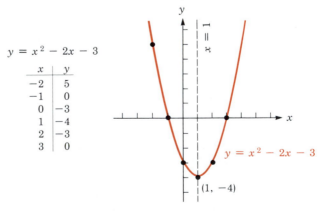

$y = x^2 - 2x - 3$

x	y
-2	5
-1	0
0	-3
1	-4
2	-3
3	0

$y = x^2 - 2x - 3$

$(1, -4)$

Figure 2–25

Solution   Plot several points whose coordinates satisfy the equation. Connect them with a smooth curve as shown in Figure 2–25.   ■

Example 2   Graph the quadratic function $y = -\dfrac{1}{3}x^2 + 3$.

Solution   Plot several points whose coordinates satisfy the equation. Connect them with a smooth curve as shown in Figure 2–26.

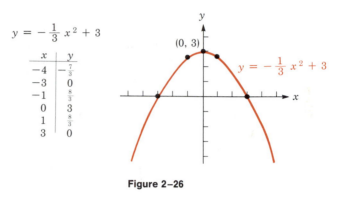

$$y = -\frac{1}{3}x^2 + 3$$

x	y
-4	$-\frac{7}{3}$
-3	0
-1	$\frac{8}{3}$
0	3
1	$\frac{8}{3}$
3	0

Figure 2–26   ■

Graphs of quadratic functions are called **parabolas**. They open upward when $a > 0$ (as in Example 1) and downward when $a < 0$ (as in Example 2). The "bottom" point of a parabola that opens upward or the "top" point of a parabola that opens downward is called the **vertex** of the parabola. The vertex of the parabola in Example 1 is the point $(1, -4)$, and the vertex of the parabola in Example 2 is the point $(0, 3)$. The vertical line that passes through the vertex is called the **axis of symmetry** because it divides the parabola into two congruent halves. The axis of symmetry of the parabola in Example 1 is the line $x = 1$, and in Example 2 it is the line $x = 0$ (the y-axis).

If a, h, and k are constants and $a \neq 0$, the equation

$$y - k = a(x - h)^2$$

defines a quadratic function because it takes the form $y = ax^2 + bx + c$ when its right side is expanded and the equation is solved for y. The graph of the equation $y - k = a(x - h)^2$ is a parabola opening upward when $a > 0$ and downward when $a < 0$. This special form of the equation of the parabola displays the coordinates of the vertex of its parabolic graph, as the following discussion will show.

Suppose that the graph of $y - k = a(x - h)^2$ is a parabola opening upward. The vertex of this parabola is that point on the graph that has the least possible y-coordinate. Because a is positive, the smallest value attainable by the right side is 0, and this occurs when $x = h$. When $x = h$, the left side of the equation attains its smallest value also, which is 0. Thus, the smallest possible value of y occurs when $y - k = 0$, or when $y = k$. The parabola's vertex, therefore, is the point (h, k). A similar argument holds if $a < 0$.

Theorem. The graph of the equation

$$y - k = a(x - h)^2$$

is a parabola with its vertex at the point (h, k). The parabola opens upward if $a > 0$ and downward if $a < 0$.

Example 3 Find the vertex of the parabola determined by $y = 2x^2 - 5x - 3$.

Solution Begin by completing the square on x:

$$y = 2x^2 - 5x - 3$$

$$y + 3 = 2x^2 - 5x \qquad \text{Add 3 to both sides.}$$

$$y + 3 = 2\left(x^2 - \frac{5}{2}x \qquad \right) \qquad \text{Factor a 2 out of the right side.}$$

$$y + 3 + \frac{25}{8} = 2\left(x^2 - \frac{5}{2}x + \frac{25}{16}\right) \qquad \text{Add } \tfrac{25}{8} \text{ to both sides.}$$

$$y + \frac{49}{8} = 2\left(x - \frac{5}{4}\right)^2 \qquad \text{Combine terms and factor.}$$

$$y - \left(-\frac{49}{8}\right) = 2\left(x - \frac{5}{4}\right)^2$$

The equation is now in the form

$$y - k = a(x - h)^2$$

with $h = \tfrac{5}{4}$ and $k = -\tfrac{49}{8}$. Thus, the vertex of the parabola is the point $(h, k) = (\tfrac{5}{4}, -\tfrac{49}{8})$. ∎

To find the vertex of any parabola given by an equation of the form $y = ax^2 + bx + c$, we can write the equation in the form $y - k = a(x - h)^2$ by completing the square.

$$y = ax^2 + bx + c$$

$$y - c = a\left(x^2 + \frac{b}{a}x \qquad \right) \qquad \text{Factor out } a \text{ on the right side and add } -c \text{ to both sides.}$$

$$y - c + \frac{b^2}{4a} = a\left(x^2 + \frac{b}{a}x + \frac{b^2}{4a^2}\right) \qquad \text{Add } \frac{b^2}{4a} \text{ to both sides.}$$

$$y - \left(c - \frac{b^2}{4a}\right) = a\left[x - \left(-\frac{b}{2a}\right)\right]^2 \qquad \text{Factor.}$$

Thus, the vertex is the point with coordinates $\left(-\dfrac{b}{2a}, c - \dfrac{b^2}{4a}\right)$, and we have the following theorem.

Theorem.   The **vertex of the parabola** determined by the quadratic function

$$y = ax^2 + bx + c \qquad (a \neq 0)$$

is the point with coordinates $\left(-\dfrac{b}{2a}, c - \dfrac{b^2}{4a}\right)$.

Example 4   A farmer plans to use 400 feet of fencing to enclose the rectangular corral shown in Figure 2–27. What dimensions would enclose the largest area?

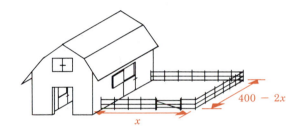

Figure 2–27

Solution   If x represents the width of the fenced area, then $400 - 2x$ represents the length. Since the area is the product of the length and width,

$$A = x(400 - 2x)$$

or

$$A = -2x^2 + 400x$$

Because the coefficient of x^2 is negative, the parabola determined by this function opens downward, and its vertex is its "highest" point. The A-coordinate of the vertex gives the largest possible area, and the x-coordinate gives the width of the corral of maximum area. The equation

$$A = -2x^2 + 400x$$

defines a quadratic function with $a = -2$, $b = 400$, and $c = 0$. The vertex of the parabola is the point

$$\left(-\frac{b}{2a}, c - \frac{b^2}{4a}\right) = \left(-\frac{400}{2(-2)}, 0 - \frac{400^2}{4(-2)}\right)$$

$$= (100,\ 20{,}000)$$

If the farmer's fence runs 100 feet out from the barn, 200 feet parallel to the barn, and 100 feet back to the barn, it will enclose the maximum area of 20,000 square feet.   ■

Symmetries of Graphs

The graph of a parabola is symmetric about its axis, making half of the parabola the mirror image of the other half. We often say that one half of the parabola is the **reflection** of its other half in this axis of symmetry. There are three other common symmetries of graphs of functions and relations.

If the point $(-x, y)$ lies on the graph of a relation whenever (x, y) does, the graph is **symmetric about the y-axis** (see Figure 2–28).

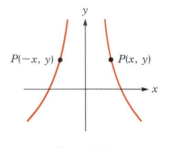

Figure 2–28

It is easy to check for y-axis symmetry of the graph of a *function* defined by $y = f(x)$. If $y = f(-x)$ is equivalent to $y = f(x)$—that is, if $y = f(x)$ remains unchanged when x is replaced with $-x$—then the graph is symmetric about the y-axis. A function with this type of symmetry is called an **even function**.

Definition.   A function f with the property that $f(-x) = f(x)$ for all x in the domain of f is called an **even function**.

A graph is **symmetric about the x-axis** if the point $(x, -y)$ lies on the graph whenever the point (x, y) does (see Figure 2–29). Except for the graph of the function $f(x) = 0$, such a graph does not pass the vertical line test and does not represent a function.

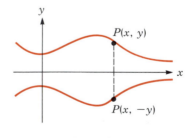

Figure 2–29

Finally, a graph is **symmetric about the origin** if the point $(-x, -y)$ lies on the graph whenever the point (x, y) does (see Figure 2–30).

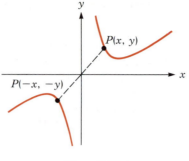

Figure 2–30

To test the graph of a function defined by $y = f(x)$ for symmetry about the origin, replace x with $-x$ and replace y with $-y$. If the equation $-y = f(-x)$ is equivalent to the equation $y = f(x)$, the graph is symmetric about the origin. A function that is symmetric about the origin is called an **odd function**.

> **Definition.**   A function f with the property that $f(-x) = -f(x)$ for all x in the domain of f is called an **odd function**.

In summary we have the following results:

> **Tests for Symmetry.**   If the point $(-x, y)$ lies on a graph whenever (x, y) does, then the graph is **symmetric about the y-axis**.
>   If $(x, -y)$ lies on the graph whenever (x, y) does, the graph is **symmetric about the x-axis**.
>   If $(-x, -y)$ lies on the graph whenever (x, y) does, the graph is **symmetric about the origin**.
>   If $y = f(x)$ and $y = f(-x)$ are equivalent, the graph of the function f is symmetric about the y-axis.
>   If $y = f(x)$ and $-y = f(-x)$ are equivalent, the graph of the function f is symmetric about the origin.

Example 5   Graph the polynomial function $y = x^3 - x$.

Solution   To check for symmetries, calculate both $f(x)$ and $f(-x)$.

$$f(x) = x^3 - x \qquad f(-x) = (-x)^3 - (-x)$$
$$= -x^3 + x$$

Because $y = f(x)$ is not equivalent to $y = f(-x)$, there is no symmetry about the y-axis. Because $y = f(x)$ is a function and not the zero function, there is no symmetry about the x-axis either. However, if x and y are replaced with $-x$ and $-y$, respectively, the resulting equation is equivalent to the original equation:

$$y = x^3 - x$$
$$-y = (-x)^3 - (-x) \qquad \text{Replace } x \text{ with } -x \text{ and } y \text{ with } -y.$$
$$-y = -x^3 + x \qquad \text{Simplify.}$$
$$y = x^3 - x \qquad \text{Multiply both sides by } -1.$$

Because the final result is identical to the original equation, the graph is symmetric with respect to the origin.

Before graphing the curve, it is useful to know the graph's x-intercepts. To find them, solve the equation $x^3 - x = 0$.

$$x^3 - x = 0$$
$$x(x^2 - 1) = 0 \qquad\qquad\qquad\qquad\qquad \text{Factor out } x.$$
$$x(x + 1)(x - 1) = 0 \qquad\qquad\qquad\qquad \text{Factor } x^2 - 1.$$
$$x = 0 \quad \text{or} \quad x + 1 = 0 \quad \text{or} \quad x - 1 = 0 \qquad \text{Set each factor equal to 0.}$$
$$x = -1 \qquad\qquad x = 1$$

Thus, the x-intercepts are 0, -1, and 1.

Finally, plot the intercepts and a few points for positive x, and use the graph's symmetry to draw the rest of the graph (see Figure 2–31).

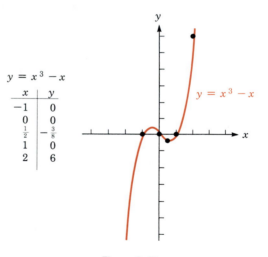

$$y = x^3 - x$$

x	y
-1	0
0	0
$\frac{1}{2}$	$-\frac{3}{8}$
1	0
2	6

Figure 2–31

Example 6   Graph the function $y = f(x) = x^4 - 5x^2 + 4$.

Solution   Because the variable x appears only with even exponents, $y = f(x)$ is equivalent to $y = f(-x)$; that is, the function's value at x is equal to its value at $-x$.

$$f(x) = x^4 - 5x^2 + 4 \qquad f(-x) = (-x)^4 - 5(-x)^2 + 4$$
$$= x^4 - 5x^2 + 4$$

Hence, this graph is symmetric about the y-axis. It is not symmetric about either the x-axis or the origin, however.

To find the x-intercepts of the graph, set y equal to 0 and solve the equation for x:

$$y = x^4 - 5x^2 + 4$$
$$0 = x^4 - 5x^2 + 4$$
$$0 = (x^2 - 4)(x^2 - 1)$$
$$0 = (x + 2)(x - 2)(x + 1)(x - 1)$$

$$x + 2 = 0 \quad \text{or} \quad x - 2 = 0 \quad \text{or} \quad x + 1 = 0 \quad \text{or} \quad x - 1 = 0$$
$$x = -2 \quad \quad x = 2 \quad \quad x = -1 \quad \quad x = 1$$

The x-intercepts are -2, 2, -1, and 1. To graph this function, plot the intercepts and several other points whose coordinates satisfy the equation, and make use of the y-axis symmetry. The graph appears in Figure 2–32.

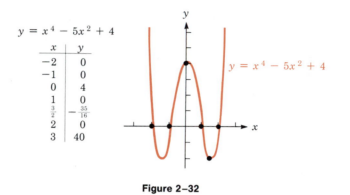

$y = x^4 - 5x^2 + 4$

x	y
-2	0
-1	0
0	4
1	0
$\frac{3}{2}$	$-\frac{35}{16}$
2	0
3	40

$y = x^4 - 5x^2 + 4$

Figure 2–32

In calculus it is shown that graphs of polynomial functions are continuous curves that rise and fall smoothly, without any sharp corners (called **cusps**) or gaps. The peaks and valleys are called **turning points**, and the graph of a polynomial function of degree n can have at most $n - 1$ turning points. Figure 2–33a could not represent the graph of a polynomial function because of the cusp and the break in the graph. Figure 2–33b could represent the graph of a polynomial function.

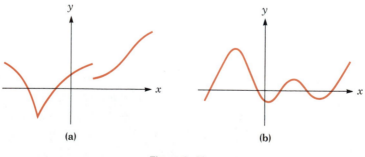

(a)

(b)

Figure 2–33

An important function in calculus is called the **difference quotient**. It represents the slope of a line, called a **secant line**, that passes through two points on the graph of a function. Consider the graph of $y = f(x)$ and the point $(a, f(a))$ on that graph (see Figure 2–34). At the point $x = a + h$, the value of the function is $f(a + h)$. The slope of the line through points $(a, f(a))$ and $(a + h, f(a + h))$ is a function of h given by

$$m(h) = \frac{\Delta y}{\Delta x} = \frac{f(a + h) - f(a)}{h}$$

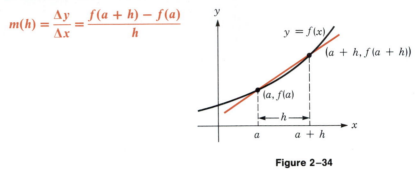

Figure 2–34

Example 7   A secant line is drawn through the points of the parabola $y = f(x) = x^2$ with x-coordinates 2 and $2 + h$, as in Figure 2–35. Find the slope of the line as a function of h.

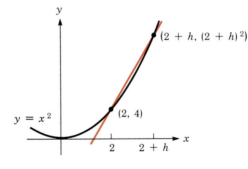

Figure 2–35

Solution   The coordinates of the points are $(2, 2^2)$ and $(2 + h, (2 + h)^2)$. Find the difference quotient and simplify.

$$\begin{aligned}
m(h) = \frac{\Delta y}{\Delta x} &= \frac{f(a + h) - f(a)}{h} \\
&= \frac{f(2 + h) - f(2)}{h} \\
&= \frac{(2 + h)^2 - 2^2}{h} \\
&= \frac{4 + 4h + h^2 - 4}{h} \\
&= \frac{4h + h^2}{h} \\
&= 4 + h \qquad \text{Factor an } h \text{ from the numerator and simplify.}
\end{aligned}$$

The slope of the secant line is $m(h) = 4 + h$.   ■

Increasing and Decreasing Functions

If the values of $f(x)$ increase as x increases on an interval, we say that the function is **increasing** on that interval (see Figure 2–36a). If the values of $f(x)$ decrease as x increases on an interval, we say the function is **decreasing** on that interval (see Figure 2–36b). If the values of $f(x)$ remain unchanged as x increases on an interval, then the function is **constant** on the interval (see Figure 2–36c).

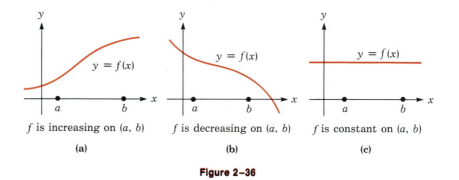

f is increasing on (a, b) f is decreasing on (a, b) f is constant on (a, b)

(a)   (b)   (c)

Figure 2–36

Definition. A function is **increasing** on (a, b) if $f(x_1) < f(x_2)$ whenever $a < x_1 < x_2 < b$.

A function is **decreasing** on (a, b) if $f(x_1) > f(x_2)$ whenever $a < x_1 < x_2 < b$.

A function f is **constant** on (a, b) if $f(x_1) = f(x_2)$ for all x_1 and x_2 on (a, b).

Example 8   Graph the function $f(x) = x^3 - 3x$ and determine the intervals on which it is increasing and on which it is decreasing.

Solution   The graph of $f(x) = x^3 - 3x$ appears in Figure 2–37. The function is increasing on $(-\infty, -1)$ and on $(1, \infty)$. The function is decreasing on $(-1, 1)$.

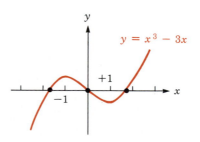

$y = x^3 - 3x$

Figure 2–37   ■

Translating and Stretching Graphs

The graph of $y = x^2 + k$ is a parabola with vertex at $(0, k)$. Figure 2–38 shows the graph of this equation for three different values of k. The graph of $y = x^2$ is the parabola that passes through the origin. The graph of $y = x^2 + 2$ is identical to the graph of $y = x^2$, except that it is shifted **2** units upward. The graph of $y = x^2 - 3$ is identical to the graph of $y = x^2$, except that it is shifted **3** units downward. Such shifts are called **vertical translations**.

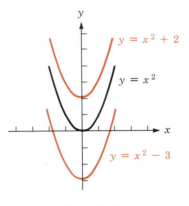

Figure 2–38

In general, we have

Vertical Translations.   The graph of $y = f(x) + |k|$ is identical to the graph of $y = f(x)$, except that it is translated $|k|$ units upward.

The graph of $y = f(x) - |k|$ is identical to the graph of $y = f(x)$, except that it is translated $|k|$ units downward.

The graph of $y = (x - h)^2$ is a parabola with vertex at $(h, 0)$. Figure 2–39 shows the graph of this equation for three different values of h. The graph of $y = (x - 2)^2$ is identical to the graph of $y = x^2$, except that it is shifted **2** units

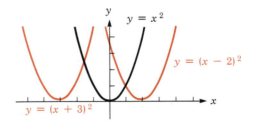

Figure 2–39

to the right. The graph of $y = (x + 3)^2$ is identical to the graph of $y = x^2$, except that it is translated **3** units to the left. Such shifts are called **horizontal translations**. In general, we have

> **Horizontal Translations.**   The graph of $y = f(x - |h|)$ is identical to the graph of $y = f(x)$, except that it is translated $|h|$ units to the right.
>   The graph of $y = f(x + |h|)$ is identical to the graph of $y = f(x)$, except that it is translated $|h|$ units to the left.

Example 9   Graph the function $y = (x - 5)^3 + 4$.

Solution   The graph of $y = (x - 5)^3 + 4$ is identical to the graph of $y = (x - 5)^3$, except it is translated **4** units upward. The graph of $y = (x - 5)^3$ is identical to the graph of $y = x^3$, except it is translated **5** units to the right. Thus, to graph $y = (x - 5)^3 + 4$, graph $y = x^3$ and translate it 4 units upward and 5 units to the right. See Figure 2–40.

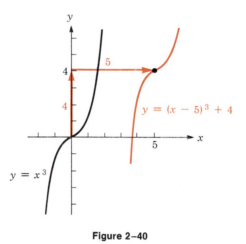

Figure 2–40

    In Figure 2–41 the graph of $y = f(x) = x^2 - 1$ is the black parabola with x-intercepts of 1 and -1. The graph of $y = 3f(x) = 3(x^2 - 1)$ appears in color. For every number x the value of y determined by $y = \mathbf{3}(x^2 - 1)$ is **3** times the value determined by $y = x^2 - 1$. The graph of $y = x^2 - 1$ has been stretched vertically by a factor of 3 to become the graph of $y = 3(x^2 - 1)$. Note that this stretching does not change the x-intercepts of the graph.

    Figure 2–42 shows the graph of $y = f(x) = x^2 - 1$ and the graph of $y = f(3x) = (3x)^2 - 1$. If the point (a, b) lies on the graph of $y = x^2 - 1$, the point $(\frac{1}{3}a, b)$ lies on the graph of $y = (3x)^2 - 1$. The graph of $y = x^2 - 1$ has been stretched horizontally by a factor of $\frac{1}{3}$ to become the graph of $y = (3x)^2 - 1$. Note that this stretching does not change the y-intercept of the graph.

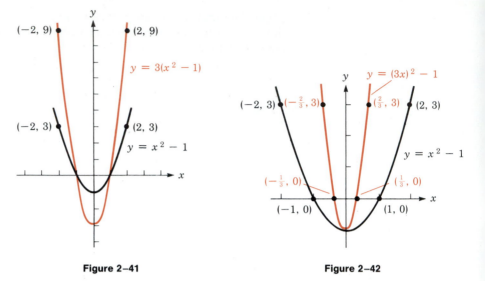

Figure 2-41   Figure 2-42

In general, we have

> **Vertical and Horizontal Stretchings.**   If $k > 0$, the graph of $y = kf(x)$ can be obtained by stretching the graph of $y = f(x)$ vertically by a factor of k. If $k < 0$, the graph of $y = kf(x)$ can be obtained by stretching the reflection of $y = f(x)$ in the x-axis by a factor of $|k|$.
>
> If $k > 0$, the graph of $y = f(kx)$ can be obtained by stretching the graph of $y = f(x)$ horizontally by a factor of $\frac{1}{k}$. If $k < 0$, the graph of $y = f(kx)$ can be obtained by stretching the reflection of $y = f(x)$ in the y-axis by a factor of $\left|\frac{1}{k}\right|$.

The effect of vertically stretching a graph by a factor of -1 is to reflect the graph in the x-axis. The effect of horizontally stretching a graph by a factor of -1 is to reflect the graph in the y-axis.

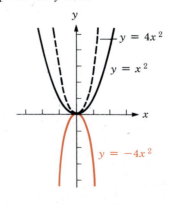

Figure 2-43

In Figure 2–43, the colored graph $y = -f(x) = -4x^2$ can be obtained by stretching the black graph of $y = x^2$ vertically by a factor of 4 to get the broken line graph of $y = 4x^2$, and then reflecting the broken line graph in the x-axis.

Exercise 2.4

In Exercises 1–8 graph each quadratic function.

1. $f(x) = x^2 - 4x + 1$
2. $f(x) = -x^2 - 4x + 1$
3. $f(x) = -(x - 1)^2$
4. $f(x) = (x + 2)^2$
5. $f(x) = x(3 - x)$
6. $f(x) = -2x(1 + x)$
7. $f(x) = (x + 2)^2 + 3$
8. $f(x) = -(x - 3)^2 - 2$

In Exercises 9–14 find the vertex of each parabola.

9. $y = x^2 - 4x + 4$
10. $y = x^2 - 10x + 25$
11. $y + 2 = -x^2 + 10x$
12. $-2x^2 + 12x - 17 - y = 0$
13. $2x^2 + 16x - y + 33 = 0$
14. $2(x + 4)^2 - x(x - 3) + y = 0$

In Exercises 15–24 find the symmetries, if any, of each curve. If the relation is a function, indicate whether the function is even, odd, or neither.

15. $y = 3x$
16. $y = x^2 + 2$
17. $y^2 = x$
18. $x^2 + y^2 = 25$
19. $y = 3x - 7$
20. $y = x^3 + 9$
21. $y = (x^3 + 1)(x^3 - 1)$
22. $y = |x|$
23. $x = |y|$
24. $|x| = |y|$

In Exercises 25–30 determine the symmetries and x- and y-intercepts, if any, of each curve. Then graph the function.

25. $y = x^2 + 1$
26. $y = 16 - x^2$
27. $y = x^3 + x$
28. $y = x^4 - x^2$
29. $y = x^6$
30. $y = x^7$

In Exercises 31–35 use this information: At a time t seconds after an object is tossed vertically upward, it reaches a height s given by the equation $s = 80t - 16t^2$.

31. In how many seconds does the object reach its maximum height?
32. In how many seconds does the object return to the point from which it was thrown?
33. What is the maximum height reached by the object?
34. Show that it takes the same time for the object to reach its maximum height as it does to fall back from that height to the point from which it was thrown.
35. Show that, if it takes the object k seconds to reach a height h on the way up, then on the way down it will reach that same height k seconds before it returns to its starting point.
36. An object is thrown from the origin of a coordinate system with x-axis along the ground and y-axis vertical. Its path, or **trajectory**, is given by the equation $y = 400x - 16x^2$. What is the object's maximum height?
37. A farmer wants to partition off a rectangular stall in a corner of his barn. The barn walls form two sides of the stall, and the farmer has 50 feet of fencing for the remaining two sides. What dimensions will maximize the area? See Illustration 1.

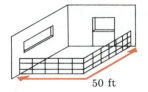

Illustration 1

38. A 24-inch-wide sheet of metal is bent into a rectangular trough; a cross section is shown in Illustration 2. What dimensions will maximize the amount of water carried? That is, what dimensions will maximize the cross-sectional area?

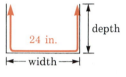

Illustration 2

39. The sum of two numbers is 6, and the sum of the squares of those two numbers is minimum. What are the two numbers?

40. A farmer will use D feet of fencing to enclose a rectangular plot of ground. Show that he will enclose the maximum area if the rectangle is a square.

41. What number most exceeds its square?

42. A rancher wishes to enclose a rectangular partitioned corral with 1800 feet of fencing (see Illustration 3). What dimensions of the corral would enclose the largest possible area? What is the maximum area?

Illustration 3

43. Find the dimensions of the rectangle of largest area that can be inscribed in the right triangle ABO shown in Illustration 4.

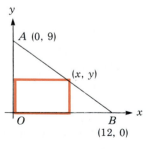

Illustration 4

44. Point P lies in the first quadrant and on the line $x + y = 1$ in such a position that the area of triangle OPA is maximum (see Illustration 5). Find the coordinates of P.

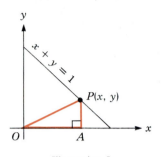

Illustration 5

In Exercises 45–46 a secant line is drawn through the points on the given curve for which $x = 2$ and $x = 2 + h$. Find the slope of the secant line as a function of h.

45. $y = f(x) = x^3$

46. $y = f(x) = x^2 - x$

In Exercises 47–48 graph the function and determine the intervals on which it is increasing and on which it is decreasing.

47. $y = \dfrac{x^3}{3} - \dfrac{x^2}{2}$

48. $y = 2x^3 + 3x^2 + 6x$

In Exercises 49–52 indicate the shifting and/or stretching of the graph of $y = f(x) = x^3$ required to produce the graph of the given equation.

49. $y = x^3 - 6$

50. $y = (x - 6)^3$

51. $y = (x + 2)^3 - 1$

52. $y = \left(\dfrac{1}{5}x\right)^3 + 2$

In Exercises 53–56 indicate the shifting and/or stretching of the graph of the first equation required to produce the graph of the second equation.

53. $y = f(x) = x^2; y = (x + 2)^2 + 3$

54. $y = |x|; y = |x - 3|$

55. $y = x^3 - x; y = -x^3 + x$

56. $y = x; y = 3x - 2$

2.5   RATIONAL FUNCTIONS

If $P(x)$ and $Q(x)$ are polynomials, the function defined by the equation

$$y = \frac{P(x)}{Q(x)}$$

is called a **rational function**. Because $Q(x)$ appears in the denominator and division by 0 is not defined, the domain of the rational function excludes the zeros of Q.

Example 1   Find the domain of the rational function defined by $y = f(x) = \dfrac{x^2 - 4}{x^2 - 1}$. Also find the symmetries of its graph.

Solution   Factor both the numerator and the denominator of the fraction:

$$y = f(x) = \frac{x^2 - 4}{x^2 - 1} = \frac{(x + 2)(x - 2)}{(x + 1)(x - 1)}$$

Observe that the numbers 1 and -1 are zeros of the denominator. Thus, the domain of the function is the set

$$\{x : x \neq 1 \text{ and } x \neq -1\}$$

The graph of the function is symmetric about the y-axis because the equations $y = f(x)$ and $y = f(-x)$ are equivalent:

$$y = f(x) = \frac{x^2 - 4}{x^2 - 1} \qquad y = f(-x) = \frac{(-x)^2 - 4}{(-x)^2 - 1}$$

$$= \frac{x^2 - 4}{x^2 - 1}$$

$$= f(x) \qquad \blacksquare$$

Example 2   Graph the rational function defined by $y = f(x) = \dfrac{x^2 - 4}{x^2 - 1}$.

Solution   In Example 1 you saw that the graph of the function is symmetric about the y-axis. You also saw that the numbers 1 and -1 are not in the domain of the function. To discover what happens when x is near 1, calculate values of y for numbers x that are close to 1:

$x < 1$		$x > 1$	
x	y	x	y
0.5	5	1.5	-1.4
0.9	16.8	1.1	-13.3
0.99	151.8	1.01	-148.3
0.999	1501.8	1.001	-1498.3

Plotting these points suggests a graph that approaches but never touches the vertical line $x = 1$ (see Figure 2–44a). Because the graph is symmetric about the y-axis, similar behavior occurs near the line $x = -1$.

The y-intercept of the graph is determined by setting x equal to 0 and computing y: the y-intercept is 4.

The x-intercepts of the graph are the zeros of the function. Because a fraction can be 0 only when its numerator is zero, the x-intercepts are $x = 2$ and $x = -2$. You can now sketch most of the curve, as shown in Figure 2–44b.

To discover the shape of the curve to the right of $x = 2$, examine the behavior of

$$y = \frac{x^2 - 4}{x^2 - 1}$$

as x gets very large. Perform a long division and write the answer in *quotient* $+ \dfrac{remainder}{divisor}$

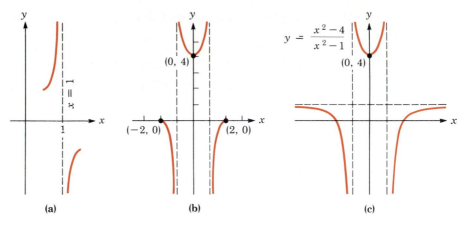

(a)   (b)   (c)

Figure 2–44

form:

$$x^2 - 1 \overline{\smash{)}\begin{array}{r} 1 \\ x^2 - 4 \\ \underline{x^2 - 1} \\ -3 \end{array}}$$

Hence,

$$y = \frac{x^2 - 4}{x^2 - 1} = 1 + \frac{-3}{x^2 - 1}$$

As x becomes very large, the denominator of the fraction $\dfrac{-3}{x^2 - 1}$ becomes very large also, and the magnitude of the fraction itself becomes very small. We indicate this by saying that the **limit** of $\dfrac{-3}{x^2 - 1}$ is 0 as x approaches infinity, and we write

$$\lim_{x \to \infty} \frac{-3}{x^2 - 1} = 0$$

Because the term $\dfrac{-3}{x^2 - 1}$ is negative and approaches 0 as x approaches infinity, corresponding values of y are less than 1 and the curve approaches the line $y = 1$ from below. Because of symmetry, the curve also approaches the line $y = 1$ as x decreases without bound (see Figure 2–44c).

The broken lines in Figure 2–44c are called **asymptotes**. The rational function of this example has two vertical asymptotes, the lines $x = 1$ and $x = -1$, and one horizontal asymptote, the line $y = 1$. The complete graph of this rational function appears in Figure 2–44c. ■

We now summarize the work done in the previous example: First you determined that the function is symmetric about the y-axis. Then you found the vertical asymptotes, $x = 1$ and $x = -1$, by setting the denominator of the rational expression equal to 0 and solving for x. You found the y-intercept by letting x equal 0 in the rational expression and evaluating y, and you found the x-intercepts by setting the numerator of the rational expression equal to 0 and solving for x. You found the horizontal asymptote by performing a long division and ignoring the remainder. Finally, you sketched the entire curve.

Example 3   Graph the rational function $y = f(x) = \dfrac{3x}{x-2}$.

Solution   First determine whether the graph has any symmetries. Because $f(-x) \neq f(x)$, the graph is not symmetric about the y-axis. Because $f(-x) \neq -f(x)$, the graph is not symmetric about the origin either.

Find the x-intercept by setting the numerator of the rational expression equal to 0 and solving for x: the x-intercept is 0. Because $y = 0$ when $x = 0$, the y-intercept is also 0 and the curve passes through the origin.

The vertical asymptotes are found by setting the denominator of the rational expression equal to 0 and solving for x: the vertical asymptote is the line $x = 2$.

The horizontal asymptote, if any, is found by dividing $3x$ by $x - 2$ and expressing the answer in $quotient + \dfrac{remainder}{divisor}$ form:

$$y = \frac{3x}{x-2} = 3 + \frac{6}{x-2}$$

Because $\lim\limits_{|x| \to \infty} \dfrac{6}{x-2} = 0$, the curve approaches the line $y = 3$. Thus, the line $y = 3$ is the horizontal asymptote. To discover the behavior of the graph to the right of $x = 2$, plot a point to the right of $x = 2$. If $x = 3$, for example, then $y = 9$. Plot the point $(3, 9)$ and use intercepts and asymptotes to sketch the graph. This curve, shown in Figure 2–45, is called a **hyperbola**.

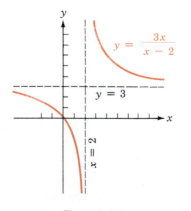

Figure 2–45

Example 4   Graph the rational function $y = \dfrac{x^2 + x - 2}{x - 3}$.

Solution   First factor the numerator and the denominator of the rational expression:

$$y = \frac{(x - 1)(x + 2)}{x - 3}$$

Most of the information is straightforward:

- The function is not symmetric to either the x-axis or the y-axis.
- The y-intercept is $\frac{2}{3}$.
- The x-intercepts are 1 and -2.
- The vertical asymptote is $x = 3$.

Perform the division and write the rational expression as

$$y = \frac{x^2 + x - 2}{x - 3} = x + 4 + \frac{10}{x - 3}$$

Because $\displaystyle\lim_{|x| \to \infty} \frac{10}{x - 3} = 0$, the graph approaches *the line $y = x + 4$*. Because this line is not horizontal, the graph has no horizontal asymptote. However, it does have a **slant** or **oblique asymptote**, the line $y = x + 4$. The graph appears in Figure 2–46.

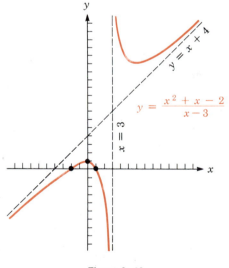

Figure 2–46

Example 5   Discuss the nature of the asymptotes of the following rational functions:

a. $y = \dfrac{x + 2}{x^2 - 1}$ **b.** $y = \dfrac{5x^2 + x + 2}{x^2 - 1}$

c. $y = \dfrac{3x^3 + 2x^2 + 2}{x^2 - 1}$ **d.** $y = \dfrac{x^4 + x + 2}{x^2 - 1}$

Solution   All four functions have vertical asymptotes at $x = 1$ and at $x = -1$ because at these values the denominators are 0. Next we consider the nature of the remaining asymptotes.

a. $y = \dfrac{x + 2}{x^2 - 1}$

Because the degree of the numerator is less than the degree of the denominator, long division is not practical. Instead, divide both numerator and denominator by x^2, which is the largest power of x in the denominator of the rational expression.

$$y = \frac{x + 2}{x^2 - 1} = \frac{\dfrac{x}{x^2} + \dfrac{2}{x^2}}{\dfrac{x^2}{x^2} - \dfrac{1}{x^2}} = \frac{\dfrac{1}{x} + \dfrac{2}{x^2}}{1 - \dfrac{1}{x^2}}$$

The three fractions in the final result, $\dfrac{1}{x}$, $\dfrac{2}{x^2}$, and $\dfrac{1}{x^2}$, all approach 0 as $|x|$ increases without bound. Hence, y approaches

$$\frac{0 + 0}{1 - 0} = 0$$

The graph of this function has a horizontal asymptote of $y = 0$. In parts **b**, **c**, and **d**, a long division can be performed.

b. $y = \dfrac{5x^2 + x + 2}{x^2 - 1} = 5 + \dfrac{6x + 2}{x^2 - 1}$

As $|x|$ increases without bound, the fraction $\dfrac{6x + 2}{x^2 - 1}$ approaches 0 (for reasons discussed in part **a**) and y approaches 5. This curve has a horizontal asymptote of $y = 5$. Note that 5 is the quotient of the lead coefficients of $P(x)$ and $Q(x)$, and that these polynomials have equal degree.

c. $y = \dfrac{3x^3 + 2x^2 + 2}{x^2 - 1} = 3x + 2 + \dfrac{3x + 4}{x^2 - 1}$

Again, the last fraction approaches 0 as $|x|$ increases without bound, and the curve approaches the slant asymptote $y = 3x + 2$.

d. $y = \dfrac{x^4 + x + 2}{x^2 - 1} = x^2 + 1 + \dfrac{x + 3}{x^2 - 1}$

As $|x|$ increases, the fractional part again approaches 0, and the curve approaches $y = x^2 + 1$. However, $y = x^2 + 1$ does not represent a line. This curve has neither horizontal nor slant asymptotes.   ■

We generalize the results of Example 5, and summarize the techniques of graphing rational functions.

Perform the following steps when you graph the rational function $y = \dfrac{P(x)}{Q(x)}$, where $\dfrac{P(x)}{Q(x)}$ is in simplified form.

Check for symmetry.   If the polynomials $P(x)$ and $Q(x)$ involve only even powers of x, the graph is symmetric about the y-axis. Otherwise, y-axis symmetry does not exist.

Look for y-intercepts.   Set x equal to 0. The corresponding value of y is the y-intercept of the graph.

Look for x-intercepts.   Set $P(x)$ equal to 0. The solutions of the equation $P(x) = 0$ (if any) are the x-intercepts of the graph.

Look for vertical asymptotes.   Set $Q(x)$ equal to 0. The solutions of the equation $Q(x) = 0$ (if any) determine the vertical asymptotes of the graph.

Look for a horizontal asymptote.   If the degree of $P(x)$ is less than the degree of $Q(x)$, then the line $y = 0$ is the horizontal asymptote.

If the degrees of $P(x)$ and $Q(x)$ are equal, then the line $y = \dfrac{p}{q}$, where p and q are the lead coefficients of $P(x)$ and $Q(x)$, is the horizontal asymptote. [Make sure that $P(x)$ and $Q(x)$ are written in descending powers of x before applying this rule.]

If the degree of $P(x)$ is greater than the degree of $Q(x)$, then there is no horizontal asymptote.

A curve might intersect a horizontal asymptote when $|x|$ is small. However, it will approach but never touch a horizontal asymptote as $|x|$ grows large.

Look for a slant asymptote.   If the degree of $P(x)$ is exactly one greater than the degree of $Q(x)$, there is a slant asymptote. To find it, perform the long division $Q(x) \overline{)\, P(x)}$ and ignore the remainder.

To find the vertical asymptotes of the graph of a rational function, we first found the zeros of the polynomial in the denominator of the function. This method works *only when the fraction* $\dfrac{P(x)}{Q(x)}$ *is in simplified form.* In the next example, the polynomials $P(x)$ and $Q(x)$ share a common factor.

Example 6   Graph the rational function $y = \dfrac{x^2 - x - 12}{x - 4}$.

Solution   Factor the numerator of the rational expression:

$$y = \frac{x^2 - x - 12}{x - 4} = \frac{(x + 3)(x - 4)}{x - 4}$$

Because the denominator is 0 when $x = 4$, the number 4 is not in the domain of the rational function. For numbers x *other than* 4, the common factor of $x - 4$ can be canceled and the fraction simplified. The result is equivalent to the original function *only if you keep the restriction that $x \neq 4$*:

$$y = x + 3 \text{ provided that } x \neq 4$$

When $x = 4$, the function is not defined. The graph of the given rational function appears in Figure 2–47. It is a line with one point missing, the point with x-coordinate of 4.

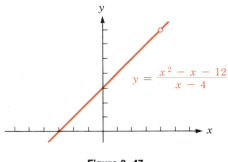

$$y = \frac{x^2 - x - 12}{x - 4}$$

Figure 2–47

Exercise 2.5

In Exercises 1–24 find all intercepts and asymptotes, and graph each rational function.

1. $y = \dfrac{1}{x - 2}$ **2.** $y = \dfrac{3}{x + 3}$ **3.** $y = \dfrac{x}{x - 1}$ **4.** $y = \dfrac{x}{x + 2}$

5. $y = \dfrac{x + 1}{x + 2}$ **6.** $y = \dfrac{x - 1}{x - 2}$ **7.** $y = \dfrac{2x - 1}{x - 1}$ **8.** $y = \dfrac{3x + 2}{x - 1}$

9. $y = \dfrac{x^2 - 9}{x^2 - 4}$ **10.** $y = \dfrac{x^2 - 4}{x^2 - 9}$ **11.** $y = \dfrac{x^2 - 5x + 6}{x^2 - 2x + 1}$ **12.** $y = \dfrac{x^2 + 7x + 12}{x^2 - 4x + 4}$

13. $y = \dfrac{3x^2}{x^2 + 1}$ **14.** $y = \dfrac{x^2 - 9}{2x^2 + 1}$ **15.** $y = \dfrac{2x^2 - 2}{x^2 - 25}$ **16.** $y = \dfrac{x^2 - 4}{3x^2 - 27}$

17. $y = \dfrac{2x^2 - 3x - 2}{x^2 + x - 2}$ **18.** $y = \dfrac{3x^2 - 4x + 1}{2x^2 + 3x + 1}$ **19.** $y = \dfrac{x^2 - 9}{2x^2 - 8}$ **20.** $y = \dfrac{3x^2 - 12}{x^2}$

21. $y = \dfrac{x^2 - 2x - 8}{x - 1}$ **22.** $y = \dfrac{x^2 + x - 6}{x + 2}$ **23.** $y = \dfrac{x^3 + x^2 + 6x}{x^2 - 1}$ **24.** $y = \dfrac{x^3 - 2x^2 + x}{x^2 - 4}$

In Exercises 25–30 the numerator and the denominator of each rational function share one or more common factors. Graph each function.

25. $y = \dfrac{x^2 - 4}{x - 2}$ **26.** $y = \dfrac{x^2 - 9}{x + 3}$ **27.** $y = \dfrac{x^2 - 4x + 3}{x - 1}$

28. $y = \dfrac{x^2 - 4x + 3}{x^2 - 1}$ **29.** $y = \dfrac{x + 3}{x^2 - 9}$ **30.** $y = \dfrac{x^2 - 1}{x^3 - x}$

2.6   ALGEBRA OF FUNCTIONS

It is possible to perform some arithmetic with functions.

Definition.   If the ranges of functions f and g are subsets of the real numbers, then:

The **sum** of f and g, denoted as $f + g$, is defined by

$$(f + g)(x) = f(x) + g(x)$$

The **difference** of f and g, denoted as $f - g$, is defined by

$$(f - g)(x) = f(x) - g(x)$$

The **product** of f and g, denoted as $f \cdot g$, is defined by

$$(f \cdot g)(x) = f(x) \cdot g(x)$$

The **quotient** of f and g, denoted as f/g, is defined by

$$(f/g)(x) = \frac{f(x)}{g(x)} \qquad \text{provided } g(x) \neq 0$$

The domain of each of these functions (unless otherwise restricted) is the set of all real numbers that are in the domains of *both* f and g. The quotient f/g has the further restriction that $g(x) \neq 0$.

Example 1   Let $f(x) = x^2 - 4$ and $g(x) = \sqrt{x}$. If all calculations involve real numbers only, find the functions **a.** $f + g$, **b.** $f \cdot g$, **c.** f/g, and **d.** g/f, and give the domain of each.

Solution   Because any real number can be squared, the domain of f is the set of real numbers. Because all calculations are to involve real numbers only, the domain of g is the set of nonnegative real numbers, $\{x : x \geq 0\}$.

a. $(f + g)(x)$ is defined as $f(x) + g(x)$. Hence,

$$(f + g)(x) = x^2 - 4 + \sqrt{x}$$

The domain is the set of all numbers common to the domains of f and g; it is the intersection of the domains of f and g. Thus, the domain of $f + g$ is $\{x : x \geq 0\}$.

b. $(f \cdot g)(x) = f(x) \cdot g(x)$
$$= (x^2 - 4) \cdot \sqrt{x}$$
$$= x^2 \sqrt{x} - 4\sqrt{x}$$

The domain of $f \cdot g$ is the intersection of the domains of f and g, or $\{x : x \geq 0\}$.

c. $(f/g)(x) = \dfrac{f(x)}{g(x)}$
$$= \frac{x^2 - 4}{\sqrt{x}}$$

The domain is the set of all numbers common to the domains of f and g, *except* 0, because division by 0 is not defined. The domain of f/g is $\{x : x > 0\}$.

d. $(g/f)(x) = \dfrac{g(x)}{f(x)}$

$\qquad\qquad = \dfrac{\sqrt{x}}{x^2 - 4}$

The domain consists of those numbers common to the domains of f and g, *except* 2, because division by 0 is not defined. The domain of g/f is $\{x : x \geq 0 \text{ and } x \neq 2\}$. ■

Example 2   Let $h(x) = x^2 + 3x + 2$. Find functions f and g such that **a.** $h = f + g$ and **b.** $h = f \cdot g$.

Solution   **a.**   Several answers are possible. One answer is $f(x) = x^2$ and $g(x) = 3x + 2$. Then

$\qquad h(x) = x^2 + 3x + 2$

$\qquad\qquad = (x^2) + (3x + 2)$

$\qquad\qquad = f(x) + g(x)$

$\qquad\qquad = (f + g)(x)$

Another possible answer is $f(x) = x^2 + 2x$ and $g(x) = x + 1$.

b.   Several answers are possible. One answer is suggested by factoring the polynomial $x^2 + 3x + 2$.

$\qquad h(x) = x^2 + 3x + 2 = (x + 1)(x + 2)$

Let $f(x) = x + 1$ and $g(x) = x + 2$. Then

$\qquad h(x) = x^2 + 3x + 2$

$\qquad\qquad = (x + 1)(x + 2)$

$\qquad\qquad = f(x) \cdot g(x)$

$\qquad\qquad = (f \cdot g)(x)$

Another possible answer is $f(x) = 3$ and $g(x) = \dfrac{x^2}{3} + x + \dfrac{2}{3}$. ■

Composition of Functions

Often one quantity depends on a second quantity that depends, in turn, on a third quantity. A farmer's income, for example, is a function of the number of bushels of grain harvested. But the harvest, in turn, is a function of the number of inches of rainfall received. Such chains of dependence can be analyzed mathematically by considering the **composition of functions**.

We suppose that $y = f(x)$ and $y = g(x)$ define two functions. To any number x in the domain of g there corresponds the value $g(x)$ in the range of g. If $g(x)$ is also in the domain of f, then $g(x)$ is an acceptable input to the function f and the corresponding value $f(g(x))$ is determined. This two-step process defines a new function, called a **composite function**, denoted by $f \circ g$.

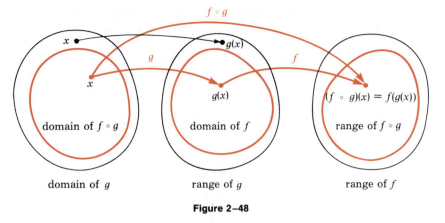

Figure 2–48

To visualize the domain of the composite function $f \circ g$, refer to Figure 2–48. The value of the independent variable x must be in the domain of the function g because x is the number used as input to the function g. However, $g(x)$ must be in the domain of f because $g(x)$ is the input to the function f. Hence, the domain of $f \circ g$ consists of all those numbers x that are permissible inputs to g and for which $g(x)$ is a permissible input to f.

> **Definition.** The **composite function $f \circ g$** is defined by
>
> $$(f \circ g)(x) = f(g(x))$$
>
> The domain of the composite function $f \circ g$ consists of all those elements in the domain of g whose images under g are in the domain of f.

Example 3   If $f(x) = x + 1$ and $g(x) = x^2$, find $(f \circ g)(2)$.

Solution   To compute $(f \circ g)(2)$, you must calculate $f(g(2))$. Because $g(x) = x^2$, you have $g(2) = 2^2 = 4$. Thus, $f(g(2)) = f(4)$. Because 4 is in the domain of f, you can calculate $f(4)$, which is $4 + 1$, or 5. Hence,

$$(f \circ g)(2) = f(g(2)) = f(4) = 5$$ ∎

Example 4   If $f(x) = 2x + 7$ and $g(x) = 4x + 1$, find **a.** $(f \circ g)(x)$ and **b.** $(g \circ f)(x)$.

Solution   **a.** $(f \circ g)(x) = f(g(x)) = f(4x + 1)$
$$= 2(4x + 1) + 7$$
$$= 8x + 9$$

b. $(g \circ f)(x) = g(f(x)) = g(2x + 7)$
$$= 4(2x + 7) + 1$$
$$= 8x + 29$$ ∎

Example 4 illustrates that composition of functions is *not* commutative. In general, $(f \circ g)(x) \neq (g \circ f)(x)$.

Example 5   Find two functions f and g such that $(f \circ g)(x) = \sqrt{x^2 + 2}$.

Solution   Let $g(x) = x^2 + 2$ and $f(x) = \sqrt{x}$. Then

$$(f \circ g)(x) = f(g(x)) = f(x^2 + 2)$$
$$= \sqrt{x^2 + 2}$$

Thus, $g(x) = x^2 + 2$ and $f(x) = \sqrt{x}$ are two functions such that $(f \circ g)(x) = \sqrt{x^2 + 2}$. ∎

Example 6   Let $f(x) = \sqrt{x}$ and $g(x) = x - 3$. Find the domain of functions **a.** f, **b.** g, **c.** $f \circ g$, and **d.** $g \circ f$.

Solution   **a.**   Because the range of a function is to be a subset of the real numbers, the domain of f is the set of nonnegative real numbers: $\{x : x \geq 0\}$.

b.   The domain of g is the set of real numbers.

c. $(f \circ g)(x)$ means $f(g(x))$. Because $g(x)$ must be in the domain of f, the value $g(x)$ must be nonnegative. Thus,

$$g(x) \geq 0$$
$$x - 3 \geq 0 \qquad \text{Because } g(x) = x - 3.$$
$$x \geq 3$$

Hence, the domain of $f \circ g$ is $\{x : x \geq 3\}$.

d. $(g \circ f)(x)$ means $g(f(x))$. The domain of $g \circ f$ contains those numbers x in the domain of f for which $f(x)$ is in the domain of g. Because the domain of g includes all real numbers, the domain of $g \circ f$ is just the domain of f itself: $\{x : x \geq 0\}$. ∎

Inverse Functions

By definition, each element x in the domain of a function has a *single* image y. For some functions, distinct x's in the domain have the *same* image in the range (see Figure 2–49). For other functions, called **one-to-one functions**, different numbers x have different images (see Figure 2–50).

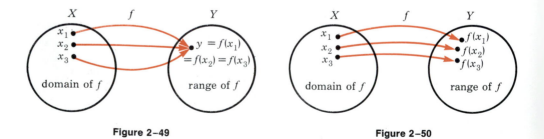

Figure 2–49   Figure 2–50

> **Definition.** A function f from a set X to a set Y is called **one-to-one** if different elements in the domain of f have different images in the range of f.
>
> In symbols, if x_1 and x_2 are two elements in the domain of a one-to-one function f and $x_1 \neq x_2$, then $f(x_1) \neq f(x_2)$.

A **horizontal line test** can be used to determine whether the graph of a function represents a one-to-one function. If any horizontal line intersects the graph more than once, the function is *not* one-to-one (see Figure 2–51). If every horizontal line that intersects the graph does so only once, the function *is* one-to-one (see Figure 2–52).

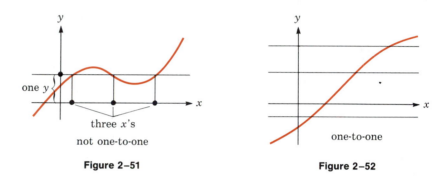

Figure 2–51   Figure 2–52

Figure 2–49 illustrates a function f from set X to set Y. Because several arrows map to a single y, the function f is not one-to-one. If the arrows of Figure 2–49 were reversed, the diagram would not represent a function, because to some y of set Y would correspond several values x in set X. If the arrows of the one-to-one function f in Figure 2–50 were reversed, the diagram would still represent a function. This "backwards function," called the **inverse** of the original function f, is denoted by the symbol f^{-1}. The "-1" indicates the inverse of function f and is *not* an exponent.

Refer to the one-to-one function f and its inverse f^{-1} shown in Figure 2–53. To each element x in the domain of f there corresponds its image $f(x)$ in the range of f. This element $f(x)$, however, is in the domain of f^{-1}. The image of

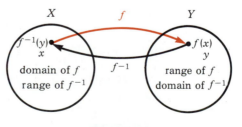

Figure 2–53

$f(x)$ under the function f^{-1} is $f^{-1}(f(x))$, which is the original number x. Thus, $(f^{-1} \circ f)(x) = f^{-1}(f(x)) = x$. Similarly, if y is an element of the domain of f^{-1}, then $(f \circ f^{-1})(y) = f(f^{-1}(y)) = y$.

Theorem. If f is a one-to-one function with domain X and range Y, then there is a one-to-one function f^{-1}, called the **inverse of f**, with domain Y and range X such that

$$(f^{-1} \circ f)(x) = x \qquad \text{and} \qquad (f \circ f^{-1})(y) = y$$

To show that one function f is the inverse of another function g, we must show that the compositions $f \circ g$ and $g \circ f$ are the **identity function**—the function that assigns x itself as the image of each real number x.

Example 7   Show that the function $f(x) = x^3$ is the inverse of $g(x) = \sqrt[3]{x}$.

Solution   Show that the composition of f and g (in both directions) is the identity function:

$$(f \circ g)(x) = f(g(x)) = f(\sqrt[3]{x}) = (\sqrt[3]{x})^3 = x$$
$$(g \circ f)(x) = g(f(x)) = g(x^3) = \sqrt[3]{x^3} = x \qquad\blacksquare$$

If f is a one-to-one function defined by $y = f(x)$, then f^{-1} reverses the correspondence of f. That is, if $f(a) = b$, then $f^{-1}(b) = a$. To determine f^{-1}, we interchange the variables x and y in the equation $y = f(x)$. The resulting equation, $x = f(y)$, defines the inverse function f^{-1}. If we can solve $x = f(y)$ for y, we will have expressed the inverse as the equation $y = f^{-1}(x)$.

Example 8   Find the inverse of the function $y = f(x) = \dfrac{3}{2}x + 2$, and verify the result.

Solution   To find the inverse of $y = \dfrac{3}{2}x + 2$, interchange the x and y to obtain

$$x = \frac{3}{2}y + 2$$

To express this equation in the form $y = f^{-1}(x)$, solve it for y:

$$x = \frac{3}{2}y + 2$$
$$2x = 3y + 4$$
$$2x - 4 = 3y$$
$$y = \frac{2x - 4}{3}$$

Thus, the inverse of the function $f(x) = \dfrac{3}{2}x + 2$ is

$$f^{-1}(x) = \frac{2x - 4}{3}$$

Verify this by showing that their composition is the identity function:

$$(f \circ f^{-1})(x) = f(f^{-1}(x)) = f\left(\frac{2x-4}{3}\right) = \frac{3}{2}\left(\frac{2x-4}{3}\right) + 2 = x - 2 + 2 = x$$

Also,

$$(f^{-1} \circ f)(x) = f^{-1}(f(x)) = f^{-1}\left(\frac{3}{2}x + 2\right) = \frac{2\left(\frac{3}{2}x + 2\right) - 4}{3} = \frac{3x + 4 - 4}{3} = x$$

■

Because x and y are interchanged when finding the inverse, the point (b, a) lies on the graph of $y = f^{-1}(x)$ whenever the point (a, b) lies on the graph of $y = f(x)$. Thus, the graphs of a function and its inverse are reflections of each other in the line $y = x$.

Example 9   Find the inverse of $y = f(x) = x^3 + 3$, and graph both the function and its inverse on a single set of coordinate axes.

Solution   The inverse of the function $y = x^3 + 3$ is found by interchanging x and y. The inverse is

$$x = y^3 + 3$$

Solve this equation for y to obtain

$$y = \sqrt[3]{x - 3}$$

Thus, $f^{-1}(x) = \sqrt[3]{x - 3}$.

The graphs appear in Figure 2–54. Note that $y = x$ is the axis of symmetry.

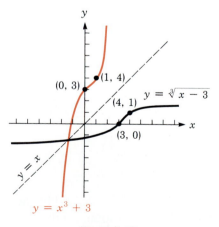

$$y = x^3 + 3$$

Figure 2–54

■

Example 10   The function $f(x) = x^2 + 3$ is not one-to-one. However, the function becomes one-to-one if the domain of f is restricted to a carefully chosen subset of the real numbers, such as $\{x : x \leq 0\}$. If the domain of f is so restricted, find **a.** the range of f, **b.** the inverse of f, and **c.** the domain and range of f^{-1}; **d.** graph both functions.

Solution   **a.**   The function f is defined by $f(x) = y = x^2 + 3$, with domain $\{x : x \leq 0\}$. If x is replaced with numbers from this domain, y ranges over the values 3 and above. Thus, the range of f is $\{y : y \geq 3\}$.

b.   To find the inverse of f, interchange x and y in the equation that defines f, and solve for y:

$$y = x^2 + 3 \qquad \text{where } x \leq 0$$
$$x = y^2 + 3 \qquad \text{where } y \leq 0 \qquad \text{Exchange } x \text{ and } y.$$
$$x - 3 = y^2 \qquad \text{where } y \leq 0$$

To solve this equation for y, take the square root of both sides. Because $y \leq 0$, you have

$$-\sqrt{x - 3} = y \qquad \text{where } y \leq 0$$

Thus, the inverse of f is defined by the equation

$$y = f^{-1}(x) = -\sqrt{x - 3}$$

c.   The domain of $f^{-1}(x)$ is $\{x : x \geq 3\}$ and the range is $\{y : y \leq 0\}$.

d.   The graphs of these two functions appear in Figure 2–55. Note that the line of symmetry is $y = x$.

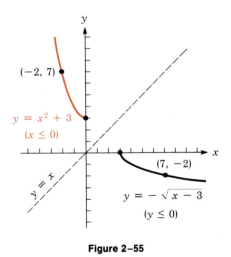

Figure 2–55

If a function is defined by the equation $y = f(x)$, we can often find the domain of f by inspection. Finding the range can be more difficult. One way to find the range of f is to find the domain of f^{-1}.

Example 11 Find the domain and the range of the function $y = f(x) = \dfrac{2}{x} + 3$.

Solution Because x cannot be 0, the domain of f is

$$\{x : x \neq 0\}$$

To find the range of f, find the domain of f^{-1}. Find f^{-1} as follows:

$$y = \frac{2}{x} + 3$$

$$x = \frac{2}{y} + 3 \qquad \text{Interchange } x \text{ and } y.$$

$$yx = 2 + 3y \qquad \text{Multiply both sides by } y.$$

$$yx - 3y = 2 \qquad \text{Add } -3y \text{ to both sides.}$$

$$y(x - 3) = 2 \qquad \text{Factor out } y.$$

$$y = \frac{2}{x - 3} \qquad \text{Divide both sides by } x - 3.$$

This final equation defines f^{-1} with domain $\{x : x \neq 3\}$. Because the range of f is equal to the domain of f^{-1}, the range of f is

$$\{y : y \neq 3\}$$ ■

Exercise 2.6

In Exercises 1–4 let $f(x) = 2x + 1$ and $g(x) = 3x - 2$. Find each function, and determine its domain.

1. $f + g$ **2.** $f - g$ **3.** $f \cdot g$ **4.** f/g

In Exercises 5–8 let $f(x) = x^2 + x$ and $g(x) = x^2 - 1$. Find each function, and determine its domain.

5. $f - g$ **6.** $f + g$ **7.** f/g **8.** $f \cdot g$

In Exercises 9–12 let $f(x) = x^2 - 1$ and $g(x) = 3x - 2$. Find each value, if possible.

9. $(f + g)(2)$ **10.** $(f \cdot g)(-3)$ **11.** $(f/g)\left(\dfrac{2}{3}\right)$ **12.** $(g/f)(-5)$

In Exercises 13–20 find two functions f and g such that the given correspondence can be expressed as the function indicated. Several answers are possible.

13. $y = 3x^2 + 2x$; $f + g$ **14.** $y = 3x^2$; $f \cdot g$

15. $y = \dfrac{3x^2}{x^2 - 1}$; f/g **16.** $y = 5x + x^4$; $f - g$

17. $y = x(3x^2 - 1)$; $f + g$ **18.** $y = (3x - 2)(3x + 2)$; $f - g$

19. $y = x^2 + 7x - 18$; $f \cdot g$ **20.** $y = 5x^5$; $f - g$

In Exercises 21–24 let $f(x) = 2x - 5$ and $g(x) = 5x - 2$. Find the following values.

21. $(f \circ g)(2)$ **22.** $(g \circ f)(-3)$ **23.** $(f \circ f)\left(-\dfrac{1}{2}\right)$ **24.** $(g \circ g)\left(\dfrac{3}{5}\right)$

In Exercises 25–28 let $f(x) = 3x^2 - 2$ and $g(x) = 4(x + 1)$. Find the following values.

25. $(g \circ f)(-3)$ **26.** $(f \circ g)(3)$ **27.** $(f \circ f)(\sqrt{3})$ **28.** $(g \circ g)(-4)$

In Exercises 29–32 let $f(x) = 3x$ and $g(x) = x + 1$. Find each composite function, and determine its domain.

29. $f \circ g$ **30.** $g \circ f$ **31.** $g \circ g$ **32.** $f \circ f$

In Exercises 33–36 let $f(x) = x^2$ and $g(x) = 2x$. Find each composite function, and determine its domain.

33. $g \circ f$ **34.** $f \circ g$ **35.** $f \circ f$ **36.** $g \circ g$

In Exercises 37–40 let $f(x) = \sqrt{x}$ and $g(x) = x^2 + 1$. Find each composite function, and determine its domain.

37. $g \circ f$ **38.** $f \circ g$ **39.** $g \circ g$ **40.** $f \circ f$

In Exercises 41–44 let $f(x) = \sqrt{x + 1}$ and $g(x) = x^2 - 1$. Find each of the following functions, and determine its domain.

41. $f \circ g$ **42.** $g \circ f$ **43.** $f \circ f$ **44.** $g \circ g$

In Exercises 45–56 find two functions f and g such that the composition $f \circ g$ expresses the given correspondence. Several answers are possible.

45. $y = 3x - 2$ **46.** $y = 7x - 5$ **47.** $y = x^2 - 2$

48. $y = x^3 + 7$ **49.** $y = (x - 2)^2$ **50.** $y = (x + 7)^3$

51. $y = \sqrt{x + 2}$ **52.** $y = \dfrac{1}{x - 5}$ **53.** $y = \sqrt{x} + 2$

54. $y = \dfrac{1}{x} - 5$ **55.** $y = x$ **56.** $y = 3$

57. Let $f(x) = 3x$. Show that $(f + f)(x) = f(x + x)$.   **58.** Let $g(x) = x^2$. Show that $(g + g)(x) \neq g(x + x)$.

59. Let $f(x) = \dfrac{x - 1}{x + 1}$. Find $(f \circ f)(x)$.   **60.** Let $g(x) = \dfrac{x}{x - 1}$. Find $(g \circ g)(x)$.

In Exercises 61–64 determine whether each function is one-to-one.

61. $y = 3x - 2$ **62.** $y = 3x^2 - 2$ **63.** $y = x^4 + x^2$ **64.** $y = x^3 - x$

In Exercises 65–68 use the horizontal line test to determine whether the graph represents a one-to-one function.

65.   **66.**   **67.**   **68.**

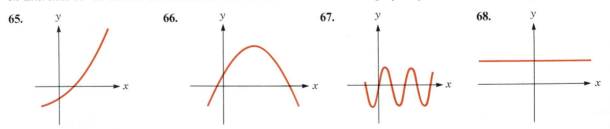

In Exercises 69–72 verify that functions f and g are inverses by determining that $f \circ g$ and $g \circ f$ are the identity function.

69. $f(x) = 3x + 2$; $g(x) = \dfrac{x - 2}{3}$ **70.** $f(x) = \dfrac{x + 1}{x}$; $g(x) = \dfrac{1}{x - 1}$

71. $f(x) = g(x) = \dfrac{x + 1}{x - 1}$ **72.** $f(x) = g(x) = \dfrac{1}{x}$

In Exercises 73–76 each equation determines a one-to-one function. Find f^{-1}.

73. $y = 3x + 2$ **74.** $y = \dfrac{1}{x + 3}$ **75.** $y = \dfrac{1}{x^3}$ **76.** $y = \dfrac{x - 1}{x}$

In Exercises 77–80 find the inverse of each one-to-one function, and graph both the function and its inverse on one set of coordinate axes.

77. $y = 5x - 3$ **78.** $y = \dfrac{1}{2x}$ **79.** $y = x^3$ **80.** $3x + 2y = 6$

In Exercises 81–84 the function defined by the given equation is one-to-one on the given domain. Find f^{-1}.

81. $f(x) = x^2 - 3 \ \{x : x \le 0\}$ **82.** $f(x) = \dfrac{1}{x^2} \ \{x : x > 0\}$

83. $f(x) = \sqrt{4 - x^2} \ \{x : 0 \le x \le 2\}$ **84.** $f(x) = \sqrt{x^2 - 1} \ \{x : x \le -1\}$

In Exercises 85–88 find the domain and the range of f. Find the range of f by finding the domain of f^{-1}.

85. $f(x) = \dfrac{x}{x - 1}$ **86.** $f(x) = \dfrac{x - 2}{x + 3}$ **87.** $f(x) = \dfrac{1}{x} - 2$ **88.** $f(x) = \dfrac{3}{x} - \dfrac{1}{2}$

REVIEW EXERCISES

In Review Exercises 1–4 graph each equation.

1. $2x - 5y = 10$ **2.** $9x + 2y = 18$ **3.** $3x + 2y = 1$ **4.** $3x - 7y = 5$

*In Review Exercises 5–8 find **a.** the length and **b.** the midpoint of the line segment PQ.*

5. $P(-3, 7); Q(3, -1)$ **6.** $P(0, 5); Q(-12, 10)$

7. $P(-\sqrt{3}, 9); Q(-\sqrt{3}, -7)$ **8.** $P(a, -a); Q(-a, a) \quad (a > 0)$

In Review Exercises 9–12 find the slope of the line PQ, if possible.

9. $P(3, -5); Q(1, 7)$ **10.** $P(2, 7); Q(-5, -7)$

11. $P(b, a); Q(a, b)$ **12.** $P(a + b, b); Q(b, b - a)$

In Review Exercises 13–22 write the equation of the line with the given properties. Express the answer in general form.

13. The line passes through the origin and the point $(-5, 7)$.

14. The line passes through $(-2, 1)$ and has a slope of -4.

15. The line passes through $(7, -5)$ and $(4, 1)$.

16. The line has a slope of $\frac{2}{3}$ and a y-intercept of 3.

17. The line has a slope of 0 and passes through $(-5, 17)$.

18. The line has no defined slope and passes through $(-5, 17)$.

19. The line passes through $(8, -2)$ and is parallel to the line segment joining $(2, 4)$ and $(4, -10)$.

20. The line passes through $(8, -2)$ and is perpendicular to the line segment joining $(2, 4)$ and $(4, -10)$.

21. The line is parallel to $3x - 4y = 7$ and passes through $(2, 0)$.

22. The line passes through $(7, 0)$ and is perpendicular to the line $3y + x - 4 = 0$.

In Review Exercises 23–28 tell whether the given equation determines y as a function of x. If it does, give the domain and range, and graph the function.

23. $y = 5x - 2$ **24.** $y = |x - 2|$ **25.** $y = \sqrt{x - 1}$ **26.** $y^2 = 2x + 7$

27. $y = x^2 - x + 1$ **28.** $y = \dfrac{x}{|x|}$

In Review Exercises 29–32 graph each polynomial function. If the graph is a parabola, find the coordinates of its vertex.

29. $y = x^2 - 3x - 2$ **30.** $x^2 + y - 2x + 4 = 0$

31. $y = x^4 + 2x^2 + 1$ **32.** $y = x^3 - 1$

In Review Exercises 33–40 graph each rational function.

33. $y = \dfrac{2}{x - 4}$ **34.** $y = \dfrac{x^2 - 9}{x^2 - 4}$ **35.** $y = \dfrac{4x}{x - 3}$

36. $y = \dfrac{x^2 + 5x + 6}{x - 1}$ **37.** $y = \dfrac{x - 2}{x^2 - 3x - 4}$ **38.** $y = \dfrac{x^2 + 4}{x^2 - 3x - 4}$

39. $y = \dfrac{3x^2 - 7x + 2}{x - 2}$ **40.** $y = \dfrac{x^3}{x}$

In Review Exercises 41–48 let $f(x) = 2x^2 - x$ and $g(x) = x + 3$. Find each value.

41. $f(2)$ **42.** $g(-3)$ **43.** $(f + g)(2)$

44. $(f - g)(-1)$ **45.** $(f \cdot g)(0)$ **46.** $(f/g)(2)$

47. $(f \circ g)(-2)$ **48.** $(g \circ f)(4)$

In Review Exercises 49–54 let $f(x) = \sqrt{x + 1}$ and $g(x) = x^2 - 1$.

49. Find $(f + g)(x)$.   **50.** Find $(f \cdot g)(x)$.   **51.** Find $(f \circ g)(x)$.   **52.** Find $(g \circ f)(x)$.

53. Determine the domain of f/g.   **54.** Determine the domain of $f \circ g$.

In Review Exercises 55–60 determine whether the given function is one-to-one. If so, determine its inverse function.

55. $y = 7x$ **56.** $y = 7x^2$ **57.** $y = \dfrac{1}{x - 1}$

58. $y = \dfrac{3}{x^3}$ **59.** $y = \dfrac{x + 2}{x - 3}$ **60.** $y = |x| + 3$

In Review Exercises 61–62 find the domain and the range of each function. Find the range by finding the domain of the function's inverse.

61. $y = f(x) = \dfrac{2x + 1}{2x - 1}$ **62.** $y = f(x) = \dfrac{2}{x} + \dfrac{1}{2}$

In Review Exercises 63–64 a secant line is drawn to the given curve through the points $(3, f(3))$ and $(3 + h, f(3 + h))$. Find the slope of the secant line as a function of h.

63. $y = f(x) = 3x + 7$ **64.** $y = f(x) = x^2 - 1$

65. Hooke's law states that the force required to stretch a spring is proportional to the amount of stretch. If a 3-pound force stretches a spring 5 inches, what force would stretch the spring 3 inches?

66. The volume of gas in a balloon varies directly as its temperature and inversely as the pressure. If the volume is 400 cubic centimeters when the temperature is 300 K and the pressure is 25 dynes per square centimeter, find the volume when the temperature is 200 K and the pressure is 20 dynes per square centimeter.

67. A moving body has a kinetic energy proportional to the square of its velocity. By what factor does the kinetic energy of an automobile increase if its speed increases from 30 miles per hour to 50 miles per hour?

68. The electrical resistance of a wire varies directly as the length of the wire and inversely as the square of its diameter. A 1000-foot length of wire, 0.05 inch in diameter, has a resistance of 200 ohms. What would be the resistance of a 1500-foot length of wire that is 0.08 inch in diameter?

3

MORE ON POLYNOMIAL FUNCTIONS AND EQUATIONS

A **polynomial function** is defined by an equation of the form $y = P(x)$, where

$$P(x) = a_n x^n + a_{n-1} x^{n-1} + a_{n-2} x^{n-2} + \cdots + a_1 x + a_0$$

is a polynomial of degree n. A **polynomial equation** has the form $P(x) = 0$. In this chapter we discuss methods for solving such equations.

3.1 THE FACTOR AND REMAINDER THEOREMS

A **zero of the polynomial** P is any real or nonreal number r for which $P(r) = 0$. It follows that a zero of the polynomial P is also a root of the polynomial equation $P(x) = 0$. For example, **2** is a zero of the polynomial

$$P(x) = 3x^2 - x - 10$$

because

$$P(\textbf{2}) = 3(\textbf{2})^2 - \textbf{2} - 10$$
$$= 12 - 2 - 10$$
$$= 0$$

Thus 2 is a root of the polynomial equation

$$3x^2 - x - 10 = 0$$

Before we attempt to find zeros of complicated polynomials, we need to know whether a given polynomial even has a zero. This question is settled by the **fundamental theorem of algebra**, first proved by Carl Friedrich Gauss (1777–1855). We state the fundamental theorem without proof.

> **The Fundamental Theorem of Algebra.** A polynomial with positive degree has at least one zero.

By the fundamental theorem, we know that polynomials such as $2x + 1$ and $32.75x^{1492} + ix^3 - (3 + 2i)x + 5$ both have zeros among the complex numbers. It may be difficult to find the zeros, and we may have to settle for approximations, but zeros do exist.

There is a relationship between a zero r of a polynomial $P(x)$ and the results of a long division of $P(x)$ by the binomial $x - r$. We illustrate this relationship with an example.

Example 1 Let $P(x) = 3x^3 - 5x^2 + 3x - 10$.

    a. Find $P(1)$ and $P(-2)$.

    b. Divide $P(x)$ by $x - 1$ and by $x + 2$.

Solution **a.**

$$P(1) = 3(1)^3 - 5(1)^2 + 3(1) - 10 \qquad P(-2) = 3(-2)^3 - 5(-2)^2 + 3(-2) - 10$$
$$= 3 - 5 + 3 - 10 \qquad\qquad\qquad = -24 - 20 - 6 - 10$$
$$= -9 \qquad\qquad\qquad\qquad\qquad = -60$$

    b.

$$
\begin{array}{r}
3x^2 - 2x + 1 \\
x - 1 \overline{\smash{)}\, 3x^3 - 5x^2 + 3x - 10} \\
\underline{3x^3 - 3x^2} \\
-2x^2 + 3x \\
\underline{-2x^2 + 2x} \\
x - 10 \\
\underline{x - 1} \\
-9
\end{array}
\qquad
\begin{array}{r}
3x^2 - 11x + 25 \\
x + 2 \overline{\smash{)}\, 3x^3 - 5x^2 + 3x - 10} \\
\underline{3x^3 + 6x^2} \\
-11x^2 + 3x \\
\underline{-11x^2 - 22x} \\
25x - 10 \\
\underline{25x + 50} \\
-60
\end{array}
$$

    ■

Note the results of Example 1. When $P(x)$ is divided by $x - 1$, the remainder is $P(1)$, or -9. When $P(x)$ is divided by $x + 2$, or $x - (-2)$, the remainder is $P(-2)$, or -60. The following theorem, called the **remainder theorem**, shows that these results are not coincidental.

The Remainder Theorem. Let $P(x)$ be a polynomial and r be any real or nonreal complex number.

    If $P(x)$ is divided by $x - r$, the remainder is $P(r)$.

Proof To divide $P(x)$ by $x - r$, we must find two polynomials: a quotient $Q(x)$ and a remainder $R(x)$ such that

    dividend = divisor · quotient + remainder

$$P(x) = (x - r) \cdot Q(x) + R(x)$$

where $R(x) = 0$, or the degree of $R(x)$ is less than the degree of the divisor $x - r$. Because the divisor is of degree 1, the remainder is 0 or a polynomial of degree 0. Thus, the remainder must be a constant, which we denote as R. Because the polynomial on the left side of the equation $P(x) = (x - r)Q(x) + R$ is identical to the polynomial on the right, the values that these polynomials assume for any replacement of the variable x must be equal. Replacing x with r, we have $P(r) = (r - r)Q(r) + R$. Because $(r - r) = 0$, it follows that $P(r) = R$; that is, the value of the polynomial P attained at $x = r$ is the remainder produced by dividing $P(x)$ by $x - r$. The theorem is proved. □

The **factor theorem**, a corollary to the remainder theorem, gives necessary and sufficient conditions for $x - r$ to be a factor of $P(x)$.

> **The Factor Theorem.** Let $P(x)$ be a polynomial and r be any real or nonreal complex number. Then,
>
> $$P(r) = 0 \qquad \text{if and only if } x - r \text{ is a factor of } P(x).$$

Proof   First, we assume that $P(r) = 0$ and prove that $x - r$ is a factor of $P(x)$. If we divide $P(x)$ by $x - r$, the remainder theorem asserts that the remainder is $P(r)$. But $P(r)$, by assumption, is 0. Thus, $P(x) = (x - r)Q(x) + 0$, and $x - r$ is a factor of $P(x)$.

Conversely, we assume that $x - r$ is a factor of $P(x)$ and prove that $P(r) = 0$. Because $x - r$ is a factor of $P(x)$, a division of $P(x)$ by $x - r$ produces a remainder of 0. By the remainder theorem, this remainder is $P(r)$. Hence, $P(r) = 0$. The theorem is proved. □

Example 2   Determine whether $x + 2$ is a factor of the polynomial $P(x) = x^4 - 7x^2 - 6x$.

Solution   Evaluate $P(-2)$. If $P(-2) = 0$, then, by the factor theorem, $x - (-2)$, or $x + 2$, will be a factor of $P(x)$.

$$P(x) = x^4 - 7x^2 - 6x$$
$$P(-2) = (-2)^4 - 7(-2)^2 - 6(-2)$$
$$= 16 - 28 + 12$$
$$= 0$$

Because $P(-2) = 0$, you know that $x - (-2)$, or $x + 2$, is a factor of $P(x)$.   ■

Example 3   Let $P(x) = 3x^3 - 5x^2 + 3x - 10$. Show that $P(2) = 0$, and use the factor theorem to factor $P(x)$.

Solution   Calculate $P(2)$:

$$P(x) = 3x^3 - 5x^2 + 3x - 10$$
$$P(2) = 3(2)^3 - 5(2)^2 + 3(2) - 10$$
$$= 3(8) - 5(4) + 6 - 10$$
$$= 0$$

Because $P(2) = 0$, it follows (by the factor theorem) that $x - 2$ is a factor of $P(x)$. To determine the other factor, divide $P(x)$ by $x - 2$:

$$
\begin{array}{r}
3x^2 +\ x\ + 5 \\
x - 2\overline{)\,3x^3 - 5x^2 + 3x - 10} \\
\underline{3x^3 - 6x^2} \\
x^2 + 3x \\
\underline{x^2 - 2x} \\
5x - 10 \\
\underline{5x - 10} \\
0
\end{array}
$$

Thus, $P(x)$ factors as

$$P(x) = (x - 2)(3x^2 + x + 5)$$ ■

Example 4   Solve the equation $3x^3 - 5x^2 + 3x - 10 = 0$.

Solution   Factor the polynomial on the left side of the equation. See Example 3.

$$3x^3 - 5x^2 + 3x - 10 = 0$$
$$(x - 2)(3x^2 + x + 5) = 0$$

To solve for x, set each factor equal to zero, and solve each resulting equation. Use the quadratic formula to solve the equation $3x^2 + x + 5 = 0$. The complete solution set is

$$\left\{ 2, \quad -\frac{1}{6} + \frac{\sqrt{59}}{6}\, i, \quad -\frac{1}{6} - \frac{\sqrt{59}}{6}\, i \right\}$$ ■

Example 5   Find a polynomial that has zeros of 2, 3, and -5.

Solution   By the factor theorem, if 2, 3, and -5 are zeros of a polynomial $P(x)$, then $x - 2$, $x - 3$, and $x - (-5)$ are factors of $P(x)$. Hence, the polynomial

$$\begin{aligned} P(x) &= (x - 2)(x - 3)(x + 5) \\ &= (x^2 - 5x + 6)(x + 5) \\ &= x^3 - 19x + 30 \end{aligned}$$

has zeros of 2, 3, and -5. ■

Example 6   Find the three cube roots of -1.

Solution   The three cube roots of -1 are the three roots of the cubic equation $x^3 = -1$, or $x^3 + 1 = 0$. One cube root of -1 is -1. By the factor theorem, $x - (-1)$, or $x + 1$, must be a factor of $x^3 + 1$. Obtain the other factor of $x^3 + 1$ by long division:

$$
\begin{array}{r}
x^2 - x + 1 \\
x + 1 \overline{\smash{)}\, x^3 + 1} \\
\underline{x^3 + x^2 } \\
-x^2 \\
\underline{-x^2 - x } \\
x + 1 \\
\underline{x + 1} \\
0
\end{array}
$$

Thus, the equation $x^3 + 1 = 0$ factors as

$$(x + 1)(x^2 - x + 1) = 0$$

Setting each factor equal to 0 and solving for x gives the three cube roots of -1:

$$\left\{ -1, \quad \frac{1}{2} + \frac{\sqrt{3}}{2}\, i, \quad \frac{1}{2} - \frac{\sqrt{3}}{2}\, i \right\}$$ ■

■ Exercise 3.1 ■

In Exercises 1–6 let $P(x) = 2x^4 - 2x^3 + 5x^2 - 1$. Evaluate the polynomial by substituting the given value of x and simplifying. Then evaluate the polynomial by using the remainder theorem.

1. $P(2)$ **2.** $P(-1)$ **3.** $P(0)$ **4.** $P(1)$ **5.** $P(-4)$ **6.** $P(4)$

7. Let $P(x) = x^5 - 1$. Find $P(2)$ by using the remainder theorem.

8. Let $P(x) = x^5 + 4x^2 - 1$. Find $P(-3)$ by using the remainder theorem.

In Exercises 9–18 use the factor theorem to decide whether each statement is true.

9. $x - 1$ is a factor of $x^7 - 1$.

10. $x - 2$ is a factor of $x^3 - x^2 + 2x - 8$.

11. $x - 1$ is a factor of $3x^5 + 4x^2 - 7$.

12. $x + 1$ is a factor of $3x^5 + 4x^2 - 7$.

13. $x + 3$ is a factor of $2x^3 - 2x^2 + 1$.

14. $x - 3$ is a factor of $3x^5 - 3x^4 + 5x^2 - 13x - 6$.

15. $x - 1$ is a factor of $x^{1984} - x^{1776} + x^{1492} - x^{1066}$.

16. $x + 1$ is a factor of $x^{1984} + x^{1776} - x^{1492} - x^{1066}$.

17. $x + 2i$ is a factor of $x^3 - x^2 + 4x - 4$.

18. $x - (1 - i)$ is a factor of $x^2 - 2x + 2$.

19. Completely solve $x^3 + 3x^2 - 13x - 15 = 0$, given that -1 is a root.

20. Completely solve $x^4 + 4x^3 - 10x^2 - 28x - 15 = 0$, given that -1 is a double root.

21. Completely solve $x^4 - 5x^3 + 7x^2 - 5x + 6 = 0$, given that $x = 2$ and $x = 3$ are roots.

22. Completely solve $x^4 + 2x^3 - 3x^2 - 4x + 4 = 0$, given that $x = 1$ and $x = -2$ are roots.

In Exercises 23–30 find the polynomial of lowest degree that has the indicated zeros.

23. $1, 1, 1$

24. $1, 0, -1$

25. $2, 4, 5$

26. $7, 6, 3$

27. $-1, 1, -\sqrt{2}, \sqrt{2}$

28. $0, 0, 0, \sqrt{3}, -\sqrt{3}$

29. $\sqrt{2}, i, -i$

30. $1 + i, 1 - i, 0$

In Exercises 31–34 find the three cube roots of each number.

31. 1

32. 64

33. -125

34. -216

35. If 0 is a zero of $P(x) = a_n x^n + a_{n-1} x^{n-1} + \cdots + a_1 x + a_0$, what is a_0?

36. If 0 occurs twice as a zero of $P(x) = a_n x^n + a_{n-1} x^{n-1} + \cdots + a_1 x + a_0$, what are a_0 and a_1?

37. Explain why the fundamental theorem of algebra guarantees that every polynomial equation of positive degree has at least one root.

38. Explain why the fundamental theorem of algebra guarantees that a first-degree polynomial equation has exactly one root.

39. The fundamental theorem of algebra demands that the polynomial be of positive degree. Would the theorem still be true if the polynomial were of degree 0? Explain.

40. For what value(s) of k is $x - 2$ a factor of $x^2 + kx + 6$?

41. For what value(s) of k is $x + k$ a factor of $2x^3 + (2k + 1)x^2 + 2x - 3$?

42. Prove: If n is a positive integer, then $x - 1$ is a factor of $x^n - 1$.

43. Prove: If n is an even positive integer, then $x + 1$ is a factor of $x^n - 1$.

44. Prove: If n is an odd positive integer, then $x + 1$ is a factor of $x^n + 1$.

3.2 SYNTHETIC DIVISION

If a polynomial is to be divided by a binomial of the form $x - r$, a method known as **synthetic division** is more efficient than long division. To see how this method works, we consider the long division of $2x^3 + 4x^2 - 3x + 10$ by $x - 3$:

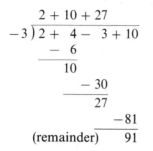

On the left is the complete long division. On the right is a modified version of the long division in which the variables have been removed. We can shorten the work on the right even further by omitting the numbers printed in color:

$$
\begin{array}{r}
2 + 10 + 27 \\
-3\,\overline{)\,2 + \ 4 - \ \ 3 + 10} \\
- \ 6 \\
\hline
10 \\
- 30 \\
\hline
27 \\
- 81 \\
\hline
\text{(remainder)} \quad 91
\end{array}
$$

We can compress the work vertically to obtain

$$
\begin{array}{r}
2 + 10 + 27 \\
-3\,\overline{)\,2 + \ 4 - \ \ 3 + 10} \\
- \ 6 - 30 - 81 \\
\hline
10 \quad 27 \quad 91
\end{array}
$$

There is no reason why the quotient, represented by the numbers 2, 10, and 27, must appear above the long division symbol. If we write the 2 on the bottom line, then the bottom line gives both quotient and remainder. The top line can then be eliminated, and the division appears as

$$
\begin{array}{r}
-3\,\overline{)\,2 + \ 4 - \ \ 3 + 10} \\
- \ 6 - 30 - 81 \\
\hline
2 \quad 10 \quad 27 \quad 91
\end{array}
$$

The bottom line was obtained by subtracting the middle line from the top line. If we replace the -3 in the divisor with $+3$, the signs of each entry in the middle line will be reversed. Then the bottom line can be obtained by addition.

Thus, we have the following final form of the synthetic division:

$$\begin{array}{r|rrrr}
+3 & 2 + & 4 - & 3 + & 10 \\
& & + 6 + & 30 + & 81 \\
\hline
& 2 & 10 & 27 & | \; 91
\end{array}$$

The coefficients of the dividend.

The coefficients of the quotient, and the remainder.

The indicated division is often written in the form $quotient + \dfrac{remainder}{divisor}$:

$$\frac{2x^3 + 4x^2 - 3x + 10}{x - 3} = 2x^2 + 10x + 27 + \frac{91}{x - 3}$$

Example 1   Use synthetic division to divide $3x^4 - 8x^3 + 10x + 3$ by $x - 2$.

Solution   Begin by writing the coefficients of the dividend and the 2 from the divisor in the following form:

$$\underline{2\,|}\; 3 \quad -8 \quad \mathbf{0} \quad 10 \quad 3$$

Write 0 for the coefficient of the missing term of x^2.

Then follow these steps:

$$\underline{2\,|}\; 3 \quad -8 \quad 0 \quad 10 \quad 3$$
$$\mathbf{3}$$

Bring down the 3.

$$\underline{2\,|}\; 3 \quad -8 \quad 0 \quad 10 \quad 3$$
$$6$$
$$3 \quad \mathbf{-2}$$

Multiply 2 and 3. Add that product to -8 to get -2.

$$\underline{2\,|}\; 3 \quad -8 \quad 0 \quad 10 \quad 3$$
$$6 \quad -4$$
$$3 \quad -2 \quad \mathbf{-4}$$

Multiply 2 and -2. Add that product to 0 to get -4.

$$\underline{2\,|}\; 3 \quad -8 \quad 0 \quad 10 \quad 3$$
$$6 \quad -4 \quad 8$$
$$3 \quad -2 \quad -4 \quad \mathbf{2}$$

Multiply 2 and -4. Add that product to 10 to get 2.

$$\underline{2\,|}\; 3 \quad -8 \quad 0 \quad 10 \quad 3$$
$$6 \quad -4 \quad -8 \quad 4$$
$$3 \quad -2 \quad -4 \quad 2 \; | \; \mathbf{7}$$

Multiply 2 and 2. Add that product to 3 to get 7.

Thus,

$$\frac{3x^4 - 8x^3 + 10x + 3}{x - 2} = 3x^3 - 2x^2 - 4x + 2 + \frac{7}{x - 2}$$

■

Example 2   Use synthetic division to find $P(-2)$ if $P(x) = 5x^3 + 3x^2 - 21x - 1$.

Solution   Because of the remainder theorem, $P(-2)$ will be the remainder when $P(x)$ is divided by $x - (-2)$. Use synthetic division, as follows:

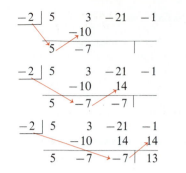

Because the remainder is 13, $P(-2) = 13$. ∎

Example 3 Find $P(i)$, where $P(x) = x^3 - x^2 + x - 1$, and $i = \sqrt{-1}$.

Solution Use synthetic division:

$$
\begin{array}{c|rrrr}
i & 1 & -1 & 1 & -1 \\
  &   & i & -1-i & 1 \\
\hline
  & 1 & i-1 & -i & 0
\end{array}
$$

Because the remainder is 0, $P(i) = 0$. ∎

Example 4 Graph $y = f(x) = x^3 + x^2 - 2x$.

Solution Let $x = 0$ to find the graph's y-intercept: if $x = 0$, then $y = 0$. Then use synthetic division to help find the coordinates of other points on the graph. For example, if $x = 1$, then

$$
\begin{array}{c|rrrr}
1 & 1 & 1 & -2 & 0 \\
  &   & 1 & 2 & 0 \\
\hline
  & 1 & 2 & 0 & 0
\end{array}
$$

The point with coordinates $(1, 0)$ lies on the graph. As another example, if $x = -1$, then

$$
\begin{array}{c|rrrr}
-1 & 1 & 1 & -2 & 0 \\
   &   & -1 & 0 & 2 \\
\hline
   & 1 & 0 & -2 & 2
\end{array}
$$

The point with coordinates $(-1, 2)$ lies on the graph. These coordinates and others that satisfy the equation appear in the graph in Figure 3–1.

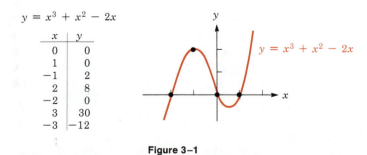

$y = x^3 + x^2 - 2x$

x	y
0	0
1	0
-1	2
2	8
-2	0
3	30
-3	-12

$y = x^3 + x^2 - 2x$

Figure 3–1 ∎

Example 5   Graph $y = f(x) = x^4 - 5x^2 + 4$.

Solution   Because $y = 4$ when $x = 0$, the y-intercept of the graph is 4. Use synthetic division to help find the coordinates of other points on the graph. For example, if $x = 3$, then

$$
\begin{array}{r|rrrr}
3 & 1 & 0 & -5 & 0 & 4 \\
  &   & 3 & 9 & 12 & 36 \\
\hline
  & 1 & 3 & 4 & 12 & \underline{\mid 40} \\
\end{array}
$$

Because x appears only to even powers, the graph is symmetric about the y-axis. Plot several points for positive numbers x to determine the shape of the graph to the right of the y-axis, and use this symmetry to determine the graph as shown in Figure 3–2.

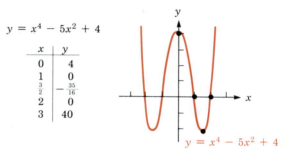

$y = x^4 - 5x^2 + 4$

x	y
0	4
1	0
$\frac{3}{2}$	$-\frac{35}{16}$
2	0
3	40

$y = x^4 - 5x^2 + 4$

Figure 3–2   ∎

Exercise 3.2

In Exercises 1–8 let $P(x) = 5x^3 + 2x^2 - x + 1$. Use synthetic division to find each value of $P(x)$.

1. $P(2)$ **2.** $P(-2)$ **3.** $P(-5)$ **4.** $P(3)$

5. $P(0)$ **6.** $P(5)$ **7.** $P(-i)$ **8.** $P(i)$

In Exercises 9–16 let $P(x) = 2x^4 - x^2 + 2$. Use synthetic division to find each value of $P(x)$.

9. $P(1)$ **10.** $P(-1)$ **11.** $P(-2)$ **12.** $P(3)$

13. $P\left(\dfrac{1}{2}\right)$ **14.** $P\left(\dfrac{1}{3}\right)$ **15.** $P(i)$ **16.** $P(-i)$

In Exercises 17–24 let $P(x) = x^4 + 8x - 8x^3 + 14x^2 - 15$. Write the terms of $P(x)$ in descending powers of x and use synthetic division to find each value of $P(x)$.

17. $P(1)$ **18.** $P(0)$ **19.** $P(-3)$ **20.** $P(-1)$

21. $P(3)$ **22.** $P(5)$ **23.** $P(2)$ **24.** $P(-5)$

In Exercises 25–32 let $P(x) = 8 - 8x^2 + x^5 - x^3$. Write the terms of $P(x)$ in descending powers of x and use synthetic division to find each value of $P(x)$.

25. $P(0)$ **26.** $P(1)$ **27.** $P(2)$ **28.** $P(-2)$

29. $P(i)$ **30.** $P(-i)$ **31.** $P(-2i)$ **32.** $P(2i)$

In Exercises 33–38 use synthetic division to determine the quotient of $P(x) = 3x^3 - 2x^2 - 6x - 4$ *and the given divisor.*
Express $P(x)$ *in the form* (divisor)(quotient) + remainder.

33. $x + 1$ **34.** $x - 1$ **35.** $x - 2$ **36.** $x + 2$

37. x **38.** $x - 7$

In Exercises 39–44 use synthetic division to express each answer in the form $quotient + \dfrac{remainder}{divisor}$.

39. $(7x^3 - 3x^2 - 5x + 1) \div (x + 1)$ **40.** $(2x^3 + 4x^2 - 3x + 8) \div (x - 3)$

41. $\dfrac{4x^4 - 3x^3 - x + 5}{x - 3}$ **42.** $\dfrac{x^4 + 5x^3 - 2x^2 + x - 1}{x + 1}$

43. $\dfrac{3x^5 - 768x}{x - 4}$ **44.** $\dfrac{x^5 - 4x^2 + 4x + 4}{x + 3}$

In Exercises 45–48 use synthetic division to find each power. [*Hint: If* $f(x) = x^5$, *then* $f(3)$ *is* 3^5.]

45. 3^5 **46.** 4^5 **47.** 2^7 **48.** 3^6

In Exercises 49–54 use synthetic division to help graph each equation.

49. $y = x^3 - 4x$ **50.** $y = x^3 - x^2 - 2x$ **51.** $y = x^5 - 2x^3$ **52.** $y = x^4 + 3x^2$

53. $y = x^5 - 3x^4 - 5x^3 + 15x^2 + 4x - 12$ **54.** $y = x^4 + 2x^3 - 5x^2 - 6x$

55. Let $P(x) = 5x^3 - 10x^2 + kx - 9$. For what value of k will the number 3 be a zero of $P(x)$?

56. Let $P(x) = 3x^3 + 8x^2 + kx - 6$. For what value of k will the number -2 be a zero of $P(x)$?

57. Use a calculator and synthetic division to find $P(1.3)$ if $P(x) = 2.5x^3 - 0.78x^2 - 2.7x + 4.3$.

58. Use a calculator and synthetic division to find $P(0.13)$ if $P(x) = 2.1x^3 - 1.2x^2 - 3.5x - 1.8$.

3.3   DESCARTES' RULE OF SIGNS AND BOUNDS ON ROOTS

The remainder theorem and synthetic division provide a way of verifying that a particular number is a root of a polynomial equation, but they do not provide the solutions of the equation. Selecting numbers at random, checking to see whether they work, and hoping for the best is not an efficient technique. This section presents some guidelines that indicate how many solutions to expect, what kind of solutions to expect, and where they can be found. The first two theorems tell how many solutions to expect when solving a polynomial equation.

> **Theorem.**   A polynomial equation of positive degree n with real or non-real coefficients has at most n distinct roots among the complex numbers.

Proof   Let $P(x)$ be a polynomial equation of degree $n > 0$ with lead coefficient a_n. The fundamental theorem of algebra asserts that $P(x)$ has a zero among the complex numbers, which we will call r_1. The factor theorem guarantees that $x - r_1$ is a factor of $P(x)$. Thus,

$$P(x) = (x - r_1)Q_1(x)$$

where $Q_1(x)$ is a polynomial of degree $n - 1$ whose lead coefficient is also a_n. If $n - 1$ is still positive, the fundamental theorem of algebra asserts that $Q_1(x)$ also has a zero, which we will call r_2. According to the factor theorem, $x - r_2$ is a factor of $Q_1(x)$, and

$$P(x) = (x - r_1)(x - r_2)Q_2(x)$$

where $Q_2(x)$ is a polynomial of degree $n - 2$ with a lead coefficient a_n.

This process can continue only to n factors of the form $x - r_i$, and the final quotient $Q_n(x)$ is a polynomial of degree $n - n$, or 0. A polynomial of degree 0 with the lead coefficient a_n is simply the constant a_n. The original polynomial $P(x)$ factors completely as

1. $P(x) = a_n(x - r_1)(x - r_2)(x - r_3) \cdots (x - r_n)$

Each value r_i is a zero of $P(x)$, and each is a root of the polynomial equation $P(x) = 0$.

We use an indirect argument to prove that $P(x) = 0$ has no other roots than the numbers $r_1, r_2, \ldots, r_n$ determined above. Suppose to the contrary that the number r is distinct from the numbers $r_1, r_2, \ldots, r_n$ and is also a root of the equation $P(x) = 0$. Then r is a zero of the polynomial $P(x)$. We substitute r for x in Equation 1 to obtain

$$P(r) = a_n(r - r_1)(r - r_2)(r - r_3) \cdots (r - r_n)$$

By assumption, $r \neq r_i$ for $i = 1, 2, 3, \ldots, n$, so none of the n factors $r - r_i$ on the right side of this equation is equal to zero. By the zero factor theorem, their product is not equal to zero, either. Thus, the right side of Equation 1 is *not* equal to zero. Also by assumption, however, r is a zero of $P(x)$, so the left side of Equation 1 *is* equal to zero. This contradiction is brought about by the assumption that r is a root of the equation $P(x) = 0$ distinct from the n roots already found. Thus, the polynomial equation $P(x) = 0$ has at most n distinct roots. $\square$

The r_i in the preceding proof need not be distinct. Any zero of a polynomial that occurs k times is called a **zero of multiplicity k**. If multiple roots are counted individually, we have the following result:

Theorem.   Let $P(x)$ be a polynomial of positive degree n with real or nonreal coefficients. If multiple roots are counted individually, the nth-degree polynomial equation $P(x) = 0$ has *exactly* n roots among the complex numbers.

Example 1   Factor the polynomial $P(x) = 2x^4 + x^3 - 6x^2$ into the form

$$P(x) = a_n(x - r_1)(x - r_2)(x - r_3) \cdots (x - r_n)$$

and determine the four roots of the fourth-degree polynomial equation $P(x) = 0$.

Solution   First factor the polynomial:

$$P(x) = 2x^4 + x^3 - 6x^2$$
$$= x^2(2x^2 + x - 6)$$
$$= x^2(2x - 3)(x + 2)$$

Write each factor of x as $x - 0$, and factor out 2, the lead coefficient of $2x - 3$:

$$P(x) = 2(x - \mathbf{0})(x - \mathbf{0})\left(x - \frac{\mathbf{3}}{\mathbf{2}}\right)(x + \mathbf{2})$$

The solution set is

$$\left\{0, 0, \frac{3}{2}, -2\right\}$$

If you count the double root 0 twice, the polynomial equation $P(x) = 0$ has exactly four roots. ∎

The next theorem points out that, if the polynomial $P(x)$ has real coefficients, there is a pattern to the complex roots of the equation $P(x) = 0$.

Theorem.   If a polynomial equation $P(x) = 0$ with real coefficients has a complex root $a + bi$ with $b \neq 0$, then the conjugate $a - bi$ is a root also.

(This theorem is often stated as "Complex roots of a real polynomial equation occur in conjugate pairs.")

Proof   Let $z = a + bi$ be a nonreal root of the polynomial equation $P(x) = a_n x^n + a_{n-1} x^{n-1} + a_{n-2} x^{n-2} + \cdots + a_1 x + a_0 = 0$, where the coefficients a_i are real numbers. Then

$$a_n z^n + a_{n-1} z^{n-1} + a_{n-2} z^{n-2} + \cdots + a_1 z + a_0 = 0$$

We must show that $\bar{z}$, the conjugate of z, is also a root. Because complex conjugates of equal numbers are equal, we have

$$\overline{a_n z^n + a_{n-1} z^{n-1} + a_{n-2} z^{n-2} + \cdots + a_1 z + a_0} = \bar{0}$$

Exercise 103 in Exercise Set 1.6 pointed out that the conjugate of the sum of two complex numbers is equal to the sum of their conjugates. Thus, the previous equation can be written as

$$\overline{a_n z^n} + \overline{a_{n-1} z^{n-1}} + \overline{a_{n-2} z^{n-2}} + \cdots + \overline{a_1 z} + \overline{a_0} = \bar{0}$$

Exercise 104 pointed out that the conjugate of the product of complex numbers is also equal to the product of the conjugates of those numbers. Thus,

$$\overline{a_n}\,\overline{z^n} + \overline{a_{n-1}}\,\overline{z^{n-1}} + \overline{a_{n-2}}\,\overline{z^{n-2}} + \cdots + \overline{a_1}\,\overline{z} + \overline{a_0} = \bar{0}$$

By assumption, the coefficients a_i are real numbers. Since the conjugate of a real number is that real number itself,

$$a_n \bar{z}^n + a_{n-1} \bar{z}^{n-1} + a_{n-2} \bar{z}^{n-2} + \cdots + a_1 \bar{z} + a_0 = 0$$

This final equation implies that $P(\bar{z}) = 0$, and that $\bar{z}$ is a root of the polynomial equation $P(x) = 0$. □

Example 2   Form a polynomial equation of lowest possible degree that has real coefficients and a double root of i.

Solution   Because i is to be a root of a polynomial equation with real coefficients, the conjugate $-i$ must also be a root. Because i is a root of multiplicity two, $-i$ must also be a root of multiplicity two. Thus, the complete solution set is

$$\{i, -i, i, -i\}$$

and the equation is

$$(x - i)(x + i)(x - i)(x + i) = 0$$
$$(x^2 + 1)(x^2 + 1) = 0$$
$$x^4 + 2x^2 + 1 = 0$$

■

Example 3   Find a quadratic equation that has a double root of i.

Solution   Complex roots of a quadratic equation with *real* coefficients will be conjugate. A quadratic equation can have two nonreal, nonconjugate roots if the equation has one or more coefficients that are not real. The number i is a double root of the quadratic equation

$$(x - i)(x - i) = 0$$

or

$$x^2 - 2ix - 1 = 0$$

■

If $P(x)$ is a polynomial of positive degree with real coefficients and the number r is a zero of $P(x)$, the *linear* polynomial $x - r$ is a factor of $P(x)$. If the nonreal number $a + bi$ is a zero of $P(x)$, its conjugate $a - bi$ is also a zero, and the product

$$[x - (a + bi)][x - (a - bi)]$$

or

$$x^2 - (a + bi)x - (a - bi)x + (a + bi)(a - bi)$$

or

$$x^2 - 2ax + (a^2 + b^2)$$

is a factor of $P(x)$. Note that this factor is a quadratic polynomial *with real coefficients*. Because the zeros of this quadratic factor are not real numbers, it is said to be **irreducible over the reals**. Thus we have the following result:

Theorem.   A polynomial with positive degree and with real coefficients can be written as the product of linear and irreducible quadratic polynomial factors.

Example 4   Express the polynomial $P(x) = x^5 + 5x^3 - 36x$ as the product of linear and irreducible quadratic polynomials.

Solution   Factor out the common factor of x. Then the remaining factor is quadratic in the variable x^2 and can be factored by inspection.

$$P(x) = x^5 + 5x^3 - 36x$$
$$= x(x^4 + 5x^2 - 36)$$
$$= x(x^2 - 4)(x^2 + 9)$$
$$= x(x + 2)(x - 2)(x^2 + 9)$$

The first three factors of the preceding expression are linear polynomials. The factor $x^2 + 9$ is a quadratic polynomial that is irreducible over the reals because its zeros are the complex conjugates $3i$ and $-3i$. ∎

René Descartes (1596– 1650) Descartes merged algebra and geometry into a single subject called analytical geometry.

Descartes' Rule of Signs

René Descartes is credited with a theorem known as **Descartes' rule of signs** that enables us to look at a polynomial equation and estimate the number of positive, negative, and nonreal roots.

If a polynomial is written in descending powers of x and we scan it from left to right, a "variation in sign" occurs whenever successive terms have opposite signs. For example,

$$P(x) = 3x^5 - 2x^4 - 5x^3 + x^2 - x - 9$$

has three variations in sign, and

$$P(-x) = 3(-x)^5 - 2(-x)^4 - 5(-x)^3 + (-x)^2 - (-x) - 9$$
$$= -3x^5 - 2x^4 + 5x^3 + x^2 + x - 9$$

has two variations in sign.

> **Descartes' Rule of Signs.**   If $P(x)$ is a polynomial with *real* coefficients, the number of positive roots of $P(x) = 0$ either is equal to the number of variations in sign of $P(x)$ or is less than that by an even number.
>   The number of negative roots of $P(x) = 0$ is either equal to the number of variations in sign of $P(-x)$ or is less than that by an even number.

The proof of this theorem is omitted.

Example 5   What possible roots can be expected for the polynomial equation

$$P(x) = x^8 + x^6 + x^4 + x^2 + 1 = 0$$

Solution   Because the equation is an eighth-degree polynomial equation, it must have eight roots. Because there are no variations in sign for $P(x)$, none of the roots can be positive. Now calculate $P(-x)$:

$$P(-x) = (-x)^8 + (-x)^6 + (-x)^4 + (-x)^2 + 1$$
$$= x^8 + x^6 + x^4 + x^2 + 1$$

Because there are no variations in sign for $P(-x)$ either, none of the roots can be negative. Zero is not a root. Thus, all eight roots are nonreal, and they occur in four conjugate pairs. ■

Example 6   Discuss the possibilities for the roots of $P(x) = 3x^3 - 2x^2 + x - 5 = 0$.

Solution   There are three variations in sign of $P(x)$, so there could be three positive roots, or only one (because 1 is less than 3 by an even number). Because $P(-x) = -3x^3 - 2x^2 - x - 5 = 0$ has no variations in sign, there are no negative roots. Furthermore, 0 is not a root.

If there are three positive roots, then all the roots are accounted for. If there is only one positive root, the remaining two roots must be nonreal. The following chart indicates these two possibilities.

Number of positive roots	Number of negative roots	Number of nonreal roots
3	0	0
1	0	2

The number of nonreal roots is the number needed to bring the total number of roots up to three. ■

Example 7   Discuss the possibilities for the roots of $P(x) = 5x^5 - 3x^3 - 2x^2 + x - 1 = 0$.

Solution $P(x)$ has three variations in sign. Thus, there are either three positive solutions or only one. Because $P(-x) = -5x^5 + 3x^3 - 2x^2 - x - 1 = 0$ has two variations in sign, there are either two negative roots or none. Each line of the following chart indicates a possible combination of positive, negative, and nonreal roots.

Number of positive roots	Number of negative roots	Number of nonreal roots
1	0	4
3	0	2
1	2	2
3	2	0

Note that in each case the number of nonreal roots is even. This result is expected, because complex zeros of a polynomial with real coefficients occur in conjugate pairs. ■

Bounds for Roots

A final theorem provides a way of finding **bounds** for the roots of the polynomial equation $P(x) = 0$, enabling us to concentrate our root-finding efforts on those regions where roots can be found. This theorem is also presented without proof.

Theorem. Let $P(x)$ be a polynomial with real coefficients, and let its lead coefficient be positive. If $P(x)$ is divided synthetically by the *positive* number c and each term in the last row is nonnegative, then no number greater than c can be a root of $P(x) = 0$. (c is called an **upper bound** of the real roots.)

If $P(x)$ is divided synthetically by the *negative* number d and the signs of the last row alternate, then no number less than d can be a root of the equation $P(x) = 0$. (d is called a **lower bound** of the real roots.) A zero that appears in the last row can be assigned either a $+$ or a $-$ sign to help the signs alternate.

Example 8   Establish the best positive integer bounds for the positive roots of $18x^3 - 3x^2 - 37x + 12 = 0$ and the best negative integer bounds for the negative roots.

Solution   Try several synthetic divisions, looking for a nonnegative last row (if you synthetically divide by a positive number) or the alternating-sign last row (if you synthetically divide by a negative number). Try 1 first to get

$$
\begin{array}{r|rrrr}
1 & 18 & -3 & -37 & 12 \\
  &   & 18 & 15 & -22 \\
\hline
  & +18 & +15 & -22 & -10 \\
\end{array}
$$

Because some of the signs in the last row are negative, 1 is not an upper bound. Now try 2:

$$
\begin{array}{r|rrrr}
2 & 18 & -3 & -37 & 12 \\
  &   & 36 & 66 & 58 \\
\hline
  & +18 & +33 & +29 & +70 \\
\end{array}
$$

Because the last row is entirely positive, no number greater than 2 can be a root. Thus, the smallest, or best, positive integer upper bound is 2.

Now try a negative divisor such as -2:

$$
\begin{array}{r|rrrr}
-2 & 18 & -3 & -37 & 12 \\
   &   & -36 & 78 & -82 \\
\hline
   & +18 & -39 & +41 & -70 \\
\end{array}
$$

The alternating signs in the last row indicate that no number less than -2 can be a root. To determine whether -2 is the largest, or best, negative integer bound, try -1 next.

$$
\begin{array}{r|rrrr}
-1 & 18 & -3 & -37 & 12 \\
   &   & -18 & 21 & 16 \\
\hline
   & +18 & -21 & -16 & +28 \\
\end{array}
$$

Because the signs in the last row do not alternate, -1 is not a lower bound. Thus, the largest, or best, negative lower bound is -2.

All of the real roots of the given equation lie between -2 and 2.   ■

▬ Exercise 3.3 ▬

1. How many roots does $x^{10} = 1$ have?   2. How many roots does $x^{40} = 1$ have?

3. One root of $x(3x^4 - 2) = 12x$ is 0. How many other roots are there?

4. One root of $3x^2(x^7 - 14x + 3) = 0$ is 0. How many other roots are there?

In Exercises 5–14 form a polynomial equation with real coefficients and lowest possible degree that has the roots indicated.

5. $0, -3, 5$ 6. $0, -3, 1$ 7. 3, and a double root of 1

8. 7, and a triple root of 0   9. a double root of $2i$ 10. 3, and a double root of i

11. $2 + i, 2 - i$ 12. $3 + 2i, 3 - 2i$ 13. $i, -2i$

14. $-i$, and a double root of 2

In Exercises 15–20 express each polynomial as a product of linear and irreducible quadratic polynomials.

15. $x^5 - x^4 - 2x^3$ 16. $x^4 - 8x^3 + 15x^2$ 17. $x^3 + x$ 18. $x^4 + 2x^2 + 1$

19. $x^4 + 5x^3 + 7x^2 + 5x + 6$, given that i is a zero.   20. $x^6 - 2x^4 - 7x^2 - 4$, given that i is a zero twice.

*In Exercises 21–34 use Descartes' rule of signs to find the number of positive, negative, and nonreal roots of each equation. **Do not attempt to find the roots.***

21. $3x^3 + 5x^2 - 4x + 3 = 0$ 22. $3x^3 - 5x^2 - 4x - 3 = 0$ 23. $2x^3 + 7x^2 + 5x + 4 = 0$

24. $-2x^3 - 7x^2 - 5x - 4 = 0$ 25. $8x^4 = -5$ 26. $-3x^3 = -5$

27. $x^4 + 8x^2 - 5x = 10$ 28. $5x^7 + 3x^6 - 2x^5 + 3x^4 + 9x^3 + x^2 + x + 1 = 0$

29. $-x^{10} - x^8 - x^6 - x^4 - x^2 - 1 = 0$ 30. $x^{10} + x^8 + x^6 + x^4 + x^2 + 1 = 0$

31. $x^9 + x^7 + x^5 + x^3 + x = 0$ (Is 0 a root?)   32. $-x^9 - x^7 - x^5 - x^3 = 0$ (Is 0 a root?)

33. $-2x^4 - 3x^2 + 2x + 3 = 0$ 34. $-7x^5 - 6x^4 + 3x^3 - 2x^2 + 7x - 4 = 0$

In Exercises 35–42 find the best integer bounds for the roots of each equation.

35. $x^2 - 5x - 6 = 0$ 36. $6x^2 + x - 1 = 0$

37. $6x^2 - 13x - 110 = 0$ 38. $3x^2 + 12x + 24 = 0$

39. $x^5 + x^4 - 8x^3 - 8x^2 + 15x + 15 = 0$ 40. $12x^3 + 20x^2 - x - 6 = 0$

41. $2x^3 + 9x^2 - 5x = 41$ 42. $x^4 - 34x^2 = -225$

43. Prove that any odd-degree polynomial equation with real coefficients must have at least one real root.

44. If a, b, c, and d are positive numbers, prove that $ax^4 + bx^2 + cx - d = 0$ has exactly two nonreal roots.

3.4   FINDING RATIONAL ROOTS OF POLYNOMIAL EQUATIONS

In this section we consider a method for finding rational roots of polynomial equations with integral coefficients.

> **Theorem.**   If the polynomial equation
>
> $$P(x) = a_n x^n + a_{n-1}x^{n-1} + a_{n-2}x^{n-2} + \cdots + a_1 x + a_0 = 0$$
>
> has integral coefficients and the rational number $\frac{p}{q}$ (in lowest terms) is a root of the equation, then p is a factor of a_0 and q is a factor of a_n.

Proof   Let the rational number $\frac{p}{q}$ (in lowest terms) be a root of

$$P(x) = a_n x^n + a_{n-1} x^{n-1} + a_{n-2} x^{n-2} + \cdots + a_1 x + a_0 = 0$$

where each coefficient is an integer. Because $\frac{p}{q}$ is a root, the equation is satisfied by $\frac{p}{q}$:

$$a_n \left(\frac{p}{q}\right)^n + a_{n-1}\left(\frac{p}{q}\right)^{n-1} + a_{n-2}\left(\frac{p}{q}\right)^{n-2} + \cdots + a_1 \left(\frac{p}{q}\right) + a_0 = 0$$

By multiplying both sides of the equation by q^n, we clear the equation of fractions. (Remember that p, q, and each of the a_i are integers.)

1. $a_n p^n + a_{n-1} p^{n-1} q + a_{n-2} p^{n-2} q^2 + \cdots + a_1 p q^{n-1} + a_0 q^n = 0$

Because all terms but the last share a common factor of p, we can rewrite the equation in the form

$$p(a_n p^{n-1} + a_{n-1} p^{n-2} q + a_{n-2} p^{n-3} q^2 + \cdots + a_1 q^{n-1}) = -a_0 q^n$$

Since p is a factor of the left side, it must also be a factor of the right side. It cannot be a factor of q^n, because the fraction $\frac{p}{q}$ is in lowest terms, and p and q share no common factor. Therefore, p and q^n share no common factor either. It follows that p must be a factor of a_0.

We return to Equation 1. Note that all terms but the first share a common factor of q, and rewrite Equation 1 as

$$q(a_{n-1} p^{n-1} + a_{n-2} p^{n-2} q + a_{n-3} p^{n-3} q^2 + \cdots + a_0 q^{n-1}) = -a_n p^n$$

Because q is a factor of the left side, it must be a factor of the right side as well. Because q cannot be a factor of p^n, it must be a factor of a_n.   □

Example 1   Determine the possible rational roots of the equation

$$\frac{1}{2} x^4 + \frac{2}{3} x^3 + 3x^2 - \frac{3}{2} x + 3 = 0$$

Solution   Since the previous theorem applies only to polynomial equations with *integral* coefficients, clear this equation of its fractional coefficients. To do so, multiply both sides by 6 to get

$$3x^4 + 4x^3 + 18x^2 - 9x + 18 = 0$$

The only possible numerators available for a rational root are the factors of the constant term 18:

$$\pm 1, \quad \pm 2, \quad \pm 3, \quad \pm 6, \quad \pm 9, \quad \text{and} \quad \pm 18$$

The only possible denominators are the factors of the lead coefficient 3:

$$\pm 1 \quad \text{and} \quad \pm 3$$

You can form the list of possible rational solutions by listing all the combinations of values from these two sets.

$$\pm\frac{1}{1}, \quad \pm\frac{2}{1}, \quad \pm\frac{3}{1}, \quad \pm\frac{6}{1}, \quad \pm\frac{9}{1}, \quad \pm\frac{18}{1}, \quad \pm\frac{1}{3}, \quad \pm\frac{2}{3}, \quad \pm\frac{3}{3}, \quad \pm\frac{6}{3}, \quad \pm\frac{9}{3}, \quad \pm\frac{18}{3}$$

Because several of these are duplicates, you can condense the list to obtain the possible rational roots:

$$\pm 1, \quad \pm 2, \quad \pm 3, \quad \pm 6, \quad \pm 9, \quad \pm 18, \quad \pm\frac{1}{3}, \quad \pm\frac{2}{3}$$

■

Example 2   Prove that $\sqrt{2}$ is irrational.

Solution $\sqrt{2}$ is a real root of the polynomial equation $x^2 - 2 = 0$. Any rational solution of the equation must have ± 1 or ± 2 as its numerator, and ± 1 as its denominator. The only possible rational solutions, therefore, are ± 1 and ± 2. Since none of these satisfies the equation, the solution $\sqrt{2}$ must be irrational.   ∎

Example 3   Solve the equation $P(x) = x^3 + x^2 - 10x + 8 = 0$.

Solution   Because the equation is of third degree, it must have three roots. According to Descartes' rule of signs, there are two possible combinations of positive, negative, and nonreal roots.

Number of positive roots	Number of negative roots	Number of nonreal roots
2	1	0
0	1	2

The only possible rational roots are

$$\pm \frac{8}{1}, \quad \pm \frac{4}{1}, \quad \pm \frac{2}{1}, \quad \pm \frac{1}{1}$$

or

$$-8, \quad -4, \quad -2, \quad -1, \quad 1, \quad 2, \quad 4, \quad 8$$

Check each one, crossing out those that do not satisfy the equation. Start, for example, with $x = 4$:

$$\begin{array}{r|rrrr} 4 & 1 & 1 & -10 & 8 \\ & & 4 & 20 & 40 \\ \hline & 1 & 5 & 10 & | \ 48 \end{array}$$

Because the remainder is not 0, the number 4 is not a root and can be crossed off the list. Because the last row in the synthetic division is entirely positive, 4 is an upper bound of the equation's roots. Thus 8 cannot be a root either and can also be crossed off the list.

$$-8, \quad -4, \quad -2, \quad -1, \quad 1, \quad 2, \quad \cancel{4}, \quad \cancel{8}$$

Now try $x = 2$:

$$\begin{array}{r|rrrr} 2 & 1 & 1 & -10 & 8 \\ & & 2 & 6 & -8 \\ \hline & 1 & 3 & -4 & | \ 0 \end{array}$$

Because the remainder is 0, the number 2 is a root, and the binomial $x - 2$ is a factor of $P(x)$. Any remaining roots must be supplied by the remaining factor, which is the quotient $1x^2 + 3x - 4$. The other roots can be found by solving the equation

$$x^2 + 3x - 4 = 0$$

This equation, called the **depressed equation**, is a quadratic equation that can be solved by factoring:

$$x^2 + 3x - 4 = 0$$
$$(x - 1)(x + 4) = 0$$
$$x - 1 = 0 \quad \text{or} \quad x + 4 = 0$$
$$x = 1 \qquad\qquad x = -4$$

The solution set is $\{2, 1, -4\}$. Note that two solutions are positive, one is negative, and none are nonreal. This is one of the possibilities predicted by Descartes' rule of signs. ■

Example 4 Solve the equation $x^8 - 2x^7 - 5x^6 + 6x^5 - x^4 + 2x^3 + 5x^2 - 6x = 0$.

Solution Being of eighth degree, the equation must have eight roots. Because x is a factor of the polynomial, 0 must be a root of the equation. The remaining seven roots must come from the depressed equation

$$x^7 - 2x^6 - 5x^5 + 6x^4 - x^3 + 2x^2 + 5x - 6 = 0$$

According to Descartes' rule of signs, there are six possible combinations of positive, negative, and nonreal roots for this equation.

Number of positive roots	Number of negative roots	Number of nonreal roots
5	2	0
3	2	2
1	2	4
5	0	2
3	0	4
1	0	6

The only possible rational roots are

$$-6, \quad -3, \quad -2, \quad -1, \quad 1, \quad 2, \quad 3, \quad 6$$

Check each one, crossing off those that do not satisfy the equation. Begin with -3:

```
-3 | 1   -2    -5     6    -1      2     5      -6
   |      -3    15   -30    72   -213   633   -1914
   ----------------------------------------------------
     1   -5    10   -24    71   -211   638 | -1920
```

Because the remainder is not 0, -3 is not a root and can be crossed off the list. Because the signs in the last row alternate, -3 is a lower bound. Thus, you can cross off -6 as well:

$$\cancel{-6}, \quad \cancel{-3}, \quad -2, \quad -1, \quad 1, \quad 2, \quad 3, \quad 6$$

Now try -2:

$$
\begin{array}{r|rrrrrrrr}
-2 & 1 & -2 & -5 & 6 & -1 & 2 & 5 & -6 \\
   &   & -2 & 8 & -6 & 0 & 2 & -8 & 6 \\
\hline
   & 1 & -4 & 3 & 0 & -1 & 4 & -3 & \,|\;0
\end{array}
$$

Because the remainder is 0, -2 is a root.

This root is negative, so you can revise the chart of positive/negative/nonreal possibilities. Until now, zero negative roots was a possibility.

Number of positive roots	Number of negative roots	Number of nonreal roots
5	2	0
3	2	2
1	2	4

Because -2 is a root, the factor theorem asserts that $x - (-2)$, or $x + 2$, is a factor of $P(x)$. Any remaining roots can be found by solving the depressed equation

$$x^6 - 4x^5 + 3x^4 - x^2 + 4x - 3 = 0$$

Because the constant term of this equation is -3, you can cross off -2, 2, and 6 from the list of possible rational roots, because none is a factor of -3. The list of candidates is now

$$-\cancel{6}, \quad -3, \quad -\cancel{2}, \quad -1, \quad 1, \quad \cancel{2}, \quad 3, \quad \cancel{6}$$

The only solutions found thus far are 0 and -2. Try -1 next because you know there must be one more negative root. The coefficients to use in the synthetic division are those of the depressed equation. Don't forget the missing x^3.

$$
\begin{array}{r|rrrrrr}
-1 & 1 & -4 & 3 & 0 & -1 & 4 & -3 \\
   &   & -1 & 5 & -8 & 8 & -7 & 3 \\
\hline
   & 1 & -5 & 8 & -8 & 7 & -3 & \,|\;0
\end{array}
$$

Because the remainder is 0, the number -1 is a root. The roots found so far are 0, -2, and -1. The root -1 cannot appear again because there can be only two negative roots and you have found them both. The current list of candidates is now

$$-\cancel{6}, \quad -3, \quad -\cancel{2}, \quad -\cancel{1}, \quad 1, \quad \cancel{2}, \quad 3, \quad \cancel{6}$$

Other roots can be found by solving the depressed equation

$$x^5 - 5x^4 + 8x^3 - 8x^2 + 7x - 3 = 0$$

Try 1 next:

$$
\begin{array}{r|rrrrrr}
1 & 1 & -5 & 8 & -8 & 7 & -3 \\
  &   & 1 & -4 & 4 & -4 & 3 \\
\hline
  & 1 & -4 & 4 & -4 & 3 & \,|\;0
\end{array}
$$

The solution 1 joins the growing list of solutions. To see whether 1 is a multiple solution, try it again in the depressed equation.

$$\begin{array}{r|rrrrr}
1 & 1 & -4 & 4 & -4 & 3 \\
& & 1 & -3 & 1 & -3 \\
\hline
& 1 & -3 & 1 & -3 & 0
\end{array}$$

Again, 1 is a root. Try 1 again to see if it will work a third time.

$$\begin{array}{r|rrrr}
1 & 1 & -3 & 1 & -3 \\
& & 1 & -2 & -1 \\
\hline
& 1 & -2 & -1 & -4
\end{array}$$

Because the remainder is nonzero, 1 is only a double root.

Solutions found thus far are 0, -2, -1, 1, and 1. Now try 3:

$$\begin{array}{r|rrrr}
3 & 1 & -3 & 1 & -3 \\
& & 3 & 0 & 3 \\
\hline
& 1 & 0 & 1 & 0
\end{array}$$

The solution 3 is added to the list of roots.

The depressed equation is now the quadratic equation $x^2 + 1 = 0$, which has the conjugate roots i and $-i$.

The eight roots of the original eighth-degree equation are

$$0, \quad -2, \quad -1, \quad 1, \quad 1, \quad 3, \quad i, \quad \text{and} \quad -i$$

Note that this list contains three positive, two negative, and two conjugate complex numbers. This combination was one of the possibilities predicted by Descartes' rule of signs. ∎

Exercise 3.4

In Exercises 1–22 find all roots for each equation.

1. $x^3 - 2x^2 - x + 2 = 0$
2. $x^3 + 2x^2 - x - 2 = 0$
3. $x^4 - 10x^3 + 35x^2 - 50x + 24 = 0$
4. $x^4 + 4x^3 + 6x^2 + 4x + 1 = 0$
5. $x^5 - 2x^4 - 2x^3 + 4x^2 + x - 2 = 0$
6. $x^5 - x^3 - 8x^2 + 8 = 0$
7. $x^4 + 3x^3 - 13x^2 - 9x + 30 = 0$
8. $x^4 - 8x^3 + 14x^2 + 8x - 15 = 0$
9. $x^7 - 12x^5 + 48x^3 - 64x = 0$
10. $x^7 + 7x^6 + 21x^5 + 35x^4 + 35x^3 + 21x^2 + 7x + 1 = 0$
11. $6x^5 - 7x^4 - 48x^3 + 81x^2 - 4x - 12 = 0$
12. $x^6 - 3x^5 - x^4 + 9x^3 - 10x^2 + 12x - 8 = 0$
13. $4x^5 - 12x^4 + 15x^3 - 45x^2 - 4x + 12 = 0$
14. $12x^4 + 20x^3 - 41x^2 + 20x - 3 = 0$
15. $3x^4 - 14x^3 + 11x^2 + 16x - 12 = 0$
16. $2x^4 - x^3 - 2x^2 - 4x - 40 = 0$
17. $4x^4 - 8x^3 - x^2 + 8x - 3 = 0$
18. $3x^3 - 2x^2 + 12x - 8 = 0$
19. $30x^3 - 47x^2 - 9x + 18 = 0$
20. $x^{-5} - 8x^{-4} + 25x^{-3} - 38x^{-2} + 28x^{-1} - 8 = 0$
21. $1 - x^{-1} - x^{-2} - 2x^{-3} = 0$
22. $x^3 - \dfrac{19}{6}x^2 + \dfrac{1}{6}x + 1 = 0$

23. Prove that $\sqrt{3}$ is irrational.
24. Prove that $\sqrt[3]{2}$ is irrational.
25. If n is an even positive integer and c is a positive constant, prove that the equation $x^n - c = 0$ has exactly two real roots.
26. If n is a positive integer that is not the square of an integer, prove that $\sqrt{n}$ is irrational.

3.5   APPROXIMATING ROOTS OF POLYNOMIAL EQUATIONS

First-degree equations are easy to solve, and second-degree equations can be solved by the quadratic formula. Formulas exist for solving third- and fourth-degree polynomial equations, although they are complicated. There are, however, no explicit algebraic formulas for solving polynomial equations of degree 5 or greater.

To solve a polynomial equation of high degree with integer coefficients, we could use the methods of Section 3.4 to find all rational roots. If the remaining depressed equation were of degree 3 or more, we might not be able to find the other roots *exactly*. There are, however, many techniques available for *approximating* the real roots of equations of high degree, even though exact roots cannot be found. In this section we develop one such method.

This method, called **binary chopping**, depends on the fact that polynomial functions are *continuous*, which means, roughly, that the graph of $y = P(x)$ can be drawn without lifting the pencil from the paper. Because polynomials are continuous functions, they have the following property, whose proof is beyond the level of this course.

The Intermediate Value Theorem for Polynomials.   Let $P(x)$ be a polynomial with real coefficients. If $a \neq b$ and C lies between $P(a)$ and $P(b)$, then there is a number c between a and b for which $P(c) = C$.

The intermediate value theorem implies that, if $P(a) \neq P(b)$, then $P(x)$ takes on all values between $P(a)$ and $P(b)$.

The graph in Figure 3–3a could be the graph of a polynomial. Note that, if C lies between $P(a)$ and $P(b)$, there is always at least one point c on the x-axis, between a and b, for which $P(c) = C$. The graph in Figure 3–3b could not be the graph of a polynomial, for there is no number c on the x-axis for which $P(c)$ is equal to the indicated value C.

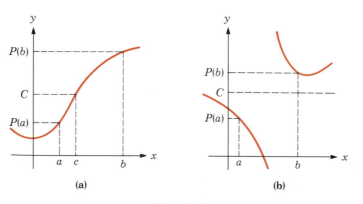

(a)   (b)

Figure 3–3

Example 1   If $P(x) = 3x^2 - 2x + 2$, then $P(1) = 3$ and $P(2) = 10$. The number 7 lies between 3 and 10. By the intermediate value theorem, there must be a number c, where $1 < c < 2$, for which $P(c) = 7$. Find c.

Solution   Solve the equation $P(c) = 7$:

$$P(c) = 7$$
$$3c^2 - 2c + 2 = 7$$
$$3c^2 - 2c - 5 = 0$$
$$(3c - 5)(c + 1) = 0$$
$$3c - 5 = 0 \quad \text{or} \quad c + 1 = 0$$
$$c = \frac{5}{3} \quad \bigg| \quad c = -1$$

The intermediate value theorem demands that c lie between 1 and 2. Because $1 < \frac{5}{3} < 2$, the number c must be $\frac{5}{3}$.   ■

If the polynomial $P(x)$ has real coefficients and $P(a)$ and $P(b)$ have opposite signs, then 0 lies between $P(a)$ and $P(b)$. The intermediate value theorem guarantees that $P(x) = 0$ for some number between a and b. Thus, we have the following theorem:

Theorem.   Let $P(x)$ be a polynomial with real coefficients. If $P(a)$ and $P(b)$ have opposite signs, there is at least one number r between a and b for which $P(r) = 0$.

The previous theorem provides a method for finding the roots of $P(x) = 0$ to any degree of accuracy desired. Suppose we find, by trial and error, the numbers x_L and x_R (for left and right) that straddle a root; that is, $x_L < x_R$ and $P(x_L)$ and $P(x_R)$ have opposite signs. For purposes of discussion, $P(x_L)$ will be negative and $P(x_R)$ will be positive. We compute a number c that is halfway between x_L and x_R (c is the average of x_L and x_R) and then evaluate $P(c)$. If $P(c)$ is 0, we have found a root. More likely, however, $P(c)$ will not be 0.

If $P(c)$ is negative, the root r lies between c and x_R, as shown in Figure 3–4a. In this case, let c become the new x_L, and repeat the process.

(a)   (b)

Figure 3–4

If $P(c)$ is positive, however, the root lies between x_L and c, as shown in Figure 3–4b. In this case, let c become the new x_R, and repeat the process.

At any stage in this procedure, the root is contained between the current numbers x_L and x_R. If the original bounds were, say, 1 unit apart, then after 10 repetitions of this procedure the root would be contained between fences that are 2^{-10}, or about 0.001 unit apart. After 20 repetitions, the bounds are only 0.000001 unit apart. The actual zero of $P(x)$ is within 0.000001 of either x_L or x_R. The method of binary chopping is well suited for solving equations by computer.

Example 2   Find $\sqrt{2}$ to two decimal places.

Solution $\sqrt{2}$ is a root of the polynomial equation $P(x) = x^2 - 2 = 0$. Note that the values $P(1) = 1^2 - 2 = -1$ and $P(2) = 2^2 - 2 = 2$ have opposite signs. Set x_L equal to 1 and x_R equal to 2 and compute the midpoint:

$$c = \frac{1+2}{2} = 1.5$$

Because $P(c) = P(1.5) = 0.25$ is a positive number, let c become the new x_R and calculate a new c. Tabulate the information in a chart to keep things straight:

Step	x_L	c	x_R	$P(x_L)$	$P(c)$	$P(x_R)$
0	1	1.5	2	negative	positive	positive
1	1	1.25	1.5	negative	negative	positive

At this point, $P(c)$ and $P(x_R)$ are of opposite signs, so c becomes the new x_L. The process continues, but without a hand calculator it can be quite difficult.

Step	x_L	c	x_R	$P(x_L)$	$P(c)$	$P(x_R)$
0	1	1.5	2	negative	positive	positive
1	1	1.25	1.5	negative	negative	positive
2	1.25	1.375	1.5	negative	negative	positive
3	1.375	1.4375	1.5	negative	positive	positive
4	1.375	1.40625	1.4375	negative	negative	positive
5	1.40625	1.421875	1.4375	negative	positive	positive
6	1.40625	1.4140625	1.421875	negative	negative	positive
7	1.4140625	1.41796875	1.421875	negative	positive	positive
8	1.4140625	1.416015625	1.41796875	negative	positive	positive

In Step 7, the bounds x_R and x_L agree to only one decimal place; our approximation of $\sqrt{2}$ is 1.4. In Step 8, x_R and x_L agree to two decimal places; $\sqrt{2} \approx 1.41$. ∎

Exercise 3.5

In Exercises 1–2 verify that $P(a) < 5 < P(b)$. Then find a number c between a and b for which $P(c) = 5$.

1. $P(x) = x^2 - x + 3; a = 1, b = 4$

2. $P(x) = x^2 - 6x + 13; a = -5, b = 3$

In Exercises 3–8 show that each equation has at least one real root between the numbers specified.

3. $x^4 - 8x^2 + 15 = 0; 1, 2$

4. $x^4 - 8x^2 + 15 = 0; 2, 3$

5. $30x^3 + 10 = 61x^2 + 39x; 2, 3$

6. $30x^3 + 10 = 61x^2 + 39x; -1, 0$

7. $30x^3 + 10 = 61x^2 + 39x; 0, 1$

8. $5x^3 - 9x^2 - 4x + 9 = 0; -1, 2$

9. Use binary chopping to evaluate $\sqrt{3}$ to one decimal place.

10. Use binary chopping to evaluate $\sqrt[3]{53}$ to one decimal place.

11. Use binary chopping to evaluate $\sqrt{5}$ to two decimal places.

12. Use binary chopping to evaluate $\sqrt[3]{102}$ to two decimal places.

13. Use binary chopping to find the only real root of $P(x) = 7x^3 - 3x^2 + 14x - 6 = 0$ to one decimal place. [*Hint:* Find $P(0)$ and $P(1)$.]

14. Use binary chopping to find a root of $35x^3 + 12x^2 + 8x + 1 = 0$ to one decimal place.

15. Use binary chopping to find a root of $2x^3 - x^2 + 2x - 1 = 0$ to two decimal places.

16. Graph the equation $y = P(x) = x^2 + 2x + 1$ and explain why binary chopping cannot be used to find the roots of $P(x) = 0$.

REVIEW EXERCISES

In Review Exercises 1–4 use the remainder theorem to evaluate the given polynomial at the number indicated.

1. $P(x) = 3x^3 - 5x^2 + 7x - 2; x = 2$

2. $P(x) = 2x^5 - 17x^3 + 3x^2 + x - 5; x = -3$

3. $P(x) = 4x^4 + 2x^3 - 3x - 2; x = 2$

4. $P(x) = 4x^4 + 2x^3 - 3x - 2; x = \frac{1}{2}$

In Review Exercises 5–10 use the factor theorem to decide whether each statement is true.

5. $x - 3$ is a factor of $5x^3 - 13x^2 - 9x - 9$.

6. $x + 5$ is a factor of $3x^3 + 12x^2 - 17x - 10$.

7. $x + 3$ is a factor of $2x^4 + 10x^3 + 4x^2 + 7x + 21$.

8. $x - 6$ is a factor of $x^5 - 6x^4 - 4x + 24$.

9. $x + i$ is a factor of $x^5 + x^3 + x^2 + 1$.

10. $x - \frac{2}{3}$ is a factor of $3x^2 - x - 2$.

11. Find the three cube roots of -64.

12. Find the three cube roots of 343.

13. Find the polynomial of lowest degree with zeros of $-1, 2$, and $\frac{3}{2}$.

14. Find the polynomial equation of lowest degree with roots of $i, -i$, and a double root of -1.

15. Use synthetic division to find the quotient when the polynomial $3x^4 + 2x^2 + 3x + 7$ is divided by $x - 3$.

16. Use synthetic division to find the quotient when the polynomial $5x^5 - 4x^4 + 3x^3 - 2x^2 + x - 1$ is divided by $x + 2$.

17. How many roots does the equation $3x^6 - 4x^5 + 3x + 2 = 0$ have?

18. How many roots does the equation $x^{1984} - 3^{2001} = 0$ have?

In Review Exercises 19–22 use Descartes' rule of signs to find the number of possible negative, positive, and nonreal roots.

19. $3x^4 + 2x^3 - 4x + 2 = 0$

20. $4x^5 + 3x^4 + 2x^3 + x^2 + x = 7$

21. $x^4 + x^2 + 24{,}576 = 0$

22. $-x^7 - 5 = 0$

23. Find all roots of the equation $2x^3 + 17x^2 + 41x + 30 = 0$.

24. Find all roots of the equation $3x^3 + 2x^2 + 2x = 1$.

25. Show that $5x^3 + 37x^2 + 59x + 18 = 0$ has a root between $x = 0$ and $x = -1$.

26. Show that $6x^3 - x^2 - 10x - 3 = 0$ has a root between $x = 1$ and $x = 2$.

27. Use binary chopping to find $\sqrt{7}$ to the nearest hundredth.

28. Use binary chopping to find an approximation of the root of the equation $0 = 3x - 1$. What is the exact root?

4

EXPONENTIAL AND LOGARITHMIC FUNCTIONS

In this chapter we discuss two functions that are of theoretical and practical importance in mathematics. One of these, the **exponential function**, is used to compute compound interest and to model population growth and radioactivity. The other, the **logarithmic function**, is used to simplify calculations, measure the acidity of a solution or the intensity of an earthquake, and determine safe noise levels for factory workers.

4.1 EXPONENTIAL FUNCTIONS

In the discussion of exponential functions, we will consider expressions such as 3^x, where x is any *real* number. In Chapter 1 we defined 3^x where x is a *rational* number. We now give meaning to 3^x where x is an *irrational* number.

To do so we consider the expression $3^{\sqrt{2}}$, where $\sqrt{2}$ is the irrational number $1.414213562\ldots$. Because $1 < \sqrt{2} < 2$, it can be shown that $3^1 < 3^{\sqrt{2}} < 3^2$. Similarly, $1.4 < \sqrt{2} < 1.5$, so $3^{1.4} < 3^{\sqrt{2}} < 3^{1.5}$.

The value of $3^{\sqrt{2}}$ is bounded by two numbers involving only rational powers of 3, as shown in the following list. As the list continues, $3^{\sqrt{2}}$ gets squeezed into a smaller and smaller interval.

$$\textbf{3.0} \qquad \approx 3^1 \quad < 3^{\sqrt{2}} < 3^2 \qquad \approx \textbf{9.0}$$
$$\textbf{4.656} \quad \approx 3^{1.4} \quad < 3^{\sqrt{2}} < 3^{1.5} \quad \approx \textbf{5.196}$$
$$\textbf{4.7070} \quad \approx 3^{1.41} \quad < 3^{\sqrt{2}} < 3^{1.42} \quad \approx \textbf{4.7590}$$
$$\textbf{4.727695} \approx 3^{1.414} < 3^{\sqrt{2}} < 3^{1.415} \approx \textbf{4.732892}$$

There is exactly one real number that is larger than any of the increasing numbers on the left of the previous list, and less than all of the decreasing numbers on the right. By definition, that number is $3^{\sqrt{2}}$.

In a similar way, it is possible to define b^x for every real number x and every positive real number b.

To find an approximation for $3^{\sqrt{2}}$, we can use a calculator and press these keys:

The display will show 4.7288044. Thus,

$$3^{\sqrt{2}} \approx 4.7288044$$

Many properties of the exponential expression b^x involve concepts from calculus and cannot be proved in this course. We shall assume the following facts involving real number exponents:

---

1. If $b > 0$ then for each x, b^x represents a single positive real number.
2. The laws of rational exponents are valid for *real* exponents.
3. If $b > 1$ and $r < s$, then $b^r < b^s$.
4. If $0 < b < 1$ and $r < s$, then $b^r > b^s$.
5. If $0 < a < b$ and $r > 0$, then $a^r < b^r$.
6. If $0 < a < b$ and $r < 0$, then $a^r > b^r$.

Because b^x is defined for real exponents, we can now define an **exponential function** with domain of the real numbers.

Definition.   The **exponential function with base b** is defined by the equation

$$y = f(x) = b^x$$

where $b > 0$ and $b \neq 1$.
   The domain of the exponential function is the set of real numbers, and the range is the set of positive real numbers.

Note that if $b > 1$ then, by Property 3 of real number exponents, you know that if $x_1 < x_2$ then $b^{x_1} < b^{x_2}$. Thus, if $b > 1$ the equation $y = b^x$ defines an *increasing* function. If $0 < b < 1$, we can conclude from Property 4 that the equation $y = b^x$ defines a *decreasing* function.

Example 1   Graph the exponential functions $y = 2^x$ and $y = 7^x$.

Solution   Calculate several pairs (x, y) that satisfy each equation. Plot the points and join them with a smooth curve. The graph of $y = 2^x$ appears in Figure 4–1a, and the graph of $y = 7^x$ appears in Figure 4–1b.
   Note that $y = 2^x$ and $y = 7^x$ are *increasing* functions, that each graph passes through the point $(0, 1)$, and that the x-axis is an **asymptote** of each graph. Note also that the graph of $y = 2^x$ passes through the point $(1, 2)$ and that the graph of $y = 7^x$ passes through the point $(1, 7)$.

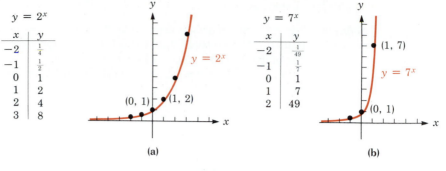

$y = 2^x$	
x	y
-2	$\frac{1}{4}$
-1	$\frac{1}{2}$
0	1
1	2
2	4
3	8

$y = 2^x$

$(0, 1)$ $(1, 2)$

(a)

$y = 7^x$	
x	y
-2	$\frac{1}{49}$
-1	$\frac{1}{7}$
0	1
1	7
2	49

$(1, 7)$

$y = 7^x$

$(0, 1)$

(b)

Figure 4–1

Example 2 Graph the exponential functions $y = \left(\dfrac{1}{2}\right)^x$ and $y = \left(\dfrac{1}{7}\right)^x$.

Solution Recall that the graphs of $y = f(x)$ and $y = f(-x)$ are reflections of each other in the y-axis. Because $y = (\frac{1}{2})^x$ is equivalent to $y = 2^{-x}$, it follows that the graphs of $y = (\frac{1}{2})^x$ and $y = 2^x$ are reflections of each other in the y-axis.

Similarly, the graphs of $y = (\frac{1}{7})^x$ and $y = 7^x$ are reflections of each other in the y-axis. The graphs appear in Figures 4–2a and 4–2b.

Note that $y = (\frac{1}{2})^x$ and $y = (\frac{1}{7})^x$ are *decreasing* functions, that each graph passes through the point $(0, 1)$, and that the x-axis is an asymptote of each graph. Note also that the graph of $y = (\frac{1}{2})^x$ passes through the point $(1, \frac{1}{2})$ and that of $y = (\frac{1}{7})^x$ passes through the point $(1, \frac{1}{7})$.

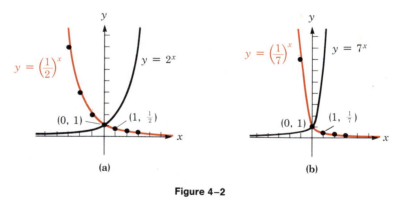

$y = \left(\dfrac{1}{2}\right)^x$ $y = 2^x$

$(0, 1)$ $(1, \frac{1}{2})$

(a)

$y = \left(\dfrac{1}{7}\right)^x$ $y = 7^x$

$(0, 1)$ $(1, \frac{1}{7})$

(b)

Figure 4–2

Because an exponential function is either increasing (for $b > 1$) or decreasing (for $0 < b < 1$), distinct real numbers x will determine distinct values b^x. Thus the exponential function is one-to-one. This fact is the basis for another property of exponents:

> **7.** If $b > 0$ and $b \neq 1$, and if $b^r = b^s$, then $r = s$.

Example 3   Graph $y = 2^{x+3}$.

Solution   The graph of $y = 2^{x+3}$ is identical to the graph of $y = f(x) = 2^x$, except that it has been translated horizontally **3** units to the left. The graph of $y = 2^{x+3}$ appears in Figure 4–3. The graph of $y = 2^x$ is shown for reference.

The graph of the colored curve can also be viewed as a *vertical stretching* of the black curve, because

$$y = 2^{x+3}$$
$$= 2^x \mathbf{2^3}$$
$$= \mathbf{8}(2^x)$$

Thus, the colored curve can also be found by stretching the black curve vertically by a factor of **8**.

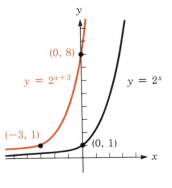

Figure 4–3

We summarize the properties of an exponential function:

The exponential function is one-to-one.
If $b > 1$, then $y = b^x$ defines an *increasing* function.
If $0 < b < 1$, then $y = b^x$ defines a *decreasing* function.
The graph of $y = b^x$ passes through the points $(0, 1)$ and $(1, b)$.
The x-axis is an asymptote of the graph of $y = b^x$.
The graphs of $y = b^x$ and $y = b^{-x}$ are reflections of each other in the y-axis.
The graph of $y = a + b^{x+k}$ is identical to the graph of $y = b^x$ except that it has been translated a distance $|a|$ in a vertical direction and a distance $|k|$ in a horizontal direction.

Example 4   Graph the function $y = 2(3^{x/2})$ and determine its domain and range.

Solution   Plot several pairs (x, y) that satisfy the equation $y = 2(3^{x/2})$ and join them with a smooth curve. The graph appears in Figure 4–4. The domain of the function is the set of real numbers, and the range is the set of positive real numbers. Note that the y-intercept is 2.

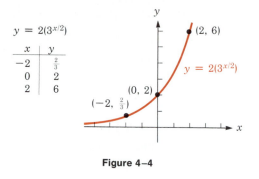

$y = 2(3^{x/2})$

x	y
-2	$\frac{2}{3}$
0	2
2	6

$y = 2(3^{x/2})$

$(2, 6)$

$(0, 2)$

$(-2, \frac{2}{3})$

Figure 4–4

Applications of Exponential Functions

A mathematical description of an observed event is called a **model** of that event. Many observed events can be modeled by functions defined by equations of the form

$$y = f(x) = ab^{kx}$$

where a, b, and k are constants. If f is an increasing function such as the one in Example 4, then y is said to **grow exponentially**. If f is a decreasing function, then y **decays exponentially**.

Radioactive decay. A decreasing function of the form $f(x) = ab^{kx}$ provides a model for a process called **radioactive decay**. The atomic structure of a radioactive material changes as it emits radiation. Uranium, for example, decays into thorium, then into radium, and eventually into lead. The rate of radioactive decay is not constant, but depends on the amount of radioactive material present. A block of radioactive material could not decay completely, because the more it decays, the less is present and the slower the rate of decay becomes.

Experiments have determined the time it takes for *half* of a given amount of radioactive material to decompose. This time, called the element's **half-life**, is constant for any given substance. The amount A of radioactive material present decays exponentially according to the model

Formula for Radioactive Decay

$$A = A_0 2^{-t/h}$$

where A_0 is the amount present at $t = 0$ and h is the material's half-life.

Example 5 The half-life of radium is 1600 years. How much of a 1-gram sample will be present in 660 years?

Solution   In this example, $A_0 = 1$, $h = 1600$, and $t = 660$. Substitute these values into the equation $A = A_0 2^{-t/h}$ and simplify.

$$A = A_0 2^{-t/h}$$
$$A = 1 \cdot 2^{-660/1600}$$
$$= 1 \cdot 2^{-0.4125} \qquad \text{Use a calculator.}$$
$$\approx 0.75$$

After 660 years, approximately 0.75 gram of radium will remain.   ■

Compound interest.   If the interest earned on money in a savings account is allowed to accumulate, then that interest also earns interest. The amount in the account grows exponentially according to the equation

Formula for Compound Interest $$A = A_0 \left(1 + \frac{r}{k} \right)^{kt}$$

where A represents the amount in the account after t years, with interest paid k times a year at an annual rate r on an initial deposit A_0.

Example 6   If \$1000 is deposited in an account that earns 12% interest compounded quarterly, how much will be in the account after 20 years?

Solution   Calculate A using the formula

$$A = A_0 \left(1 + \frac{r}{k} \right)^{kt}$$

with $A_0 = 1000$, $r = 0.12$, and $t = 20$. Because quarterly interest payments occur four times a year, $k = 4$.

$$A = A_0 \left(1 + \frac{r}{k} \right)^{kt}$$
$$A = 1000 \left(1 + \frac{0.12}{4} \right)^{4(20)}$$
$$= 1000(1.03)^{80}$$
$$\approx 10,640.89$$

In 20 years, the account will contain \$10,640.89.   ■

Base-e Exponential Functions

In mathematical models of natural events, one number appears often as the base of an exponential function. That number is represented by e, a symbol introduced by Leonhard Euler (1707–1783). We introduce this important number by allowing k in the compound interest formula

$$A = A_0 \left(1 + \frac{r}{k} \right)^{kt}$$

Leonhard Euler (1707–1783) Euler was one of the most prolific mathematicians of all time. He did much of his work after he became blind.

to become very large. To see what happens, we let $k = rp$, where p is a new variable, and proceed as follows:

$$A = A_0\left(1 + \frac{r}{k}\right)^{kt}$$

$$= A_0\left(1 + \frac{r}{rp}\right)^{rpt} \qquad \text{Substitute } rp \text{ for } k.$$

$$= A_0\left(1 + \frac{1}{p}\right)^{rpt} \qquad \text{Simplify } \frac{r}{rp}.$$

$$= A_0\left[\left(1 + \frac{1}{p}\right)^{p}\right]^{rt} \qquad \text{Remember that } x^{mn} = (x^m)^n.$$

Because r is a positive constant and $k = rp$, it follows that, as k becomes very large, so does p. The question of what happens to the value of A as k becomes very large is tied to the question: What happens to the value of $(1 + \frac{1}{p})^p$ as p becomes very large?

Some results calculated for increasing values of p appear in Table 4–1.

Table 4–1

p	$\left(1 + \dfrac{1}{p}\right)^p$
1	2
10	2.5937
1,000	2.7169
1,000,000	2.7182805
$\vdots$	$\vdots$

The results in the table suggest that, as p increases, the value of $(1 + \frac{1}{p})^p$ approaches a fixed number. This number is e, an irrational number with a decimal representation of $2.71828182845904\ldots$.

If interest on an amount A_0 is compounded more and more often, the number p grows large without bound, and the formula

$$A = A_0\left[\left(1 + \frac{1}{p}\right)^{p}\right]^{rt}$$

becomes

Formula for Continuous Compound Interest

$$A = A_0 e^{rt}$$

When the amount invested grows exponentially according to the formula $A = A_0 e^{rt}$, interest is said to be **compounded continuously**.

Example 7 If $1000 accumulates interest at an annual rate of 12%, compounded continuously, how much money will be in the account in 20 years?

Solution $A = A_0 e^{rt}$

$A = 1000 e^{(0.12)(20)}$

Evaluate A with a calculator by finding the product of 0.12 and 20, pressing the $\boxed{e^x}$ key, and multiplying that result by 1000. Round the answer to the nearest penny. In 20 years, the account will contain \$11,023.18.

If your calculator does not have an $\boxed{e^x}$ key, try pressing $\boxed{\text{INV}}$ $\boxed{\ln x}$, or consult your owner's manual. ∎

The exponential function $y = e^x$ is so important that it is often called *the* exponential function.

Example 8   Graph the exponential function.

Solution   Use a calculator to find several pairs (x, y) that satisfy the equation $y = e^x$. Plot them, and join them with a smooth curve. The graph appears in Figure 4–5.

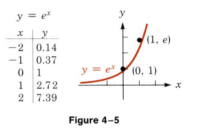

$y = e^x$	
x	y
-2	0.14
-1	0.37
0	1
1	2.72
2	7.39

Figure 4–5 ∎

Applications of Base-e Exponential Functions

Population growth.   A population of living organisms changes with time. Many factors—such as birth and death rates, immigration, pollution, diet, wars, plagues, and famines—affect the population. Models of population growth that account for several factors can be very complex. The **Malthusian model of population growth** assumes constant birth rate B and death rate D and neglects all other factors. In that model, the population P grows exponentially according to the formula

Formula for Population Growth $P = P_0 e^{kt}$

where P_0 is the population at $t = 0$, and $k = B - D$. If t is measured in years, then k is called the **annual growth rate**.

Example 9   The annual birth rate in a certain country is 19.0 per 1000 population, and the death rate is 7.0 per 1000. Based on a current population of 2.3 million, what would a Malthusian model predict the population to be in 50 years?

Solution   Use the Malthusian model of population growth

$P = P_0 e^{kt}$

The number k is the difference between the birth and death rates. The birth rate B is $\frac{19}{1000}$, or 0.019. The death rate D is $\frac{7}{1000}$, or 0.007. Thus,

$$k = B - D$$
$$= 0.019 - 0.007$$
$$= 0.012$$

Substitute $\mathbf{2.3 \times 10^6}$ for P_0, $\mathbf{50}$ for t, and $\mathbf{0.012}$ for k in the equation $P = P_0 e^{kt}$ and simplify.

$$P = \mathbf{P_0 e^{kt}}$$
$$P = \mathbf{(2.3 \times 10^6)}e^{(\mathbf{0.012})(\mathbf{50})}$$
$$= (2.3 \times 10^6)(1.82)$$
$$= 4.2 \times 10^6$$

After 30 years, the population will exceed 4 million.   ■

Example 10   A population of 1000 bacteria doubles in 8 hours. Assuming a Malthusian model, what will be the population in 12 hours?

Solution   The population P grows according to the formula $P = P_0 e^{kt}$. Let $P_0 = \mathbf{1000}$, $P = \mathbf{2000}$, and $t = \mathbf{8}$ Then solve the resulting equation for e^k.

$$\mathbf{P = P_0 e^{kt}}$$
$$\mathbf{2000 = 1000}e^{k \cdot \mathbf{8}}$$
$$2 = e^{k8}$$

or

$$2 = (e^k)^8 \qquad \text{Remember that } x^{mn} = (x^m)^n.$$
$$2^{1/8} = \left[(e^k)^8\right]^{1/8} \qquad \text{Raise both sides to the 1/8 power.}$$
$$\mathbf{2^{1/8}} = \mathbf{e^k}$$

We know that the population grows according to the formula

$$P = 1000e^{kt}$$

or

$$P = 1000(e^k)^t \qquad \text{Remember that } x^{mn} = (x^m)^n.$$
$$= 1000(\mathbf{2^{1/8}})^t \qquad \text{Substitute } 2^{1/8} \text{ for } e^k.$$
$$= 1000(2^{t/8})$$

To find the population after 12 hours, substitute $\mathbf{12}$ for t in the formula $P = 1000(2^{t/8})$ and simplify.

$$P = 1000(2^{t/8})$$
$$= 1000(2^{12/8})$$
$$= 1000(2^{3/2})$$
$$\approx 1000(2.8284) \qquad \text{Use a calculator.}$$
$$\approx 2800$$

After 12 hours, there are approximately 2800 bacteria.   ■

Exercise 4.1

In Exercises 1–8 graph each exponential function.

1. $y = 5^x$

2. $y = \left(\dfrac{1}{5}\right)^x$

3. $y = \left(\dfrac{1}{3}\right)^x$

4. $y = 3^x$

5. $y = 2.5^x$

6. $y = (0.4)^x$

7. $y = 10^x$

8. $y = (0.1)^x$

In Exercises 9–18 graph the function defined by each equation.

9. $y = 2^{x+1}$

10. $y = 2^{x-3}$

11. $y = -3^{x-2}$

12. $y = -3^{x+2}$

13. $y = 5(2^x)$

14. $y = 2(5^x)$

15. $y = 2 + 3^x$

16. $y = 3^x - 3$

17. $y = 2^{|x|}$

18. $y = 3^{-x}$

In Exercises 19–24 find the value of b, if any, that would cause the graph of $y = b^x$ to look like the graph indicated.

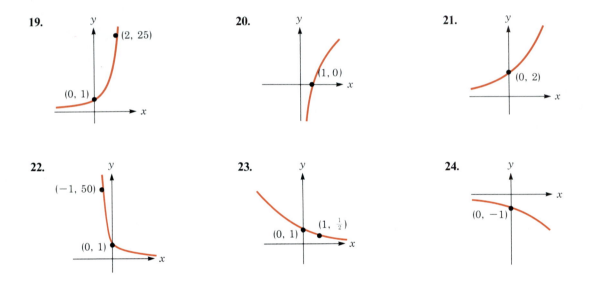

19.   (2, 25)   (0, 1)

20.   (1, 0)

21.   (0, 2)

22.   (−1, 50)   (0, 1)

23.   (0, 1)   (1, ½)

24.   (0, −1)

In Exercises 25–30 graph each function. **Use a calculator.**

25. $f(x) = -e^x$

26. $f(x) = e^{-x}$

27. $f(x) = e^{x/2}$

28. $f(x) = e^{2x}$

29. $f(x) = e^x + 1$

30. $f(x) = e^{x+1}$

31. A radioactive material decays according to the formula $A = A_0 \left(\frac{2}{3}\right)^t$, where A_0 is the amount present initially, and t is measured in years. What amount will be present in 5 years?

32. Tritium, a radioactive isotope of hydrogen, has a half-life of 12.4 years. From an initial sample of 0.05 gram, how much will remain after 100 years?

33. The half-life of radioactive carbon ^{14}C is 5700 years. How much of an initial sample will remain after 3000 years?

34. The biological half-life of triazolam, a drug used for treating insomnia, is 2.3 hours. What percent of an initial dose will remain after 5 hours?

35. The biological half-life of the asthma medication theophylline is 4.5 hours for smokers and 8 hours for non-smokers. Twelve hours after administration of equal doses, what is the ratio of drug retained in a smoker's system to that in a nonsmoker's system?

In Exercises 36–44 assume there are no deposits or withdrawals.

36. An initial deposit of $500 earns 10% interest, compounded quarterly. How much will be in the account in 10 years?

37. An initial deposit of $1000 earns 12% interest, compounded monthly. How much will be in the account in $4\frac{1}{2}$ years?

38. If $1 had been invested in 1776 at 5% interest, compounded annually, what would it be worth in 2076?

39. Some financial institutions pay daily interest, compounded by the **360/365 method**, by using the formula

$$A = P\left(1 + \frac{r}{360}\right)^{365t} \qquad (t \text{ is in years})$$

Using this method, how large will an initial investment of $1000 be in 5 years, at a 12% annual interest rate?

40. An initial deposit of $5000 grows at a continuously compounded annual rate of 14% interest. What will the investment be worth in 10 years?

41. An initial investment of $5000 earns 11.2% interest, compounded continuously. What will the investment be worth in 12 years?

42. An initial deposit of $2000 earns 8% interest, compounded continuously. How much will be in the account in 15 years?

43. An account now contains $11,180. It has been accumulating interest at the rate of 13%, compounded continuously, for 7 years. What was the initial deposit?

44. An account now contains $3610. It has been continuously accumulating interest at the rate of $10\frac{1}{2}\%$. How much was in the account one year ago?

45. A colony of 6 million bacteria is growing in a culture medium. The population P after t hours is modeled by the formula $P = (6 \times 10^6)(2.3)^t$. What is the population after 4 hours?

46. The growth of a population is modeled by $P = 173e^{0.03t}$. How large will the population be when $t = 20$?

47. The decline of a population is modeled by $P = 1.2 \times 10^6 e^{-0.008t}$. How large will the population be when $t = 30$?

48. The population of North Rivers is growing exponentially according to the model $P = 375(1.3)^t$, where t is measured in years from the present date. What will be the population in 3 years?

49. A bacteria culture doubles in size every 24 hours. By what factor will it have increased in 36 hours?

50. An initial deposit of $5000 grows at a continuously compounded annual rate of 14%. What will the investment be worth in 100 years?

51. The world population is approximately 4.2 billion and is growing at an annual rate of 1.9%. Assuming a Malthusian growth model, what will be the world's population in 30 years?

52. The charge remaining in a battery is decreasing exponentially according to the formula $C = (3 \times 10^{-4})(0.7)^t$, where C is the charge after t days. What is the charge after 5 days?

53. The definition of the exponential function $y = b^x$ excluded the possibility that $b = 1$. Graph the function $y = 1^x$. Is this function one-to-one?

54. The definition of the exponential function $y = b^x$ excluded the possibility that $b < 0$. Attempt to graph the equation $y = (-2)^x$ by determining the value of y for the following numbers x: 3, 2, 1, and $\frac{1}{2}$. Explain why $y = (-2)^x$ does not define an exponential function.

55. Graph the function $y = \frac{1}{2}(e^x + e^{-x})$ for values of x between -2 and 2. The graph looks like a parabola, but it is not. It is called a **catenary** and represents the shape of a cable hanging between two supports.

56. The value of e can be calculated to any degree of accuracy by adding the first several terms of the following list:

$$1, \quad 1, \quad \frac{1}{2}, \quad \frac{1}{2\cdot 3}, \quad \frac{1}{2\cdot 3\cdot 4}, \quad \frac{1}{2\cdot 3\cdot 4\cdot 5}, \quad \frac{1}{2\cdot 3\cdot 4\cdot 5\cdot 6}, \ldots$$

The more terms that are added, the closer the sum is to e. Calculate an approximation to the value of e by adding the first ten terms of the list. To how many decimal places is your sum accurate?

57. If the graphs of $y = b^x$ and $y = e^{kx}$ are identical, show that $b = e^k$.

58. If the graphs of $y = e^x$ and $y = b^{kx}$ are identical, to what is b^k equal?

59. Graph the function

$$f(x) = \frac{1}{\sqrt{2\pi}} e^{-x^2/2}$$

This bell-shaped curve is the **normal probability density function** used in statistics.

60. Find a function $f(x)$ with the property that $f(a + b) = f(a)f(b)$ for all real numbers a and b.

61. Find a function $f(x)$ with the property that $f(ab) = [f(a)]^b$ for all real numbers a and b.

62. What number x is a solution to the equation $e^{2x} - 2e^x + 1 = 0$? (*Hint:* $e^{2x} = (e^x)^2$.)

In Exercises 63–64 assume that f is a function with the property that $f(x + y) = f(x)f(y)$ for all real numbers x and y, and that $f(y)$ is never 0.

63. Prove that $f(0) = 1$.   **64.** Prove that $f(-x) = \dfrac{1}{f(x)}$.

4.2 LOGARITHMIC FUNCTIONS

Because the exponential function defined by $y = b^x$ is one-to-one, it has an inverse that is defined by the equation $x = b^y$. To express this inverse function in the form $y = f^{-1}(x)$, we must solve the equation $x = b^y$ for y. To do so, we need to introduce a new function.

> **Definition.**   The **logarithmic function with base b** is defined by the equation
>
> $$y = \log_b x$$
>
> where $b > 0$ and $b \neq 1$. This equation is equivalent to the exponential equation
>
> $$b^y = x$$

Because the logarithmic function $y = \log_b x$ is the inverse of the one-to-one exponential function $y = b^x$, the logarithmic function is one-to-one also. The domain of the logarithmic function is the range of the exponential function: it is the set of positive real numbers. The range of the logarithmic function is the domain of the exponential function: it is the set of real numbers.

The previous definition implies that any pair (x, y) that satisfies the equation $x = b^y$ also satisfies the equation $\log_b x = y$. Thus,

$$\log_5 25 = 2 \quad \text{because} \quad 5^2 = 25$$
$$\log_7 1 = 0 \quad \text{because} \quad 7^0 = 1$$
$$\log_{16} 4 = \frac{1}{2} \quad \text{because} \quad 16^{1/2} = 4$$

and

$$\log_2\left(\frac{1}{8}\right) = -3 \quad \text{because} \quad 2^{-3} = \frac{1}{8}$$

Note that in each case the logarithm of a number is an exponent.

Because the domain of the logarithmic function is the set of positive real numbers, it is impossible to find the logarithm of zero or the logarithm of a negative number.

Example 1   Find the value of y in each of the following equations.

    a. $\log_5 1 = y$ **b.** $\log_2 8 = y$ **c.** $\log_7(\frac{1}{7}) = y$

Solution   **a.** Change the equation $\log_5 1 = y$ into the equivalent exponential form $5^y = 1$. Because $5^0 = 1$, it follows that $y = 0$. Hence, $\log_5 1 = 0$.

    b. $\log_2 8 = y$ is equivalent to $2^y = 8$. Because $2^3 = 8$, it follows that $y = 3$. Hence, $\log_2 8 = 3$.

    c. $\log_7(\frac{1}{7}) = y$ is equivalent to $7^y = \frac{1}{7}$. Because $7^{-1} = \frac{1}{7}$, it follows that $y = -1$. Hence, $\log_7(\frac{1}{7}) = -1$.   ■

Example 2   Find the value of a in each equation:

    a. $\log_3(\frac{1}{9}) = a$ **b.** $\log_a 32 = 5$ **c.** $\log_9 a = -\frac{1}{2}$

Solution   **a.** $\log_3(\frac{1}{9}) = a$ is equivalent to $3^a = \frac{1}{9}$. Because $3^{-2} = \frac{1}{9}$, it follows that $a = -2$.

    b. $\log_a 32 = 5$ is equivalent to $a^5 = 32$. Because $2^5 = 32$, it follows that $a = 2$.

    c. $\log_9 a = -\frac{1}{2}$ is equivalent to $9^{-1/2} = a$. Because $9^{-1/2} = \frac{1}{3}$, it follows that $a = \frac{1}{3}$.   ■

Example 3   Graph the functions defined by **a.** $y = \log_2 x$ and **b.** $y = \log_2(x - 3)$.

Solution   **a.** The equation $y = \log_2 x$ is equivalent to the equation $2^y = x$. Calculate and plot pairs (x, y) that satisfy the equation $x = 2^y$, and connect them with a smooth curve as in Figure 4–6.

    b. The graph of $y = \log_2(x - 3)$ is identical to that of $y = \log_2 x$, except that it is translated 3 units to the right.

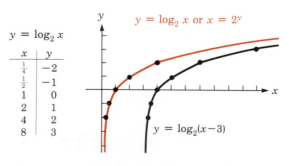

Figure 4–6   ■

Example 4   Graph the logarithmic function defined by $y = \log_{1/2} x$.

Solution   Proceed as in Example 3. The graph appears in Figure 4–7.

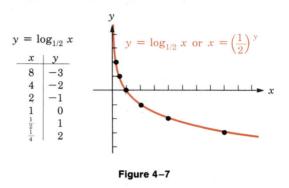

$y = \log_{1/2} x$

x	y
8	-3
4	-2
2	-1
1	0
$\frac{1}{2}$	1
$\frac{1}{4}$	2

$y = \log_{1/2} x$ or $x = \left(\dfrac{1}{2}\right)^{y}$

Figure 4–7

Graphs of logarithmic functions are always similar to those in Figure 4–8. If $b > 1$, the logarithmic function is *increasing* as in Figure 4–8a; if $0 < b < 1$, the logarithmic function is *decreasing* as in Figure 4–8b. Note that the graphs of $y = \log_b x$ pass through the points $(1, 0)$ and $(b, 1)$ and that the y-axis is an asymptote to the curve.

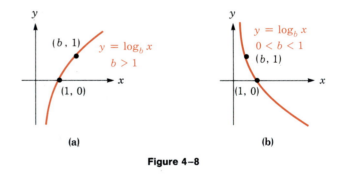

(a)

(b)

Figure 4–8

The exponential and logarithmic functions to base b are inverses of each other and, therefore, have symmetry about the line $y = x$. The graphs of $y = \log_b x$ and $y = b^x$ appear as in Figure 4–9a if $b > 1$ and as in Figure 4–9b if $0 < b < 1$.

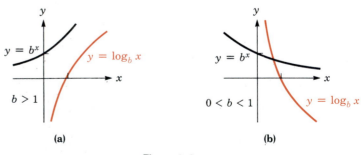

(a)

(b)

Figure 4–9

Several properties of logarithms can be found by expressing the properties of exponents in logarithmic form.

Properties of Logarithms.   If M, N, p, and b are positive numbers, and $b \neq 1$, then

1. $\log_b 1 = 0$
2. $\log_b b = 1$
3. $\log_b b^x = x$
4. $b^{\log_b x} = x$
5. $\log_b MN = \log_b M + \log_b N$
6. $\log_b \left(\dfrac{M}{N} \right) = \log_b M - \log_b N$
7. $\log_b M^P = p \log_b M$
8. If $\log_b r = \log_b s$, then $r = s$.

Proof   Properties 1 and 2 follow directly from the definition of the logarithmic function. Properties 3 and 4 reflect the fact that the logarithmic and the exponential functions are inverses of each other and that their composition is the identity function. To prove Property 5, let $m = \log_b M$ and $n = \log_b N$. Using the definition of logarithm gives these equations in the form

$$M = b^m \quad \text{and} \quad N = b^n$$

Then, we have

$$MN = b^m b^n$$

and, using a property of exponents, we obtain

$$MN = b^{m+n}$$

By the definition of logarithm, this equation is equivalent to

$$\log_b MN = m + n$$

We substitute the values of m and n to complete the proof of Property 5.

$$\log_b MN = \log_b M + \log_b N$$

The proofs of Properties 6 and 7 are left as exercises. Property 8 is a result of the fact that the logarithm function is one-to-one.   □

Note that Property 5 states that the logarithm of the *product* of M and N is the sum of the logarithms of M and N. The logarithm of the *sum* of two numbers is *not* the sum of their logarithms:

$$\log_b(M + N) = \log_b M + \log_b N$$

Example 5   If A, B, C, and b are positive numbers, and $b \neq 1$, then

a. $\log_b(ABC) = \log_b(AB) + \log_b C$ Use Property 5.
$$= \log_b A + \log_b B + \log_b C \quad \text{Use Property 5.}$$

b. $\log_b(A^3 B^2) = \log_b A^3 + \log_b B^2$ Use Property 5.

$= 3\log_b A + 2\log_b B$ Use Property 7.

c. $\log_b\left(\dfrac{A}{BC}\right) = \log_b A - \log_b(BC)$ Use Property 6.

$= \log_b A - (\log_b B + \log_b C)$ Use Property 5.

$= \log_b A - \log_b B - \log_b C$

d. $\log_b b^2 = 2\log_b b$ Use Property 7.

$= 2(1)$ Use Property 2.

$= 2$ ■

Logarithms are easiest to find by using a calculator. Most scientific calculators provide logarithms to two bases: base 10 and base e.

For computational purposes, base-10 logarithms, called **common logarithms**, are the most convenient. In this book, if the base b is not indicated in the notation $\log_b x$, assume that b is 10:

$$\log x \quad \text{means} \quad \log_{10} x$$

Because the number e appears so often in mathematical models of events in nature, base-e logarithms are called **natural logarithms**. They are also called **Napierian logarithms** after the inventor of logarithms, John Napier (1550–1617). Natural logarithms are usually denoted by the symbol ln x (pronounced "ell-en"), rather than $\log_e x$:

$$\ln x \quad \text{means} \quad \log_e x$$

John Napier (1550–1617)
Napier invented a
device, called **Napier's
rods**, that did
multiplications
mechanically. His device
was a forerunner of
modern-day computers.

Example 6   Use a calculator to find **a.** log 2.34 and **b.** ln 2.34.

Solution   **a.**   To find log 2.34 enter the number 2.34 and press the ⬚log⬚ key. (You may have to press a ⬚2nd⬚ function key first.) The display should read .3692158574. Hence,

$$\log 2.34 \approx 0.3692$$

b.   To find ln 2.34, enter the number 2.34 and press the ⬚ln x⬚ key. The display shows .850150929. Hence,

$$\ln 2.34 \approx 0.8502$$ ■

Example 7   Find the value of x in each equation:

a. $\log x = 0.7482$ **b.** $\ln x = 1.335$

Solution   **a.** $\log x = 0.7482$ is equivalent to $10^{0.7482} = x$. To find x, enter the number 10, press the ⬚y^x⬚ key, enter the number .7482, and press the ⬚=⬚ key. The display reads

5.6001544. Hence,

$$x \approx 5.6$$

If your calculator has a $\boxed{10^x}$ key, enter .7482 and press that key to get the same result.

b. $\ln x = 1.335$ is equivalent to $e^{1.335} = x$. To find x, enter the number 1.335 and press the $\boxed{e^x}$ key. If your calculator does not have an $\boxed{e^x}$ key, press $\boxed{\text{INV}}\ \boxed{\ln x}$. The display reads 3.79999595. Hence,

$$x \approx 3.8$$ ■

Example 8 Solve the equation $e^{2x} - 5e^x + 6 = 0$.

Solution Because e^{2x} can be written as $(e^x)^2$, the given equation is quadratic in the variable e^x. It can be solved as follows:

$$e^{2x} - 5e^x + 6 = 0$$
$$(e^x)^2 - 5e^x + 6 = 0$$
$$(e^x - 2)(e^x - 3) = 0$$

$e^x - 2 = 0$	or	$e^x - 3 = 0$	
$e^x = 2$		$e^x = 3$	
$x = \ln 2$		$x = \ln 3$	$e^x = y$ is equivalent to $x = \ln y$.
$x \approx 0.6931$		$x \approx 1.0986$	

■

Just as $y = e^x$ represents a special exponential function, so does $y = \ln x$ represent a special logarithmic function.

Example 9 Graph the logarithmic function $f(x) = \ln x$.

Solution Find and plot several pairs (x, y) that satisfy the equation $y = \ln x$, and join them with a smooth curve. The graph appears in Figure 4–10.

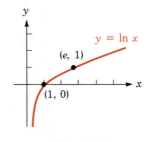

Figure 4–10 ■

Applications of Logarithms

Isothermal expansion. Suppose some amount E of energy is added to one mole of a gas, causing it to expand from an initial volume V_i to a final volume V_f

without changing its temperature. Then

Formula for Isothermal Expansion

$$E = RT \ln\left(\frac{V_f}{V_i}\right)$$

where E is measured in joules, T is the temperature of the system measured in degrees Kelvin, and R is the **universal gas constant**, 8.314 joules/mole/K.

Example 10   What energy must be supplied to triple the volume of one mole of gas at a constant temperature of 300 K?

Solution   Substitute **8.314** for R and **300** for T in the equation $E = RT \ln\left(\frac{V_f}{V_i}\right)$. Because the final volume is to be three times the initial volume, also substitute $3V_i$ for V_f. Then determine E as follows:

$$E = RT \ln\left(\frac{V_f}{V_i}\right)$$

$$E = (8.314)(300) \ln\left(\frac{3V_i}{V_i}\right)$$

$$= 2494 \ln 3$$

$$\approx 2740$$

Approximately 2740 joules of energy must be added to the gas to triple its volume. ■

pH scale.   In chemistry, common logarithms are used to express the acidity of solutions. The more acidic a solution, the greater is the concentration of hydrogen ions. This concentration is indicated indirectly by the **pH scale**, or the **hydrogen-ion index**. The pH of a solution is defined by the equation

Formula for pH

$$pH = -\log[H^+]$$

where $[H^+]$ is the hydrogen-ion concentration in gram-ions per liter. Pure water has a few free hydrogen ions—$[H^+]$ is approximately 10^{-7} gram-ions per liter. The pH of pure water is

$$pH = -\log[H^+]$$
$$= -\log 10^{-7}$$
$$= -(-7)\log 10 \qquad \text{Use Property 7.}$$
$$= -(-7) \qquad \text{Use Property 2.}$$
$$= 7$$

The pH of pure water is 7.

Example 11   Seawater is slightly alkaline, with a pH of approximately 8.5. What is the hydrogen-ion concentration of seawater?

Solution   Substitute **8.5** for pH in the logarithmic equation $pH = -\log[H^+]$. Then change it to exponential form and solve for $[H^+]$:

$$\mathbf{pH} = -\log[H^+]$$
$$\mathbf{8.5} = -\log[H^+]$$
$$-8.5 = \log[H^+]$$
$$[H^+] = 10^{-8.5}$$
$$[H^+] \approx 3.2 \times 10^{-9} \qquad \text{Use a calculator.}$$

The hydrogen-ion concentration of seawater is approximately 3.2×10^{-9} gram-ions per liter.   ■

Exercise 4.2

In Exercises 1–24 find the value of x. A calculator is of no value.

1. $\log_2 8 = x$

2. $\log_{1/2}\left(\frac{1}{8}\right) = x$

3. $\log_{1/2} 8 = x$

4. $\log_{25} 5 = x$

5. $\log_5 25 = x$

6. $\log_8 x = 2$

7. $\log_x 8 = 3$

8. $\log_7 x = 0$

9. $\log_7 x = 1$

10. $\log_4 x = \frac{1}{2}$

11. $\log_x\left(\frac{1}{16}\right) = -2$

12. $\log_{125} x = \frac{2}{3}$

13. $\log_{100}\left(\frac{1}{1000}\right) = x$

14. $\log_{5/2}\left(\frac{4}{25}\right) = x$

15. $\log_{27} 9 = x$

16. $\log_{12} x = 0$

17. $\log_x 5^3 = 3$

18. $\log_x 5 = 1$

19. $\log_x\left(\frac{9}{4}\right) = 2$

20. $\log_x\left(\frac{\sqrt{3}}{3}\right) = \frac{1}{2}$

21. $\log_{\sqrt{3}} x = -4$

22. $\log_\pi x = 3$

23. $\log_{2\sqrt{2}} x = 2$

24. $\log_4 8 = x$

In Exercises 25–32 graph each function.

25. $y = \log_5 x$

26. $y = \log_{1/5} x$

27. $y = \log_{1/3} x$

28. $y = \log_3 x$

29. $y = \ln x^2$

30. $y = \ln 2x$

31. $y = (\ln x) - 1$

32. $y = \ln(x - 1)$

In Exercises 33–38 graph each pair of equations on one set of coordinate axes.

33. $y = \log_3 x$ and $y = \log_3(3x)$

34. $y = \log_3 x$ and $y = \log_3\left(\frac{x}{3}\right)$

35. $y = \log_2 x$ and $y = \log_2(x + 1)$

36. $y = \log_2 x$ and $y = \log_2(-x)$

37. $y = \log_5 x$ and $y = 5^x$

38. $y = \ln x$ and $y = e^x$

In Exercises 39–44 find the value of b, if any, that would cause the graph of $y = \log_b x$ to look like the graph indicated.

39.   **40.**   **41.**

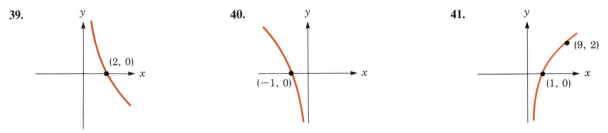

(2, 0)   (−1, 0)   (9, 2)   (1, 0)

42.

43.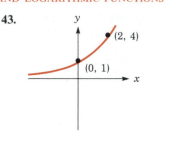

44.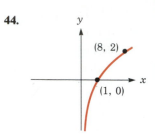

In Exercises 45–66 tell whether the given statement is true.

45. $\log_b ab = \log_b + 1$

46. $\log_b\left(\dfrac{1}{a}\right) = -\log_b a$

47. $\log_b 0 = 1$

48. $\log_b 2 = \log_2 b$

49. $\log_b(x + y) \neq \log_b x + \log_b y$

50. $\log_b xy = (\log_b x)(\log_b y)$

51. If $\log_a b = c$, then $\log_b a = c$.

52. $\log_b b^b = b$

53. $b^{\log_b b} = b$

54. $\log_b(-x) = -\log_b x$

55. If $\log_b a = c$, then $\log_b a^p = pc$.

56. $\dfrac{\log A}{\log B} = \log A - \log B$

57. $\log(A - B) = \dfrac{\log A}{\log B}$

58. $n \log_b \sqrt[n]{a} = \log_b a$

59. $\dfrac{1}{3} \log_b a^3 = \log_b a$

60. The base of a system of logarithms cannot be negative.

61. A logarithm cannot be negative.

62. The logarithm of a negative number is negative.

63. If $0 < x < 1$, then $\log_b x < 0$.

64. $\log_b y + \log_{1/b} y = 0$

65. $\log_b y + \log_b\left(\dfrac{1}{y}\right) = 0$

66. $\log_b\left(\dfrac{1}{b}\right) + 1 = 0$

In Exercises 67–76 assume that $\log_{10} 4 \approx 0.6021$, $\log_{10} 7 \approx 0.8451$, and $\log_{10} 9 \approx 0.9542$. Use these values and the properties of logarithms to find approximate values of each quantity.

67. $\log_{10} 28$

68. $\log_{10}\left(\dfrac{7}{4}\right)$

69. $\log_{10} 2.25$

70. $\log_{10} 36$

71. $\log_{10}\left(\dfrac{63}{4}\right)$

72. $\log_{10}\left(\dfrac{4}{63}\right)$

73. $\log_{10} 3$

74. $\log_{10} 5$

75. $\log_{10} 3.6$

76. $\log_{10} 2.5$

In Exercises 77–82 use a calculator to find the value of the variable. Express all answers to four decimal places.

77. $\log 3.25 = x$

78. $\ln 3.25 = x$

79. $\ln y = 4.24$

80. $\log y = 3.87$

81. $\log M = \ln 5$

82. $\ln M = \log 5$

83. One mole of gas expands isothermally to double its volume. If the gas temperature is 273 K, what energy is absorbed?

84. One thousand joules is absorbed by one mole of gas at 400 K. If the temperature does not change, by what factor does the volume increase?

85. The hydrogen-ion concentration of sour pickles is 6.31×10^{-4}. What is their pH?

86. The pH of gastric juice is approximately 5. What is its hydrogen-ion concentration?

87. The pH of apples can range from 2.9 to 3.3. What is the range in the hydrogen-ion concentration?

88. Sodium alizarin sulfate changes color from yellow to violet as the pH of a solution changes from 4.5 to 6.5. For what range of hydrogen-ion concentrations does this indicator test?

89. The coaxial power cable of Illustration 1 has a central wire with radius $R_1 = 0.25$ centimeter. It is insulated from a surrounding shield with inside radius $R_2 = 2$ centimeters. The maximum voltage the cable can withstand is the **breakdown voltage** V of the insulation. V is given by the formula

$$V = ER_1 \ln \frac{R_2}{R_1}$$

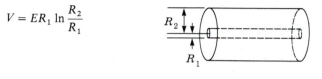

Illustration 1

where E is the **dielectric strength** of the insulation. If $E = 400{,}000$ volts/centimeter, calculate V.

90. If the inside diameter of the shield in Exercise 89 were doubled, what voltage could the cable withstand?

91. In electrical engineering, the voltage gain of an amplifier is specified in **decibels**, given by

$$\text{decibel voltage gain} = 20 \log \frac{E_O}{E_I}$$

where E_O is the output voltage of the amplifier and E_I is the input voltage. If an output of 40 volts results for an input voltage of 0.5 volt, what is the amplifier's decibel voltage gain?

92. The intensity of an earthquake is measured on the **Richter scale**. The intensity R is given by

$$R = \log \frac{A}{P}$$

where A is the amplitude of the tremor (measured in micrometers) and P is the period of the tremor (the time of one oscillation of the earth's surface, measured in seconds). What does an earthquake measure on the Richter scale if the amplitude is 10,000 micrometers (1 centimeter) and the period is 0.1 second?

93. Prove: If $\log_a b = c$, then $\log_b a = \frac{1}{c}$.

94. Prove: $\log_{1/b} x = -\log_b x$.

In Exercises 95–98 assume that f is a function with the property that $f(xy) = f(x) + f(y)$ for all real numbers x and y.

95. Prove that $f(1) = 0$.

96. Prove that $f(x^2) = 2f(x)$.

97. Prove that $f\left(\frac{1}{x}\right) = -f(x)$.

98. Prove that $f\left(\frac{x}{y}\right) = f(x) - f(y)$.

99. Prove Property 6 of logarithms.

100. Prove Property 7 of logarithms.

4.3 EXPONENTIAL AND LOGARITHMIC EQUATIONS

An **exponential equation** is one that contains a variable in an exponent. A **logarithmic equation** is one that involves logarithms of expressions that contain the variable.

Example 1   Solve the exponential equation $3^x = 5$.

Solution   Because the logarithms of equal numbers are equal, you can take the common logarithm of both sides of the equation. Property 7 of logarithms then provides a means for moving the variable x from its position as an exponent to a position as a factor:

$$3^x = 5$$

$$\log 3^x = \log 5 \qquad \text{Take the common logarithm of both sides.}$$

$$x \log 3 = \log 5 \qquad \text{Use Property 7 of logarithms.}$$

$$x = \frac{\log 5}{\log 3} \qquad \text{Divide both sides by log 3.}$$

$$\approx 1.465 \qquad \text{Use a calculator.}$$

Thus, $x \approx 1.465$.   ■

Example 2   Solve the exponential equation $6^{x-3} = 2^x$.

Solution

$$6^{x-3} = 2^x$$

$$\log 6^{x-3} = \log 2^x \qquad \text{Take the common logarithm of both sides.}$$

$$(x-3) \log 6 = x \log 2 \qquad \text{Use Property 7 of logarithms.}$$

$$x \log 6 - 3 \log 6 = x \log 2 \qquad \text{Remove parentheses.}$$

$$x \log 6 - x \log 2 = 3 \log 6 \qquad \text{Add } 3 \log 6 - x \log 2 \text{ to both sides.}$$

$$x(\log 6 - \log 2) = 3 \log 6 \qquad \text{Factor out } x \text{ from the left side.}$$

$$x = \frac{3 \log 6}{\log 6 - \log 2} \qquad \text{Divide both sides by log 6 } - \text{ log 2.}$$

$$x \approx 4.893 \qquad \text{Use a calculator.}$$ ■

Example 3   The **hyperbolic sine** of x, denoted as $\sinh x$, is defined by the equation

$$\sinh x = \frac{e^x - e^{-x}}{2}$$

Solve the equation $\sinh x = 2$ for x.

Solution   Use the definition of the hyperbolic sine to write the equation $\sinh x = 2$ in the form

$$\frac{e^x - e^{-x}}{2} = 2$$

Multiplying both sides by $2e^x$ transforms this equation into a quadratic equation in the variable e^x:

$$2e^x \left(\frac{e^x - e^{-x}}{2} \right) = 2(2e^x)$$

$$e^x e^x - e^x e^{-x} = 4e^x$$

$$e^{2x} - 1 = 4e^x$$

$$(e^x)^2 - 1 = 4e^x \qquad\qquad a^{mn} = (a^m)^n$$

$$(e^x)^2 - 4e^x - 1 = 0$$

This is a quadratic equation of the form $az^2 + bz + c = 0$, with $z = e^x$, $a = 1$, $b = -4$, and $c = -1$. Use the quadratic formula to find its solutions.

$$z = \frac{-b \pm \sqrt{b^2 - 4ac}}{2a}$$

$$e^x = \frac{-(-4) \pm \sqrt{(-4)^2 - 4(1)(-1)}}{2(1)}$$

$$= \frac{4 \pm \sqrt{16 + 4}}{2}$$

$$= \frac{4 \pm 2\sqrt{5}}{2}$$

$$e^x = 2 + \sqrt{5} \qquad \text{or} \qquad e^x = 2 - \sqrt{5}$$

$$\approx 4.236 \qquad\qquad\qquad \approx -0.236$$

To solve the first exponential equation, $e^x \approx 4.236$, first convert it to logarithmic form, and then use a calculator:

$$x \approx \ln 4.236$$

$$x \approx 1.444$$

Because the range of the exponential function is the set of positive real numbers, e^x cannot be negative. Thus, the second exponential equation, $e^x \approx -0.236$, has no solution.

Thus, $x \approx 1.444$ is the only solution of the equation $\sinh x = 2$. ■

Example 4   Solve the logarithmic equation $\log x + \log(x - 3) = 1$.

Solution

$$\log x + \log(x - 3) = 1$$

$$\log x(x - 3) = 1 \qquad \text{Use Property 5 of logarithms.}$$

$$x(x - 3) = 10^1 \qquad \text{Use the definition of logarithm to change equation to exponential form.}$$

$$x^2 - 3x - 10 = 0 \qquad \text{Remove parentheses and add } -10 \text{ to both sides.}$$

$$(x + 2)(x - 5) = 0 \qquad \text{Factor } x^2 - 3x - 10.$$

$$x + 2 = 0 \qquad \text{or} \qquad x - 5 = 0$$

$$x = -2 \qquad\qquad\qquad x = 5$$

Check: Because a negative number does not have a logarithm, the number -2 is not a solution. Check the remaining number, 5:

$$\log x + \log(x - 3) = 1$$

$$\log 5 + \log(5 - 3) \overset{?}{=} 1 \qquad \text{Substitute 5 for } x.$$

$$\log 5 + \log 2 \overset{?}{=} 1$$

$$\log (5 \cdot 2) \overset{?}{=} 1 \qquad \text{Use Property 3 of logarithms.}$$

$$\log 10 \overset{?}{=} 1$$

$$1 = 1 \qquad \text{Use Property 2 of logarithms.}$$

Thus, 5 is a solution. ■

Example 5   Solve the logarithmic equation $\log_b(3x + 2) - \log_b(2x - 3) = 0$.

Solution $\log_b(3x + 2) - \log_b(2x - 3) = 0$

$\log_b(3x + 2) = \log_b(2x - 3)$

$3x + 2 = 2x - 3$ Use Property 8 of logarithms.

$x = -5$ Add $-2x - 2$ to both sides.

Check:

$$\log_b(3x + 2) - \log_b(2x - 3) = 0$$

$$\log_b[3(-5) + 2] - \log_b[2(-5) - 3] \stackrel{?}{=} 0$$

$$\log_b(-13) - \log_b(-13) \stackrel{?}{=} 0$$

Because the logarithm of a negative number does not exist, the number -5 is not a solution. The given equation has no solutions.   ∎

If we know the base-a logarithm of a number, we can find the logarithm of that number to some other base, b, by using the **change-of-base formula**.

Change-of-Base Formula.

$$\log_b x = \frac{\log_a x}{\log_a b}$$

Proof   We begin with the equation $\log_b x = y$ and proceed as follows.

1. $\log_b x = y$

$b^y = x$ Change the equation from logarithmic to exponential form.

$\log_a b^y = \log_a x$ Take the base-a logarithm of both sides.

$y \log_a b = \log_a x$ Use Property 7 of logarithms.

$y = \dfrac{\log_a x}{\log_a b}$ Divide both sides by $\log_a b$.

$\log_b x = \dfrac{\log_a x}{\log_a b}$ Refer to Equation 1 and substitute $\log_b x$ for y.   □

If we know logarithms to base a (for example, $a = 10$), we can find the logarithm of x to a new base, b. To do so, we divide the base-a logarithm of x by the base-a logarithm of the new base b.

Example 6   Find $\log_3 5$ by using the change-of-base formula.

Solution   Use the change-of-base formula with $b = 3$, $a = 10$, and $x = 5$.

$$\log_b x = \frac{\log_a x}{\log_a b}$$

$$\log_3 5 = \frac{\log_{10} 5}{\log_{10} 3}$$

$$\approx 1.465   \text{Use a calculator.}$$ ∎

Applications of Exponential and Logarithmic Equations

Example 7   When a living organism dies, the oxygen/carbon dioxide cycle common to all living things ceases, and carbon-14—a radioactive isotope with a half-life of 5700 years—is no longer absorbed. By using a process called carbon dating, archeologists can answer such questions as: How old is a wooden statue that contains only $\frac{1}{3}$ of the carbon-14 it had originally?

Solution   The amount A of radioactive material present at time t is given by the model

$$A = A_0 2^{-t/h}$$

where A_0 is the amount present initially, and h is the material's half-life.

To determine the time t when A is one-third of A_0, substitute $\dfrac{A_0}{3}$ for A and **5700** for h. Then solve for t:

$$A = A_0 2^{-t/h}$$

$$\frac{A_0}{3} = A_0 2^{-t/5700} \qquad \text{Substitute } \frac{A_0}{3} \text{ for } A \text{ and 5700 for } h.$$

$$1 = 3 \cdot 2^{-t/5700} \qquad \text{Multiply both sides by } \frac{3}{A_0}.$$

$$\log 1 = \log (3 \cdot 2^{-t/5700}) \qquad \text{Take the common logarithm of both sides.}$$

$$0 = \log 3 + \log 2^{-t/5700} \qquad \text{Use Properties 1 and 5 of logarithms.}$$

$$-\log 3 = -\frac{t}{5700} \log 2 \qquad \begin{array}{l}\text{Add } -\log 3 \text{ to both sides and}\\ \text{use Property 7 of logarithms.}\end{array}$$

$$t = 5700 \frac{\log 3}{\log 2} \qquad \text{Multiply both sides by } -\frac{5700}{\log 2}.$$

$$\approx 9034.29$$

The wooden statue is approximately 9000 years old.   ■

Example 8   When there is sufficient food supply and space, populations of living organisms tend to increase exponentially according to the Malthusian model

Formula for Population Growth

$$P = P_0 e^{kt}$$

where P_0 is the initial population (at $t = 0$), and k depends on the rate of growth.

The bacterial population in a laboratory culture medium increased from an initial population of 500 to 1500 in 3 hours. By the Malthusian model, when will the population reach 10,000?

Solution   First determine k as follows:

$$P = P_0 e^{kt}$$

$$1500 = 500 e^{k \cdot 3} \qquad \text{Substitute 1500 for } P, 500 \text{ for } P_0, \text{ and 3 for } t.$$

$$3 = e^{3k}$$

$$3k = \ln 3 \qquad \text{Change the equation from exponential to logarithmic form.}$$

$$k = \frac{\ln 3}{3}$$

To find when the population will reach 10,000 substitute **10,000** for P, **500** for P_0, and $\dfrac{\ln 3}{3}$ for k in the equation $P = P_0 e^{kt}$ and solve for t:

$$P = P_0 e^{kt}$$

$$10{,}000 = 500 e^{(\ln 3/3)t}$$

$$20 = e^{(\ln 3/3)t}$$

$$\frac{\ln 3}{3} t = \ln 20 \qquad\qquad \text{Change the equation from exponential to logarithmic form.}$$

$$t = \frac{3 \ln 20}{\ln 3}$$

$$\approx 8.18$$

The culture will reach the 10,000 mark in approximately 8 hours.   ∎

Example 9   In electrical engineering, common logarithms are used to express the voltage gain (or loss) of an electronic device such as an amplifier or a length of transmission line. The unit of gain (or loss), called the **decibel**, is defined by the equation

Formula for Decibel Gain   **decibel voltage gain** $= 20 \log \dfrac{E_o}{E_I}$

where E_O is the output voltage of the device and E_I is the input voltage.

If the output of an amplifier is 40 volts and its decibel voltage gain is 38, what is the amplifier's input voltage?

Solution   Substitute **38** for the decibel voltage gain and **40** for E_O into the formula, and solve the resulting logarithmic equation for E_I:

$$\text{decibel voltage gain} = 20 \log\left(\frac{E_o}{E_I}\right)$$

$$38 = 20 \log\left(\frac{40}{E_I}\right)$$

$$\frac{38}{20} = \log\left(\frac{40}{E_I}\right)$$

$$1.9 = \log\left(\frac{40}{E_I}\right)$$

$$\frac{40}{E_I} = 10^{1.9} \qquad\qquad \text{Change equation from exponential to logarithmic form.}$$

$$E_I = \frac{40}{10^{1.9}}$$

$$E_1 \approx 0.5 \qquad\qquad \text{Use a calculator.}$$

This input voltage is approximately 0.5 volt.   ∎

Example 10   In seismology, common logarithms are used to measure the intensity of earthquakes on the **Richter scale**. The intensity R is given by

Formula for Richter Scale

$$R = \log\left(\frac{A}{P}\right)$$

where A is the amplitude of the tremor (measured in micrometers) and P is the period of the tremor (the time of one oscillation of the earth's surface, measured in seconds). Given a constant period, by what factor must the amplitude of a tremor change to cause an increase of 1 on the Richter scale?

Solution   Let A be the amplitude of a tremor that registers R on the Richter scale. Given that P is a constant, you must find a constant k such that a tremor of amplitude kA registers $R + 1$ on the Richter scale. To do so, solve the logarithmic equation

$$\log\left(\frac{A}{P}\right) + 1 = \log\left(\frac{kA}{P}\right)$$

for k, as follows:

$$\log A - \log P + 1 = \log k + \log A - \log P \qquad \text{Use Properties 5 and 6 of logarithms.}$$
$$1 = \log k$$
$$k = 10$$

If the period of a tremor remains constant, but the amplitude increases by a factor of 10, the Richter scale value increases by 1. Note that this result is independent of the amplitude.   ■

Example 11   Experiments in physiology suggest that the relationship between loudness and intensity of sound is a logarithmic one known as the **Weber–Fechner law**: the apparent loudness L of a sound is proportional to the natural logarithm of its actual intensity I:

Formula for Weber–Fechner Law

$$L = k \ln I$$

What actual increase in intensity will cause a doubling of the apparent loudness of a sound?

Solution   If the original loudness is L_0, caused by an actual intensity of I_0, then $L_0 = k \ln I_0$. To double the apparent loudness, multiply both sides of the equation by 2 and use Property 7 of logarithms:

$$2L_0 = 2k \ln I_0$$
$$= k \ln(I_0)^2$$

Thus, to double the apparent volume of a sound, the actual intensity must be squared.   ■

Exercise 4.3

In Exercises 1–18 solve each exponential equation.

1. $4^x = 5$
2. $7^x = 12$
3. $13^{x-1} = 2$
4. $5^{x+1} = 3$
5. $2^{x+1} = 3^x$
6. $5^{x-3} = 3^{2x}$
7. $2^x = 3^x$
8. $3^{2x} = 4^x$

9. $7^{x^2} = 10$ **10.** $8^{x^2} = 11$ **11.** $8^{x^2} = 9^x$ **12.** $5^{x^2} = 2^{5x}$

13. $3^{2x} - 3^x - 2 = 0$ **14.** $2(5^{2x}) - 5(5^x) + 2 = 0$ **15.** $2^{2x} - 1 = 0$

16. $e^{4x} - e^{2x} = 0$ **17.** $e^x + e^{-x} = 2$ **18.** $e^x - e^{-x} = 5$

In Exercises 19–36 solve each logarithmic equation.

19. $\log(2x - 3) = \log(x + 4)$ **20.** $\log(3x + 5) - \log(2x + 6) = 0$

21. $\log\left(\dfrac{4x + 1}{2x + 9}\right) = 0$ **22.** $\log\left(\dfrac{5x + 2}{2(x + 7)}\right) = 0$

23. $\log x^2 = 2$ **24.** $\log x^3 = 3$

25. $\log x + \log(x - 48) = 2$ **26.** $\log x - \log(x + 9) = 1$

27. $\log x + \log(x - 15) = 2$ **28.** $\log x + \log(x + 21) = 2$

29. $\log(x + 90) = 3 - \log x$ **30.** $\log(x - 6) - \log(x - 2) = \log\left(\dfrac{5}{x}\right)$

31. $\log(x - 1) - \log 6 = \log(x - 2) - \log x$ **32.** $\log(2x - 3) - \log(x - 1) = 0$

33. $\log_{10} x^2 = (\log_{10} x)^2$ **34.** $\log_{10}(\log_{10} x) = 1$

35. $\log_3 x = \log_3\left(\dfrac{1}{x}\right) + 4$ **36.** $\log_5(7 + x) + \log_5(8 - x) - \log_5 2 = 2$

In Exercises 37–40 find the logarithm with the indicated base.

37. $\log_3 7$ **38.** $\log_7 3$ **39.** $\log_{\sqrt{2}} \sqrt{5}$ **40.** $\log_\pi e$

41. A parchment fragment found in an ancient tomb contains 60% of the carbon-14 that it is assumed to have had initially. Approximately how old is the fragment?

42. Only 10% of the carbon-14 in a small wooden bowl remains. How old is the bowl?

43. The half-life of tritium is 12.4 years. How long will it take for 25% of a sample to decompose?

44. An isotope of lead, lead-201, has a half-life of 8.4 hours. How many hours ago was there 30% more of the isotope?

45. An isotope of thorium, thorium-227, has a half-life of 18.4 days. How long will it take 80% of a sample to decompose?

46. In 2 years 20% of a newly discovered radioactive material decays. What is its half-life?

47. A bacteria culture grows according to the formula

$$P = P_0 a^t$$

If it takes 5 days for the culture to triple in size, how long does it take to double in size?

48. The population of Willard County is now 140,000 and is expected to double in 30 years. Assuming Malthusian growth, when might the population be expected to reach 200,000?

49. The rodent population in a small farm community is 10% less this year than last. Assuming an exponential decrease in population, when might health officials expect the population to be one-half of last year's?

50. If an exponentially growing population will increase its present size by 50% in 5 years, when was it one-half of its current size?

51. The decibel voltage gain of an amplifier is 29. If the output is 20 volts, what is the input voltage?

52. The decibel voltage gain of an amplifier is 35. If the input signal is 0.05 volt, what is the output voltage?

53. An earthquake has an amplitude of 5000 micrometers and a period of 0.2 second. What does it measure on the Richter scale?

54. An earthquake with an amplitude of 8000 micrometers measures 6 on the Richter scale. What is its period?

55. An earthquake with a period of 0.25 second measures 4 on the Richter scale. What is its amplitude?

56. If the intensity of a sound is doubled, what is the apparent change in the loudness?

57. What increase in the intensity of a sound will cause an apparent tripling of the loudness?

58. If the intensity of a sound is tripled, what is the apparent change in the loudness?

In Exercises 59–62 recall that an initial amount of A_0 dollars left on deposit t years at an annual interest rate r compounded k times a year will grow to A dollars, where

$$A = A_0\left(1 + \frac{r}{k}\right)^{kt}$$

59. If $500 is deposited into an account paying 12% interest compounded semiannually, how long will it take the account to increase to $800?

60. If $1300 is deposited into an account paying 14% interest compounded quarterly, how long will it take to increase the amount to $2100?

61. A sum of $5000 deposited in an account grows to $7000 in 5 years. Assuming annual compounding, what interest rate is paid?

62. What annual interest rate, compounded annually, will provide the same growth to a single initial deposit that 13%, compounded quarterly, will provide?

63. When a battery is being charged, its charge C at time t is modeled by the formula

$$C = M(1 - e^{-kt})$$

where M is the theoretical maximum charge, and k is a constant that depends on the charger and the battery. Show that the time t required to charge a battery to a given level C is

$$t = -\frac{1}{k}\ln\left(1 - \frac{C}{M}\right)$$

64. A quick rule-of-thumb for determining how long it takes an investment to double is known as the **rule of seventy**: Divide 70 by the rate (as a percent). At 5%, for example, it requires $\frac{70}{5}$, or 14, years to double the capital. At 10%, it takes $\frac{70}{10}$, or 7, years. Why does this formula work?

65. The intensity I of light a distance x meters beneath the surface of a lake decreases exponentially. If the light intensity at 6 meters is 70% of the intensity at the surface, at what depth will the intensity be 20%?

66. The power output (or input) of an amplifier is directly proportional to the square of the voltage output (or input). Show that the formula for decibel voltage gain is

$$\text{decibel voltage gain} = 10 \log\left(\frac{P_O}{P_I}\right)$$

where P_O is the power output and P_I is the power input.

In Exercises 67–70 solve for x in terms of y.

67. $y = \dfrac{e^x + e^{-x}}{2}$

68. $y = \dfrac{e^x - e^{-x}}{2}$

69. $y = \dfrac{e^x + e^{-x}}{e^x - e^{-x}}$

70. $y = \dfrac{e^x - e^{-x}}{e^x + e^{-x}}$

REVIEW EXERCISES

In Review Exercises 1–4 graph the function defined by each equation.

1. $y = \left(\dfrac{6}{5}\right)^x$ **2.** $y = \left(\dfrac{3}{4}\right)^x$ **3.** $y = \log x$ **4.** $y = \ln x$

In Review Exercises 5–8 graph each pair of equations on one set of coordinate axes.

5. $y = \left(\dfrac{1}{3}\right)^x$ and $y = \log_{1/3} x$ **6.** $y = \left(\dfrac{2}{5}\right)^x$ and $y = \log_{2/5} x$

7. $y = 4^x$ and $y = \log_4 x$ **8.** $y = 3^x$ and $y = \log_3 x$

In Review Exercises 9–30 solve each equation for x.

9. $\log_2 x = 3$ **10.** $\log_3 x = -2$ **11.** $\log_x 9 = 2$ **12.** $\log_x 0.125 = -3$

13. $\log_7 7 = x$ **14.** $\log_3 \sqrt{3} = x$ **15.** $\log_8 \sqrt{2} = x$ **16.** $\log_6 36 = x$

17. $\log_{1/3} 9 = x$ **18.** $\log_{1/2} 1 = x$ **19.** $\log_x 3 = \dfrac{1}{3}$ **20.** $\log_x 25 = -2$

21. $\log_2 x = 5$ **22.** $\log_{\sqrt{3}} x = 4$ **23.** $\log_{\sqrt{3}} x = 6$ **24.** $\log_{0.1} 10 = x$

25. $\log_x 2 = -\dfrac{1}{3}$ **26.** $\log_x 32 = 5$ **27.** $\log_{0.25} x = -1$ **28.** $\log_{0.125} x = -\dfrac{1}{3}$

29. $\log_{\sqrt{2}} 32 = x$ **30.** $\log_{\sqrt{5}} x = -4$

In Review Exercises 31–34 use a calculator to find the value of the variable, if possible. Express each answer to four decimal places.

31. $\log 735.4 = x$ **32.** $\log(-0.002345) = y$ **33.** $\ln\left(\dfrac{2}{15}\right) = z$ **34.** $\ln M = 5.345$

In Review Exercises 35–44 solve for x, if possible.

35. $3^x = 7$ **36.** $1.2 = (3.4)^{5.6x}$

37. $2^x = 3^{x-1}$ **38.** $\log x + \log(29 - x) = 2$

39. $\log_2 x + \log_2(x - 2) = 3$ **40.** $\log_2(x + 2) + \log_2(x - 1) = 2$

41. $e^{x \ln 2} = 9$ **42.** $\ln x = \ln(x - 1)$

43. $\ln x = \ln(x - 1) + 1$

44. $\ln x = \log_{10} x$ (*Hint:* Use the change-of-base formula.)

45. A wooden statue excavated from the sands of Egypt has a carbon-14 content that is $\frac{1}{3}$ of that found in living wood. The half-life of carbon-14 is 5700 years. How old is the statue?

46. The pH of grapefruit juice is approximately 3.1. What is its hydrogen-ion concentration?

47. Some chemistry texts define the pH of a solution as the common logarithm of the reciprocal of the hydrogen-ion concentration:

$$pH = \log_{10} \dfrac{1}{[H^+]}$$

Show that this definition is equivalent to the one given in the text.

48. What is the half-life of a radioactive material if $\frac{1}{3}$ of it decays in 20 years?

5

TRIGONOMETRIC FUNCTIONS

Many events in nature tend to repeat themselves in a predictable way. For example, average noon temperatures in Chicago go up and down depending on the seasons. To model many such events, called **periodic events**, we need six new functions called the **trigonometric functions**.

5.1 THE TRIGONOMETRIC FUNCTIONS

In geometry, an angle is defined as a figure formed by two rays originating from a common point, called the angle's **vertex**. A common measure of the size of an angle is the **degree**, denoted by the symbol °. An angle has a measure of 1° if it represents $\frac{1}{360}$ of one complete rotation.

We shall usually indicate fractional parts of a degree by using decimal fractions. For example, $\frac{1}{2}° = 0.5°$, and $33\frac{3}{4}° = 33.75°$. Another method for expressing parts of a degree makes use of **minutes**, denoted by ′, and **seconds**, denoted by ″, where

$$1' = \frac{1}{60}(1°) \qquad \text{and} \qquad 1'' = \frac{1}{60}(1') = \frac{1}{3600}(1°)$$

The angle 37°24′18″, for example, is read as "37 degrees, 24 minutes, and 18 seconds."

Example 1 Change 37°24′18″ to decimal degrees.

Solution Because $24' = \frac{24}{60}(1°)$, it follows that $24' = 0.4°$. Because $18'' = \frac{18}{3600}(1°)$, it follows that $18'' = 0.005°$. Thus,

$$37°24'18'' = 37° + 0.4° + 0.005°$$
$$= 37.405° \qquad ■$$

Example 2 Change 83.41° to degrees–minutes–seconds.

Solution   You must change 0.41° to minutes and seconds. Because 1° is 60′, and 1′ is 60″, you have

$$0.41° = (0.41)60′$$
$$= 24.6′$$
$$= 24′ + (0.6)60″$$
$$= 24′ + 36″$$

Thus, 83.41° = 83°24′36″.   ■

In trigonometry, we define an angle differently.

> **Definition.**   Consider two rays with a common initial point O. A **trigono-metric angle** is the rotation required to move one ray, called the **initial side**, into coincidence with the other, called the **terminal side**.
>   Angles that are rotations in a counterclockwise direction are considered to be positive; those in a clockwise direction are considered to be negative.

To draw a trigonometric angle, we must show a curved arrow, drawn from the angle's initial side to its terminal side, to indicate the direction of rotation. Trigonometric angles of 40°, 135°, and −135° are shown in Figure 5–1.

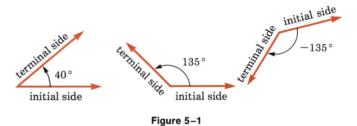

Figure 5–1

> **Definition.**   A trigonometric angle is in **standard position** if it is drawn on a rectangular coordinate system with its vertex at the origin and its initial side on the positive x-axis.
>   An angle is called a **first-**, **second-**, **third-**, or **fourth-quadrant angle** de-pending on whether its terminal side lies in the first, second, third, or fourth quadrant. If the terminal side lies on the x- or y-axis, the angle is called a **quadrantal angle**.

Trigonometric angles are often denoted by Greek letters. Figure 5–2 illus-trates angles θ (theta), ϕ (phi), α (alpha), and β (beta). Angles θ and ϕ are in standard position because each one has its vertex at the origin and its initial

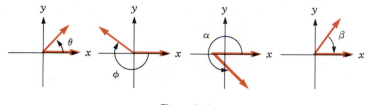

Figure 5–2

side on the positive *x*-axis. Angle θ is a first-quadrant angle, and angle ϕ is a second-quadrant angle. Angles α and β are not in standard position; the vertex of angle α is not the origin, and the initial side of angle β is not on the positive *x*-axis.

Definition.   If the terminal sides of two trigonometric angles in standard position coincide, the angles are called **coterminal angles**.

Example 3   Find two positive angles and two negative angles that are coterminal with an angle in standard position that measures 100°.

Solution   A trigonometric angle in standard position that measures 100° is shown in Figure 5–3. Two positive angles that are coterminal with it are shown in Figure 5–4. They are

$$100° + 360° \quad \text{or} \quad 460° \qquad (100° \text{ plus one revolution})$$
$$100° + 2(360°) \quad \text{or} \quad 820° \qquad (100° \text{ plus two revolutions})$$

Two negative angles that are coterminal with a 100° angle are also shown in Figure 5–4. They are

$$100° - 360° \quad \text{or} \quad -260° \qquad (100° \text{ minus one revolution})$$
$$100° - 2(360°) \quad \text{or} \quad -620° \qquad (100° \text{ minus two revolutions})$$

As this example suggests, if *n* is a positive integer, then any angle of $100° \pm n360°$ is coterminal with an angle of 100°.

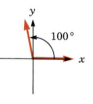

Figure 5–3

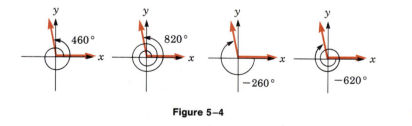

Figure 5–4

Most of the work in trigonometry is based on the following trigonometric functions.

Hipparchus (2nd century B.C.) Hipparchus is given credit for inventing trigonometry. He defined the trigonometric functions as ratios of a chord of a circle to the radius of the circle.

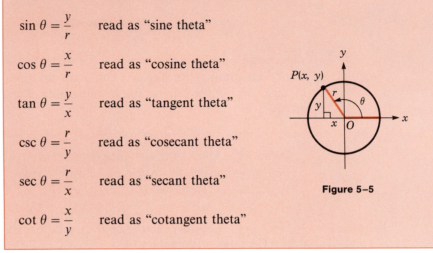

Definition. Let θ be a trigonometric angle in standard position, as shown in Figure 5–5. Let $P(x, y)$ be any point (except the origin) on the terminal side of angle θ. Then the length of segment OP is $r = \sqrt{x^2 + y^2}$. The **six trigonometric functions of angle θ** are:

$\sin \theta = \dfrac{y}{r}$ read as "sine theta"

$\cos \theta = \dfrac{x}{r}$ read as "cosine theta"

$\tan \theta = \dfrac{y}{x}$ read as "tangent theta"

$\csc \theta = \dfrac{r}{y}$ read as "cosecant theta"

$\sec \theta = \dfrac{r}{x}$ read as "secant theta"

$\cot \theta = \dfrac{x}{y}$ read as "cotangent theta"

Figure 5–5

Remember that division by 0 is not permitted. Thus, if $x = 0$, then $\tan \theta$ and $\sec \theta$ are undefined. If $y = 0$, then $\csc \theta$ and $\cot \theta$ are undefined.

The previous definition implies that the values of trigonometric functions of coterminal angles are equal.

Example 4   Suppose that the point with coordinates $(-3, 4)$ is on the terminal side of angle θ, which is in standard position. Calculate values for the six trigonometric functions of θ.

Solution   Refer to Figure 5–6 and use the distance formula to calculate r.

$$r = \sqrt{(-3)^2 + 4^2} = \sqrt{25} = 5$$

From the definitions of the trigonometric functions,

$\sin \theta = \dfrac{y}{r} = \dfrac{4}{5}$

$\cos \theta = \dfrac{x}{r} = \dfrac{-3}{5} = -\dfrac{3}{5}$

$\tan \theta = \dfrac{y}{x} = \dfrac{4}{-3} = -\dfrac{4}{3}$

$\csc \theta = \dfrac{r}{y} = \dfrac{5}{4}$

$\sec \theta = \dfrac{r}{x} = \dfrac{5}{-3} = -\dfrac{5}{3}$

$\cot \theta = \dfrac{x}{y} = \dfrac{-3}{4} = -\dfrac{3}{4}$

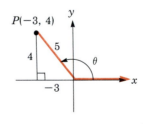

Figure 5–6

Any angle that is coterminal with angle θ will produce the same value for each of the trigonometric functions. ■

The following theorem enables us to use *any* point on the terminal side of angle θ to compute values for the six trigonometric functions of θ.

> **Theorem.** If angle θ is in standard position and if P is some point, other than the origin, on its terminal side, the values of the trigonometric functions of θ are independent of the choice of point P.

Proof   Let $P(x, y)$ and $P'(x', y')$ be two points on the terminal side of angle θ. Furthermore, let $d(OP) = r$ and $d(OP') = r'$. Then points P, O, and Q and points P', O, and Q' determine the similar triangles POQ and $P'OQ'$ shown in Figure 5–7. Because all ratios of corresponding sides of similar triangles are equal,

$$\sin \theta = \frac{y}{r} = \frac{y'}{r'}$$

and

$$\cos \theta = \frac{x}{r} = \frac{x'}{r'}$$

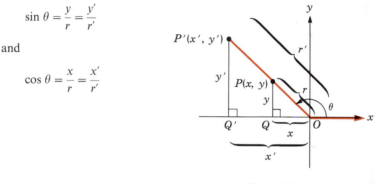

Figure 5–7

Thus, the values of $\sin \theta$ and $\cos \theta$ are independent of the choice of point P. It can be shown in a similar manner that the values of the remaining trigonometric functions are also independent of the choice of point P. □

Example 5   Suppose that $\sin \alpha = \frac{12}{13}$ and that α is in quadrant II (QII). Find $\cos \alpha$.

Solution   Because any point P on the terminal side of α can be used to compute $\cos \alpha$, choose a point with a y-coordinate of 12 and $r = 13$, for then $\sin \alpha = \frac{y}{r} = \frac{12}{13}$ (see Figure 5–8). Then, substitute **12** for y and **13** for r in the equation $x^2 + y^2 = r^2$ and solve for x.

$$x^2 + y^2 = r^2$$
$$x^2 + 12^2 = 13^2$$
$$x^2 = 169 - 144$$
$$x^2 = 25$$
$$x = \pm 5$$

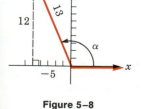

Figure 5–8

Because α is in QII, the x-coordinate of P must be negative, so $x = -5$. Thus,

$$\cos \alpha = \frac{x}{r} = \frac{-5}{13} = -\frac{5}{13}$$

■

Because r is always positive, the sine and cosecant functions are positive whenever y is positive. This occurs in QI and QII. The cosine and secant functions are positive whenever x is positive, and this occurs in QI and QIV. The tangent and cotangent functions are positive whenever x and y have the same sign, and this occurs in QI and QIII. Figure 5–9 will help you remember this information.

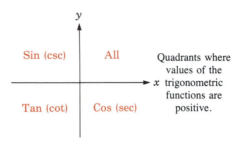

Figure 5–9

Example 6   If angle β is not in QII and $\cos \beta = -\frac{3}{4}$, find $\sin \beta$ and $\tan \beta$.

Solution   Because $\cos \beta = -\frac{3}{4}$, β must be in QII or QIII. Because β is not in QII, it must be in QIII (see Figure 5–10). Choose point $P(x, y)$ so that $x = -3$ and $r = 4$. Then,

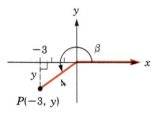

Figure 5–10

$$x^2 + y^2 = r^2$$
$$(-3)^2 + y^2 = 4^2$$
$$y^2 = 16 - 9$$
$$y = \pm\sqrt{7}$$

Choose y to be $-\sqrt{7}$ because y is negative in QIII. Thus,

$$\sin \beta = \frac{y}{r} = \frac{-\sqrt{7}}{4} = -\frac{\sqrt{7}}{4}$$

$$\tan \beta = \frac{y}{x} = \frac{-\sqrt{7}}{-3} = \frac{\sqrt{7}}{3}$$

The Fundamental Identities

The trigonometric functions can be grouped into reciprocal pairs. For example, $\cos \theta$ and $\sec \theta$ are reciprocals because

$$\frac{x}{r} \quad \text{and} \quad \frac{r}{x}$$

are reciprocals. Similarly, $\sin \theta$ and $\csc \theta$ are reciprocals and $\tan \theta$ and $\cot \theta$ are reciprocals. Thus,

$$\csc \theta = \frac{1}{\sin \theta} \qquad \sec \theta = \frac{1}{\cos \theta} \qquad \cot \theta = \frac{1}{\tan \theta}$$

Each of the previous equations is true for all values of θ for which each expression is defined. Recall that such equations are called **identities**. There are many other identities that involve the trigonometric functions. For example,

$$\frac{\sin \theta}{\cos \theta} = \frac{\dfrac{y}{r}}{\dfrac{x}{r}} = \frac{\left(\dfrac{y}{r}\right)r}{\left(\dfrac{x}{r}\right)r} = \frac{y}{x} = \tan \theta$$

Thus,

$$\frac{\sin \theta}{\cos \theta} = \tan \theta$$

In a similar way, we can show that

$$\frac{\cos \theta}{\sin \theta} = \cot \theta$$

The trigonometric functions involve the variables x, y, and r, which are related by the formula $y^2 + x^2 = r^2$. This formula enables us to establish three more important identities.

$$y^2 + x^2 = r^2$$

$$\frac{y^2}{r^2} + \frac{x^2}{r^2} = \frac{r^2}{r^2} \qquad \text{Divide both sides by } r^2.$$

$$\left(\frac{y}{r}\right)^2 + \left(\frac{x}{r}\right)^2 = 1$$

$$\sin^2 \theta + \cos^2 \theta = 1 \qquad \text{Use the definition of } \sin \theta \text{ and } \cos \theta.$$

Thus,

$$\sin^2 \theta + \cos^2 \theta = 1$$

By dividing both sides of $y^2 + x^2 = r^2$ by x^2, we can show that

$$\tan^2 \theta + 1 = \sec^2 \theta$$

and, by dividing by y^2, show that

$$\cot^2 \theta + 1 = \csc^2 \theta$$

The eight fundamental identities discussed so far are summarized as follows:

The Eight Fundamental Identities.   For any trigonometric angle θ for which the functions are defined,

$$\csc \theta = \frac{1}{\sin \theta} \qquad \tan \theta = \frac{\sin \theta}{\cos \theta} \qquad \sin^2 \theta + \cos^2 \theta = 1$$

$$\sec \theta = \frac{1}{\cos \theta} \qquad \cot \theta = \frac{\cos \theta}{\sin \theta} \qquad \tan^2 \theta + 1 = \sec^2 \theta$$

$$\cot \theta = \frac{1}{\tan \theta} \qquad\qquad\qquad\qquad \cot^2 \theta + 1 = \csc^2 \theta$$

Example 7   If $\cos \theta = \frac{3}{5}$, use the identity $\sin^2 \theta + \cos^2 \theta = 1$ to find $\sin \theta$.

Solution $\sin^2 \theta + \cos^2 \theta = 1$

$\sin^2 \theta + \left(\dfrac{3}{5}\right)^2 = 1$ Substitute $\frac{3}{5}$ for $\cos \theta$.

$\sin^2 \theta = 1 - \dfrac{9}{25}$

$\sin^2 \theta = \dfrac{16}{25}$

$\sin \theta = \pm\dfrac{4}{5}$

Thus, $\sin \theta = \frac{4}{5}$ or $\sin \theta = -\frac{4}{5}$.   ■

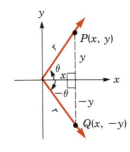

Figure 5–11

Suppose angles θ and $-\theta$ are drawn in standard position as in Figure 5–11. Further, suppose that points $P(x, y)$ and $Q(x, -y)$ lie on the terminal sides of their respective angles. Then, the values of the trigonometric functions of θ and $-\theta$ are

$$\begin{cases} \sin(-\theta) = \dfrac{-y}{r} \\[2ex] \sin \theta = \dfrac{y}{r} \end{cases} \quad \begin{cases} \cos(-\theta) = \dfrac{x}{r} \\[2ex] \cos \theta = \dfrac{x}{r} \end{cases} \quad \begin{cases} \tan(-\theta) = \dfrac{-y}{x} \\[2ex] \tan \theta = \dfrac{y}{x} \end{cases}$$

$$\begin{cases} \csc(-\theta) = \dfrac{r}{-y} \\[2ex] \csc \theta = \dfrac{r}{y} \end{cases} \quad \begin{cases} \sec(-\theta) = \dfrac{r}{x} \\[2ex] \sec \theta = \dfrac{r}{x} \end{cases} \quad \begin{cases} \cot(-\theta) = \dfrac{x}{-y} \\[2ex] \cot \theta = \dfrac{x}{y} \end{cases}$$

Because these results are true for angles in any quadrant, we have

---

Trigonometric Functions of $-\theta$.

$\sin(-\theta) = -\sin \theta \qquad \csc(-\theta) = -\csc \theta$

$\cos(-\theta) = \ \ \cos \theta \qquad \sec(-\theta) = \ \ \sec \theta$

$\tan(-\theta) = -\tan \theta \qquad \cot(-\theta) = -\cot \theta$

Recall that f is an even function if $f(-x) = f(x)$ for all x, and that f is an odd function if $f(-x) = -f(x)$ for all x. Thus, the cosine and secant functions are even functions and the sine, tangent, cosecant, and cotangent functions are odd functions.

Example 8   Show that the function determined by $f(\theta) = \sin \theta + \tan \theta$ is an odd function.

Solution The given function is an odd function if and only if $f(-\theta) = -f(\theta)$ for all θ. Show that this is true by evaluating $f(-\theta)$ and proceeding as follows:

$$f(\theta) = \sin\theta + \tan\theta$$
$$f(-\theta) = \sin(-\theta) + \tan(-\theta) \qquad \text{Substitute } -\theta \text{ for } \theta.$$
$$= -\sin\theta - \tan\theta \qquad \text{Use the rules for trigonometric functions of } -\theta.$$
$$= -(\sin\theta + \tan\theta) \qquad \text{Factor out } -1.$$
$$= -f(\theta)$$

Because $f(-\theta) = -f(\theta)$ for all θ, the given function is an odd function. ∎

Exercise 5.1

In Exercises 1–2 change each angle to decimal degrees. Give each answer to the nearest thousandth.

1. $84°36'24''$

2. $25°48'6''$

In Exercises 3–4 change each angle to degrees–minutes–seconds.

3. $23.14°$

4. $73.87°$

In Exercises 5–8 tell whether the angle is in standard position, and indicate whether the angle is positive or negative.

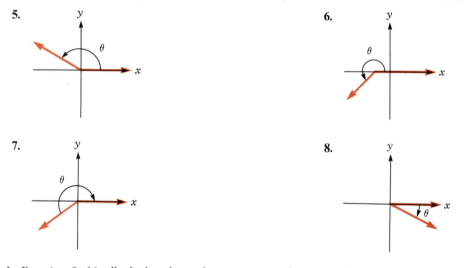

5.

6.

7.

8.

In Exercises 9–14 tell whether the angles are coterminal. Assume that all angles are in standard position.

9. $40°, 400°$

10. $90°, -270°$

11. $135°, 270°$

12. $135°, -135°$

13. $740°, 380°$

14. $-340°, -700°$

In Exercises 15–22 point P is on the terminal side of angle θ, which is to be in standard position. Draw a positive angle θ and calculate the sine, cosine, and tangent of θ, if possible.

15. $P(3, 4)$

16. $P(-5, -12)$

17. $P(-9, 40)$

18. $P(9, -40)$

19. $P(1, 1)$

20. $P(-3, 3)$

21. $P(-3, 4)$

22. $P(-1, -1)$

In Exercises 23–26 point P is on the terminal side of angle θ, which is to be in standard position. Draw a positive angle θ and calculate the cosecant, secant, and cotangent of θ, if possible.

23. $P(3, 5)$ **24.** $P(24, 10)$ **25.** $P(24, -10)$ **26.** $P(-3, -4)$

In Exercises 27–32 find the values of the remaining trigonometric functions of θ. Assume that θ is positive and in standard position.

27. $\sin \theta = \dfrac{3}{5}$; θ in QI   **28.** $\tan \theta = 1$; θ not in QI

29. $\cot \theta = \dfrac{5}{12}$; $\cos \theta = -\dfrac{5}{13}$ **30.** $\cos \theta = \dfrac{\sqrt{5}}{5}$; $\csc \theta = -\dfrac{\sqrt{5}}{5}$

31. $\sec \theta = -\dfrac{5}{3}$; $\csc \theta = \dfrac{5}{4}$ **32.** $\tan \theta = -\dfrac{40}{9}$; $\cos \theta = \dfrac{9}{41}$

In Exercises 33–36 use the letters x, y, and r and the definitions of the trigonometric functions to verify the following identities.

33. $\tan^2 \theta + 1 = \sec^2 \theta$ **34.** $\cot^2 \theta + 1 = \csc^2 \theta$

35. $\cot^2 \theta + \sin^2 \theta = \csc^2 \theta - \cos^2 \theta$ **36.** $\tan^2 \theta + \cos^2 \theta = \sec^2 \theta - \sin^2 \theta$

37. Let θ be a second-quadrant angle. Draw a figure similar to Figure 5–11 and show that $\sin(-\theta) = -\sin \theta$, $\cos(-\theta) = \cos \theta$, and $\tan(-\theta) = -\tan \theta$.

38. Let θ be a third-quadrant angle. Draw a figure similar to Figure 5–11 and show that $\sin(-\theta) = -\sin \theta$, $\cos(-\theta) = \cos \theta$, and $\tan(-\theta) = -\tan \theta$.

39. Let θ be a fourth-quadrant angle. Draw a figure similar to Figure 5–11 and show that $\sin(-\theta) = -\sin \theta$, $\cos(-\theta) = \cos \theta$, and $\tan(-\theta) = -\tan \theta$.

40. If θ is a second-quadrant angle, is $\sin(-\theta)$ positive or negative?

41. If θ is a third-quadrant angle, is $\cos(-\theta)$ positive or negative?

42. Identify the product $\sin(-\theta) \cos(-\theta) \tan(-\theta)$ as positive or negative if θ is a second-quadrant angle.

43. Identify the product $\sin(-\theta) \cos(-\theta) \tan(-\theta)$ as positive or negative if θ is a third-quadrant angle.

44. Identify the product $\sin(-\theta) \cos(-\theta) \tan(-\theta)$ as positive or negative if θ is a fourth-quadrant angle.

In Exercises 45–50 tell whether each function is odd or even. If it is neither, so indicate.

45. $f(\theta) = \sin \theta + \tan \theta$ **46.** $f(\theta) = \cot \theta + \sin \theta$ **47.** $f(\theta) = \sec \theta + \cos \theta$ **48.** $f(\theta) = \csc \theta + \sec \theta$
49. $f(\theta) = \sin \theta - \tan \theta$ **50.** $f(\theta) = \cos \theta - \sec \theta$

5.2   TRIGONOMETRIC FUNCTIONS OF ANGLES

It is easy to find the values of the trigonometric functions for certain angles. We begin by considering the quadrantal angles.

An angle of $0°$ placed in standard position has both its initial side and its terminal side on the positive x-axis. See Figure 5–12. The point $P(1, 0)$ lies on the terminal side, and r, which is $d(OP)$, is 1 unit. Four of the values of the six

trigonometric functions of 0° are

$$\sin 0° = \frac{y}{r} = \frac{0}{1} = 0$$

$$\cos 0° = \frac{x}{r} = \frac{1}{1} = 1$$

$$\tan 0° = \frac{y}{x} = \frac{0}{1} = 0$$

$$\sec 0° = \frac{r}{x} = \frac{1}{1} = 1$$

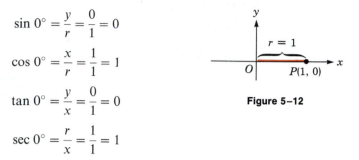

Figure 5–12

Johannes Müller
(Regiomontanus) (1436–
1476) Müller first
developed trigonometry
into a separate subject.

Values for cot 0° and csc 0° are undefined because division by 0 is undefined.

As another illustration, we consider an angle of 270° in standard position and the point $P(0, -1)$ on its terminal side. See Figure 5–13. Again, four functions have values and two do not.

$$\sin 270° = \frac{y}{r} = \frac{-1}{1} = -1$$

$$\cos 270° = \frac{x}{r} = \frac{0}{1} = 0$$

$$\cot 270° = \frac{x}{y} = \frac{0}{-1} = 0$$

$$\csc 270° = \frac{r}{y} = \frac{1}{-1} = -1$$

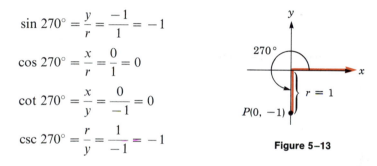

Figure 5–13

Values for tan 270° and sec 270° are undefined.

Example 1   Find sin 990°.

Solution   Because 990° is greater than 360°, find an angle that is less than 360° but coterminal with 990°. This angle can be found by repeatedly subtracting 360° from 990° until an angle less than 360° is found. Then, find the sine of that angle.

$$\sin 990° = \sin[990° - 2(360°)]$$
$$= \sin 270°$$
$$= -1$$

We now refer to the 30°–60° right triangle OPA shown in Figure 5–14 with a hypotenuse r with length of 2 units. Because the hypotenuse of a 30°–60° right triangle is twice as long as the side opposite the 30° angle, it follows that $d(PA)$ is 1 unit. By using the Pythagorean theorem, we can determine that

$d(OA)$ is $\sqrt{3}$ units. Thus,

$$\sin 30° = \frac{y}{r} = \frac{1}{2}$$

$$\cos 30° = \frac{x}{r} = \frac{\sqrt{3}}{2}$$

$$\tan 30° = \frac{y}{x} = \frac{1}{\sqrt{3}} = \frac{\sqrt{3}}{3}$$

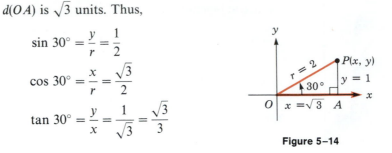

Figure 5–14

Because the values of the cosecant, secant, and cotangent functions are the reciprocals of the values of the sine, cosine, and tangent functions, we have

$$\csc 30° = 2 \qquad \sec 30° = \frac{2}{\sqrt{3}} = \frac{2\sqrt{3}}{3} \qquad \cot 30° = \sqrt{3}$$

A similar argument determines the values of the trigonometric functions of a 60° angle. See Figure 5–15.

$$\sin 60° = \frac{y}{r} = \frac{\sqrt{3}}{2}$$

$$\cos 60° = \frac{x}{r} = \frac{1}{2}$$

$$\tan 60° = \frac{y}{x} = \sqrt{3}$$

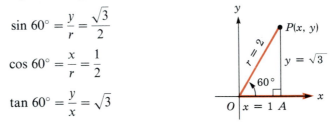

Figure 5–15

The values of the remaining three functions are the corresponding reciprocals.

$$\csc 60° = \frac{2\sqrt{3}}{3} \qquad \sec 60° = 2 \qquad \cot 60° = \frac{\sqrt{3}}{3}$$

We now refer to the isosceles right triangle OPA shown in Figure 5–16 with each of its equal sides of length 1 unit. By using the Pythagorean theorem, we can determine that the length of its hypotenuse r is $\sqrt{2}$ units. Thus,

$$\sin 45° = \frac{y}{r} = \frac{1}{\sqrt{2}} = \frac{\sqrt{2}}{2}$$

$$\cos 45° = \frac{x}{r} = \frac{1}{\sqrt{2}} = \frac{\sqrt{2}}{2}$$

$$\tan 45° = \frac{y}{x} = 1$$

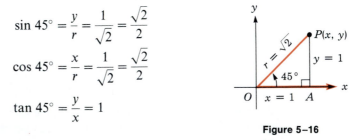

Figure 5–16

We can use the reciprocal relationships to determine values for the other trigonometric functions.

We summarize the previous results and others as follows:

θ	$\sin \theta$	$\cos \theta$	$\tan \theta$	$\csc \theta$	$\sec \theta$	$\cot \theta$
0°	0	1	0	undefined	1	undefined
30°	$\dfrac{1}{2}$	$\dfrac{\sqrt{3}}{2}$	$\dfrac{\sqrt{3}}{3}$	2	$\dfrac{2\sqrt{3}}{3}$	$\sqrt{3}$
45°	$\dfrac{\sqrt{2}}{2}$	$\dfrac{\sqrt{2}}{2}$	1	$\sqrt{2}$	$\sqrt{2}$	1
60°	$\dfrac{\sqrt{3}}{2}$	$\dfrac{1}{2}$	$\sqrt{3}$	$\dfrac{2\sqrt{3}}{3}$	2	$\dfrac{\sqrt{3}}{3}$
90°	1	0	undefined	1	undefined	0
180°	0	-1	0	undefined	-1	undefined
270°	-1	0	undefined	-1	undefined	0

We can now determine the values of the trigonometric functions of certain angles in any quadrant.

Example 2   Find **a.** sin 120°, **b.** cos 120°, and **c.** tan 120°.

Solution   Draw an angle of 120° in standard position and mark a point $P(x, y)$ on its terminal side at a distance 2 units from the origin. See Figure 5–17. Then draw segment PA perpendicular to the x-axis to form the 30°–60° triangle OPA, called the **reference triangle**. The lengths of the sides of triangle OPA are shown in the figure. Thus,

a. $\sin 120° = \dfrac{y}{r} = \dfrac{\sqrt{3}}{2}$

b. $\cos 120° = \dfrac{x}{r} = -\dfrac{1}{2}$

c. $\tan 120° = \dfrac{y}{x} = -\sqrt{3}$

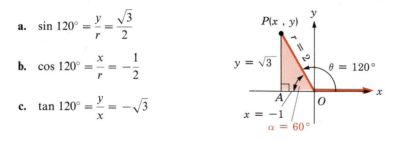

Figure 5–17

Note that angle α, called the **reference angle**, is formed by the terminal side of θ and the x-axis. Also note that the values of the trigonometric functions of 120° are equal to the values of the functions of the reference angle of 60°, except for sign.   ■

Example 3   Find **a.** csc 225°, **b.** sec 225°, and **c.** cot 225°.

Solution   Draw an angle of 225° in standard position and mark a point $P(x, y)$ on its terminal side at a distance $\sqrt{2}$ units from the origin (see Figure 5–18). Then draw segment PA perpendicular to the x-axis to form the reference triangle OPA, which is an isosceles right triangle. The lengths of the sides of triangle OPA are shown in the figure. Thus,

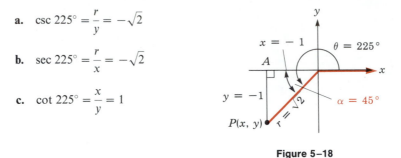

a. $\csc 225° = \dfrac{r}{y} = -\sqrt{2}$

b. $\sec 225° = \dfrac{r}{x} = -\sqrt{2}$

c. $\cot 225° = \dfrac{x}{y} = 1$

Figure 5–18

Note that the acute angle α between the terminal side of angle θ and the x-axis is 45°. Also note that the values of the functions of 225° are equal to the values of the functions of 45°, except for sign. ∎

The results of Examples 2 and 3 suggest the following fact.

> The trigonometric functions of any angle θ are equal to those of the reference angle of θ, except possibly for sign.

The appropriate sign to use for the value of a trigonometric function of θ can be determined by considering the quadrant in which the terminal side of θ lies.

Example 4   If $\theta = -405°$, find **a.** $\sin \theta$ and **b.** $\cos \theta$.

Solution   Draw an angle θ of $-405°$ in standard position, as in Figure 5–19. Angle θ is a QIV angle with a reference angle of 45°. In QIV, $\sin \theta$ will be negative, and $\cos \theta$ will be positive. Thus,

a. $\sin(-405°) = -\sin 45° = -\dfrac{\sqrt{2}}{2}$

b. $\cos(-405°) = \cos 45° = \dfrac{\sqrt{2}}{2}$

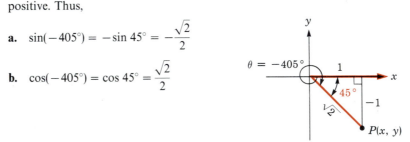

Figure 5–19   ∎

Example 5   If $\sin \theta = -\dfrac{1}{2}$ and $\cos \theta = \dfrac{\sqrt{3}}{2}$, find θ.

Solution   Because sin θ is negative and cos θ is positive, θ must be a QIV angle (see Figure 5–20). Because

$$\sin 30° = \frac{1}{2} \quad \text{and} \quad \cos 30° = \frac{\sqrt{3}}{2}$$

the reference angle of θ must be 30°. There are infinitely many QIV angles that have reference angles of 30°. Two examples are $-30°$ and 330°. All such angles are of the form

$$330° \pm n360°$$

where n is a nonnegative integer.

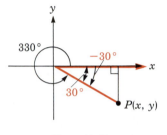

Figure 5–20

To find approximate values for the trigonometric functions of general angles, we use either tables or a calculator.

Using Tables

Example 6   Use Table A in Appendix II to find the sine, cosine, tangent, and cotangent of 43.5°.

Solution   A portion of Table A is presented below. In the **Degrees** column on the left side of the table, locate the number 43.5. Move to the right in that row and read the entries in the columns headed by **Sin**, **Cos**, **Tan**, and **Cot**.

Radians	Degrees	Sin	Cos	Tan	Cot		
.7575	43.4°	.6871	.7266	.9457	1.057	46.6°	.8133
.7592	43.5°	.6884	.7254	.9490	1.054	46.5°	.8116
.7610	43.6°	.6896	.7242	.9523	1.050	46.4°	.8098
		Cos	**Sin**	**Cot**	**Tan**	**Degrees**	**Radians**

$$\sin 43.5° \approx 0.6884 \qquad \cos 43.5° \approx 0.7254$$
$$\tan 43.5° \approx 0.9490 \qquad \cot 43.5° \approx 1.054$$

Example 7   Use Table A in Appendix II to find the sine, cosine, tangent, and cotangent of 46.5°.

Solution   The **Degrees** column on the left side of the table ends at 45°. However, 46.5° can be found on the right side in a column that is footed by **Degrees**. See the table in Example 6. Move to the left in the row containing 46.5° and read the entries in the columns footed by **Cos**, **Sin**, **Cot**, and **Tan**.

$$\sin 46.5° \approx 0.7254 \qquad \cos 46.5° \approx 0.6884$$
$$\tan 46.5° \approx 1.054 \qquad \cot 46.5° \approx 0.9490$$

∎

Note that the same row of the table was used in both Examples 6 and 7; it did double duty for the complementary angles 43.5° and 46.5°. (Remember that, if the sum of two acute angles is 90°, they are called **complementary angles**.) Note also that

$$\sin 43.5° = \cos 46.5° \qquad \text{and} \qquad \tan 43.5° = \cot 46.5°$$

are true statements. Because $\sin 43.5° = \cos 46.5°$, their reciprocals must be equal also. Therefore,

$$\csc 43.5° = \sec 46.5°$$

Definition. The trigonometric functions of sine and cosine are called **cofunctions**. The tangent and cotangent functions are cofunctions, as are the secant and cosecant functions.

It is always true that the sine of an acute angle θ is equal to the cosine of the complement of θ. In like manner, the tangent of an acute angle θ is equal to the cotangent of the complement of θ, and the secant of an acute angle θ is equal to the cosecant of the complement of θ.

Theorem. If θ is any acute angle, any trigonometric function of θ is equal to the cofunction of the complement of θ:

$$\sin \theta = \cos(90° - \theta) \qquad \text{and} \qquad \cos \theta = \sin(90° - \theta)$$
$$\tan \theta = \cot(90° - \theta) \qquad \text{and} \qquad \cot \theta = \tan(90° - \theta)$$
$$\csc \theta = \sec(90° - \theta) \qquad \text{and} \qquad \sec \theta = \csc(90° - \theta)$$

Later, you will see that these relationships can be extended to all values of θ.

Example 8   Use Table A in Appendix II to find the values of the sine, cosine, tangent, and cotangent of an angle of 107°.

Solution   Although an angle of 107° does not appear in the table, its reference angle, 73°, does. Remember that any trigonometric function of an angle in standard position can differ only in sign from that same trigonometric function of its reference angle. Because a 107°

angle is a QII angle, only the sine and cosecant functions are positive; the rest are negative:

$$\sin 107° = +\sin 73° \approx \quad 0.9563$$
$$\cos 107° = -\cos 73° \approx -0.2924$$
$$\tan 107° = -\tan 73° \approx -3.271$$
$$\cot 107° = -\cot 73° \approx -0.3057 \qquad \blacksquare$$

Using Calculators

Example 9   Use a calculator to find the values of **a.** sin 313.27°, **b.** cos(−28.2°), and **c.** tan 90°.

Solution   Set your calculator for degree measure of angles.

a.   To evaluate sin 313.27°, enter the number 313.27 and press the $\boxed{\text{SIN}}$ key. The display should read −0.7281317515. To the nearest ten-thousandth, sin 313.27° = −0.7281.

b.   To evaluate cos(−28.2°), enter the number 28.2. Press the $\boxed{+/-}$ and $\boxed{\text{COS}}$ keys. The display should read 0.8813034521. To the nearest ten-thousandth, cos(−28.2°) = 0.8813.

c.   Finally, to attempt to evaluate tan 90°, enter 90 and press the $\boxed{\text{TAN}}$ key. The display will either blink 9's at you or read ERROR. Either way, it is telling you that tan 90° is undefined. $\qquad \blacksquare$

Example 10   Find the value of sec 43°.

Solution   Neither tables nor calculators will allow you to evaluate the secant function directly. You must use the property that sec 43° is the reciprocal of cos 43°. To find sec 43° on a calculator, set your calculator for degrees, enter the number 43, and then press in order the $\boxed{\text{COS}}$ and $\boxed{1/x}$ keys. This gives the reciprocal of cos 43°, which is sec 43° ≈ 1.367327461. To the nearest ten-thousandth, sec 43° = 1.3673. $\qquad \blacksquare$

Example 11   If θ is an acute angle and cos θ = 0.7660, find angle θ.

Solution   Angle θ can be found by using either a calculator or Table A.
  If you use a calculator, be sure it is set for degrees. Enter the number .7660 and press the $\boxed{\text{INV}}$ and $\boxed{\text{COS}}$ keys. If your calculator does not have an $\boxed{\text{INV}}$ key, consult your owner's manual. To the nearest tenth, θ = 40.0°.
  If you use Table A, find the column headed by **Cos** at the top of the page. Run your finger down the column, moving to successive pages if necessary, until you find the number .7660. Move to the left in that row to find the value of 40.0° in the **Degree** column. $\qquad \blacksquare$

Example 12   If θ is between 180° and 270° and sin θ = −0.9397, find angle θ.

Solution   See Figure 5–21. Any trigonometric function of θ has the same value as that same trigonometric function of θ's reference angle α, except possibly for sign. Enter the number

.9397 in your calculator and press the ⟨INV⟩ and ⟨SIN⟩ keys. The display will give the value of the acute reference angle α. To the nearest tenth, $\alpha = 70.0°$. Because θ is a QIII angle, add 180° to α to find angle θ. Thus, $\theta = 180° + 70.0° = 250.0°$. Table A can be used to find angle α if you do not have a calculator.

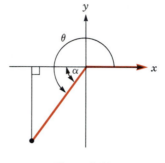

Figure 5–21

Example 13   If θ is a negative angle in QII and cot $\theta = -2.9042$, find angle θ.

Solution   Find the tangent of θ by entering -2.9042 in your calculator and pressing the ⟨1/x⟩ key. Then press ⟨INV⟩ ⟨TAN⟩. The display reads $-19°$, which is an angle in QIV. To obtain the required negative angle in QII, add $-180°$. Thus, $\theta = -199°$.   ■

⬛ Exercise 5.2

In Exercises 1–8 find the exact value of the sine, cosine, and tangent of each angle, if possible. ***Do not use a calculator.***

1. 135°   **2.** 630°   **3.** −450°   **4.** −30°

5. −240°   **6.** 300°   **7.** 540°   **8.** −315°

In Exercises 9–16 find the exact value of the cosecant, secant, and cotangent of each angle, if possible. ***Do not use a calculator.***

9. 225°   **10.** 1260°   **11.** −1080°   **12.** −225°

13. −210°   **14.** −480°   **15.** 585°   **16.** 150°

In Exercises 17–28 evaluate each expression. ***Do not use a calculator.***

17. sin 0° + cos 0° tan 45°   **18.** sin² 90° + cos 180° tan 0°

19. cos² 90° + cos 90° sin² 180°   **20.** cos² 0° + sin² 90° + cot² 90°

21. sin² 270° + csc² 270° + cot² 270°   **22.** cos 180° sin 180° − tan² 180°

23. sin 30° cos 60° − 2 tan² 60°   **24.** sin² 120° cos 45° + tan 45° sin 90°

25. sin 45° cos 330° − tan 150° tan 60°   **26.** cos 30° tan 60° + cos³ 45° tan 45°

27. csc² 210° sec 30° − sec 315° cot 60°   **28.** csc 90° csc 210° + csc 45° sin 135°

In Exercises 29–40 use the given information to find values of θ, where $0° \le \theta < 360°$. ***Do not use a calculator.***

29. $\tan \theta = \dfrac{\sqrt{3}}{3}$; $\sin \theta = \dfrac{1}{2}$ **30.** $\tan \theta = -1$; $\cos \theta = -\dfrac{\sqrt{2}}{2}$

31. $\tan \theta = -\sqrt{3}$; $\cos \theta = \dfrac{1}{2}$

32. $\cot \theta = \sqrt{3}$; $\cos \theta = -\dfrac{\sqrt{3}}{2}$

33. $\sin \theta = -\dfrac{1}{2}$; $\sec \theta = -\dfrac{2\sqrt{3}}{3}$

34. $\cos \theta = \dfrac{\sqrt{3}}{2}$; $\csc \theta = 2$

35. $\tan \theta = -1$; $\sec \theta = \sqrt{2}$

36. $\tan \theta = -\sqrt{3}$; $\cos \theta = -\dfrac{1}{2}$

37. $\sec \theta = -\sqrt{2}$; $\cot \theta = -1$

38. $\sin \theta = -1$

39. $\tan \theta$ is undefined

40. $\csc \theta = -\sqrt{2}$; $\cot \theta = -1$

In Exercises 41–46 use Table A in Appendix II to find the values of the sine, cosine, and tangent of the given angle.

41. 17°   **42.** 55.2°   **43.** 73°   **44.** 34.8°   **45.** 89°   **46.** 0.6°

In Exercises 47–58 use a calculator to find each value to four decimal places.

47. sin 23.1°   **48.** sin 57.8°   **49.** cos 133.7°   **50.** cos 211.7°

51. tan 223.5°   **52.** tan(−223.5°)   **53.** csc 312.4°   **54.** csc 129.2°

55. sec(−47.4°)   **56.** sec 11.3°   **57.** cot 640.6°   **58.** cot 302.2°

In Exercises 59–70 a positive angle θ is in a given quadrant and the value of a trigonometric function is given. Use a calculator to find θ to the nearest tenth of a degree.

59. QI; tan θ = 0.2493   **60.** QI; sin θ = 0.9986   **61.** QII; cos θ = −0.3420

62. QII; cos θ = −0.9063   **63.** QIII; sin θ = −0.4540   **64.** QIII; cos θ = −0.7193

65. QIV; tan θ = −5.6713   **66.** QIV; sin θ = −0.1908   **67.** QI; csc θ = 1.3250

68. QII; sec θ = −57.2987   **69.** QIII; cot θ = 1.1918   **70.** QIV; csc θ = −11.4737

5.3   RADIAN MEASURE AND THE CIRCULAR FUNCTIONS

Although degree measure of angles is common, a different unit of angular measure, called the **radian**, is more convenient in many mathematical applications.

Definition.   Consider the circle with radius r shown in Figure 5–22. The **radian measure** of the central angle θ is given by the formula

$$\theta = \frac{s}{r}$$

Figure 5–22

where s is the length of the intercepted arc, and s and r are in the same units.

Example 1   Suppose a central angle θ in a circle with a radius of 6 inches intercepts an arc of 2 feet. Find the radian measure of θ.

Solution   Since s and r need to be in the same units, express 2 feet as 24 inches. Then substitute **24** for s and **6** for r in the formula $\theta = \frac{s}{r}$ to obtain

$$\theta = \frac{s}{r} = \frac{24}{6} = 4$$

The radian measure of θ is 4 radians.   ■

To determine how the radian measure of an angle is related to its degree measure, we consider the circle in Figure 5–23 with radius r and central angle of 180°. Because the length of an arc intercepted by a 180° angle is one-half of the circumference of the circle, or πr, the radian measure of a 180° angle is

$$180° = \frac{\pi r}{r} \text{ radians}$$

or

$$\textbf{180°} = \boldsymbol{\pi} \textbf{ radians}$$

Figure 5–23

This result determines two equations that can be used to convert angular measure from one system to the other.

$$\textbf{1°} = \frac{\boldsymbol{\pi}}{\textbf{180}} \textbf{ radian} \qquad \text{and} \qquad \textbf{1 radian} = \frac{\textbf{180°}}{\boldsymbol{\pi}}$$

Example 2   Change to degrees: **a.** $\frac{5\pi}{6}$ radians and **b.** 7 radians.

Solution   **a.** $1 \text{ radian} = \frac{180°}{\pi}$ **b.** $1 \text{ radian} = \frac{180°}{\pi}$

$\frac{5\pi}{6} \cdot 1 \text{ radian} = \frac{5\pi}{6} \cdot \frac{180°}{\pi}$ $7 \cdot 1 \text{ radian} = 7 \cdot \frac{180°}{\pi}$

$\frac{5\pi}{6} \text{ radians} = 150°$ $7 \text{ radians} = \frac{1260°}{\pi}$ ■

Example 3   Change to radians: **a.** 120° and **b.** $-315°$.

Solution   **a.** $1° = \frac{\pi}{180} \text{ radian}$ **b.** $1° = \frac{\pi}{180} \text{ radian}$

$\textbf{120} \cdot 1° = \textbf{120} \cdot \frac{\pi}{180} \text{ radians}$ $\textbf{-315} \cdot 1° = \textbf{-315} \cdot \frac{\pi}{180} \text{ radians}$

$120° = \frac{2\pi}{3} \text{ radians}$ $-315° = -\frac{7\pi}{4} \text{ radians}$ ■

The following table gives the degree measure and the corresponding radian measure of five common angles:

Degree measure	0	30°	45°	60°	90°
Radian measure	0	$\dfrac{\pi}{6}$	$\dfrac{\pi}{4}$	$\dfrac{\pi}{3}$	$\dfrac{\pi}{2}$

The information in this table enables us to convert angles such as 120° and 225° from degree measure to radian measure in the following way:

$$120° = 2(\mathbf{60°}) = 2\left(\frac{\pi}{\mathbf{3}}\right) = \frac{2\pi}{3} \qquad 225° = 5(\mathbf{45°}) = 5\left(\frac{\pi}{\mathbf{4}}\right) = \frac{5\pi}{4}$$

We can also use this information to change angles such as $\frac{7\pi}{4}$ and $\frac{11\pi}{6}$ from radian measure to degree measure:

$$\frac{7\pi}{4} = 7\left(\frac{\pi}{\mathbf{4}}\right) = 7(\mathbf{45°}) = 315° \qquad \frac{11\pi}{6} = 11\left(\frac{\pi}{\mathbf{6}}\right) = 11(\mathbf{30°}) = 330°$$

Thus, we can find values of the trigonometric functions of angles measured in radians. For example,

$$\sin\frac{\pi}{3} = \sin 60° = \frac{\sqrt{3}}{2} \quad \text{and} \quad \tan\left(-\frac{7\pi}{4}\right) = -\tan 7(45°) = -\tan 315° = 1$$

By using radian measure, arc length of a circle can be calculated. If we multiply both sides of the formula $\theta = \frac{s}{r}$ by r, the following formula results:

Formula for Arc Length of a Circle.   If a central angle θ in a circle with radius r is measured in radians, then the length s of the intercepted arc is given by the formula

$$s = r\theta$$

See Figure 5–24.

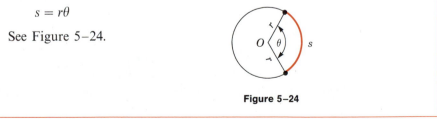

Figure 5–24

Example 4   What is the length of arc intercepted by a central angle of 150° in a circle with a radius of 18 meters?

Solution   First change 150° to radians. The result is $\frac{5\pi}{6}$ (see Example 2). Then substitute $\frac{5\pi}{6}$ for θ and 18 for r in the formula $s = r\theta$ and simplify.

$$s = 18\left(\frac{5\pi}{6}\right) = 15\pi \approx 47$$

The arc length is approximately 47 meters.   ■

The shaded area of the circle in Figure 5–25 is called a **sector of the circle**. To find the area of the sector, we first suppose that θ is measured in radians. We then set up a proportion indicating that the area A of the sector is to the area of the entire circle as the length of arc s is to the entire circumference of the circle.

$$\frac{A}{\pi r^2} = \frac{s}{2\pi r}$$

Figure 5–25

We can solve this proportion for A to obtain

$$A = \frac{1}{2}rs$$

Because $s = r\theta$, we can substitute $r\theta$ for s in the previous formula to obtain

$$A = \frac{1}{2}r^2\theta$$

Thus, we have

Formula for the Area of a Sector of a Circle.   If a sector of a circle with radius r has a central angle θ measured in radians, the area of the sector is given by the formula

$$A = \frac{1}{2}r^2\theta$$

Example 5   A sector of a circle has a central angle of 50.0° and an area of 605 square centimeters. Find the radius of the circle to the nearest tenth of a centimeter.

Solution   First change the measure of 50.0° to radians:

$$1° = \frac{\pi}{180} \text{ radian}$$

$$50.0° = \frac{5\pi}{18} \text{ radian} \qquad \text{Multiply both sides by 50.0 and simplify.}$$

Now substitute **605** for A and $\dfrac{5\pi}{18}$ for θ in the formula for the area of a sector and solve for r:

$$A = \frac{1}{2}r^2\theta$$

$$605 = \frac{1}{2}r^2\frac{5\pi}{18}$$

$$\frac{605(2)(18)}{5\pi} = r^2$$

$$1386.56 \approx r^2$$

$$r \approx \sqrt{1386.56}$$

$$r \approx 37.2$$

The radius of the circle is approximately 37.2 centimeters.   ■

Linear and Angular Velocity

The question "How fast is that train moving?" might be answered "60 miles per hour." This answer indicates the train's **linear velocity**, a measure of how far the train will travel per unit of time. The question "How fast is that phonograph record turning?" might be answered "$33\frac{1}{3}$ revolutions per minute." This answer indicates the record's **angular velocity**, a measure of the angle through which the record rotates per unit of time.

   Linear and angular velocity are often related. Because a wheel of a car, for example, moves with the car, it has linear velocity. Because the wheel is turning, it also has angular velocity. To find the relationship between the linear velocity of a point on the wheel and the angular velocity of the wheel, we consider the formula for arc length

$$s = r\theta$$

where s represents the length of the arc that the point has traveled during some time interval t, and θ represents the angle through which the wheel has turned during that same time interval. We then divide both sides of the formula by t to obtain

$$\frac{s}{t} = r\frac{\theta}{t}$$

Because $\frac{s}{t}$ represents a change in distance with respect to time, it represents the linear velocity of the point on the wheel. Because $\frac{\theta}{t}$ represents a change in rotation with respect to time, it represents the angular velocity of the wheel. If we denote the linear velocity $\frac{s}{t}$ by v, and the angular velocity $\frac{\theta}{t}$ by ω (the Greek letter omega), the previous equation becomes

$$v = r\omega$$

where ω is in units of radians per unit time.

Example 6   What is the linear velocity of a point on the Equator in miles per hour?

Solution   The radius of the earth is approximately 3960 miles. Because the earth completes one rotation (2π radians) every 24 hours, its angular velocity is

$$\omega = \frac{2\pi}{24}\frac{\text{rad}}{\text{hr}}$$

$$= \frac{\pi}{12}\frac{\text{rad}}{\text{hr}}$$

To find the linear velocity of a point on the Equator, substitute **3960** for r and $\frac{\pi}{12}$ for ω in the formula for linear velocity and simplify:

$$v = r\omega$$

$$v = 3960\left(\frac{\pi}{12}\right)$$

$$\approx 1040$$

The linear velocity of a point on the Equator is approximately 1040 miles per hour.

■

Example 7   An 8-inch-diameter pulley drives a 6-inch-diameter pulley. The larger pulley makes 15 revolutions per second. What is the angular velocity of the smaller pulley in revolutions per second? (See Figure 5–26.)

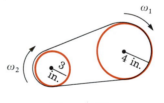

Figure 5–26

Solution   The angular velocity of the drive pulley is

$$15\frac{\text{rev}}{\text{sec}} \cdot 2\pi\frac{\text{rad}}{\text{rev}} = 30\pi\frac{\text{rad}}{\text{sec}}$$

Assuming that the belt that connects the two pulleys does not slip, the linear velocities of points on either circumference are the same—the product $r_1\omega_1$ for one pulley is equal to the product $r_2\omega_2$ for the second pulley. Thus

$$r_1\omega_1 = r_2\omega_2$$
$$4(30\pi) = 3(\omega_2) \qquad \text{Substitute 4 for } r_1, 30\pi \text{ for } \omega_1, \text{ and 3 for } r_2.$$
$$\frac{4(30\pi)}{3} = \omega_2$$
$$40\pi = \omega_2$$

The angular velocity of the smaller pulley is 40π rad/sec. To convert to revolutions per second, multiply by $1/(2\pi)$ rev/rad and simplify.

$$40\pi \, \frac{\text{rad}}{\text{sec}} \cdot \frac{1}{2\pi} \frac{\text{rev}}{\text{rad}} = 20 \, \frac{\text{rev}}{\text{sec}}$$

The angular velocity of the smaller pulley is 20 rev/sec. ∎

The Unit Circle

All of the algebraic, exponential, and logarithmic functions previously studied had domains and ranges that were subsets of the real number system. Thus far, however, the domains of the trigonometric functions have been sets of angles. We shall now show that the trigonometric functions can also be thought of as functions with real number domains.

The argument begins by associating every real number with the length of exactly one arc on the **unit circle**, the circle that is centered at the origin and has a radius of 1 unit. If t is a positive number, we can find the arc of length t by measuring a distance t in a counterclockwise direction along an arc of the unit circle beginning at the point $C(1, 0)$ (see Figure 5–27a). This determines arc CP of length t. If t is a negative number, we can find the arc of length t by measuring a distance $|t|$ in a clockwise direction along an arc of the unit circle beginning at the point $C(1, 0)$ (see Figure 5–27b). This determines arc CP of length t.

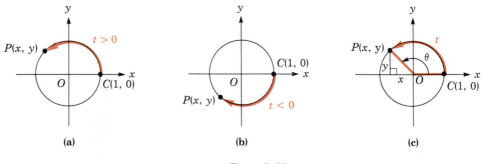

(a)   (b)   (c)

Figure 5–27

In either case, marking off the arc determines a single point P with coordinates (x, y) that corresponds to the real number t. If t is any real number, we can place the arc determined by t on the unit circle, as in Figure 5–27c. If this arc is intercepted by a central angle θ measured in radians, we know by the formula

$$s = r\theta$$

that

$$t = 1\theta \qquad \text{Substitute } t \text{ for } s \text{ and } 1 \text{ for } r.$$

or that

$$\theta = t$$

Thus, when the measure of an arc on the unit circle is the real number t, then t is also the radian measure of the central angle θ determined by that arc, and

$$\sin \theta = \sin t \qquad \csc \theta = \csc t$$
$$\cos \theta = \cos t \qquad \sec \theta = \sec t$$
$$\tan \theta = \tan t \qquad \cot \theta = \cot t$$

Hence, we can think of each trigonometric expression as being either a trigonometric function of an angle measured in radians or as a trigonometric function of a real number. The important point is this: *The trigonometric functions can now be thought of as functions that have domains and ranges that are subsets of the real numbers.* Because the real number t is associated with a point on the unit circle, the trigonometric functions of a real number are often called the **circular functions**.

The Circular Functions.   If $P(x, y)$ is the point on the unit circle of Figure 5–28 that corresponds to the real number t, then

$$\sin t = y$$
$$\cos t = x$$
$$\tan t = \frac{y}{x} \quad (x \neq 0)$$
$$\csc t = \frac{1}{y} \quad (y \neq 0)$$
$$\sec t = \frac{1}{x} \quad (x \neq 0)$$
$$\cot t = \frac{x}{y} \quad (y \neq 0)$$

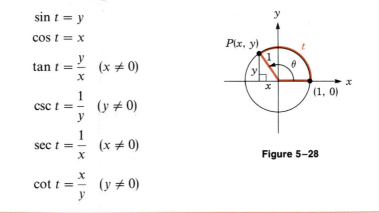

Figure 5–28

Example 8   Find **a.** $\sin\left(\dfrac{\pi}{3}\right)$ and **b.** $\cos\left(\dfrac{5\pi}{2}\right)$.

Solution   **a.** $\sin\left(\dfrac{\pi}{3}\right) = \sin\left(\dfrac{\pi}{3} \text{ rad}\right)$

$$= \frac{\sqrt{3}}{2}$$

b. $\cos\left(\dfrac{5\pi}{2}\right) = \cos\left(\dfrac{5\pi}{2}\,\text{rad}\right)$

$\qquad\qquad = \cos\left(\dfrac{\pi}{2}\,\text{rad}\right) \qquad \dfrac{\pi}{2}$ is coterminal with $\dfrac{5\pi}{2}$.

$\qquad\qquad = 0$ ◼

Example 9   Find **a.** $\tan\left(\dfrac{3\pi}{4}\right)$ and **b.** $\csc\left(\dfrac{7\pi}{6}\right)$.

Solution   **a.** $\tan\left(\dfrac{3\pi}{4}\right) = \tan\left(\dfrac{3\pi}{4}\,\text{rad}\right)$

$\qquad\qquad\qquad = -\tan\left(\dfrac{\pi}{4}\,\text{rad}\right) \qquad \dfrac{3\pi}{4}$ is a QII angle with reference angle of $\dfrac{\pi}{4}$.

$\qquad\qquad\qquad = -1$

b. $\csc\left(\dfrac{7\pi}{6}\right) = \csc\left(\dfrac{7\pi}{6}\,\text{rad}\right)$

$\qquad\qquad\qquad = -\csc\left(\dfrac{\pi}{6}\,\text{rad}\right) \qquad \dfrac{7\pi}{6}$ is a QIII angle with reference angle of $\dfrac{\pi}{6}$.

$\qquad\qquad\qquad = -2$ ◼

Example 10   Find the coordinates (x, y) of the point P on the unit circle that corresponds to the real numbers **a.** $\dfrac{\pi}{6}$ and **b.** $-\dfrac{5\pi}{4}$.

Solution   **a.**   From the definition of the circular functions, you have

$$x = \cos t \qquad \text{and} \qquad y = \sin t$$

Hence the coordinates (x, y) of the point P that corresponds to $\dfrac{\pi}{6}$ are

$$x = \cos\dfrac{\pi}{6} \qquad \text{and} \qquad y = \sin\dfrac{\pi}{6}$$

$$= \dfrac{\sqrt{3}}{2} \qquad\qquad\qquad\quad = \dfrac{1}{2}$$

Thus, point P has coordinates $\left(\dfrac{\sqrt{3}}{2}, \dfrac{1}{2}\right)$.

b.   The coordinates (x, y) of the point P that corresponds to $-\dfrac{5\pi}{4}$ are

$$x = \cos\left(-\dfrac{5\pi}{4}\right) \qquad \text{and} \qquad y = \sin\left(-\dfrac{5\pi}{4}\right)$$

$$= -\dfrac{\sqrt{2}}{2} \qquad\qquad\qquad\qquad = \dfrac{\sqrt{2}}{2}$$

Thus, the point P has coordinates $\left(-\dfrac{\sqrt{2}}{2}, \dfrac{\sqrt{2}}{2}\right)$. ◼

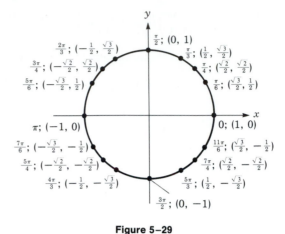

Figure 5–29

On the unit circle shown in Figure 5–29, points (cos t, sin t) are displayed for various values of t.

▬ Exercise 5.3 ▬

1. A central angle θ in a circle with a radius of 5 inches intercepts an arc that is 1 foot long. Find the radian measure of θ.

2. A central angle ϕ in a circle with a radius of 18 inches intercepts an arc that is 1.5 yards long. Find the radian measure of ϕ.

3. A central angle ϕ in a circle with a diameter of 4 meters intercepts an arc that is 2 meters long. Find the radian measure of ϕ.

4. A central angle θ in a circle with a diameter of 2.30 meters intercepts an arc that is 25 centimeters long. Find the radian measure of θ.

In Exercises 5–12 change each angle to radians.

5. $15°$	6. $75°$	7. $120°$	8. $210°$
9. $330°$	10. $780°$	11. $-520°$	12. $-880°$

In Exercises 13–20 each angle is expressed in radians. Express each one in degrees.

13. $\dfrac{3\pi}{4}$	14. 3π	15. $\dfrac{5\pi}{2}$	16. $\dfrac{7\pi}{3}$
17. $-\dfrac{4\pi}{3}$	18. $-\dfrac{11\pi}{6}$	19. 6	20. -8

21. Find the length of an arc of a circle with radius of 18 meters that is intercepted by a central angle of $120°$.

22. Find the length of an arc of a circle with radius of 25 centimeters that is intercepted by a central angle of $315°$.

23. Find the radius of a circle if a central angle of $25°$ intercepts an arc of 17 centimeters.

24. Find the central angle of a circle in radians that intercepts an arc of 10 centimeters if the circle has a diameter of 10 centimeters.

In Exercises 25–26 use 3960 miles for the radius of the earth.

25. The latitude of Manchester, New Hampshire, is 43.0° N. How far is Manchester from the Equator? See Illustration 1. Give the answer to the nearest 10 miles.

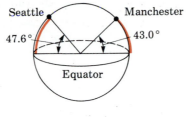

Illustration 1

26. The latitude of Seattle, Washington, is 47.6° N. How far is Seattle from the Equator? See Illustration 1. Give the answer to the nearest 10 miles.

27. Find the area, to the nearest hundredth, of a sector of a circle if the sector has a central angle of 30° and the circle has a radius of 20 units.

28. If a circle contains a sector with a central angle of 120° and an area of 30 square meters, find the diameter of the circle. Give the answer to the nearest hundredth.

29. Find the angular velocity of the minute hand of a clock in radians per hour.

30. Find the angular velocity of the earth in its orbit in radians per month.

31. A car is moving at 88 ft/sec. How fast are its 1.5-foot-radius tires spinning in radians per second?

32. A car is moving at 40 ft/sec. Find the angular velocity of its 2-foot-diameter tires in radians per second.

33. The 30-inch-diameter tires of a truck are turning at 400 rpm. What is the linear velocity of the truck in feet per minute?

34. The 27-inch tires of a bicycle are turning at the rate of 125 rpm. What is the linear velocity of the bicycle in inches per minute?

*In Exercises 35–42 evaluate each expression. **Do not use a calculator.***

35. $\sin \dfrac{\pi}{6}$ **36.** $\cos \dfrac{\pi}{6}$ **37.** $\cos\left(-\dfrac{5\pi}{6}\right)$ **38.** $\sin\left(-\dfrac{5\pi}{6}\right)$

39. $\tan \dfrac{5\pi}{4}$ **40.** $\cot \dfrac{5\pi}{4}$ **41.** $\sec \dfrac{5\pi}{3}$ **42.** $\csc \dfrac{7\pi}{3}$

In Exercises 43–50 use a calculator to evaluate each expression to four decimal places. Remember to set the calculator in radian mode.

43. $\sin 2$ **44.** $\cos 8$ **45.** $\tan 5$ **46.** $\sec 5$

47. $\sin\left(-\dfrac{3}{\pi}\right)$ **48.** $\cos\left(-\dfrac{3}{\pi}\right)$ **49.** $\csc(3 + \pi)$ **50.** $\cos(3 - \pi)$

*In Exercises 51–58 find the coordinates of the point P on the unit circle that correspond to each real number. **Do not use a calculator.***

51. $\dfrac{3\pi}{2}$ **52.** $-\dfrac{\pi}{2}$ **53.** $-\pi$ **54.** 2π

55. $-\dfrac{\pi}{4}$ **56.** $-\dfrac{3\pi}{4}$ **57.** $-\dfrac{2\pi}{3}$ **58.** $\dfrac{2\pi}{3}$

59. Show that the area of a sector of the unit circle is given by the formula $A = \frac{1}{2}\theta$, where θ is the central angle of the sector measured in radians.

60. Show that the area of triangle OAR in Illustration 2 is given by the formula

$$A = \frac{1}{2}\cos\theta\sin\theta$$

Illustration 2

61. Show that the area of triangle OTP in Illustration 2 is given by the formula

$$A = \frac{1}{2}\tan\theta$$

62. Use the inequality

area of $\triangle OAR \leq$ area of sector $OTR \leq$ area of $\triangle OTP$

to show that the ratio $\dfrac{\sin\theta}{\theta}$ approaches 1 as θ approaches 0. Note that θ must be a real number for this ratio to have meaning. This fact is very important in calculus. (*Hint:* Refer to Exercises 59–61 and Illustration 2.)

5.4   GRAPHS OF FUNCTIONS INVOLVING SIN X AND COS X

Recall that the sine of an angle θ is equal to the sine of any angle that is co-terminal with θ. A similar statement is true for the cosine of angle θ. Therefore, for any number x,

$$\sin x = \sin(x \pm 2\pi) = \sin(x \pm 4\pi) = \cdots$$
$$\cos x = \cos(x \pm 2\pi) = \cos(x \pm 4\pi) = \cdots$$

The sine and cosine functions are called **periodic functions** because, as x increases, the values of $\sin x$ and $\cos x$ repeat in a predictable way.

> **Definition.**   A function f is said to be **periodic with period p** if p is the smallest positive number for which
>
> $$f(x) = f(x + p)$$
>
> for all x in the domain of f.

The sine function has a period of 2π because $\sin x = \sin(x + 2\pi)$ for all x, and $\sin x = \sin(x + p)$ is true for no positive number that is less than 2π. To graph the sine function, we can find points (x, y) that satisfy the equation $y = \sin x$ and plot them on a rectangular coordinate system. Remember that x, often called the **argument of the function**, can be thought of either as an angle measured in radians or as a real number.

The table of values in Figure 5–30 shows that

As x increases from	The values of sin x	Creating the graph
0 to $\dfrac{\pi}{2}$	increase from 0 to 1	
$\dfrac{\pi}{2}$ to π	decrease from 1 to 0	
π to $\dfrac{3\pi}{2}$	decrease from 0 to -1	
$\dfrac{3\pi}{2}$ to 2π	increase from -1 to 0	

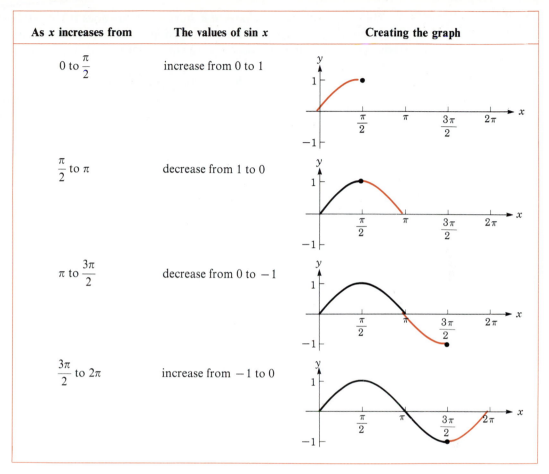

$y = \sin x$

x	y
0	0
$\frac{\pi}{6}$	.5
$\frac{\pi}{3}$	.87
$\frac{\pi}{2}$	1
$\frac{2\pi}{3}$	.87
$\frac{5\pi}{6}$	.5
π	0
$\frac{7\pi}{6}$	$-.5$
$\frac{4\pi}{3}$	$-.87$
$\frac{3\pi}{2}$	-1
$\frac{5\pi}{3}$	$-.87$
$\frac{11\pi}{6}$	$-.5$
2π	0

If we plot the pairs of values $(x, \sin x)$ shown in the figure, we can draw the graph of $y = \sin x$, often called a **sine wave**. We note that, for all x,

$$-1 \le \sin x \le 1$$

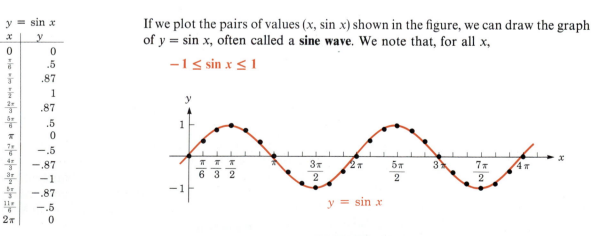

$y = \sin x$

Figure 5–30

The graph of $y = \sin x$ does not start at 0, nor does it end at 2π. The sine wave continues on forever in both directions. Because the function is periodic, however, if we know its behavior through one period, we know its behavior everywhere.

The cosine function also has a period of 2π because $\cos x = \cos(x + 2\pi)$ for all x, and $\cos x = \cos(x + p)$ is true for no positive number less than 2π. Because $\cos 0 = 1$, we shall draw its graph beginning at the point (0, 1). The table of values in Figure 5–31 shows that

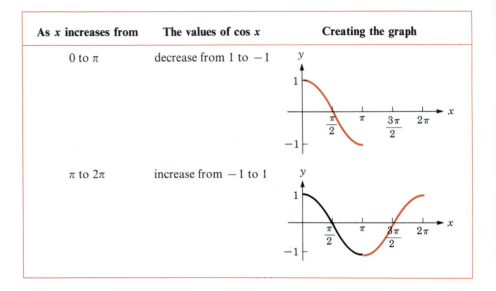

As x increases from	The values of $\cos x$	Creating the graph
0 to π	decrease from 1 to -1	
π to 2π	increase from -1 to 1	

The graph of $y = \cos x$ appears in Figure 5–31. We note that, for all x,

$$-1 \le \cos x \le 1$$

$y = \cos x$

x	y
0	1
$\frac{\pi}{6}$	.87
$\frac{\pi}{3}$	.5
$\frac{\pi}{2}$	0
$\frac{2\pi}{3}$	$-.5$
$\frac{5\pi}{6}$	$-.87$
π	-1
$\frac{7\pi}{6}$	$-.87$
$\frac{4\pi}{3}$	$-.5$
$\frac{3\pi}{2}$	0
$\frac{5\pi}{3}$	.5
$\frac{11\pi}{6}$	.87
2π	1

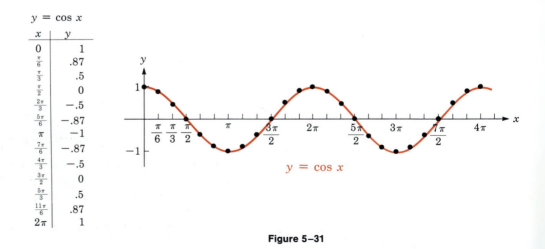

Figure 5–31

Example 1   Graph the function defined by $y = 3 \sin x$.

Solution   The values of $3 \sin x$ can be found by multiplying each value of $\sin x$ by 3. Thus, the values of $3 \sin x$ must be between -3 and 3 as x increases from 0 to 2π.

As x increases from	The values of $3 \sin x$
0 to $\dfrac{\pi}{2}$	increase from 0 to 3
$\dfrac{\pi}{2}$ to $\dfrac{3\pi}{2}$	decrease from 3 to -3
$\dfrac{3\pi}{2}$ to 2π	increase from -3 to 0

Note that the graph of $y = 3 \sin x$, as shown in Figure 5–32, can be found by stretching the graph of $y = \sin x$ vertically by a factor of 3.

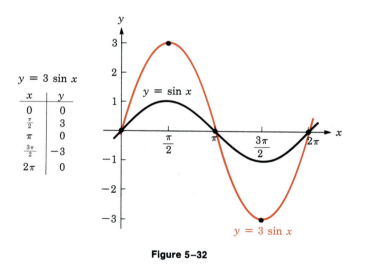

x	y
0	0
$\frac{\pi}{2}$	3
π	0
$\frac{3\pi}{2}$	-3
2π	0

Figure 5–32

Example 2   Graph $y = 2 \cos x$ and $y = -2 \cos x$ on the same set of coordinate axes.

Solution   The values of $2 \cos x$ can be found by multiplying each value of $\cos x$ by 2. As x increases from 0 to π, the values of $2 \cos x$ decrease from 2 to -2; as x increases from π to 2π, they increase from -2 back to 2. Thus, the graph of $y = 2 \cos x$ can be found by stretching the graph of $y = \cos x$ vertically by a factor of 2.

The graph of $y = -2 \cos x$ can be found by reflecting the graph of $y = 2 \cos x$ in the x-axis. Both graphs appear in Figure 5–33.

$y = 2 \cos x$ $y = -2 \cos x$

x	y
0	2
$\frac{\pi}{2}$	0
π	-2
$\frac{3\pi}{2}$	0
2π	2

x	y
0	-2
$\frac{\pi}{2}$	0
π	2
$\frac{3\pi}{2}$	0
2π	-2

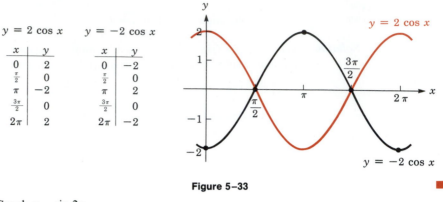

Figure 5–33

Example 3   Graph $y = \sin 2x$.

Solution   Because the graph of $y = \sin x$ completes one cycle as x increases from 0 to 2π, the graph of $y = \sin 2x$ must complete one cycle as $2x$ increases from 0 to 2π, or as x increases from 0 to π. Thus, the period of the graph of $y = \sin 2x$ is π, and the sine wave will oscillate twice as fast as the graph of $y = \sin x$. The graph of $y = \sin 2x$ appears in Figure 5–34. Note that the coefficient of x is a number greater than 1 and that the curve is compressed horizontally.

$y = \sin 2x$

x	y
0	0
$\frac{\pi}{4}$	1
$\frac{\pi}{2}$	0
$\frac{3\pi}{4}$	-1
π	0

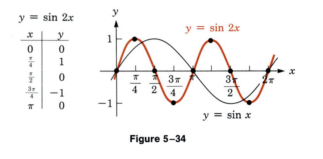

Figure 5–34

Example 4   Graph $y = \cos \frac{1}{3} x$.

Solution   Because the graph of $y = \cos x$ completes one cycle as x increases from 0 to 2π, the graph of $y = \cos \frac{1}{3}x$ must complete one cycle as $\frac{1}{3}x$ increases from 0 to 2π, or as x increases from 0 to 6π. Thus, the period of the graph of $y = \cos \frac{1}{3}x$ is 6π, and the graph will oscillate one-third as fast as the graph of $y = \cos x$. The graph of $y = \cos \frac{1}{3}x$ appears in Figure 5–35. Note that the coefficient of x is a positive number less than 1 and that the curve is stretched horizontally.

$y = \cos \frac{1}{3}x$

x	y
0	1
$\frac{3\pi}{2}$	0
3π	-1
$\frac{9\pi}{2}$	0
6π	1

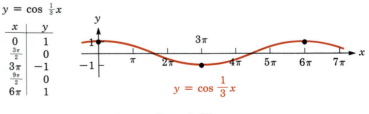

Figure 5–35

Example 5   Graph $y = \cos \pi x$.

Solution   Because the graph of $y = \cos x$ completes one cycle as x increases from 0 to 2π, the graph of $y = \cos \pi x$ must complete one cycle as πx increases from 0 to 2π, or as x increases from 0 to 2. Thus, the period of the graph of $y = \cos x$ is 2. The graph of $y = \cos \pi x$ appears in Figure 5–36. Note that the coefficient of x is a number greater than 1 and that the curve is compressed.

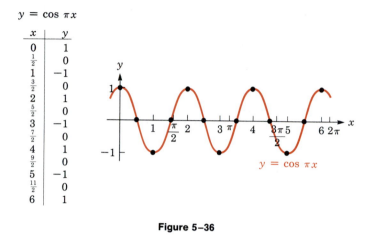

$y = \cos \pi x$

x	y
0	1
$\frac{1}{2}$	0
1	-1
$\frac{3}{2}$	0
2	1
$\frac{5}{2}$	0
3	-1
$\frac{7}{2}$	0
4	1
$\frac{9}{2}$	0
5	-1
$\frac{11}{2}$	0
6	1

Figure 5–36

Period and Amplitude

One cycle of the graph of $y = a \sin bx$ is completed as $|bx|$ increases from 0 to 2π, or as x increases from 0 to $\left|\frac{2\pi}{b}\right|$. Thus, the period of the graph of $y = a \sin bx$ is $\left|\frac{2\pi}{b}\right|$.

Because $-1 \leq \sin bx \leq 1$, the largest value that can be attained by $y = a \sin bx$ is $|a| \cdot 1$, or just $|a|$. This value is called the **amplitude** of the graph of $y = a \sin bx$.

A similar argument applies to determine the period and the amplitude of the graph of $y = a \cos bx$.

Period and Amplitude.   The **period** of the graph of $\left\{ \begin{array}{l} y = a \sin bx \\ y = a \cos bx \end{array} \right\}$ is $\left|\dfrac{2\pi}{b}\right|$, and the **amplitude** is $|a|$.

Example 6   Graph $y = 5 \sin 7x$.

Solution   The amplitude is 5 and the period is $\frac{2\pi}{7}$. One cycle of the graph of $y = 5 \sin 7x$ appears in Figure 5–37.

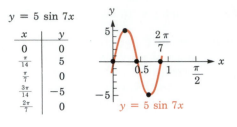

$y = 5 \sin 7x$

x	y
0	0
$\frac{\pi}{14}$	5
$\frac{\pi}{7}$	0
$\frac{3\pi}{14}$	-5
$\frac{2\pi}{7}$	0

Figure 5–37

Example 7   Graph $y = 2 \cos \frac{1}{2} x$.

Solution   The amplitude is 2 and the period is $\frac{2\pi}{\frac{1}{2}}$, or 4π. Because $2 \cos \frac{1}{2}x$ is zero when $x = \ldots, -\pi, \pi, 3\pi, \ldots$, the graph intersects the x-axis at these points. One cycle of the graph appears in Figure 5–38.

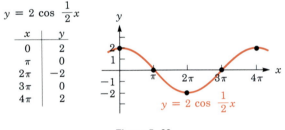

$y = 2 \cos \frac{1}{2}x$

x	y
0	2
π	0
2π	-2
3π	0
4π	2

Figure 5–38

Exercise 5.4

In Exercises 1–16 find the amplitude and period of each function. **Do not construct a graph.**

1. $y = 2 \sin x$ **2.** $y = 3 \cos x$ **3.** $y = \cos 9x$ **4.** $y = -\sin 11x$

5. $y = \sin \frac{1}{3} x$ **6.** $y = \cos \frac{1}{4} x$ **7.** $y = -\cos 0.2x$ **8.** $y = \sin 0.25x$

9. $y = 3 \sin \frac{1}{2} x$ **10.** $y = \frac{1}{2} \cos \frac{1}{3} x$ **11.** $y = -\frac{1}{2} \cos \pi x$ **12.** $y = 17 \sin 2\pi x$

13. $y = 3 \sin 2\pi x$ **14.** $y = 8 \cos \pi x$ **15.** $y = -\frac{1}{3} \sin \frac{3x}{\pi}$ **16.** $y = -\frac{5}{3} \cos \frac{x}{\pi}$

In Exercises 17–26 graph each pair of functions over the indicated interval.

17. $y = \sin x$ and $y = 2 \sin x$, $0 \le x \le 2\pi$

18. $y = \cos x$ and $y = -3 \cos x$, $0 \le x \le 2\pi$

19. $y = \cos x$ and $y = -\frac{1}{3} \cos x$, $0 \le x \le 2\pi$

20. $y = \sin x$ and $y = \frac{1}{2} \sin x$, $0 \le x \le 2\pi$

21. $y = \sin x$ and $y = \sin 2x$, $0 \le x \le 2\pi$

22. $y = \cos x$ and $y = \cos 3x$, $0 \le x \le 2\pi$

23. $y = \cos x$ and $y = \cos \frac{1}{3} x$, $0 \le x \le 6\pi$

24. $y = \sin x$ and $y = \sin \frac{1}{2} x$, $0 \le x \le 4\pi$

25. $y = \sin x$ and $y = \sin \pi x$, $0 \le x \le 2\pi$

26. $y = \cos x$ and $y = \cos \pi x$, $0 \le x \le 2\pi$

In Exercises 27–40 graph each function over an interval that is at least one period long.

27. $y = 3 \cos x$ **28.** $y = 4 \sin x$ **29.** $y = -\sin x$ **30.** $y = -\cos x$

31. $y = \cos 2x$ **32.** $y = \sin 3x$ **33.** $y = -\sin \dfrac{x}{4}$ **34.** $y = \cos \dfrac{x}{4}$

35. $y = 3 \sin \pi x$ **36.** $y = -2 \cos \pi x$ **37.** $y = \dfrac{1}{2} \cos 4x$ **38.** $y = -4 \sin 3x$

39. $y = -4 \sin \dfrac{x}{2}$ **40.** $y = \dfrac{1}{3} \cos \dfrac{x}{2}$

5.5 GRAPHS OF FUNCTIONS INVOLVING TAN X, COT X, CSC X, AND SEC X

Because the reference angles of x and $x + \pi$ are equal (see Figure 5–39), and because the tangents of angles in nonadjacent quadrants have the same sign, we have

$$\tan x = \tan(x + \pi)$$

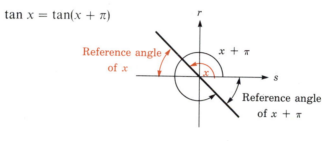

Figure 5–39

for all x. Furthermore, π is the smallest positive number p for which $\tan x = \tan(x + p)$. Thus, the period of the tangent function is π.

Because $\tan 0 = 0$, the graph of $y = \tan x$ passes through the origin, as in Figure 5–40. Because $\tan \frac{\pi}{2}$ is undefined, the graph cannot intersect the line $x = \frac{\pi}{2}$. However, as x gets close to $\frac{\pi}{2}$, the value of $|\tan x|$ becomes large without bound, and the line $x = \frac{\pi}{2}$ is a vertical asymptote. Other asymptotes are the lines $x = \frac{\pi}{2} \pm n\pi$, where n is a nonnegative integer.

$y = \tan x$

x	y
0	0
$\frac{\pi}{4}$	1
$\frac{3\pi}{4}$	-1
π	0
$\frac{5\pi}{4}$	1
$\frac{7\pi}{4}$	-1

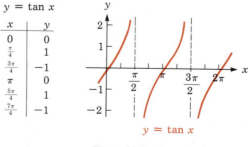

$y = \tan x$

Figure 5–40

For all x between $\frac{\pi}{2}$ and π, the values of tan x are negative, returning to 0 when $x = \pi$. The graph appears in Figure 5–40. Because $|\tan x|$ can grow large without bound, the tangent function has no amplitude.

The cotangent function also has a period of π, and its vertical asymptotes are the lines $x = 0 \pm n\pi$, where n is a nonnegative integer. The graph appears in Figure 5–41. The cotangent function has no amplitude.

The graph of $y = \csc x$ has a period of 2π because csc x is the reciprocal of sin x and sin x has a period of 2π. Values for csc x are undefined whenever sin $x = 0$, and this occurs at $x = 0 \pm n\pi$, where n is a nonnegative integer. These values determine vertical asymptotes for the graph of $y = \csc x$. The graph appears in Figure 5–42. The cosecant function has no amplitude.

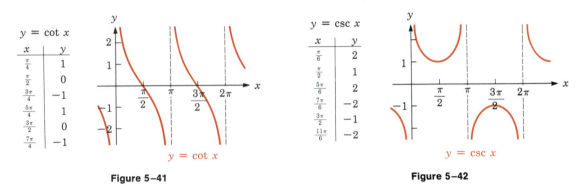

$y = \cot x$

x	y
$\frac{\pi}{4}$	1
$\frac{\pi}{2}$	0
$\frac{3\pi}{4}$	-1
$\frac{5\pi}{4}$	1
$\frac{3\pi}{2}$	0
$\frac{7\pi}{4}$	-1

$y = \cot x$

Figure 5–41

$y = \csc x$

x	y
$\frac{\pi}{6}$	2
$\frac{\pi}{2}$	1
$\frac{5\pi}{6}$	2
$\frac{7\pi}{6}$	-2
$\frac{3\pi}{2}$	-1
$\frac{11\pi}{6}$	-2

$y = \csc x$

Figure 5–42

The graph of $y = \sec x$ has a period of 2π because sec x is the reciprocal of cos x and cos x has a period of 2π. Values for sec x are undefined whenever cos $x = 0$, and this occurs at $x = \frac{\pi}{2} \pm n\pi$, where n is a nonnegative integer. These values determine vertical asymptotes for the graph of $y = \sec x$. The graph appears in Figure 5–43. The secant function has no amplitude.

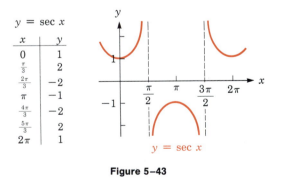

$y = \sec x$

x	y
0	1
$\frac{\pi}{3}$	2
$\frac{2\pi}{3}$	-2
π	-1
$\frac{4\pi}{3}$	-2
$\frac{5\pi}{3}$	2
2π	1

$y = \sec x$

Figure 5–43

Example 1   Graph $y = \tan 3x$ on the interval from 0 to π.

Solution   The graph of $y = \tan 3x$ will intersect the x-axis whenever $\tan 3x = 0$. In the interval from 0 to π, this occurs when $x = 0$, $x = \frac{\pi}{3}$, $x = \frac{2\pi}{3}$, and $x = \pi$.

Because the period of the tangent function is π, the graph of $y = \tan 3x$ will complete one cycle as $3x$ increases from 0 to π, or as x increases from 0 to $\frac{\pi}{3}$. Thus, the period of $y = \tan 3x$ is $\frac{\pi}{3}$.

In the interval from 0 to π, no value for $\tan 3x$ is defined when $x = \frac{\pi}{6}$, $x = \frac{\pi}{2}$, and $x = \frac{5\pi}{6}$. These equations determine vertical asymptotes for the graph of $y = \tan 3x$.

Using the previous information, you can draw the graph as shown in Figure 5–44.

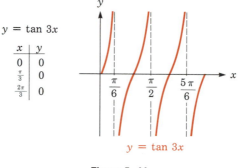

$y = \tan 3x$

x	y
0	0
$\frac{\pi}{3}$	0
$\frac{2\pi}{3}$	0

$y = \tan 3x$

Figure 5–44

Example 2   Graph $y = 2 \csc x$ on the interval from 0 to 2π.

Solution   Because $\csc x$ is the reciprocal of $\sin x$, its period is 2π. Thus, the period of $y = 2 \csc x$ is also 2π. Furthermore, no value for $2 \csc x$ is defined when $\sin x = 0$. In the interval from 0 to 2π, this occurs at $x = 0$, $x = \pi$, and $x = 2\pi$. These equations determine the vertical asymptotes for the graph of $y = 2 \csc x$. A table of values and the graph appear in Figure 5–45.

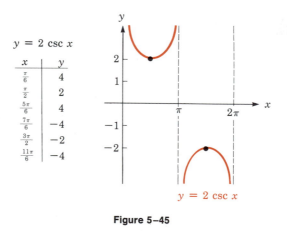

$y = 2 \csc x$

x	y
$\frac{\pi}{6}$	4
$\frac{\pi}{2}$	2
$\frac{5\pi}{6}$	4
$\frac{7\pi}{6}$	-4
$\frac{3\pi}{2}$	-2
$\frac{11\pi}{6}$	-4

$y = 2 \csc x$

Figure 5–45

Example 3   Graph $y = -2 \sec 2x$ on the interval from 0 to 2π.

Solution   Because $\sec 2x$ is the reciprocal of $\cos 2x$, it has the same period as $\cos 2x$. Thus, the period of $y = -2 \sec 2x$ is $\frac{2\pi}{2}$, or just π. Furthermore, no value for $-2 \sec 2x$ is defined when $\cos 2x$ is 0. In the interval from 0 to 2π, this occurs at $x = \frac{\pi}{4}$, at $x = \frac{3\pi}{4}$, at $x = \frac{5\pi}{4}$, and at $x = \frac{7\pi}{4}$. These equations determine vertical asymptotes for the graph of $y = -2 \sec 2x$. A table of values and the graph appear in Figure 5–46.

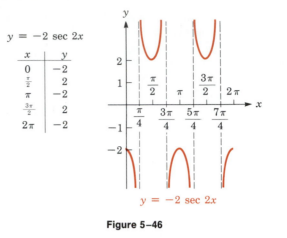

$y = -2 \sec 2x$

x	y
0	-2
$\frac{\pi}{2}$	2
π	-2
$\frac{3\pi}{2}$	2
2π	-2

$y = -2 \sec 2x$

Figure 5–46

We summarize the results of this section.

The **period** of $\left\{ \begin{matrix} y = a \tan bx \\ y = a \cot bx \end{matrix} \right\}$ is $\left| \dfrac{\pi}{b} \right|$.

The **period** of $\left\{ \begin{matrix} y = a \csc bx \\ y = a \sec bx \end{matrix} \right\}$ is $\left| \dfrac{2\pi}{b} \right|$.

The tangent, cotangent, cosecant, and secant functions have no amplitude.

Exercise 5.5

In Exercises 1–18 give the period of each function. **Do not construct the graph.**

1. $y = 3 \tan x$

2. $y = 2 \csc x$

3. $y = \dfrac{1}{2} \sec x$

4. $y = \dfrac{1}{3} \cot x$

5. $y = \dfrac{1}{3} \tan 3x$

6. $y = \dfrac{1}{2} \sec 2x$

7. $y = -2 \csc \pi x$

8. $y = -3 \tan \pi x$

9. $y = 3 \sec \dfrac{x}{3}$

10. $y = -2 \tan \dfrac{\pi x}{3}$

11. $y = \dfrac{7}{2} \cot \dfrac{2\pi x}{3}$

12. $y = -\dfrac{2}{3} \csc \dfrac{\pi x}{3}$

13. $y = 3 \csc \dfrac{\pi x}{2}$

14. $y = -4 \sec \dfrac{2\pi x}{5}$

15. $y = -\cot \dfrac{x}{2\pi}$

16. $y = 7 \csc \dfrac{x}{4\pi}$

17. $y = -\dfrac{2}{5} \sec \dfrac{3x}{\pi}$

18. $y = \dfrac{7}{9} \cot \dfrac{2x}{\pi}$

In Exercises 19–30 graph each equation over the indicated interval.

19. $y = 2 \tan x, \ -\dfrac{\pi}{2} < x < \dfrac{3\pi}{2}$

20. $y = 2 \csc x, \ 0 < x \leq 2\pi$

21. $y = -3 \sec x, \ 0 \leq x \leq 2\pi$

22. $y = -\csc 2x, \ 0 < x < \dfrac{3\pi}{2}$

23. $y = \cot 2x, 0 < x < \pi$

24. $y = \sec 3x, 0 \le x \le \dfrac{2\pi}{3}$

25. $y = -2 \tan \dfrac{x}{2}, 0 < x < 2\pi$

26. $y = -3 \csc \dfrac{x}{2}, 0 < x < 4\pi$

27. $y = 2 \sec 2x, 0 \le x \le 2\pi$

28. $y = 2 \csc 2x, 0 < x < 2\pi$

29. $y = -2 \cot \dfrac{\pi}{4} x, 0 < x < 4$

30. $y = -2 \sec \dfrac{\pi}{4} x, 0 \le x \le 8$

31. Explain why the tangent, cotangent, secant, and cosecant functions have no amplitude.

32. For what values of x, if any, are the values of $\sin x$ and $\csc x$ equal?

33. For what values of x, if any, are the values of $\cos x$ and $\sec x$ equal?

34. For what values of x, if any, are the values of $\tan x$ and $\cot x$ equal?

5.6 TRANSLATIONS, ADDITION OF ORDINATES, AND MISCELLANEOUS GRAPHS

In applied work, we often encounter graphs of trigonometric functions that have been translated from standard position in either a vertical or a horizontal direction. Recall that the graph of a function is translated vertically by adding a constant to the function, and horizontally by adding a constant to the argument of the function.

Example 1 Graph $y = 2 + \cos x$.

Solution The values of $2 + \cos x$ will be similar to the values of $\cos x$, except that each one will be increased by 2. Thus, the graph of $y = 2 + \cos x$ has a period of 2π and has y values between 1 and 3. The graph will intersect the line $y = 2$ at $x = \frac{\pi}{2} \pm n\pi$, where n is a non-negative integer. The graph appears in Figure 5–47.

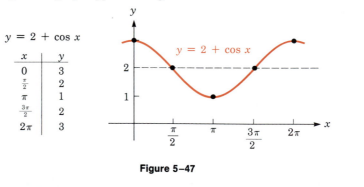

x	y
0	3
$\frac{\pi}{2}$	2
π	1
$\frac{3\pi}{2}$	2
2π	3

Figure 5–47

Example 2 Graph $y = -3 + \tan \dfrac{x}{2}$.

Solution The function $y = \tan \frac{x}{2}$ has a period of $\dfrac{\pi}{\frac{1}{2}}$, or 2π, and has vertical asymptotes at $x = \pi \pm n2\pi$, where n is a nonnegative integer. The values of $-3 + \tan \frac{x}{2}$ will be similar to the values of $\tan \frac{x}{2}$, except that each value of $\tan \frac{x}{2}$ will be decreased by 3. The graph of $y = -3 + \tan \frac{x}{2}$ appears in Figure 5–48.

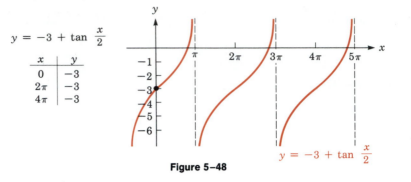

Figure 5–48

The preceding examples suggest the following facts:

> **Vertical Translations.**   The graph of $y = k + a \sin bx$ is identical to the graph of $y = a \sin bx$ except that it is translated $|k|$ units
>
> $$\left. \begin{matrix} \text{up} \\ \text{down} \end{matrix} \right\} \quad \text{if } k \text{ is} \quad \left\{ \begin{matrix} \text{positive} \\ \text{negative} \end{matrix} \right.$$
>
> A similar statement is true for each of the other trigonometric functions.

Example 3   Graph $y = \sin\left(x + \dfrac{\pi}{6}\right)$.

Solution   One complete cycle of $y = \sin(x + \frac{\pi}{6})$ is described as $x + \frac{\pi}{6}$ increases from 0 to 2π, or as x increases from $-\frac{\pi}{6}$ to $(2\pi - \frac{\pi}{6})$. At $x = -\frac{\pi}{6}$, the value of $\sin(x + \frac{\pi}{6})$ is zero. The graph of $y = \sin(x + \frac{\pi}{6})$ looks like the graph of $y = \sin x$, except that it is translated to the left by a distance of $\frac{\pi}{6}$. The graph appears in Figure 5–49. The graph of $y = \sin x$ is included for reference.

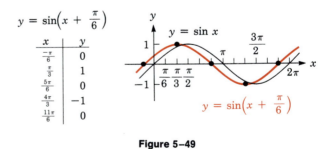

Figure 5–49

The distance that a graph of a trigonometric function is translated to the left or right is called the **phase shift** of the graph.

Example 4   Graph $y = 3 \cos\left(2x - \dfrac{\pi}{3}\right)$.

Solution Rewrite $y = 3 \cos (2x - \frac{\pi}{3})$ in the form $y = a \cos b(x + c)$ by factoring out a 2 from the binomial $2x - \frac{\pi}{3}$. Then

$$y = 3 \cos\left(2x - \frac{\pi}{3} \right)$$

$$= 3 \cos 2\left(x - \frac{\pi}{6} \right)$$

$$= 3 \cos 2\left[x + \left(-\frac{\pi}{6} \right) \right]$$

From the equation $y = 3 \cos 2[x + (-\frac{\pi}{6})]$, you can see that the amplitude is 3, the period is $\frac{2\pi}{2} = \pi$, and the graph looks like the graph of $y = 3 \cos 2x$. However, it is translated $\frac{\pi}{6}$ units to the right. The graph appears in Figure 5–50. The graph of $y = 3 \cos 2x$ is included for reference.

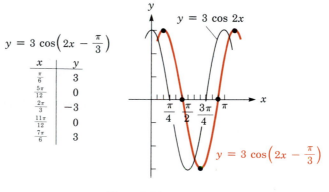

$$y = 3 \cos\left(2x - \frac{\pi}{3} \right)$$

x	y
$\frac{\pi}{6}$	3
$\frac{5\pi}{12}$	0
$\frac{2\pi}{3}$	-3
$\frac{11\pi}{12}$	0
$\frac{7\pi}{6}$	3

Figure 5–50

Example 5 Graph $y = \tan\left(x - \frac{\pi}{4} \right)$.

Solution One cycle of the graph of $y = \tan(x - \frac{\pi}{4})$ is completed as $x - \frac{\pi}{4}$ increases from 0 to π, or as x increases from $\frac{\pi}{4}$ to $\frac{5\pi}{4}$. At $x = \frac{\pi}{4}$, the value of $\tan(x - \frac{\pi}{4})$ is zero. The graph of $y = \tan(x - \frac{\pi}{4})$ looks like the graph of $y = \tan x$, except that it is translated $\frac{\pi}{4}$ units to the right. The graph appears in Figure 5–51.

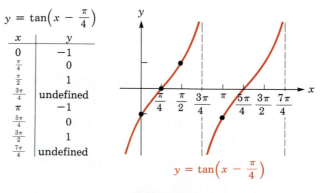

$$y = \tan\left(x - \frac{\pi}{4} \right)$$

x	y
0	-1
$\frac{\pi}{4}$	0
$\frac{\pi}{2}$	1
$\frac{3\pi}{4}$	undefined
π	-1
$\frac{5\pi}{4}$	0
$\frac{3\pi}{2}$	1
$\frac{7\pi}{4}$	undefined

$$y = \tan\left(x - \frac{\pi}{4} \right)$$

Figure 5–51

Example 6   Graph $y = \csc\left(3x - \dfrac{\pi}{2}\right)$.

Solution   Factor out a 3 from $3x - \dfrac{\pi}{2}$ to write that equation in the form $y = \csc\, b(x + c)$.

$$y = \csc\left(3x - \frac{\pi}{2}\right)$$

$$= \csc 3\left[x + \left(-\frac{\pi}{6}\right)\right]$$

The period of the cosecant function is $\left|\dfrac{2\pi}{b}\right|$ or in this case $\left|\dfrac{2\pi}{3}\right|$. The graph looks like the graph of $y = \csc 3x$, except that it is translated $\dfrac{\pi}{6}$ units to the right. The graph appears in Figure 5–52.

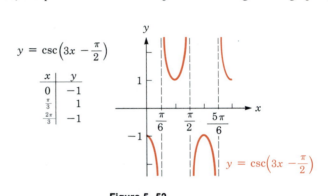

$y = \csc\left(3x - \dfrac{\pi}{2}\right)$

x	y
0	-1
$\frac{\pi}{3}$	1
$\frac{2\pi}{3}$	-1

Figure 5–52

The following statements summarize the facts involving phase shifts of the trigonometric functions.

Phase Shift.   The graph of

$$\begin{cases} y = a \sin b(x + c) \\ y = a \cos b(x + c) \\ y = a \tan b(x + c) \end{cases}$$

is identical to the graph of

$$\begin{cases} y = a \sin bx \\ y = a \cos bx \\ y = a \tan bx \end{cases}$$

except that it is translated $|c|$ units to the

$$\begin{Bmatrix} \text{left} \\ \text{right} \end{Bmatrix} \text{ if } c \text{ is } \begin{Bmatrix} \text{positive} \\ \text{negative} \end{Bmatrix}$$

The number $|c|$ is called the **phase shift** of the graph. A similar statement is true for the other trigonometric functions.

Addition of Ordinates

We can often graph the sum of two functions by graphing each function separately and then adding their corresponding y-values. This technique is called **addition of ordinates**.

Example 7   Graph $y = x + \cos x$.

Solution   It is possible to make an extensive table of values, plot the points, and draw the curve. However, there is an easier way. Use your knowledge of the functions $y = x$ and $y = \cos x$, and draw each one separately, as in Figure 5–53. Then pick several numbers x, such as x_1, and add the values of x_1 and $\cos x_1$ together to obtain the y-value of a point on the graph of $y = x + \cos x$. For example, if $x_1 = 0$, then the value of $\cos x_1 = 1$. Thus, the point $(0, 0 + 1)$ or $(0, 1)$ is on the graph of $y = x + \cos x$.

As another example, if $x_1 = 1$, the value of $\cos x_1 \approx 0.54$. Thus, the point $(1, 1 + 0.54)$ or $(1, 1.54)$ is on the graph.

It is easy to use a compass to add these values of y. (Refer to Figure 5–53.) Set your compass so that it spans segment BA. Transfer that length to form segment CD, locating point D on the desired graph. However, if $x_1 = \pi$, the value of $\cos x_1$ is negative. In this case, you must subtract the magnitudes of the y-values. Use your compass to transfer length MN to form segment PQ, locating point Q on the desired graph. Repeat this process of adding y-values to determine several points. Then join them with a smooth curve to obtain the graph of $y = x + \cos x$, as in Figure 5–53.

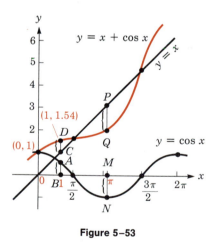

Figure 5–53

Example 8   Graph $y = \cos x + 2 \sin 2x$.

Solution   For values of x between 0 and $\frac{\pi}{2}$, the values of $\cos x$ and $2 \sin 2x$ are both positive. In this case, add the y-values by transferring segments such as MN to new positions such as PQ (see Figure 5–54). This locates point Q, which is on the graph of $y = \cos x + 2 \sin 2x$.

For values of x that are between π and $\frac{3\pi}{2}$, the value of $2 \sin 2x$ is positive, whereas the value of $\cos x$ is negative. In this case, subtract the magnitudes of the y-values of the two functions by transferring segments such as AB to new positions such as CD. This locates point D on the graph of $y = \cos x + 2 \sin 2x$. The graph appears in Figure 5–54.

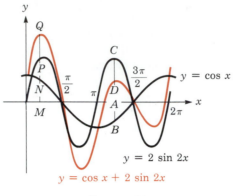

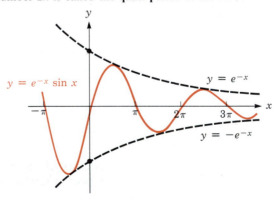

Figure 5–54

Suppose that the amplitude A of the graph of $y = A \sin bx$ is not a constant but, rather, a function of x such as $A(x)$:

$$y = A(x) \sin bx$$

Then we say that the graph has **varying amplitude**. In the next example, the amplitude decreases exponentially, and the graph illustrates a **damped oscillation**.

Example 9   Graph $y = e^{-x} \sin x$.

Solution   Determine the x-intercepts of the graph of $y = e^{-x} \sin x$. Because $e^{-x} \neq 0$ for any x, the only intercepts of the graph are the zeros of the sine function: $x = n\pi$, where n is an integer.

Because $\sin x$ varies between -1 and 1, and $e^{-x} > 0$ for all x, we can establish bounds on the function $y = e^{-x} \sin x$:

$$-1 \leq \quad \sin x \quad \leq 1$$
$$-1e^{-x} \leq e^{-x} \sin x \leq 1e^{-x}$$
$$-e^{-x} \leq e^{-x} \sin x \leq e^{-x}$$

Thus, the graph of the given function lies between the two curves $y = -e^{-x}$ and $y = e^{-x}$. The graph of $y = e^{-x} \sin x$ appears in Figure 5–55. The broken-line graphs of $y = -e^{-x}$ and $y = e^{-x}$ are called **envelopes** of the curve. Because the graph of $y = e^{-x} \sin x$ is not periodic, the number 2π is called the **quasi-period** of the function.

Figure 5–55

Exercise 5.6

In Exercises 1–12 give the number of units and the direction (either up or down) that the graph of each function has been shifted. Also give the period. **Do not draw the graph.**

1. $y = 2 + \sin x$
2. $y = -4 + \cos x$
3. $y = \tan x - 1$
4. $y = \csc x + 3$
5. $y = 7 + 9 \sec 5x$
6. $y = 7 + 9 \cot 5x$
7. $y = 3 - \sin x$
8. $y = -2 - \cos x$
9. $y + 5 = \csc 2x$
10. $y + 5 = \cot 2x$
11. $y = 2(3 + \tan \pi x)$
12. $y = -4(1 - \sec \pi x)$

In Exercises 13–30 give the period and the phase shift (including direction), if any, of each function. **Do not draw the graph.**

13. $y = \sin\left(x - \dfrac{\pi}{3}\right)$
14. $y = \cos\left(x + \dfrac{\pi}{4}\right)$
15. $y = \cos\left(x + \dfrac{\pi}{6}\right)$

16. $y = -\sin\left(x - \dfrac{\pi}{2}\right)$
17. $y + 2 = 3 \cos 2\pi x$
18. $y = 3 \sin \dfrac{2x}{\pi}$

19. $y = \tan(x - \pi)$
20. $y = \csc\left(x + \dfrac{\pi}{6}\right)$
21. $y = -\sec\left(x + \dfrac{\pi}{4}\right)$

22. $y = -2 \sec\left(x - \dfrac{\pi}{3}\right)$
23. $y = \sin(2x + \pi)$
24. $y = \cos(2x - \pi)$

25. $y = \tan\left(\dfrac{\pi x}{2} + \dfrac{\pi}{4}\right)$
26. $y = \csc\left(\dfrac{2\pi x}{3} + \dfrac{\pi}{9}\right)$
27. $y = 2 \sec\left(\dfrac{1}{3}x - 6\pi\right)$

28. $y = 2 \cot\left(\dfrac{\pi}{10} + \dfrac{x}{5}\right)$
29. $2y = 3 \cot\left(7x - \dfrac{21}{2}\pi\right)$
30. $17y = \sec\left(\dfrac{x}{5} + \dfrac{\pi}{4}\right)$

In Exercises 31–46 graph each equation through at least one period.

31. $y = -4 + \sin x$
32. $y + 2 = \tan \dfrac{x}{2}$
33. $y = 3 - \sec x$
34. $y = 1 + \csc x$

35. $y = \cot \dfrac{x}{2} - 2$
36. $y = 1 - 2 \cos x$
37. $y = \sin\left(x + \dfrac{\pi}{2}\right)$
38. $y = -\cos\left(x - \dfrac{\pi}{2}\right)$

39. $y = \tan\left(x - \dfrac{\pi}{2}\right)$
40. $y = \csc\left(x + \dfrac{\pi}{4}\right)$
41. $y = \cos(2x + \pi)$
42. $y = \sin(3x - \pi)$

43. $y = \sec\left(3x + \dfrac{\pi}{2}\right)$
44. $y = \tan\left(\dfrac{x}{2} - \dfrac{\pi}{2}\right)$

45. $y - 1 = \csc\left(2x - \dfrac{\pi}{6}\right)$
46. $y + 2 = \cot\left(\dfrac{x}{3} - \dfrac{\pi}{2}\right)$

In Exercises 47–52 use the method of addition of ordinates to graph each equation on the interval from 0 to 4π.

47. $y = \sin x + \cos x$
48. $y = 2 \sin x + \cos x$
49. $y = \sin \dfrac{x}{2} + \sin x$

50. $y = x + \sin x$
51. $y = -x + \cos x$
52. $y = x - \cos x$

In Exercises 53–62 graph each equation.

53. $y = \sin |x|$
54. $y = \cos |x|$
55. $y = \sin^2 x$
56. $y = \tan^2 x$
57. $y = 2^x \sin x$
58. $y = 2^{-x} \sin x$
59. $y = |x| \cos x$
60. $y = |x| \sin x$
61. $y = \dfrac{\sin x}{x}$
62. $y = 2^{-x} \cos x$

REVIEW EXERCISES

In Review Exercises 1–4 tell whether each angle is in standard position.

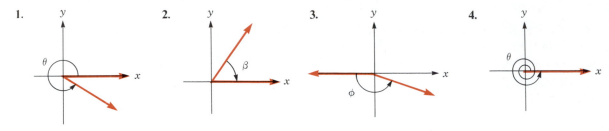

1.   **2.**   **3.**   **4.**

5.   Are angles of $310°$ and $-410°$ coterminal?   **6.**   Are angles of $190°$ and $820°$ coterminal?

In Review Exercises 7–10 assume that angle θ is in standard position and find the values of the remaining trigonometric functions of θ.

7. $\sin \theta = -\dfrac{7}{10}$; θ in QIII   **8.** $\tan \theta = \dfrac{7}{9}$; θ not in QI

9. $\cos \theta = -\dfrac{7}{10}$; θ in QII   **10.** $\cot \theta = -\dfrac{9}{8}$; θ in QIV

*In Review Exercises 11–14 evaluate each expression. **Do not use a calculator.***

11. $\sin 45° \cos 30°$ **12.** $\cos 120° \tan 135°$

13. $\tan^2 225° \cos^2 30° \sin^2 300°$ **14.** $\sec 30° \csc 30° + \sec 330° \csc 330°$

*In Review Exercises 15–18 find the value of the sine, cosine, and tangent of the given angle. **Do not use a calculator.***

15. $930°$ **16.** $1380°$ **17.** $-300°$ **18.** $-585°$

In Review Exercises 19–22 use a calculator or Table A in Appendix II to evaluate the sine, cosine, and tangent of the given angle. Give the answer to four decimal places.

19. $15°$ **20.** $160°$ **21.** $265°$ **22.** $-340°$

In Review Exercises 23–28 use a calculator or Table A in Appendix II to find angle ϕ $(0° \le \phi < 360°)$ in the given quadrant. Check your work with a calculator.

23. $\sin \phi = 0.8746$; ϕ in QII   **24.** $\tan \phi = 0.6009$; ϕ in QIII

25. $\cos \phi = 0.7314$; ϕ in QIV   **26.** $\sec \phi = 1.871$; ϕ in QI

27. $\cot \phi = -0.1763$; ϕ in QII   **28.** $\csc \phi = -1.046$; ϕ in QIV

In Review Exercises 29–32 change each angle to radians.

29. $105°$ **30.** $325°$ **31.** $318°$ **32.** $-105°$

In Review Exercises 33–36 each angle is expressed in radians. Express each one in degrees.

33. $\dfrac{19\pi}{6}$ **34.** $-\dfrac{5\pi}{6}$ **35.** 7π **36.** 8

In Review Exercises 37–40 find the values of each function without using a calculator. Then check your work with a calculator.

37. $\sin \dfrac{5\pi}{6}$

38. $\cos\left(-\dfrac{13\pi}{6}\right)$

39. $\tan\left(-\dfrac{\pi}{3}\right)$

40. $\csc \dfrac{\pi}{6}$

41. The latitude of Springfield, Illinois, is 39.8° N. How far is Springfield from the Equator? Use 3960 miles as the radius of the earth.

42. Des Moines, Iowa, is approximately 2870 miles north of the Equator. If the radius of the earth is about 3960 miles, find the latitude of Des Moines.

43. Find the area of a sector of a circle if the sector has a central angle of 15° and the circle has a radius of 12 centimeters.

44. Find the angular velocity of the earth in radians per second.

45. A truck is moving at 50 ft/sec. How fast are its 32-inch-diameter tires turning in revolutions per minute?

46. A vehicle has 40-inch-diameter tires that are making 100 rpm. How fast is the vehicle going in feet per minute?

In Review Exercises 47–48 find the coordinates of the point P on the unit circle that corresponds to each real number. Do not use a calculator.

47. $\dfrac{7\pi}{6}$

48. $\dfrac{13\pi}{4}$

In Review Exercises 49–52 use a calculator to find the values of each function, if possible. Give all answers to the nearest ten-thousandth.

49. $\cos 7$

50. $\tan(2 + \pi)$

51. $\csc 3\pi$

52. $\sin(\pi^3)$

In Review Exercises 53–56 find the amplitude and the period of each function.

53. $y = 4 \sin 3x$

54. $y = \dfrac{\cos 4x}{8}$

55. $y = -\dfrac{1}{3} \cos \dfrac{x}{3}$

56. $y = 0.875 \sin \dfrac{x}{4}$

In Review Exercises 57–60 find both the vertical shift and the phase shift, if any, for the graph of each function.

57. $y = 2 + \tan x$

58. $y = \csc\left(2x + \dfrac{\pi}{3}\right)$

59. $y - 4 = 3 \sin\left(\dfrac{x}{7} + \dfrac{3}{2}\right)$

60. $y = -\cos\left(\dfrac{x}{5} - \dfrac{\pi}{2}\right) - 1$

In Review Exercises 61–70 graph each equation.

61. $y = 4 \sin x$

62. $y = 0.5 \cos x$

63. $y = \cos \dfrac{x}{4}$

64. $y = \tan \dfrac{\pi x}{3}$

65. $y = 3 + \sin x$

66. $y = -2 + \tan x$

67. $y = 2 \sin\left(x - \dfrac{5\pi}{6}\right)$

68. $y = \tan\left(x - \dfrac{2\pi}{3}\right)$

69. $y = 2 \cos x + \sin \dfrac{x}{2}$

70. $y = \dfrac{x}{2} + \cos \dfrac{x}{2}$

6
APPLICATIONS OF TRIGONOMETRY

In this chapter, we shall discuss several applications of the trigonometric functions.

6.1 RIGHT TRIANGLE TRIGONOMETRY

The word *trigonometry* means "measuring triangles (trigons)." The ability to determine all sides and angles of a right triangle when only some are known has made trigonometry indispensible in astronomy, navigation, and surveying.

To solve right triangles, we view the definitions of the trigonometric functions in a different way. We place right triangle ABC on a coordinate system so it is the reference triangle for the acute angle A. See Figure 6–1. We let a be the length of BC, the side opposite angle A, let b be the length of AC, the side adjacent to angle A, and let c be the length of the hypotenuse. The trigonometric functions of the acute angle A can be defined as ratios involving the sides of right triangle ABC.

Definition. If angle A is an acute angle in the right triangle of Figure 6–1, then

$$\sin A = \frac{\text{opposite side}}{\text{hypotenuse}} \qquad \csc A = \frac{\text{hypotenuse}}{\text{opposite side}}$$

$$\cos A = \frac{\text{adjacent side}}{\text{hypotenuse}} \qquad \sec A = \frac{\text{hypotenuse}}{\text{adjacent side}}$$

$$\tan A = \frac{\text{opposite side}}{\text{adjacent side}} \qquad \cot A = \frac{\text{adjacent side}}{\text{opposite side}}$$

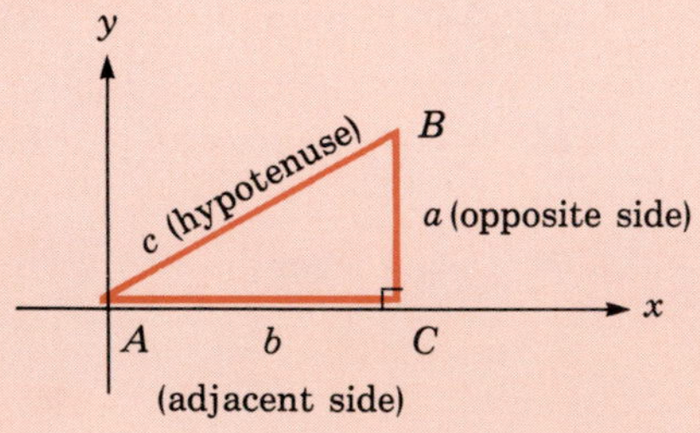

Figure 6–1

Example 1   In the right triangle shown in Figure 6–2, $a = 5$ and $c = 13$. Find **a.** side b and **b.** csc B.

Solution   **a.**   Use the Pythagorean theorem to find the length of side b:

$$a^2 + b^2 = c^2$$
$$5^2 + b^2 = 13^2$$
$$b^2 = 144$$
$$b = 12$$

The length of side b is 12 units.

b. $\text{csc } B = \dfrac{\text{hypotenuse}}{\text{side opposite angle } B}$

$$= \dfrac{c}{b}$$

$$= \dfrac{13}{12}$$

Thus, $\text{csc } B = \dfrac{13}{12}$.

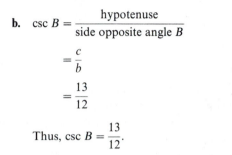

Figure 6–2

If an observer looks up at an object such as an airplane, the angle that the observer's line of sight makes with the horizontal is called the **angle of elevation**. If the observer looks down to see the object, the angle made with the horizontal is called the **angle of depression**. See Figure 6–3.

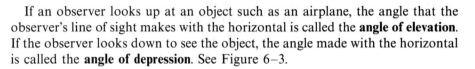

Figure 6–3

Example 2   A circus tightrope walker ascends from the ground to a platform 75 feet above an arena by walking a taut cable. How far must the performer walk if the cable makes an angle of 37°30′ with the ground?

Solution   In Figure 6–4, the height of the platform represents the side opposite the 37°30′ angle, and x represents the length of the hypotenuse, which is the distance the performer must walk. Because the sine function involves the opposite side and the hypotenuse, proceed

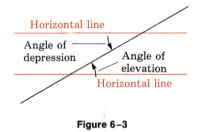

Figure 6–4

as follows:

$$\sin 37.5° = \frac{\text{opposite side}}{\text{hypotenuse}} \qquad \text{Change } 37°30' \text{ to } 37.5°.$$

$$\sin 37.5° = \frac{75}{x}$$

$$x \sin 37.5° = 75$$

$$x = \frac{75}{\sin 37.5°}$$

$$\approx 123.2010$$

$$\approx 123$$

The tightrope artist must walk approximately 123 feet.   ■

You might wonder why the result in Example 2 was rounded off to three significant digits. If the angle were exactly 37.5° and the opposite side were exactly 75 feet long, rounding off would be unnecessary. More likely, however, the measurements are not exact, but only approximate. In that case, an answer to the nearest ten-thousandth is unwarranted.

Calculators provide answers to 8, 10, or 12 digits. You must decide how many of them are significant. A good rule of thumb for determining acceptable accuracy is provided in the following table.

Accuracy in measurement of sides	Accuracy in measurement of angles
Two significant digits	Nearest degree
Three significant digits	Nearest tenth of a degree, or the nearest multiple of 10 minutes
Four significant digits	Nearest hundredth of a degree, or the nearest minute

It is important to remember that answers can be only as accurate as the least accurate of the given data. However, if a calculation requires several intermediate steps, keep at least one extra digit until you round the final answer.

If a number is not written in scientific notation, it can be difficult to decide how many significant digits the number has. For example, is the number 140 accurate to two or three significant digits? If the number is rounded to the nearest ten, the zero is merely a placeholder and 140 has two significant digits. However, if the number has been rounded to the nearest unit, the zero is significant and 140 has three significant digits. In this book, we shall assume the greatest possible number of significant digits unless it is stated otherwise—therefore, 140 has three significant digits.

3234 has four significant digits

104 has three significant digits

140.00 has five significant digits

0.00012 has two significant digits

0.000120 has three significant digits

1.0003 has five significant digits

In nautical navigation and surveying, the **bearing** of a point, such as point A in Figure 6–5, from a point O (the observer) is the acute angle measured from the north–south line to the line segment OA. This bearing is denoted as N 30° E and is read as "north 30° east" or "30° east of north." The bearing of point B from point O is N 75° W, the bearing of point C from O is S 20° W, and the bearing of point D from O is S 80° E.

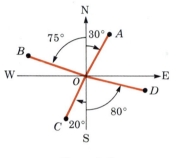

Figure 6–5

Example 3   Perryville is 25.0 miles due south of Rock City, and Prairie Town is due east. What is the bearing of Prairie Town from Perryville if they are 94.0 miles apart?

Solution   To find the bearing, find angle θ, shown in Figure 6–6, by using the cosine ratio:

$$\cos \theta = \frac{\text{adjacent side}}{\text{hypotenuse}}$$

$$\cos \theta = \frac{25.0}{94.0}$$

$$\approx 0.2660$$

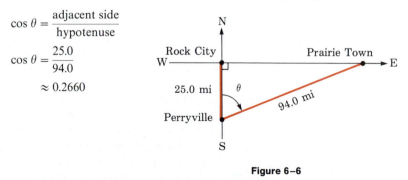

Figure 6–6

Now look up .2660 in the body of Table A in Appendix II or enter .2660 and press INV COS on a calculator set in degree mode, and you will find that $\theta \approx 74.6°$. Because the distances between the towns are accurate to three significant digits, the angle is given to the nearest tenth of a degree. Thus, the bearing of Perryville from Prairie Town is N 74.6° E. ■

Example 4   Two buildings are separated by a distance of d meters. From the top of the shorter building, the angle of elevation to the top of the taller building is θ, and the angle of depression to the base of the taller building is ϕ. How tall are the two buildings?

Solution   Let the height of the shorter building be a meters and the height of the taller building be b meters, as in Figure 6–7. Further suppose that point C on the taller building is level with the top of the shorter building. From the figure, it follows that

$$\tan \phi = \frac{a}{d}$$

or

$$a = d \tan \phi$$

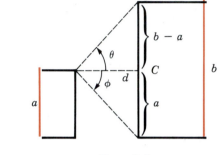

Also,

$$\tan \theta = \frac{b - a}{d}$$

or

$$b - a = d \tan \theta$$

Figure 6–7

Solving the previous equation for b and substituting gives

$$b = a + d \tan \theta$$
$$= d \tan \phi + d \tan \theta \qquad \text{Substitute } d \tan \phi \text{ for } a.$$
$$= d(\tan \phi + \tan \theta)$$

The height of the shorter building is $d \tan \phi$, and the height of the taller building is $d(\tan \phi + \tan \theta)$. ■

Example 5   An observer A notices that the Space Needle of Seattle is due north and that the angle of elevation to its top is 44.4°. A second observer B, 706 feet due east of A, notices that the bearing of the Space Needle is N 48.8° W. How tall is the Space Needle?

Solution   Refer to Figure 6–8, in which triangle PAB lies on the ground, and triangle APQ sits on its edge with the Space Needle as one of its sides. Not enough information is given about triangle APQ to determine the height h from that triangle alone. However, if the distance x were known, the height h could be computed. Note that

$$\tan 44.4° = \frac{h}{x}$$

or

$$h = x \tan 44.4°$$

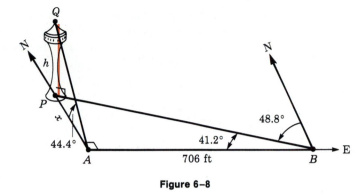

Figure 6–8

The distance x can be found by working with the triangle on the ground. Triangle PAB is a right triangle with a right angle at A. Angle PBA is $90.0° - 48.8°$, or $41.2°$. Form the equation

$$\tan 41.2° = \frac{x}{706}$$

or

$$x = 706 \tan 41.2°$$

Putting these facts together gives

$$h = {}_x\tan 44.4°$$
$$h = \mathbf{706\ tan\ 41.2°}\tan 44.4° \qquad \text{Substitute } 706 \tan 41.2° \text{ for } x.$$
$$\approx 605$$

Thus, the Space Needle is approximately 605 feet tall.   ■

Vectors

Quantities that have only magnitude are called **scalar quantities**. However, some quantities have both magnitude and direction. For example, the flight of an airplane is described by both a speed and a direction. Quantities with both magnitude and direction are called **vector quantities** and are represented by mathematical entities called **vectors**.

> **Definition.**   A **vector** is a directed line segment. The direction of the vector is indicated by the angle it makes with some convenient reference line.
>   The **norm**, or **magnitude**, of a vector is the length of the line segment. If a vector is denoted by **v**, the norm of the vector is denoted by $|\mathbf{v}|$.

Any two directed line segments that have the same length and the same direction are regarded as **equal vectors**. When drawing diagrams containing vectors, it is common to position the vectors in the most convenient locations. For example, a force of 30 pounds exerted in a northwesterly direction can be represented by the directed line segment in Figure 6–9. The length of the vector is 30 units, and, because it is the terminal side of a 135° angle, it points northwest.

A 20-mph wind blowing from the east can be represented by the vector shown in Figure 6–10. An airplane flying 350 mph on a **heading** (an intended direction of travel measured clockwise from the north line) of 240° can be represented by the vector shown in Figure 6–11.

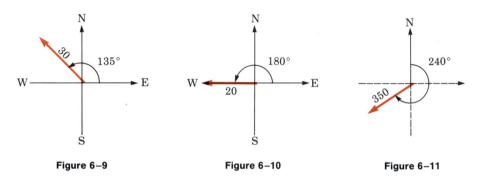

Figure 6–9   Figure 6–10   Figure 6–11

Vector quantities can be added, but the process must take into account both their norms and their directions. For example, two forces of 40 pounds exerted on the same object might not combine to be an 80-pound force. If they acted in opposite directions, the combined force would be zero.

Vector quantities can be added by using the **parallelogram law**. If two vectors originating from a common point are adjacent sides of a parallelogram, then their **vector sum**, called the **resultant vector**, is the vector represented by the diagonal of the parallelogram that is drawn from the common point. In Figure 6–12, the sum of vectors **AB** and **AD** is the vector **AC**.

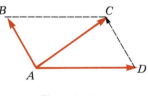

Figure 6–12

Note that vector **DC** has the same magnitude and direction as vector **AB**.

Example 6   A boat, capable of a speed of 8.0 mph in still water, attempts to go directly across a river with a current of 3.0 mph. By what angle is the boat pushed off its intended path? What is the effective speed of the boat?

Solution   The given velocities can be represented by vectors as shown in Figure 6–13. The direction the boat travels, called its **course**, is represented by a vector sum, which is the diagonal of the rectangle. Angle θ specifies the direction in which the boat is forced to travel, and $d(OP)$ specifies the effective speed of the boat. Because $d(AP)$ is also 3.0 units, $\tan \theta = \frac{3}{8}$.

From this relationship, you can calculate θ and find that

$$\theta \approx 21^\circ$$

Thus, the river pushes the boat approximately 21° off its intended path.

Then use the Pythagorean theorem to find $d(OP)$ and, hence, the speed of the boat:

$$\begin{aligned} d(OP) &= \sqrt{8^2 + 3^2} \\ &= \sqrt{73} \\ &\approx 8.5 \end{aligned}$$

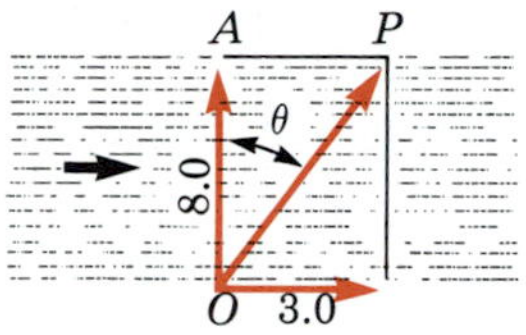

Figure 6–13

The effective speed of the boat is approximately 8.5 mph. ■

Example 7 An airplane, capable of a speed of 270 mph in still air, sets a heading of 75°. A very strong wind, blowing in the direction of 165°, forces the plane onto a course that is due east. What is the velocity of the wind, and what is the **ground speed** (the speed relative to the ground) of the plane?

Solution The velocities are represented by the vectors in Figure 6–14. Vector **w** represents the velocity of the wind, and vector **v** represents the ground speed of the plane. Because $165^\circ - 75^\circ = 90^\circ$, the vector parallelogram is a rectangle with angle α equal to 90°. Because each triangle formed by the diagonal is a right triangle,

$$\tan 15^\circ = \frac{|\mathbf{w}|}{270}$$

Multiplying both sides by 270 gives

$$\begin{aligned} |\mathbf{w}| &= 270 \tan 15^\circ \\ &\approx 72 \end{aligned}$$

Also,

$$\cos 15^\circ = \frac{270}{|\mathbf{v}|}$$

or

$$\begin{aligned} |\mathbf{v}| \cos 15^\circ &= 270 \\ |\mathbf{v}| &= \frac{270}{\cos 15^\circ} \\ &\approx 280 \end{aligned}$$

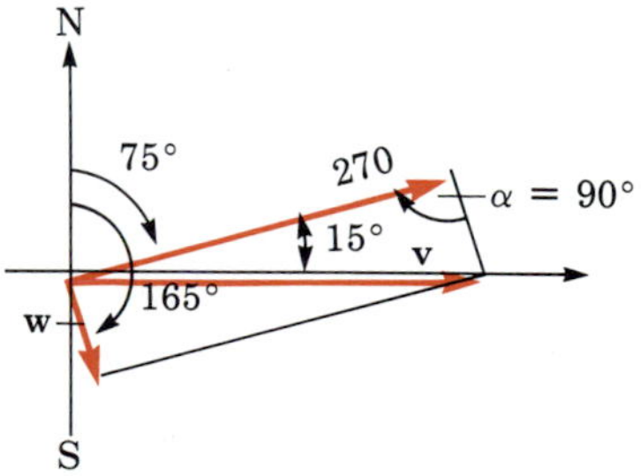

Figure 6–14

The wind velocity is approximately 72 mph, and the ground speed of the plane is approximately 280 mph. ■

Instead of two vectors combining to form a resultant vector, we often separate a vector into its component parts. For example, we suppose that the car shown

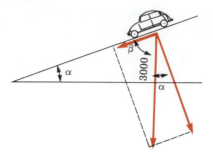

Figure 6–15

in Figure 6–15, weighing 3000 pounds, is parked on a hill. It is pulled downward by gravity with a force of 3000 pounds. Part of this force appears as a tendency for the car to roll down the hill, and another part presses the car against the hill. Just how the 3000 pounds is apportioned depends on the angle of the hill. If there were no hill, there would be no tendency to roll; if the hill were very steep, only a small force would hold the car on the road. The weight of the car is said to be **resolved** into two components—one directed down the hill, and the other directed into the hill. These vectors obey the parallelogram law. Note that the angle α the road makes with the horizontal is equal to the angle between the two vectors because both of these angles are complementary to angle β.

Example 8   A 3000-pound car sits on a 23.0° incline. What force is required to prevent the car from rolling down the hill? With what force is the car held to the roadway?

Solution   You must find the norms of vectors **t** and **n** in Figure 6–16. Because $OABC$ is a rectangle, the opposite sides are equal and angle $C = 90°$. Thus,

$$\sin 23.0° = \frac{|t|}{3000}$$

$$|t| = 3000 \sin 23.0°$$

$$\approx 1170$$

Also,

$$\cos 23.0° = \frac{|n|}{3000}$$

$$|n| = 3000 \cos 23.0°$$

$$\approx 2760$$

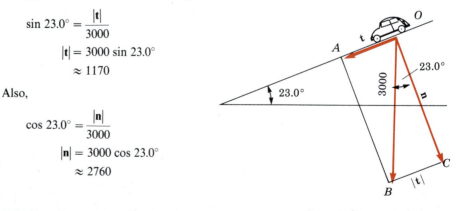

Figure 6–16

A force of approximately 1170 pounds is required to prevent rolling, and a force of approximately 2760 pounds holds the car on the road. ∎

Exercise 6.1

In Exercises 1–2 solve each triangle by finding the unknown sides and angles.

1.

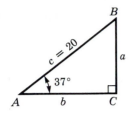

2.

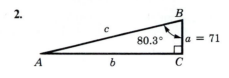

3. If the angle of elevation to the top of a flagpole from a point 40 feet from its base is $20°$, find the height of the flagpole.

4. A person on the edge of a cliff looks down at a boat on a lake. The angle of depression of the person's line of sight is $10.0°$, and the line-of-sight distance from the person to the boat is 555 feet. How high is the cliff?

5. A car drives up a long hill on a road that makes an angle of $7°30'$ with the horizontal. How far will the car travel to reach the top if the horizontal distance traveled is 715 feet?

6. On an approach to an airport, a plane is descending at an angle of $3°30'$. How much altitude is lost as the plane travels a horizontal distance of 21.2 miles?

7. A plane loses 2750 feet in altitude as it travels a horizontal distance of 39,300 feet. What is its angle of descent?

8. A train rises 230 feet as it travels 1 mile up a steep grade. What is its angle of ascent? (*Hint:* 5280 feet = 1 mile.)

9. The angle of depression from the top of a tree to a point of the ground is $57°45'$. Find the height of the tree if the line-of-sight distance from the top of the tree to the point on the ground is 34.23 feet.

10. An observer noted that the angle of elevation to a plane passing over a landmark was $32°40'$. If the landmark was 1530 meters from the observer, what was the altitude of the plane?

11. A ship leaves from a port of call bearing S $12.7°$ E. How far south has the ship traveled during a trip of 327 miles?

12. The bearing of Madison, Wisconsin, from Stevens Point, Wisconsin, is S $4.1°$ E. The distance between the cities is 108 miles. How much farther east is Madison than Stevens Point?

13. A ship leaves port and sails 8800 kilometers due west. It then sails 4500 kilometers due south. To the nearest tenth of a degree, what is the ship's bearing from its port?

14. A ship is 3.3 miles from a lighthouse. It is also due north of a buoy that is 2.5 miles due east of the lighthouse. Find the bearing of the ship from the lighthouse.

15. Two lighthouses are on an east–west line. The bearing of a ship from one lighthouse is N $59°$ E, and the bearing from the other lighthouse is N $31°$ W. How far apart are the lighthouses if the ship is 5.0 miles from the first lighthouse?

16. Two lookout stations are on a north–south line. The bearing of a forest fire from one lookout is S $67°$ E, and the bearing of the fire from the second lookout is N $23°$ E. If the fire is 3.5 kilometers from the second lookout station, how far is the fire from the first lookout station?

17. A plane flying at 18,100 feet passes directly over an observer. Thirty seconds later, the observer notes that the plane's angle of elevation is $31.0°$. How fast is the plane going in miles per hour? (*Hint:* 5280 feet = 1 mile.)

18. A plane flying horizontally at 650 miles per hour passes directly over a small city. One minute later, the pilot notes that the angle of depression to that city is $13°$. What is the plane's altitude in feet?

19. Use the information in Illustration 1 to compute the height of George Washington's face on Mount Rushmore.

20. A boat is 537 meters from a lighthouse and has a bearing from the lighthouse of N 33.7° W. A second boat is 212 meters from the same lighthouse and has a bearing from the lighthouse of S 20.1° W. How many meters north of the second boat is the first?

21. A plane is flying at an altitude of 5120 feet. As it approaches an island, the navigator determines the angles of depression as in Illustration 2. What is the length of the island in feet and in miles?

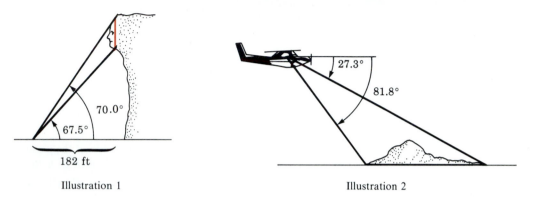

Illustration 1   Illustration 2

22. From an observation point 6500 kilometers from a launch site, an observer watches the vertical flight of a rocket. At one instant, the angle of elevation of the rocket is 15°. How far will the rocket ascend in the time it takes the angle of elevation to increase by 57°?

23. Refer to Illustration 3 and find θ.

24. Refer to Illustration 3 and find ϕ.

25. Compute the height h of the Sears Tower using the information given in Illustration 4.

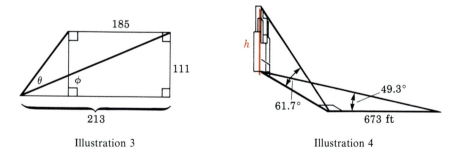

Illustration 3   Illustration 4

26. Compute the height h of the Empire State Building using the information given in Illustration 5.

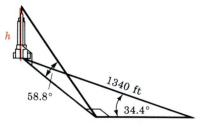

Illustration 5

27. Bill and Paula, standing on the same side of and in line with the Washington Monument, are looking at its top. The angle of elevation from Bill's position is 34.1°, and the angle of elevation from Paula's position is 60.0°. If Bill and Paula stand on level ground and are 500 feet apart, how tall is the monument?

28. Use the information given in Illustration 6 to compute the height h of the figure part of the Statue of Liberty.

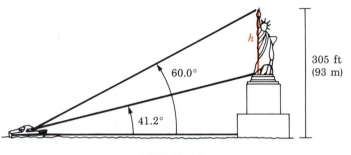

305 ft
(93 m)

60.0°

41.2°

h

Illustration 6

29. Use the information given in Illustration 7 to compute the height h of the Gateway Arch in St. Louis.

30. The bearing of point C from point A is N 25.0° E, and the bearing of point C from point B is N 40.0° W. If A and B are on an east–west line and 200 kilometers apart, how far is B from C?

31. Gail is standing 152 feet due east of a television tower. She then travels along a bearing of S 60.0° W. How tall is the tower if the angle of elevation to its top from her closest position to the tower is 58.2°?

32. Jeffrey stands due south of the Eiffel Tower and notes that the angle of elevation to its top is 60.0°. His brother Grant, standing 298 feet due east of Jeff, notes an angle of elevation of 56.9°. How tall is the tower? (*Hint:* First consider a triangle on the ground and use the Pythagorean theorem.)

33. Two tangents t are drawn from a point P to a circle of radius r as in Illustration 8. The angle between the tangents is θ. How long is each tangent?

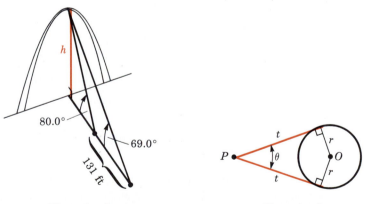

h

80.0°

69.0°

131 ft

Illustration 7

t

P

θ

r

O

r

t

Illustration 8

34. A regular polygon has n equal sides, each of length a. The radius of the inscribed circle is r. Express a as a function of n and r.

35. An isosceles triangle has a base angle of α and a side of k centimeters. Express the length b of its base in terms of α and k.

36. At noon a ship left port bearing N $\alpha°$ E and sailed k kilometers. At the same time another ship left port bearing N $\alpha°$ W and also sailed k kilometers. Express the distance d between the two ships in terms of α and k.

37. Point A is k meters from a building. From point A the angle of elevation to the top of the building is α, and the angle of depression to the base of the building is β. Express the height H of the building in terms of k, α, and β.

38. A ladder k feet long reaches a height of H feet on the side of a building, and the ladder makes an angle of α with the horizontal. The ladder slips so that it reaches a height of only h feet on the side of the building. It then makes an angle of β with the horizontal. Express the distance d that the ladder has come down the building in terms of k, α, and β.

39. From an observation point D meters from a launch site, an observer watches the vertical flight of a rocket. At one instant, the angle of elevation of the rocket is θ, and at a later instant, ϕ. How far has the rocket flown during that time?

40. Two helicopters are hovering at an altitude of 5000 feet. The pilot of one of the helicopters observes a crate being dropped from the other aircraft. At one instant, the pilot measures the angle of depression to the crate to be α. When the crate hits the ground, the pilot measures the angle of depression to be β. How far has the crate fallen in that time?

41. Two tall buildings are separated by a distance of d meters. From the top of the shorter building, the angle of elevation to the top of the taller building is α, and the angle of depression to the base of the taller building is β. Express the height a of the shorter building in terms of d and β.

42. Express the height b of the taller building described in Exercise 41 in terms of d, α, and a.

In Exercises 43–58 you will need vectors.

43. A boat capable of a speed of 6 miles per hour in still water attempts to go directly across a river. As the boat crosses the river, it drifts 30° from its intended path. How strong is the current? What is the effective speed of the boat?

44. A boat capable of a speed of 11 miles per hour in still water attempts to go directly across a river with a current of 5.6 miles per hour. By what angle is the boat pushed off its intended path? What is the effective speed of the boat?

45. Laura can row a boat $\frac{1}{2}$ mile per hour in still water. She attempts to row straight across a river that has a current of 1 mile per hour. If she must row for 2 hours to cross the river, by what angle is the current pushing her off her intended path? Give the answer to the nearest degree.

46. A boat attempts to go directly across a river with a current of 3.7 miles per hour. The current causes the boat to drift 23° from its intended path. How far will the boat travel if the trip takes 10 minutes?

47. A plane has a heading of 260.0° and is flying at 357 miles per hour. If a southerly wind causes its course (the direction it is actually going) to be due west, find the ground speed of the plane (its speed relative to the ground).

48. A plane has an airspeed of 411 miles per hour and a heading of 90.0°. A wind from the north is blowing at 31.0 miles per hour. By how many degrees is the plane blown off its heading? Find the ground speed of the plane (its speed relative to the ground).

49. A plane leaves an airport with a heading of 45.0° and an airspeed of 201 miles per hour. At the same time, another plane leaves the same airport with a heading of 135.0° and an airspeed of 305 miles per hour. At the end of 2 hours, what is the bearing of the first plane from the second? (Assume no wind.)

50. A plane leaves an airport with a heading of 170°. At the same time, a second plane leaves the same airport with a heading of 260°. One hour later, the first plane is 1300 miles directly southeast of the second plane. What is the airspeed of the first plane? (Assume no wind.)

51. What force is required to keep a 2210-pound car from rolling down a ramp that makes a 10.0° angle with the horizontal?

52. A force of 25 pounds is necessary to hold a barrel in place on a ramp that makes an angle of 7° with the horizontal. How much does the barrel weigh?

53. A vehicle presses against a roadway with a force of 1100 pounds. How much does the vehicle weigh if the roadway makes an angle of 18.0° with the horizontal?

54. A board will break if it is subjected to a force greater than 350 pounds. Will the board hold a 450-pound piano supported by a single dolly as it slides up the board and into a truck? Assume that the board makes an angle of 35.0° with the horizontal.

55. A garden tractor weighing 351 pounds is being driven up a ramp onto a trailer. If the tractor presses against the ramp with a force of 341 pounds, what angle does the ramp make with the horizontal?

56. If a force of 21.3 pounds is necessary to keep a 50.1-pound barrel from rolling down an inclined plane, what angle does the inclined plane make with the horizontal?

57. A 201-pound force is directed due east. What force, directed due north, is needed to produce a resultant force of 301 pounds? What angle is formed by the vectors representing the 201-pound force and the 301-pound force?

58. A 312-pound force is directed due west. What force, directed exactly southeast, would cause the resultant force to be directed due south?

In Exercises 59–60 you must draw some lines to create right triangles.

59. A plane leaves an airport at 12:00 noon with a heading of 60.0° and an airspeed of 451 miles per hour. One hour later, another plane leaves the same airport with a heading of 290.0° and an airspeed of 611 miles per hour. What is the bearing of the first plane from the second at 2:00 P.M.? (Assume no wind.)

60. A plane leaves an airport at 3:00 P.M. with a heading of 70.0° and an airspeed of 512 miles per hour. Two hours later, another plane leaves the same airport with a heading of 100.0° and an airspeed of 621 miles per hour. How far apart are the planes at 6:00 P.M.? (Assume no wind.)

6.2 THE LAW OF COSINES

The problem-solving techniques of the previous section cannot be applied to triangles that do not contain a 90° angle, called **oblique triangles**. To solve oblique triangles when three sides are given, or when two sides and the angle between them are given, we use a set of formulas called the **law of cosines**.

In the following discussion, capital letters will be used to name vertices of a triangle, and corresponding lowercase letters will be used to name sides opposite those vertices. For example, side b represents the side opposite angle B.

To develop the law of cosines, we refer to Figure 6–17, where the coordinates of points A, B, and C are $(0, 0)$, $(c \cos A, c \sin A)$, and $(b, 0)$, respectively. We can now use the distance formula to compute a^2.

$$a^2 = (c \cos A - b)^2 + (c \sin A - 0)^2$$
$$= c^2 \cos^2 A - 2bc \cos A + b^2 + c^2 \sin^2 A$$
$$= c^2(\cos^2 A + \sin^2 A) + b^2 - 2bc \cos A$$
$$= c^2 + b^2 - 2bc \cos A$$

Remember that $\cos^2 \theta + \sin^2 \theta = 1$.

1. $a^2 = b^2 + c^2 - 2bc \cos A$

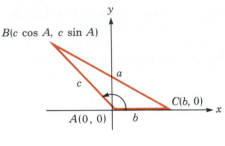

Figure 6–17

Although the triangle in Figure 6–17 is obtuse, this derivation is valid for any triangle. Also, it need not be vertex A that is placed at the origin. If point B had been placed at the origin instead of point A, a different but similar formula would occur.

2. $b^2 = c^2 + a^2 - 2ca \cos B$

If point C had been placed at the origin, a third formula would occur.

3. $c^2 = a^2 + b^2 - 2ab \cos C$

These three formulas are called the **law of cosines**.

The Law of Cosines.   The square of any side of any triangle is equal to the sum of the squares of the remaining two sides, minus twice the product of these two sides and the cosine of the angle between them.

$a^2 = b^2 + c^2 - 2bc \cos A$

$b^2 = c^2 + a^2 - 2ca \cos B$

$c^2 = a^2 + b^2 - 2ab \cos C$

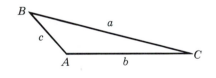

Example 1   Oblique triangle ABC has $b = 27$, $c = 14$, and $A = 43°$. Find side a.

Solution   Sides b and c and the included angle A are given. See Figure 6–18. Use the law of cosines.

$a^2 = b^2 + c^2 - 2bc \cos A$

$\quad = 27^2 + 14^2 - 2(27)(14) \cos 43°$

$\quad \approx 729 + 196 - 552.90$

$\quad \approx 372.10$

$a \approx 19.29$

$\quad \approx 19$

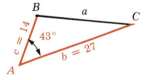

Figure 6–18

The value of a is approximately 19 units.

Example 2   In the oblique triangle ABC in Figure 6–19, $a = 5.2$, $b = 3.7$, and $c = 7.1$ units. Find angle B.

Solution   Use the form of the law of cosines that involves angle B.

$$b^2 = a^2 + c^2 - 2ac \cos B$$

Solve this formula for $\cos B$, substitute the values for a, b, and c, and calculate angle B.

$$\cos B = \frac{a^2 + c^2 - b^2}{2ac}$$

$$\cos B = \frac{5.2^2 + 7.1^2 - 3.7^2}{2(5.2)(7.1)}$$

$$\cos B \approx 0.8635$$

$$B \approx 30.29°$$

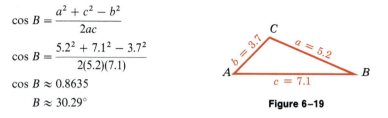

Figure 6–19

To the nearest degree, angle $B = 30°$. ∎

Example 3   A farmer uses two horses and two ropes to pull a tractor out of the mud. When pulled tight, the ropes form an angle of 27°, and each horse exerts a pull of 950 pounds. What force is applied to the tractor?

Solution   By the parallelogram law for adding vectors, the combined force is length b in the diagram of Figure 6–20. Because opposite sides of a parallelogram are equal, all four sides are 950 units in length. Because consecutive angles of a parallelogram are supplementary, the obtuse angle at B is $180° - 27° = 153°$. Apply the law of cosines to triangle ABC.

$$b^2 = a^2 + c^2 - 2ac \cos B$$

$$b^2 = 950^2 + 950^2 - 2(950)(950) \cos 153°$$

$$\approx 3{,}413{,}266.8$$

$$b \approx 1847.5$$

$$\approx 1800$$

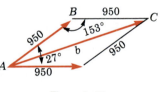

Figure 6–20

The combined pull of the horses is approximately 1800 pounds. ∎

Example 4   An airplane flies N 86.2° W for a distance of 143 kilometers. The pilot, experiencing some engine problems, alters course and flies 79.5 kilometers in the direction S 32.7° E in an attempt to find a place to land. He crash-lands safely in a cornfield. How far, and in what direction, will the rescue team need to travel?

Solution   The course diagram appears in Figure 6–21. To find distance d, use the law of cosines.

$$d^2 = 79.5^2 + 143^2 - 2(79.5)(143) \cos 53.5°$$

$$\approx 13{,}244.8$$

$$d \approx 115.1$$

$$\approx 115$$

The pilot is about 115 kilometers from the airport.

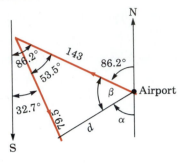

Figure 6–21

As a first step toward finding the direction in which the rescue team must travel, use the law of cosines again to find angle β.

$$79.5^2 = 143^2 + 115^2 - 2(143)(115) \cos \beta$$

$$\cos \beta = \frac{143^2 + 115^2 - 79.5^2}{2(143)(115)}$$

$$\cos \beta \approx 0.8317$$

$$\beta \approx 33.7°$$

Angle α is $180° - 86.2° - 33.7°$, or $60.1°$. The rescue team must bear S 60.1° W. ∎

Example 5   A 100-pound weight is suspended by two cables as in Figure 6–22. The tension on the left cable is 55 pounds, and on the right cable, 75 pounds. What angle does each cable make with the horizontal?

Solution   The forces in the two ropes are such that their resultant force is 100 pounds, directed upward to exactly counter the 100-pound downward pull. The force diagram appears in Figure 6–23. The angles that the cables make with the horizontal are the complements of angles α and β. Angles α and β can be found by using the law of cosines.

$$55^2 = 75^2 + 100^2 - 2(75)(100) \cos \alpha$$

$$\cos \alpha = \frac{75^2 + 100^2 - 55^2}{2(75)(100)}$$

$$\cos \alpha = 0.8400$$

$$\alpha \approx 32.9°$$

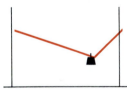

Figure 6–22

Similarly, you have

$$75^2 = 55^2 + 100^2 - 2(55)(100) \cos \beta$$

$$\cos \beta = \frac{55^2 + 100^2 - 75^2}{2(55)(100)}$$

$$\cos \beta \approx 0.6727$$

$$\beta \approx 47.7°$$

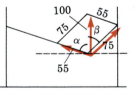

Figure 6–23

The left cable makes an angle of $90° - 47.7° \approx 42°$ with the horizontal. The right cable is $90° - 32.9° \approx 57°$ from the horizontal. ∎

Exercise 6.2

*In Exercises 1–12 refer to Illustration 1 and find the required value. **Use a calculator.***

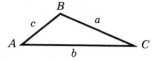

Illustration 1

1. $a = 42.9$ cm, $c = 37.2$ cm, $B = 99.0°$; find b.
2. $b = 192$ m, $c = 86.9$ m, $A = 21.2°$; find a.
3. $a = 2730$ km, $b = 3520$ km, $C = 21.7°$; find c.
4. $b = 2.1$ km, $c = 1.3$ km, $A = 14°$; find a.
5. $a = 91.1$ cm, $c = 87.6$ cm, $B = 43.2°$; find b.
6. $a = 107$ cm, $b = 205$ cm, $C = 86.5°$; find c.
7. $a = 19$ km, $b = 23$ km, $c = 18$ km; find A.
8. $a = 14.3$ km, $b = 29.7$ km, $c = 21.3$ km; find B.
9. $a = 30$ ft, $b = 40$ ft, $c = 50$ ft; find C.
10. $a = 130$ mi, $b = 50$ mi, $c = 120$ mi; find A.
11. $a = 1580$, $b = 2137$, $c = 3152$; find B.
12. $a = 0.0031$, $b = 0.0047$, $c = 0.0093$; find C.

13. Two men are pulling on ropes attached to the bumper of a car that is stuck in a snowdrift. If one man pulls with a force of 114 pounds and the other with a force of 97 pounds and the angle between the ropes is 13°, what is the force exerted on the car?

14. Two forces, one of 75 pounds and the other of 90 pounds, are exerted at an angle of 102° from each other. What is the magnitude of the resultant force?

15. In Exercise 13, what is the angle between the resultant force and the rope pulled by the stronger man?

16. In Exercise 14, what is the angle between the resultant force and the direction of the 90-pound force?

17. A donkey and a horse are tied to a large stone. The horse pulls with a force of 950 pounds; the donkey lazily tugs with a force of 150 pounds. The angle between their tethers is 19.5°. With what force do they pull on the stone?

18. A ship sails 21.2 nautical miles in a direction of N 42.0° W and then turns onto a course of S 15.0° E and sails 19.0 nautical miles. How far is the ship from its starting point?

19. A ship sails 14.3 nautical miles in a direction of S 28.0° W and then turns onto a course of S 52.0° W and sails 23.2 nautical miles. How far is the ship from its starting point?

20. To measure the length of a lake, a surveyor determines the measurements shown in Illustration 2. How long is the lake?

21. To estimate the cost of building a tunnel, a surveyor must find the distance through a hill. The surveyor determines the measurements shown in Illustration 3. How long must the tunnel be to pass through the hill?

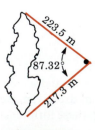

Illustration 2

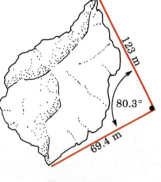

Illustration 3

22. The three circles in Illustration 4 have radii of 4.0 centimeters, 7.0 centimeters, and 9.0 centimeters. If the circles are externally tangent to one another, what are the angles of the triangle that joins their centers?

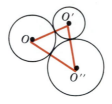

Illustration 4

23. The three circles in Illustration 4 have radii of 21.2 centimeters, 19.3 centimeters, and 31.2 centimeters. If the circles are externally tangent to one another, what are the angles of the triangle that joins their centers?

24. Show that the Pythagorean theorem is a special case of the law of cosines.

25. To determine whether two interior walls meet at a right angle, carpenters often mark a point 3 feet from the corner on one wall and a point (at the same height) on the other wall 4 feet from the corner. If the straight-line distance between those points is 5 feet, the walls are square. At what angle do the walls meet if the distance measures 4 feet 10 inches?

26. To build a counter top for a kitchen, a cabinetmaker must determine the angle at which two walls meet. The method of Exercise 25 is used. What is the angle between the walls if the measured distance is 5 feet 3 inches?

27. Triangle ABC is formed by points $A(3, 4)$, $B(1, 5)$, and $C(5, 9)$. Find angle A to the nearest tenth of a degree.

28. In Exercise 27, find angle B to the nearest tenth of a degree.

29. In Illustration 5, D is the midpoint of BC. Find angle α to the nearest tenth of a degree.

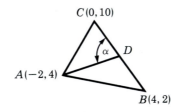

Illustration 5

30. In Exercise 29, find angle DAB to the nearest tenth of a degree.

31. A lighthouse is 15.0 nautical miles N 23.0° W of a dock. A ship leaves the dock heading due east at 26.3 knots. How long will it take for the ship to reach a distance of 35.0 nautical miles from the lighthouse?

32. Consider triangle ABC in Figure 6-17. Rotate triangle ABC so that point B is at the origin. Prove that $b^2 = c^2 + a^2 - 2ac \cos B$.

33. Consider triangle ABC in Figure 6-17. Rotate triangle ABC so that point C is at the origin. Prove that $c^2 = a^2 + b^2 - 2ab \cos C$.

6.3   THE LAW OF SINES

To solve oblique triangles when two angles and a side are given, or when two sides and an angle opposite one of them are given, we use another set of formulas called the **law of sines**. To develop these formulas, we refer to the triangles in Figures 6–24 and 6–25, in which line segment CD is drawn perpendicular to side AB or its extension. If the length of segment CD is h, then

$$h = b \sin A$$

and

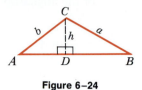

$$h = a \sin B$$

Figure 6–24

Because both $b \sin A$ and $a \sin B$ are equal to h, they are equal to each other.

$$b \sin A = a \sin B$$

or

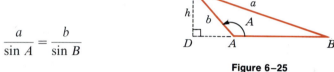

$$\frac{a}{\sin A} = \frac{b}{\sin B}$$

Figure 6–25

In the two right triangles of Figure 6–25, these relations hold:

$$h = a \sin B$$

and

$$h = b \sin(180° - A) = b \sin A$$

Because $a \sin B$ and $b \sin A$ both equal h, they are equal to each other.

$$b \sin A = a \sin B$$

or

$$\frac{a}{\sin A} = \frac{b}{\sin B}$$

In either of these triangles, the perpendicular need not be drawn from C to AB or its extension. If drawn from another vertex, similar reasoning yields

$$\frac{a}{\sin A} = \frac{c}{\sin C}$$

or

$$\frac{b}{\sin B} = \frac{c}{\sin C}$$

Because of the transitive law of equality, it follows that

$$\frac{a}{\sin A} = \frac{c}{\sin C} = \frac{b}{\sin B}$$

These results are summed up in a formula called the **law of sines**.

> **The Law of Sines.** The sides in any triangle are proportional to the sines of the angles opposite those sides.
>
> $$\frac{a}{\sin A} = \frac{b}{\sin B} = \frac{c}{\sin C}$$
>
>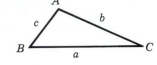

Three equations are implicit in the law of sines, each providing a relation between two angles of a triangle and the sides opposite those angles. If any three of these are known, the fourth can be calculated.

Example 1   In Figure 6–26, $a = 14$ kilometers, $A = 21°$, and $B = 35°$. Find side b.

Solution   Note that the given information fills three of the four spots in the law of sines.

$$\frac{a}{\sin A} = \frac{b}{\sin B}$$

$$\frac{14}{\sin 21°} = \frac{b}{\sin 35°}$$

$$b = \frac{14 \sin 35°}{\sin 21°}$$

$$b \approx 22.4$$

$$\approx 22$$

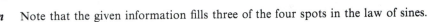

Figure 6–26

Thus, b is approximately 22 kilometers.   ■

Example 2   In Figure 6–27, $a = 29$ meters, $B = 42°$, and $C = 31°$. Find side c.

Solution   The law of sines is not convenient to use unless you know angle A. Because A is the "rest of" 180°, you have

$$A = 180° - B - C$$

$$= 180° - 42° - 31°$$

$$= 107°$$

Now use the law of sines.

$$\frac{a}{\sin A} = \frac{c}{\sin C}$$

$$\frac{29}{\sin 107°} = \frac{c}{\sin 31°}$$

$$c = \frac{29 \sin 31°}{\sin 107°}$$

$$c \approx 15.62$$

$$\approx 16$$

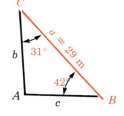

Figure 6–27

Thus, c is approximately 16 meters.   ■

Example 3   A ship is sailing due east. The skipper observes a lighthouse with a bearing of N 37.5° E. After the ship has sailed 4.70 nautical miles, the bearing to the lighthouse is N 9.0° W. How close to the lighthouse did the ship pass?

Solution   The information of the problem is contained in Figure 6–28. If you can find distance b first, then the required distance d can be obtained by solving right triangle ACD. The law of sines provides a way to compute b.

$$\frac{b}{\sin B} = \frac{c}{\sin C}$$

$$\frac{b}{\sin 81.0°} = \frac{4.70}{\sin 46.5°}$$

$$b = \frac{4.70 \sin 81.0°}{\sin 46.5°}$$

$$b \approx 6.3996$$

$$\approx 6.40$$

In right triangle ACD, it follows that

$$d = b \sin A$$

$$\approx 6.40 \sin 52.5°$$

$$\approx 5.0775$$

$$\approx 5.08$$

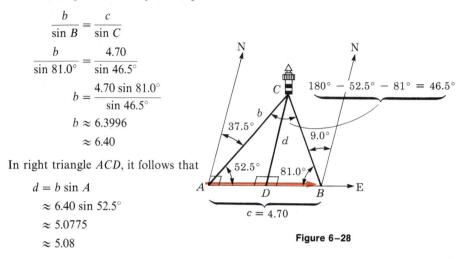

Figure 6–28

The ship's closest approach to the lighthouse is 5.08 nautical miles.   ∎

Example 4   On a bright moonlit night, the moon is directly overhead in city A. In city B, 2500 miles to the north, the moonlight strikes a pole at an angle of 36.57°. If the radius of the earth is 4000 miles (to two significant digits), find the distance from the earth to the moon.

Solution   Refer to Figure 6–29. Because 2500 miles is $\frac{1}{10}$ of the earth's circumference, angle BOA is $\frac{1}{10}$ of a complete revolution, or 36°. Angle OBM is supplementary to 36.57°. Hence, angle OBM is 143.43°. Because the sum of the angles in any triangle must be 180°, angle M is 0.57°. Use the law of sines to set up the following proportion and solve for x.

$$\frac{x}{\sin 143.43°} = \frac{4000}{\sin 0.57°}$$

$$x = \frac{4000 \sin 143.43°}{\sin 0.57°}$$

$$x \approx 239{,}518$$

$$x \approx 240{,}000$$

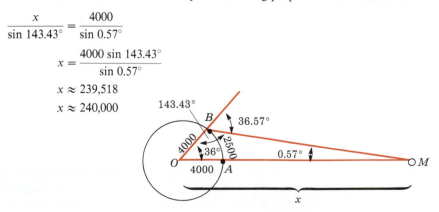

Figure 6–29

The estimate of the distance to the moon is $240,000 - 4,000 = 236,000$ miles. To two significant digits, the distance is 240,000 miles. ∎

Example 5   A pilot wishes to fly in the direction of 20.0° east of north, against a 45.0 mph wind blowing from the east. The airspeed of the plane is to be 185 mph. What should be the pilot's heading, and what will be the plane's ground speed?

Solution   Refer to Figure 6–30. The pilot's intended direction of travel is represented by the vector **OA**. The direction of vector **OA** is 20.0° east of north. The length of vector **OH** represents the plane's airspeed of 185 mph. To find the pilot's heading, you must find angle θ, which is the direction of vector **OH**.

In triangle OAH, side AH has a length of 45.0 miles. Angle OAH is $90.0° + 20.0°$, or $110.0°$. Use the law of sines on triangle OAH to find $\theta - 20.0°$. You can then find θ.

$$\frac{185}{\sin 110.0°} = \frac{45.0}{\sin(\theta - 20.0°)}$$

$$\sin(\theta - 20.0°) = \frac{45.0 \sin 110.0°}{185}$$

$$\sin(\theta - 20.0°) \approx 0.2286$$

$$\theta - 20.0° \approx 13.2°$$

$$\theta \approx 33.2°$$

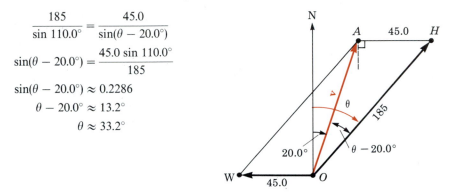

Figure 6–30

To attain his intended direction of travel, the pilot must set a heading of 33.2°.

In the figure, the ground speed of the plane is represented by the length v of vector **OA**. To determine this length, calculate angle H and use the law of cosines. Note that

$$\text{angle } H \approx 180.0° - 13.2° - 110.0°$$

$$\approx 56.8°$$

By the law of cosines, you get

$$v^2 \approx 45^2 + 185^2 - 2(45)(185) \cos 56.8°$$

$$\approx 27,133$$

$$v \approx 164.7$$

$$\approx 165$$

The plane's ground speed is approximately 165 mph. ∎

The Ambiguous Case

A triangle might or might not be determined if two sides and a nonincluded angle are given. This case is called the **ambiguous case**.

When two sides and a nonincluded angle of a triangle are given, we can carefully make a scale drawing and use common sense to determine whether two, one, or no triangles exist. If a scale drawing indicates a situation that is too close to call, we can let the law of sines decide the outcome.

Example 6   Three parts of a triangle are exactly $a = 1$, $b = 2$, and $A = 20°$. Find angles B and C to the nearest tenth of a degree and side c to the nearest hundredth.

Solution   A sketch such as Figure 6–31 indicates that there are two possible triangles. By the law of sines, it follows that

$$\frac{a}{\sin A} = \frac{b}{\sin B}$$

$$\sin B = \frac{b \sin A}{a}$$

$$= \frac{2 \sin 20°}{1}$$

$$\approx 0.6840$$

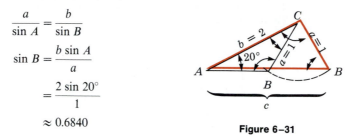

Figure 6–31

There are two possible values of B, one acute (a first-quadrant angle) and the other obtuse (a second-quadrant angle):

$$B \approx 43.2°$$

or

$$B \approx 180° - 43.2° = 136.8°$$

The third angle C has two possibilities also:

$$C = 180° - A - B$$
$$C \approx 180° - 20° - 43.2°$$
$$\approx 116.8°$$

or

$$C = 180° - A - B$$
$$C \approx 180° - 20° - 136.8°$$
$$\approx 23.2°$$

The third side c can also be found by using the law of sines. Side c has two possibilities:

$$\frac{c}{\sin C} = \frac{a}{\sin A}$$

$$c \approx \frac{1 \sin 116.8°}{\sin 20°} \approx 2.61$$

or

$$c \approx \frac{1 \sin 23.2°}{\sin 20°} \approx 1.15$$

The two triangles are shown in Figure 6–32.

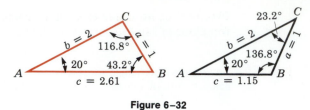

Figure 6–32

Example 7   In triangle ABC, $a = 1$, $b = 2$, and $A = 45°$. If possible, find the length of side c.

Solution   By the law of sines, it follows that

$$\frac{a}{\sin A} = \frac{b}{\sin B}$$

$$\frac{1}{\sin 45°} = \frac{2}{\sin B}$$

$$\sin B = 2 \sin 45°$$

$$= 2\left(\frac{\sqrt{2}}{2}\right)$$

$$= \sqrt{2}$$

Because $\sqrt{2}$ is greater than 1, and $\sin B$ cannot be greater than 1, there is no triangle that satisfies the given conditions.

Example 8   A vertical tower 255 feet tall stands on a hill. From a point 700 feet down the hill, the angle between the hill and the observer's line of sight to the top of the tower is 12.0°. What is the angle of inclination of the hill (the angle the ground makes with the horizontal)?

Solution   Use the information to draw Figure 6–33. By the law of sines, it follows that

$$\frac{255}{\sin 12.0°} = \frac{700}{\sin B}$$

$$\sin B = \frac{700 \sin 12.0°}{255}$$

$$\approx 0.5707$$

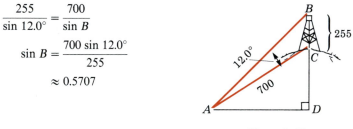

Figure 6–33

There are two possibilities for angle B—one acute and one obtuse. However, because of the diagram, you want only the acute angle. Because $\sin B \approx 0.5707$,

$$B \approx 34.8°$$

Once you know the measure of angle B, you can compute the measure of angle C:

$$C = 180° - A - B$$

$$\approx 180° - 12.0° - 34.8°$$

$$\approx 133.2°$$

The angle the hill makes with the horizontal is $C - 90°$, which is approximately $43.2°$. See Figure 6–34.

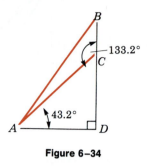

Figure 6–34 ■

Exercise 6.3

In Exercises 1–10 refer to Illustration 1. Use the law of sines and a calculator to find the required value.

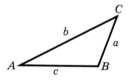

Illustration 1

1. $A = 12°$, $B = 97°$, $a = 14$ km; find b.
2. $A = 19°$, $C = 102°$, $c = 37$ ft; find a.
3. $A = 21.3°$, $B = 19.2°$, $a = 143$ m; find c.
4. $B = 8.6°$, $C = 9.2°$, $c = 2.73$ m; find b.
5. $A = 99.8°$, $C = 43.2°$, $b = 186$ m; find a.
6. $A = 14.61°$, $B = 87.10°$, $c = 1437$ ft; find b.
7. $B = 32.0°$, $b = 120$ in., $a = c$; find a.
8. $C = 12°$, $c = 5.4$ in., $b = a$; find b.
9. $A = x°$, $C = 2x°$, $B = 3x°$, $b = 7.93$ m; find a.
10. $A = x°$, $C = 3x°$, $B = 5x°$, $c = 12.5$ m; find a.

11. To measure the distance up a steep hill, a woman determines the measurements shown in Illustration 2. What is the distance d?

12. A ship sails 3.2 nautical miles on a bearing of N 33° E. After reaching a lighthouse, the ship turns and sails 6.7 nautical miles to a position that is due east of its starting point. What is the bearing of the lighthouse from the ship's final position?

13. Points A and B shown in Illustration 3 are on opposite sides of a river. A tree at point C is 310 feet from point A. If angle $A = 125°$ and angle $C = 32°$, how wide is the river?

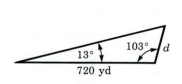

Illustration 2

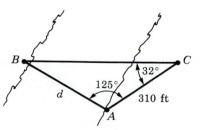

Illustration 3

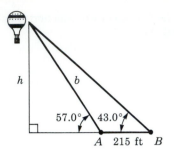

Illustration 4

14. Observers at points A and B shown in Illustration 4 are directly in line with a hot-air balloon and are themselves 215 feet apart. The angles of elevation of the balloon from points A and B are as shown in the illustration. How high is the balloon? (*Hint:* First use the law of sines to find b.)

15. A radio tower 175 feet tall is located on the top of a hill. At a point 800 feet down the hill, the angle of elevation to the top of the tower is 19.0°. What angle θ does the hill make with the horizontal? See Illustration 5.

16. Two children are on a riverbank, 120 feet apart. They each sight the same tree on the opposite bank. See Illustration 6. If angle $A = 79°$ and angle $B = 63°$, how wide is the river?

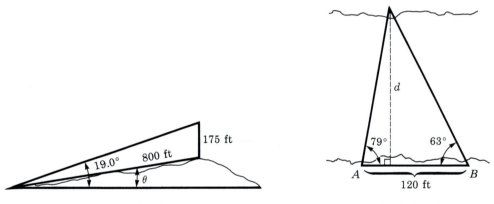

Illustration 5   Illustration 6

17. A ship sails due north at 3.2 knots (nautical miles per hour). At 2:00 P.M. the captain sights a lighthouse in the direction of N 43° W. One hour later, the lighthouse bears S 78° W. How close to the lighthouse did the ship sail?

18. A ship sails on a course bearing N 21° E at a speed of 14 knots. At noon, the first mate sights an island in the direction of N 35° E and one hour later sights the same island due east. If the ship continues on its course, how close will it come to the island?

In Exercises 19–26 calculate all possibilities for the indicated value. If no triangle exists, so indicate.

19. $A = 42.0°$, $a = 123$ ft, $b = 96.0$ ft; find B.

20. $B = 56.2°$, $b = 13.5$ yd, $c = 15.3$ yd; find C.

21. $C = 98.6°$, $a = 42.1$ cm, $c = 47.3$ cm; find A.

22. $B = 17.5°$, $a = 0.063$ m, $b = 0.152$ m; find A.

23. $A = 57°$, $b = 13$ m, $a = 12$ m; find c.

24. $C = 48°$, $b = 29$ km, $c = 26$ km; find a.

25. $B = 87°$, $a = 35$ cm, $b = 32$ cm; find c.

26. $C = 38°$, $a = 12$ cm, $b = 40$ cm; find c

27. A triangular piece of land owned by Farmer Brown is bounded by three straight highways. The angle between two of them—U.S. 45 and County M—is 43°. The farmer's property runs for 2500 feet along County M and for 2000 feet along the third highway—scenic Silo Drive. How much frontage could Farmer Brown own on U.S. 45?

28. From the roof of Farmer Brown's barn, the angle of elevation to the top of a lookout tower is 17°. From the barn's ground level 43 feet below, the angle of elevation to the top of the tower is 21°. How far above ground level is the top of the tower?

29. A 210-foot television tower stands on the top of an office building. From a point on level ground, the angles of elevation to the top and the base of the tower are 25.2° and 21.1°, respectively. How tall is the building?

30. A pilot leaves point A and flies 800 kilometers with a heading of 320° to point B. From B, she flies due south to a point C, which is 700 kilometers from point A. How long is the BC leg of the trip?

6.4   SIMPLE HARMONIC MOTION; SNELL'S LAW

Many types of motion can be modeled by the trigonometric functions. Perhaps the most important of these is an oscillating type of motion called *simple harmonic motion*.

Simple Harmonic Motion

One type of motion, known as **simple harmonic motion**, is produced by forces that vary periodically. For example, simple harmonic motion describes the vibrations of a guitar string or a column of air in an organ pipe, and the oscillations of the balance wheel of a watch or the pendulum of a clock.

We refer to the circle shown in Figure 6–35, which is centered at the origin O and has a radius of A. We let $P(x, y)$ be a point on the circle, $X(x, 0)$ be the projection of point P on the x-axis, and $Y(0, y)$ be the projection of P on the y-axis. If point P travels around the circle with a constant angular velocity, then point X will move back and forth between points Q and R, and point Y will oscillate between points S and T. The motions of points X and Y are examples of simple harmonic motion.

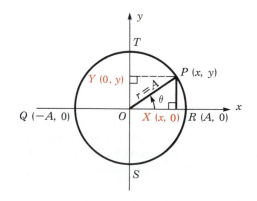

Figure 6–35

To develop equations that model such oscillatory motion, we assume that point $P(x, y)$ starts at point $R(A, 0)$ and moves in a counterclockwise direction with a constant angular velocity ω measured in radians per unit time. If the radius OP rotates through an angle θ in a time interval t, then

$$\omega = \frac{\theta}{t}$$

or

$$\theta = \omega t$$

By the definitions of the trigonometric functions, we have

$$\cos \theta = \frac{x}{A} \qquad \text{and} \qquad \sin \theta = \frac{y}{A}$$

or

$$x = A \cos \theta \qquad \text{and} \qquad y = A \sin \theta$$

Because $\theta = \omega t$, we have

$$x = A \cos \omega t \qquad \text{and} \qquad y = A \sin \omega t$$

Thus, as point P travels around the circle with a constant angular velocity ω, the equation $x = A \cos \omega t$ gives the position on the x-axis of point $X(x, 0)$ at any given time t. Similarly, the equation $y = A \sin \omega t$ gives the position on the y-axis of point $Y(0, y)$ at any given time t.

Definition. Any motion that can be described by an equation of the form

$$x = A \cos \omega t \qquad \text{or} \qquad y = A \sin \omega t$$

is called **simple harmonic motion**.
The **amplitude** of the oscillation is $|A|$, and the **period** is $\left| \frac{2\pi}{\omega} \right|$.

The **frequency** of the oscillation is $\left| \frac{\omega}{2\pi} \right|$, the reciprocal of the period.

Note that the period of an oscillation indicates the time necessary to complete one oscillation. The frequency of oscillation indicates the number of oscillations that are completed per unit time.

If an object suspended by a spring is moved a distance A units above its equilibrium position and then released, it will start to bounce. See Figure 6–36. Its position above (positive) or below (negative) the equilibrium, or starting, position is given by the formula

$$y = A \cos \left(\sqrt{\frac{k}{m}}\, t \right)$$

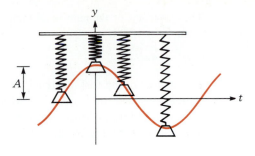

Figure 6–36

where k, called the **spring constant**, is determined by the construction of the spring, m is the mass of the object, and t is the time in seconds since the object was released.

The period of oscillation p of the described motion is

$$p = \frac{2\pi}{\sqrt{\dfrac{k}{m}}} = 2\pi \sqrt{\frac{m}{k}}$$

The frequency of oscillation is the reciprocal of p:

$$f = \frac{1}{2\pi} \sqrt{\frac{k}{m}}$$

Example 1   A mass of 4 grams, attached to a spring with a spring constant of 9 g/cm, is pulled 10 centimeters below the equilibrium position and released. Find the amplitude, the period, and the frequency of the oscillation.

Solution   The position of the mass relative to its equilibrium position is given by

$$y = A \cos\left(\sqrt{\frac{k}{m}}\, t\right)$$

where $k = 9$ and $m = 4$. Because the mass is brought 10 centimeters below equilibrium, $A = -10$. Thus,

$$y = -10 \cos\left(\sqrt{\frac{9}{4}}\, t\right)$$

$$= -10 \cos\left(\frac{3}{2}\, t\right)$$

The amplitude of the oscillation is $|A| = |-10| = 10$. The period, or the time necessary to complete one cycle, is

$$p = \frac{2\pi}{\dfrac{3}{2}} = \frac{4\pi}{3}$$

Figure 6–37

The frequency, or the number of cycles per second, is the reciprocal of the period:

$$f = \frac{3}{4\pi}$$

■

When a tuning fork is struck, its tongs vibrate at a rate that depends on its dimensions and the material of which it is made. The vibrations of the fork set the surrounding air into vibration, as shown in Figure 6–37. These vibrations are described by the equation

$$y = A \sin(2\pi f t)$$

where f is the frequency, t is the time, and A is the amplitude of the oscillations.

Example 2   One tine of a tuning fork takes 0.0035 second to complete one vibration. What is the frequency of the tone?

Solution   The period of $y = A \sin(2\pi f t)$ is

$$\frac{2\pi}{2\pi f} = \frac{1}{f}$$

Because the period is given to be 0.0035 second, you have

$$\frac{1}{f} = 0.0035$$

or

$$f \approx 285.714$$

The frequency is about 286 cycles per second, or 286 **hertz**.   ■

A pendulum swinging through a small arc makes an angle θ with the vertical, where θ is given by the formula

$$\theta = A \cos\left(\sqrt{\frac{g}{l}} \, t\right)$$

where $|A|$ is the amplitude of the swing, l is the length of the pendulum, and g is a constant related to the force of gravity. The period and frequency of the motion are

$$p = \frac{2\pi}{\sqrt{\dfrac{g}{l}}} = 2\pi \sqrt{\frac{l}{g}} \quad \text{and} \quad f = \frac{1}{p} = \frac{1}{2\pi} \sqrt{\frac{g}{l}}$$

Snell's Law

Light travels more rapidly in air than it does in glass. When light passes from one substance into another in which it slows down, its path is bent. This is called **refraction**.

We suppose that a beam of light is passing from air into glass, as shown in Figure 6–38. Angle i is called the **angle of incidence**, and angle r is the **angle of refraction**. The two are related by **Snell's law**, named after the Dutch astronomer Willebrord Snell (1591–1629):

$$n = \frac{\sin i}{\sin r}$$

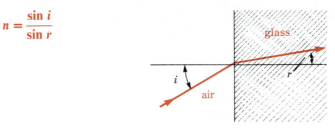

Figure 6–38

where the constant n, called the **index of refraction**, depends on the optical properties of the glass.

Example 3   Find the index of refraction of light striking a piece of glass at a $10°$ angle if the angle of refraction is $6°$.

Solution   See Figure 6–39. Substitute $10°$ for i and $6°$ for r in Snell's law and simplify:

$$n = \frac{\sin i}{\sin r} = \frac{\sin 10°}{\sin 6°} \approx \frac{0.1736}{0.1045} \approx 1.7$$

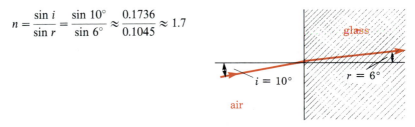

Figure 6–39

The index of refraction is approximately 1.7.   ∎

Exercise 6.4

In Exercises 1–2 each equation determines the position of a point X that moves harmonically along the x-axis. Give the coordinates of point X when t = 5.

1. $x = 4\cos(3\pi t)$

2. $x = -3\cos\left(\frac{\pi}{15}t\right)$

In Exercises 3–4 each equation determines the position of a point Y that moves harmonically along the y-axis. Give the coordinates of point Y when t = 2.

3. $y = -2\sin\left(\frac{5\pi}{3}t\right)$

4. $y = 7\sin\left(\frac{\pi}{2}t\right)$

In Exercises 5–8 each equation describes simple harmonic motion. Find the amplitude, period, and frequency of each oscillation.

5. $x = \dfrac{3}{2}\cos\left(\dfrac{\pi}{2}t\right)$ **6.** $y = -\dfrac{2}{5}\sin\left(\dfrac{2\pi}{3}t\right)$ **7.** $y = 5\sin\left(\dfrac{3\pi}{4}t\right)$ **8.** $x = -5\cos\left(\dfrac{11\pi}{6}t\right)$

9. A spring with a spring constant of 6 newtons per meter hangs from a ceiling and supports a mass of 24 kilograms. If the mass is pulled 12 centimeters below the equilibrium position and released, find the amplitude, period, and frequency of the oscillation.

10. A spring supports a mass of 12 grams. The frequency of its oscillation is $\frac{1}{\pi}$ hertz (cycles per second). What is the spring constant? The answer will be in dynes per centimeter.

11. What is the effect on the period of oscillation if a mass suspended by a spring is doubled?

12. What is the effect on the frequency of oscillation if a mass suspended by a spring is doubled?

13. One tine of a tuning fork takes 0.0042 second to complete one vibration. What is the frequency of the tone?

14. A tuning fork vibrates at a frequency of 256 cycles per second. How long does it take to complete one cycle?

15. A pendulum 1 meter long is set swinging through a small arc. If $g = 9.8$ m/sec², what is the period and the frequency of the oscillation?

16. A pendulum of a clock is set swinging. If $g = 9.8$ m/sec² and the frequency of the oscillation is 2 cycles per second, how long is the pendulum?

In Exercises 17–22 assume that a wave travels at a velocity of v m/sec. The distance that a wave of frequency f travels in the time of 1 cycle is v/f meters. This distance is called the **wavelength**.

17. What is the wavelength of international musical A, 435 hertz, traveling 335 m/sec in air?

18. What is the wavelength of a sound of 435 hertz traveling in water, where its velocity is 1500 m/sec?

19. If the frequency of a tone is doubled, its sound is raised by one octave. By what factor is the wavelength changed?

20. If the wavelength in steel of a 1000 hertz sound is 5.8 meters, what is the sound's velocity in steel?

21. A radio station operates on a federally assigned frequency of 1400 kilohertz (1 kilohertz = 1000 hertz). If radio waves travel at 300 million m/sec, what is the wavelength of the radio station's signal?

22. The wavelength of visible light lies between 4000 and 7000 angstroms. What is the range of frequencies of light, which has velocity of 300 million m/sec? (*Hint:* 1 angstrom = 1×10^{-10} meters.)

23. Graph the equation $y = A\sin(2\pi ft)$.   **24.** Graph the equation $y = A\cos kt$.

25. A beam of light enters a block of ice with an angle of incidence of 8.0° and is refracted to 6.1°. What is the index of refraction of the ice?

26. A clear liquid, tetrachloroethylene, has an index of refraction equal to that of a certain brand of glass. Would you be able to see a piece of that glass if it were submerged in the liquid? How could you distinguish between real diamonds and a handful of fakes?

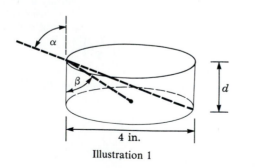

4 in.

Illustration 1

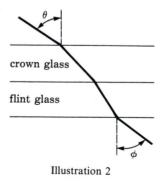

crown glass

flint glass

Illustration 2

27. A man looks over the rim of his empty 4-inch-diameter soup bowl and sees the edge of its bottom. See Illustration 1. When the bowl is filled with a fluid with an index of refraction of 1.2, he sees the center of its bottom. How deep is the bowl?

28. Plates of crown glass (index of refraction 1.517) and flint glass (index of refraction 1.656) are cemented together as in Illustration 2. Find angle ϕ in terms of θ.

REVIEW EXERCISES

1. From a location 32.1 meters from the base of a flagpole, the angle of elevation to its top is α. Find α if the flagpole is 10.0 meters tall.

2. The angle of depression from a window in a building to a point on the ground is 17.7°. If the point on the ground is 187 feet from the base of the building, how high is the observer?

3. Owatonna, Minnesota, is approximately 55 miles due south of Minneapolis, and the bearing of Winona, Minnesota, from Minneapolis is about S 45° E. How far is Winona from Owatonna if Owatonna is due west of Winona?

4. Assume that the bearing of South Bend, Indiana, from Fort Wayne, Indiana, is N 48° W and that the distance between the cities is about 71 miles. Further assume that South Bend is due north of Indianapolis and that the bearing of Fort Wayne is N 21° E. How far is Fort Wayne from Indianapolis?

5. A barrel weighing 60 pounds rests on a ramp that makes an angle of 10° with the horizontal. How much force is necessary to keep the barrel from rolling down the ramp?

6. A 2500-pound car rests on a hill. A force of 500 pounds is required to keep the car from rolling down the hill. What angle does the hill make with the horizontal?

In Review Exercises 7–18 consider the given parts of triangle ABC. Use a calculator to solve for the required value, if possible. If more than one value is possible, give both.

7. $a = 12$, $c = 15$, $B = 30°$; find b.
8. $b = 23$, $a = 13$, $C = 125°$; find c.
9. $c = 0.5$, $b = 0.8$, $A = 50°$; find a.
10. $a = 28.7$, $b = 37.8$, $C = 11.2°$; find c.
11. $a = 12$, $c = 18$, $C = 40°$; find A.
12. $b = 17$, $a = 12$, $A = 25°$; find B.
13. $c = 31.5$, $b = 27.5$, $B = 16.2°$; find C.
14. $a = 315.2$, $b = 457.8$, $A = 32.51°$; find B.
15. $A = 24.3°$, $B = 56.8°$, $a = 32.3$; find b.
16. $B = 10.3°$, $C = 59.4°$, $c = 341$; find b.
17. $b = 17$, $c = 21$, $B = 42°$; find A.
18. $c = 189$, $a = 150$, $C = 85.3°$; find B.

19. Two airplanes leave an airport at 2:00 P.M., one with a heading of 30.0° and a ground speed of 425 miles per hour and the other with a heading of 85.0° and a ground speed of 375 miles per hour. How far apart are they at 3:30 P.M.?

20. Find the largest angle of the triangle with vertices at (0, 0), (5, 0), and (7, 8).

21. From a point 542 feet away from the base of the Leaning Tower of Pisa, the angle of elevation to its top is 17.9°. If the tower makes an angle with the ground of 94.5° (see Illustration 1), how tall is the tower?

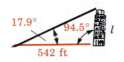

17.9°   94.5° l

542 ft

Illustration 1

22. From a point 312 feet from the base of the Great Pyramid of Khufu (Cheops) at Gizeh, the angle of elevation to its top is 25.5°. The pyramid makes an angle with the ground of 141.8° (see Illustration 2). Find its slant height h.

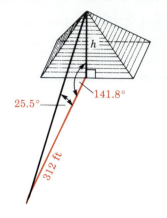

Illustration 2

23. The equation $x = -2\cos(\frac{\pi}{3}t)$ determines the position of a point X that moves harmonically along the x-axis. Give the coordinates of point X when $t = 7$, and the amplitude, period, and frequency of the oscillation.

24. What is the effect on the period of oscillation if a mass suspended by a spring is cut in half?

25. A tuning fork vibrates at a frequency of 440 cycles per second. How long does it take to complete one cycle?

26. A beam of light enters a block of ice with an angle of incidence of 9.0°. What is the angle of refraction if the index of refraction of the ice is 1.4?

7 TRIGONOMETRIC IDENTITIES AND EQUATIONS

In the previous chapter, we used trigonometry to solve applied problems such as finding heights of towers and distances across rivers. We now use trigonometry in a more theoretical way. Because the trigonometric functions are related to one another and to certain algebraic expressions, they often enable us to simplify very complicated mathematical expressions. In this chapter, we shall explore some of these relationships by verifying trigonometric identities and solving trigonometric equations.

7.1 VERIFYING IDENTITIES

Recall the following fundamental identities:

$$\csc x = \frac{1}{\sin x} \qquad \tan x = \frac{\sin x}{\cos x} \qquad \tan^2 x + 1 = \sec^2 x$$

$$\cot^2 x + 1 = \csc^2 x$$

$$\sec x = \frac{1}{\cos x} \qquad \cot x = \frac{\cos x}{\sin x} \qquad \sin(-x) = -\sin x$$

$$\cos(-x) = \cos x$$

$$\cot x = \frac{1}{\tan x} \qquad \sin^2 x + \cos^2 x = 1 \qquad \tan(-x) = -\tan x$$

These identities and the rules of algebra enable us to show that many more equations are identities. In doing so, we shall usually work with one side of the equation only, until it is transformed into the other side. Here are some suggestions to follow when verifying identities:

1. Memorize the fundamental identities. Whenever you see one side of one of these identities, the other should come to mind immediately.
2. Start with the more complicated side of the equation and try to transform it into the other side.
3. Remember that trigonometric fractions can be added, multiplied, and simplified just like algebraic fractions. Trigonometric expressions can be factored and similar terms combined.
4. If one side of an equation contains only a single trigonometric function, eliminate all other functions, if possible, from the other side.

5. It is sometimes helpful to change all functions into sine and cosine functions before proceeding.

6. If the numerator or the denominator of a fraction contains a factor of $1 \pm \sin x$, consider multiplying both the numerator and the denominator by the conjugate $1 \mp \sin x$. This will create a factor of $1 - \sin^2 x$, which can be replaced with $\cos^2 x$. The same idea applies to factors of $1 \pm \cos x$, $\sec x \pm 1$, and $\sec x \pm \tan x$.

Example 1　Verify that $\dfrac{\tan x}{\sin x} = \sec x$ is an identity.

Solution　Work on the left side because it is the more complicated.

$$\frac{\tan x}{\sin x} = \frac{\dfrac{\sin x}{\cos x}}{\dfrac{\sin x}{1}} \qquad \text{Replace } \tan x \text{ with } \frac{\sin x}{\cos x}.$$

$$= \frac{\sin x}{\cos x} \cdot \frac{1}{\sin x} \qquad \text{Invert and multiply.}$$

$$= \frac{1}{\cos x} \qquad \text{Divide out } \sin x.$$

$$= \sec x \qquad \text{Replace } \frac{1}{\cos x} \text{ with } \sec x.$$

Because the left side has been transformed into the right side, the identity has been verified. ■

Example 2　Verify the identity $\tan x + \cot x = \csc x \sec x$.

Solution　Work on the left side, change $\tan x$ and $\cot x$ into expressions containing $\sin x$ and $\cos x$, and proceed as follows:

$$\tan x + \cot x = \frac{\sin x}{\cos x} + \frac{\cos x}{\sin x}$$

$$= \frac{\sin x \, \sin x}{\sin x \, \cos x} + \frac{\cos x \, \cos x}{\sin x \, \cos x} \qquad \text{Change each fraction into a fraction with a common denominator.}$$

$$= \frac{\sin^2 x + \cos^2 x}{\sin x \cos x} \qquad \text{Add the fractions.}$$

$$= \frac{1}{\sin x \cos x} \qquad \text{Replace } \sin^2 x + \cos^2 x \text{ with 1.}$$

$$= \frac{1}{\sin x} \cdot \frac{1}{\cos x} \qquad \text{Write the fraction as two separate fractions.}$$

$$= \csc x \sec x$$

Because the left side has been transformed into the right side, the identity is verified. ■

Example 3　Verify that $1 + \tan x = \sec x(\cos x + \sin x)$.

Solution   Work on the right side.

$$\sec x(\cos x + \sin x) = \frac{1}{\cos x}(\cos x + \sin x)$$

$$= \frac{\cos x}{\cos x} + \frac{\sin x}{\cos x}$$

$$= 1 + \tan x$$

The identity is verified.   ■

Example 4   Verify that $(1 + \tan^2 x)\cos^2 x = 1$.

Solution   Change all expressions into sine and cosine functions and proceed as follows:

$$(1 + \tan^2 x)\cos^2 x = \left(1 + \frac{\sin^2 x}{\cos^2 x}\right)\cos^2 x$$

$$= \cos^2 x + \sin^2 x$$

$$= 1$$

The identity is verified.   ■

Example 5   Verify that $\cos^4 x - \sin^4 x = 1 - 2\sin^2 x$.

Solution   The left side is the difference of two squares and can be factored. Because the expression on the right side involves only the sine function, eliminate the cosine function from the left side.

$$\cos^4 x - \sin^4 x = (\cos^2 x + \sin^2 x)(\cos^2 x - \sin^2 x)$$

$$= 1 \cdot (\cos^2 x - \sin^2 x)$$

$$= \cos^2 x - \sin^2 x$$

$$= 1 - \sin^2 x - \sin^2 x \qquad \cos^2 x = 1 - \sin^2 x.$$

$$= 1 - 2\sin^2 x$$

The identity is verified.   ■

Example 6   Verify that $\dfrac{1}{\sec x - \tan x} - \dfrac{1}{\sec x + \tan x} = 2\tan x$.

Solution   Find a common denominator for the fractions on the left side and add them:

$$\frac{1}{\sec x - \tan x} - \frac{1}{\sec x + \tan x}$$

$$= \frac{\sec x + \tan x}{(\sec x - \tan x)(\sec x + \tan x)} - \frac{\sec x - \tan x}{(\sec x + \tan x)(\sec x - \tan x)}$$

$$= \frac{\sec x + \tan x - \sec x + \tan x}{\sec^2 x - \tan^2 x}$$

$$= \frac{2\tan x}{1}$$

$$= 2\tan x$$

The identity is verified.   ■

Sometimes we can work on both sides of an identity independently until each side is transformed into a common third expression. When following this strategy, it is important that each step in the process be reversible so that each side of the identity can be derived from the common third expression. It is also important that each side be worked on independently. It is incorrect to multiply or divide both sides of an equation whose truth we are trying to establish by an expression containing a variable because the resulting equation might not be equivalent to the given equation.

Example 7   Verify that $\dfrac{1 - \cos x}{1 + \cos x} = (\csc x - \cot x)^2$.

Solution   In this example, change the left side and the right side of the equation into a common third expression. The left side can be changed as follows:

$$\frac{1 - \cos x}{1 + \cos x} = \frac{(1 - \cos x)\,(\mathbf{1 - \cos x})}{(1 + \cos x)\,(\mathbf{1 - \cos x})} \qquad \text{Multiply numerator and denominator by } 1 - \cos x.$$

$$= \frac{(1 - \cos x)^2}{1 - \cos^2 x}$$

$$= \frac{(1 - \cos x)^2}{\sin^2 x}$$

$$= \left(\frac{1 - \cos x}{\sin x}\right)^2$$

The right side can be changed as follows:

$$(\csc x - \cot x)^2 = \left(\frac{1}{\sin x} - \frac{\cos x}{\sin x}\right)^2$$

$$= \left(\frac{1 - \cos x}{\sin x}\right)^2$$

Each side has been transformed independently into the expression

$$\left(\frac{1 - \cos x}{\sin x}\right)^2$$

Because each step is reversible, it follows that the given equation is an identity.   ∎

Example 8   For what values of x is $\sqrt{1 - \cos^2 x} = \sin x$?

Solution   Take the positive square root of both sides of the identity $1 - \cos^2 x = \sin^2 x$ to obtain

$$\sqrt{1 - \cos^2 x} = \sqrt{\sin^2 x}$$

$$= |\sin x|$$

The equation $\sqrt{1 - \cos^2 x} = |\sin x|$ is an identity, but, because $\sin x$ is sometimes negative, $\sqrt{1 - \cos^2 x} = \sin x$ is *not* an identity. If you think of x as a real number from 0 to 2π, $\sin x$ is nonnegative when $0 \le x \le \pi$. Thus, $\sqrt{1 - \cos^2 x} = \sin x$ only if x is a number from 0 to π, from 2π to 3π, from 4π to 5π, and so on.   ∎

When verifying identities, remember to write the variable associated with a trigonometric function. The notation "cos x," for example, represents a numerical value, but the notation "cos" is meaningless.

Exercise 7.1

In Exercises 1–10 indicate whether the statement is an identity. If so, verify it. If not, explain why. Remember that if an equation is false for one value of its variable, it cannot be an identity.

1. $\sin x + \sin x = 2 \sin x$

2. $\sec^2 x - \tan^2 x = 1$

3. $\sin x + \cos x = 1$

4. $\dfrac{1}{\tan x} = \dfrac{\cos x}{\sin x}$

5. $\tan \alpha \cos \alpha \csc \alpha = 1$

6. $\cot \beta \sec \beta \cos \beta = 1$

7. $\sqrt{1 - \sin^2 x} = \cos x$

8. $\sqrt{1 - \cos^2 x} = |\sin x|$

9. $(\sin x + 1)^2 = \sin^2 x + 2 \sin x + 1$

10. $\sin^2 x - 1 = \cos^2 x$

In Exercises 11–76 verify each identity.

11. $\dfrac{1 - \cos^2 x}{\sin x} = \sin x$

12. $\dfrac{\cot x}{\csc x} = \cos x$

13. $\dfrac{1 + \tan^2 x}{\sec^2 x} = 1$

14. $\tan x \csc x = \sec x$

15. $(\sin x + \cos x)^2 = 1 + 2 \sin x \cos x$

16. $(\sin x + 1)^2 = \sin^2 x + 2 \sin x + 1$

17. $(1 + \cos x)(1 - \cos x) = \sin^2 x$

18. $(\sin x + 1)(\sin x - 1) = -\cos^2 x$

19. $(\sec x + 1)(\sec x - 1) = \tan^2 x$

20. $\sin^4 x - \cos^4 x = \sin^2 x - \cos^2 x$

21. $\sin x(\sin x + \cot x \cos x) = 1$

22. $(\csc x - 1)(\csc x + 1) = \cot^2 x$

23. $\dfrac{1}{1 - \cos^2 x} = 1 + \cot^2 x$

24. $\cos x(\cos x + \sin x \tan x) = 1$

25. $\dfrac{\sin x}{\cos^2 x - 1} = -\csc x$

26. $\dfrac{1}{1 - \sin^2 x} = 1 + \tan^2 x$

27. $\dfrac{1 - \cos^2 x}{1 - \sin^2 x} = \tan^2 x$

28. $\sin^2 x + \cos^2 x = \cos^2 x \sec^2 x$

29. $\dfrac{1 - \sin^2 x}{1 + \tan^2 x} = \cos^4 x$

30. $\tan x \sin x = \dfrac{\csc x}{\cot x + \cot^3 x}$

31. $\dfrac{\cos x(\cos x + 1)}{\sin x} = \cos x \cot x + \cot x$

32. $\cos^2 x \csc x - \csc x = -\sin x$

33. $\sin^2 x \sec x - \sec x = -\cos x$

34. $(\sin x - \cos x)(1 + \sin x \cos x) = \sin^3 x - \cos^3 x$

35. $\sin^2 x - \tan^2 x = -\sin^2 x \tan^2 x$

36. $\dfrac{1 + \cot x}{\csc x} = \sin x + \cos x$

37. $\dfrac{1 - \csc x}{\cot x} = \tan x - \sec x$

38. $\dfrac{\cos^2 x - \tan^2 x}{\sin^2 x} = \cot^2 x - \sec^2 x$

39. $\dfrac{1}{\sec x - \tan x} = \tan x + \sec x$

40. $\csc^2 x + \sec^2 x = \csc^2 x \sec^2 x$

41. $\dfrac{\cos x}{\cot x} + \dfrac{\sin x}{\tan x} = \sin x + \cos x$

42. $\dfrac{\cos x}{1 + \sin x} = \sec x - \tan x$

43. $\dfrac{\cos x}{1 - \sin x} = \dfrac{1 + \sin x}{\cos x}$

44. $\cos^2 x + \sin x \cos x = \dfrac{\cos x(\cot x + 1)}{\csc x}$

45. $(\sin x + \cos x)^2 + (\sin x - \cos x)^2 = 2$

46. $\dfrac{\sec x + 1}{\tan x} = \dfrac{\tan x}{\sec x - 1}$

47. $\dfrac{\cos x - \cot x}{\cos x \cot x} = \dfrac{\sin x - 1}{\cos x}$

48. $\cos^4 x - \sin^4 x = 2 \cos^2 x - 1$

49. $\dfrac{1}{1 + \sin x} + \dfrac{1}{1 - \sin x} = 2 \sec^2 x$

50. $\sqrt{\dfrac{1 - \sin x}{1 + \sin x}} = \sec x - \tan x$ with $0 < x < \dfrac{\pi}{2}$

51. $\sqrt{\dfrac{1 - \sin x}{1 + \sin x}} = \dfrac{1 - \sin x}{\cos x}$ with $0 < x < \dfrac{\pi}{2}$

52. $-\sqrt{\dfrac{\sec x - 1}{\sec x + 1}} = \dfrac{1 - \sec x}{\tan x}$ with $0 < x < \dfrac{\pi}{2}$

53. $\sqrt{\dfrac{\csc x - \cot x}{\csc x + \cot x}} = \dfrac{\sin x}{1 + \cos x}$ with $0 < x < \dfrac{\pi}{2}$

54. $\dfrac{\cos x}{1 - \sin x} - \dfrac{1}{\cos x} = \tan x$

55. $\dfrac{1}{\sec x(1 + \sin x)} = \sec x(1 - \sin x)$

56. $\dfrac{\cot x - \cos x}{\cot x \cos x} = \dfrac{1 - \sin x}{\cos x}$

57. $\dfrac{1}{\tan x(\csc x + 1)} = \tan x(\csc x - 1)$

58. $\dfrac{(\cos x + 1)^2}{\sin^2 x} = 2 \csc^2 x + 2 \csc x \cot x - 1$

59. $\dfrac{\csc x}{\sec x - \csc x} = \dfrac{\cot^2 x + \csc x \sec x + 1}{\tan^2 x - \cot^2 x}$

60. $(\sin x + \cos x)^2 - (\sin x - \cos x)^2 = 4 \sin x \cos x$

61. $\dfrac{\sin^2 x \tan x + \sin^2 x}{\sec x - 2 \sin x \tan x} = \dfrac{\sin^2 x}{\cos x - \sin x}$

62. $\dfrac{\sin x + \cos x + 1}{\sin x + \cos x - 1} = \csc x \sec x + \csc x + \sec x + 1$

63. $\dfrac{\sec x + \tan x}{\csc x + \cot x} = \dfrac{\cot x - \csc x}{\tan x - \sec x}$

64. $\dfrac{\cos x + \sin x}{\cos x - \sin x} = \dfrac{\csc x + 2 \cos x}{\csc x - 2 \sin x}$

65. $\dfrac{3 \cos^2 x + 11 \sin x - 11}{\cos^2 x} = \dfrac{3 \sin x - 8}{1 + \sin x}$

66. $\dfrac{3 \sin^2 x + 5 \cos x - 5}{\sin^2 x} = \dfrac{3 \cos x - 2}{1 + \cos x}$

67. $\dfrac{\cos x + \sin x + 1}{\cos x - \sin x - 1} = \dfrac{1 + \cos x}{-\sin x}$

68. $\dfrac{1 + \sin x + \cos x}{1 + \sin x - \cos x} = \dfrac{\sin x}{1 - \cos x}$

69. $\dfrac{1 + \sin x}{\cos x} = \dfrac{\sin x - \cos x + 1}{\sin x + \cos x - 1}$

70. $\dfrac{\tan x + \sec x + 1}{\tan x + \sec x - 1} = \dfrac{1 + \cos x}{\sin x}$

71. $\dfrac{\csc x + 1 + \cot x}{\csc x + 1 - \cot x} = \dfrac{1 + \cos x}{\sin x}$

72. $\dfrac{3 \cot x \csc x - 2 \csc^2 x}{\csc^2 x + \cot x \csc x} - 3 = -5(\csc^2 x - \csc x \cot x)$

73. $\dfrac{\sin^3 x \cos x + \cos x - \sin^2 x \cos x - \cos x \sin x}{\cos^4 x} = \sec x - \tan x$

74. $\dfrac{\sin x + \sin x \cos x - \sin x \cos^2 x - \sin x \cos^3 x}{\cos^4 x - 2 \cos^2 x + 1} = \csc x + \cot x$

75. $\dfrac{1}{\sec x + \csc x - \sec x \csc x} = \dfrac{\sin x + \cos x + 1}{2}$

76. $\dfrac{2(\sin x + 1)}{1 + \cot x + \csc x} = \sin x + 1 - \cos x$

7.2   IDENTITIES INVOLVING SUMS AND DIFFERENCES OF TWO QUANTITIES

We shall often encounter expressions that contain trigonometric functions of sums or differences of two quantities. It is tempting to believe that an expression such as $\cos(u + v)$ is equal to $\cos u + \cos v$. However, this is not true, as the following work shows:

$$\cos\left(\frac{\pi}{6} + \frac{\pi}{3}\right) = \cos\frac{\pi}{2} = 0$$

but

$$\cos\frac{\pi}{6} + \cos\frac{\pi}{3} = \frac{\sqrt{3}}{2} + \frac{1}{2} = \frac{\sqrt{3} + 1}{2}$$

Thus,

$$\cos\left(\frac{\pi}{6} + \frac{\pi}{3}\right) \neq \cos\frac{\pi}{6} + \cos\frac{\pi}{3}$$

In general, a trigonometric function of the sum (or difference) of two quantities is not equal to the sum (or difference) of the trigonometric functions of each quantity. It is possible, however, to develop formulas for finding trigonometric functions of a sum (or difference). To do so, we shall use the distance formula to derive a formula to evaluate the expression $\cos(u + v)$, where u and v are real numbers.

On the unit circle in Figure 7–1a, we locate point P corresponding to u, point R corresponding to $u + v$, point Q corresponding to $-v$, and point S corresponding to 0. We then form triangles POQ and ROS as in Figure 7–1b. Because these triangles are isosceles triangles with equal vertex angles, they are congruent. Hence, $d(RS) = d(PQ)$. The coordinates of points P, Q, R, and S are:

P: $(\cos u, \sin u)$
Q: $(\cos[-v], \sin[-v]) = (\cos v, -\sin v)$
R: $(\cos[u + v], \sin[u + v])$
S: $(1, 0)$

Because $d(RS) = d(PQ)$, we have $d(RS)^2 = d(PQ)^2$ and

$$[\cos(u + v) - 1]^2 + [\sin(u + v) - 0]^2$$
$$= (\cos u - \cos v)^2 + [\sin u - (-\sin v)]^2$$

or

$$\cos^2(u + v) - 2\cos(u + v) + 1 + \sin^2(u + v)$$
$$= \cos^2 u - 2\cos u \cos v + \cos^2 v + \sin^2 u + 2\sin u \sin v + \sin^2 v$$

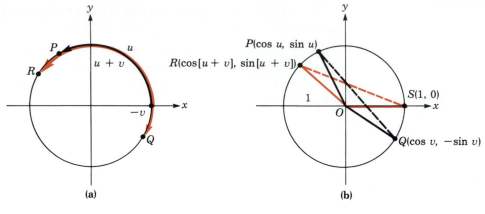

Figure 7–1

Because $\cos^2 x + \sin^2 x = 1$, the previous equation can be written as

$$1 - 2\cos(u + v) + 1 = 1 - 2\cos u \cos v + 1 + 2\sin u \sin v$$
$$2 - 2\cos(u + v) = 2 - 2\cos u \cos v + 2\sin u \sin v$$
$$-2\cos(u + v) = -2\cos u \cos v + 2\sin u \sin v$$

Dividing both sides of the previous equation by -2 gives the identity

$$\cos(u + v) = \cos u \cos v - \sin u \sin v$$

The identity for the cosine of the difference of two quantities follows from the fact that $u - v = u + (-v)$:

$$\cos(u - v) = \cos[u + (-v)]$$
$$= \cos u \cos(-v) - \sin u \sin(-v)$$

Because $\cos(-v) = \cos v$ and $\sin(-v) = -\sin v$, we have

$$\cos(u - v) = \cos u \cos v + \sin u \sin v$$

Example 1   Find the value of $\cos \frac{\pi}{12}$ without using tables or a calculator.

Solution   Because $\frac{\pi}{12} = \frac{\pi}{4} - \frac{\pi}{6}$, it follows that

$$\cos \frac{\pi}{12} = \cos\left(\frac{\pi}{4} - \frac{\pi}{6}\right)$$
$$= \cos \frac{\pi}{4} \cos \frac{\pi}{6} + \sin \frac{\pi}{4} \sin \frac{\pi}{6}$$
$$= \frac{\sqrt{2}}{2}\left(\frac{\sqrt{3}}{2}\right) + \frac{\sqrt{2}}{2}\left(\frac{1}{2}\right)$$
$$= \frac{\sqrt{6} + \sqrt{2}}{4}$$

We have seen that any trigonometric function of an acute angle is equal to the cofunction of the complement of that angle. This property can be extended to all angles. For example,

$$\cos(90° - u) = \cos 90° \cos u + \sin 90° \sin u$$
$$= 0 \cdot \cos u + 1 \cdot \sin u$$
$$= \sin u$$

Thus, we have

$$\cos(90° - u) = \sin u$$

Because $u = 90° - (90° - u)$, we can find a value for $\cos u$:

$$\cos u = \cos[90° - (90° - u)]$$
$$= \cos 90° \cos(90° - u) + \sin 90° \sin(90° - u)$$
$$= \sin(90° - u)$$

Thus, we have

$$\sin(90° - u) = \cos u$$

In the exercises, you will be asked to show that

$$\tan(90° - u) = \cot u$$

To develop a formula for $\sin(u + v)$, we can substitute $u + v$ for x in the equation $\sin x = \cos(90° - x)$ and proceed as follows:

$$\sin(u + v) = \cos[90° - (u + v)]$$
$$= \cos[(90° - u) - v]$$
$$= \cos(90° - u) \cos v + \sin(90° - u) \sin v$$

Because $\cos(90° - u) = \sin u$ and $\sin(90° - u) = \cos u$, we have

$$\sin(u + v) = \sin u \cos v + \cos u \sin v$$

To find the formula for $\sin(u - v)$, we substitute $-v$ for v in the previous equation and simplify:

$$\sin(u - v) = \sin[u + (-v)]$$
$$= \sin u \cos(-v) + \cos u \sin(-v)$$
$$= \sin u \cos v - \cos u \sin v \qquad \cos(-v) = \cos v;$$
$$\sin(-v) = -\sin v.$$

Thus,

$$\sin(u - v) = \sin u \cos v - \cos u \sin v$$

To find identities for the tangent of the sum of difference of two quantities, we make use of the identity

$$\tan x = \frac{\sin x}{\cos x}$$

and substitute $u + v$ for x:

$$\tan(u + v) = \frac{\sin(u + v)}{\cos(u + v)}$$

$$= \frac{\sin u \cos v + \cos u \sin v}{\cos u \cos v - \sin u \sin v}$$

To simplify this result, we divide both the numerator and the denominator of the fraction by $\cos u \cos v$:

$$\tan(u + v) = \frac{\dfrac{\sin u \cos v}{\cos u \cos v} + \dfrac{\cos u \sin v}{\cos u \cos v}}{\dfrac{\cos u \cos v}{\cos u \cos v} - \dfrac{\sin u \sin v}{\cos u \cos v}}$$

$$= \frac{\tan u + \tan v}{1 - \tan u \tan v}$$

Thus,

$$\tan(u + v) = \frac{\tan u + \tan v}{1 - \tan u \tan v}$$

If we substitute $-v$ for v in the previous equation and simplify, we obtain an identity involving the tangent of the difference of two quantities:

$$\tan(u - v) = \frac{\tan u - \tan v}{1 + \tan u \tan v}$$

Example 2   Given that $\sin \alpha = \frac{12}{13}$ with α in QI, and $\cos \beta = -\frac{4}{5}$, with β in QII, find $\sin(\beta - \alpha)$.

Solution   Because $\sin \alpha = \frac{12}{13}$ and α is in QI, you can draw Figure 7–2a and determine that $\cos \alpha = \frac{5}{13}$. Because $\cos \beta = -\frac{4}{5}$ and β is in QII, you can draw Figure 7–2b and determine that $\sin \beta = \frac{3}{5}$. Then substitute $\frac{3}{5}$ for $\sin \beta$, $\frac{5}{13}$ for $\cos \alpha$, $-\frac{4}{5}$ for $\cos \beta$, and $\frac{12}{13}$ for $\sin \alpha$ in the following identity and simplify:

$$\sin(\beta - \alpha) = \sin \beta \cos \alpha - \cos \beta \sin \alpha$$

$$= \frac{3}{5}\left(\frac{5}{13}\right) - \left(-\frac{4}{5}\right)\left(\frac{12}{13}\right)$$

$$= \frac{15}{65} + \frac{48}{65}$$

$$= \frac{63}{65}$$

Thus, $\sin(\beta - \alpha) = \frac{63}{65}$.

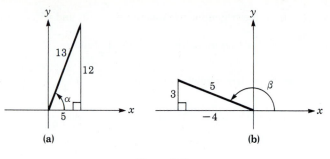

Figure 7–2

Example 3   Verify the identity $\tan\left(u + \dfrac{\pi}{4}\right) = \dfrac{1 + \tan u}{1 - \tan u}$.

Solution   Use the formula for the tangent of the sum of two angles:

$$\tan\left(u + \frac{\pi}{4}\right) = \frac{\tan u + \tan\dfrac{\pi}{4}}{1 - \tan u \tan\dfrac{\pi}{4}}$$

$$= \frac{\tan u + 1}{1 - (\tan u)(1)}$$

$$= \frac{1 + \tan u}{1 - \tan u}$$

The identity is verified.

Example 4   If $f(x) = \cos x$, verify the identity

$$\frac{f(x + h) - f(x)}{h} = \cos x\left(\frac{\cos h - 1}{h}\right) - \sin x\left(\frac{\sin h}{h}\right)$$

Solution $\dfrac{f(x + h) - f(x)}{h} = \dfrac{\cos(x + h) - \cos x}{h}$

$$= \frac{\cos x \cos h - \sin x \sin h - \cos x}{h}$$

$$= \frac{\cos x(\cos h - 1) - \sin x \sin h}{h}$$

$$= \cos x\left(\frac{\cos h - 1}{h}\right) - \sin x\left(\frac{\sin h}{h}\right)$$

Exercise 7.2

1. Find sin 195° from the trigonometric functions of 45° and 150°.
2. Find cos 195° from the trigonometric functions of 45° and 150°.
3. Find tan 195° from the trigonometric functions of 225° and 30°.

4. Find $\tan 165°$ from the trigonometric functions of $210°$ and $45°$.

5. Find $\cos \dfrac{11\pi}{12}$ from the trigonometric functions of $\dfrac{\pi}{6}$ and $\dfrac{3\pi}{4}$.

6. Find $\sin \dfrac{11\pi}{12}$ from the trigonometric functions of $\dfrac{\pi}{6}$ and $\dfrac{3\pi}{4}$.

7. Find $\cos \dfrac{19\pi}{12}$ from the trigonometric functions of $\dfrac{11\pi}{6}$ and $\dfrac{\pi}{4}$.

8. Find $\tan \dfrac{19\pi}{12}$ from the trigonometric functions of $\dfrac{11\pi}{6}$ and $\dfrac{\pi}{4}$.

In Exercises 9–16 choose two appropriate angles and use an identity of Section 7.2 to evaluate each expression. **Do not use a calculator.**

9. $\sin 255°$

10. $\cos 285°$

11. $\tan 105°$

12. $\cot 255°$

13. $\cos \dfrac{\pi}{12}$

14. $\sin \dfrac{7\pi}{12}$

15. $\sin \dfrac{5\pi}{12}$

16. $\cos \dfrac{13\pi}{12}$

17. Show that $\sin(60° + \theta) = \dfrac{\sqrt{3}}{2} \cos \theta + \dfrac{1}{2} \sin \theta$.

18. Show that $\cos\left(\dfrac{\pi}{2} + x\right) = -\sin x$.

19. Show that $\tan(\pi + x) = \tan x$.

20. Show that $\sin\left(\dfrac{3\pi}{2} - x\right) = -\cos x$.

21. Show that $\cos(\pi - x) = -\cos x$.

22. Show that $\tan\left(\dfrac{\pi}{4} - x\right) = \dfrac{1 - \tan x}{1 + \tan x}$.

In Exercises 23–30 express each quantity as a single function of one angle.

23. $\sin 10° \cos 30° + \cos 10° \sin 30°$

24. $\cos 20° \cos 30° - \sin 20° \sin 30°$

25. $\dfrac{\tan 75° + \tan 40°}{1 - \tan 75° \tan 40°}$

26. $\sin 100° \cos 80° - \cos 100° \sin 80°$

27. $\cos 120° \cos 40° + \sin 120° \sin 40°$

28. $\dfrac{\tan 37° - \tan 125°}{1 + \tan 37° \tan 125°}$

29. $\sin x \cos 2x + \sin 2x \cos x$

30. $\cos 2y \cos 3y - \sin 2y \sin 3y$

31. Given that $\sin \alpha = \frac{3}{5}$, α in QII, and $\cos \beta = -\frac{12}{13}$, β in QII, find $\sin(\alpha + \beta)$ and $\cos(\alpha - \beta)$.

32. Given that $\sin \alpha = \frac{7}{25}$, α in QI, and $\sin \beta = \frac{15}{17}$, β in QI, find $\sin(\alpha - \beta)$ and $\cos(\alpha + \beta)$.

33. Given that $\tan \alpha = -\frac{5}{12}$, α in QII, and $\tan \beta = \frac{15}{8}$, β in QI, find $\tan(\alpha + \beta)$ and $\tan(\alpha - \beta)$.

34. Given that $\sin \alpha = \frac{12}{13}$, α in QI, and $\sin(\alpha + \beta) = \frac{24}{25}$, $\alpha + \beta$ in QII, find $\sin \beta$ and $\cos \beta$.

35. Given that $\cos \beta = -\frac{15}{17}$, β in QII, and $\sin(\alpha - \beta) = -\frac{24}{25}$, $\alpha - \beta$ in QIV, find $\sin \alpha$ and $\cos \alpha$.

36. Given that $\sin(\alpha + \beta) = \frac{3}{5}$, $\alpha + \beta$ in QI, and $\cos(\alpha - \beta) = \frac{12}{13}$, $\alpha - \beta$ in QIV, find $\sin 2\alpha$ and $\cos 2\beta$.

In Exercises 37–51 verify each identity.

37. $\sin(30° + \theta) - \cos(60° + \theta) = \sqrt{3} \sin \theta$

38. $\sin(60° + \theta) - \cos(30° + \theta) = \sin \theta$

39. $\sin(30° + \theta) + \cos(60° + \theta) = \cos \theta$

40. $\sin(A + B) - \sin(A - B) = 2 \cos A \sin B$

41. $\cos(A + B) + \cos(A - B) = 2 \cos A \cos B$

42. $\sin(A + B) \sin(A - B) = \sin^2 A - \sin^2 B$

43. $\cos(A + B) \cos(A - B) = \cos^2 A + \cos^2 B - 1$

44. $\cot(A + B) = \dfrac{\cot A \cot B - 1}{\cot A + \cot B}$

45. $\cot(A - B) = \dfrac{\cot A \cot B + 1}{\cot B - \cot A}$

46. $\cos(A + B) - \cos(A - B) = -2 \sin A \cos B$

47. $\sin(A + B) + \sin(A - B) = 2 \sin A \cos B$

48. $-\tan A = \cot\left(A + \dfrac{\pi}{2}\right)$

49. $\dfrac{\tan A + \tan B}{1 + \tan A \tan B} = \dfrac{\sin(A + B)}{\cos(A - B)}$

50. $\dfrac{\cos(A + B)}{\sin(A - B)} = \dfrac{\cot A - \tan B}{1 - \cot A \tan B}$

51. $\dfrac{\cos(A - B)}{\sin(A + B)} = \dfrac{1 + \tan A \tan B}{\tan A + \tan B}$

52. Derive a formula for $\cos(A + B + C)$.

53. Derive a formula for $\sin(A + B + C)$.

54. Derive a formula for $\cos(A - B - C)$.

55. Derive a formula for $\tan(A + B + C)$.

56. Show that $\tan(90° - u) = \cot u$.

57. If $f(x) = \sin x$, verify the identity

$$\frac{f(x + h) - f(x)}{h} = (\sin x)\frac{\cos h - 1}{h} + (\cos x)\frac{\sin h}{h}$$

58. If $f(x) = -\cos x$, verify the identity

$$\frac{f(x + h) - f(x)}{h} = (\cos x)\frac{1 - \cos h}{h} + (\sin x)\frac{\sin h}{h}$$

7.3 THE DOUBLE-ANGLE IDENTITIES

If u and v represent equal quantities, the formulas for the sine, cosine, and tangent of a sum can be transformed into formulas called the **double-angle identities**.

If $v = u$ in the identity $\sin(u + v) = \sin u \cos v + \cos u \sin v$, we obtain

$$\sin(u + u) = \sin u \cos u + \cos u \sin u$$

$$\mathbf{\sin 2u = 2 \sin u \cos u}$$

Similarly, if $v = u$ in the identity $\cos(u + v) = \cos u \cos v - \sin u \sin v$, we obtain

$$\cos(u + u) = \cos u \cos u - \sin u \sin u$$

$$\mathbf{\cos 2u = \cos^2 u - \sin^2 u}$$

There are two variations of the identity for $\cos 2u$. To obtain one of them, we substitute $1 - \cos^2 u$ for $\sin^2 u$ and simplify:

$$\cos 2u = \cos^2 u - \sin^2 u$$
$$= \cos^2 u - (1 - \cos^2 u)$$

Thus,

$$\mathbf{\cos 2u = 2 \cos^2 u - 1}$$

To obtain the other, we substitute $1 - \sin^2 u$ for $\cos^2 u$ and simplify:

$$\cos 2u = \cos^2 u - \sin^2 u$$
$$= 1 - \sin^2 u - \sin^2 u$$

Thus,

$$\cos 2u = 1 - 2 \sin^2 u$$

To find the double-angle identity for the tangent function, we let $v = u$ in the identity

$$\tan(u + v) = \frac{\tan u + \tan v}{1 - \tan u \tan v}$$

and simplify:

$$\tan(u + u) = \frac{\tan u + \tan u}{1 - \tan u \tan u}$$

$$\tan 2u = \frac{2 \tan u}{1 - \tan^2 u}$$

Example 1   Simplify the expression sin 5u cos 5u.

Solution   Because the given expression is one-half of 2 sin 5u cos 5u, it is one-half of sin 2(5u):

$$\sin 5u \cos 5u = \frac{1}{2}(2 \sin 5u \cos 5u)$$

$$= \frac{1}{2} \sin 2(5u)$$

$$= \frac{1}{2} \sin 10u$$

■

Example 2   Use a double-angle identity to evaluate $\tan 2(\frac{\pi}{3})$.

Solution   Use the double-angle identity for tan 2u and substitute $\frac{\pi}{3}$ for u:

$$\tan 2u = \frac{2 \tan u}{1 - \tan^2 u}$$

$$\tan 2\left(\frac{\pi}{3}\right) = \frac{2 \tan \dfrac{\pi}{3}}{1 - \tan^2 \dfrac{\pi}{3}}$$

$$= \frac{2\sqrt{3}}{1 - 3}$$

$$= -\sqrt{3}$$

■

Example 3   Find the **a.** sine, **b.** cosine, and **c.** tangent of 2θ if $\cos \theta = \frac{12}{13}$ and θ is in QIV.

Solution   Use the information that $\cos \theta = \frac{12}{13}$ and that θ is in QIV to draw that angle in standard position, as in Figure 7–3. From the figure, it follows that

$$\sin \theta = \frac{-5}{13} = -\frac{5}{13}$$

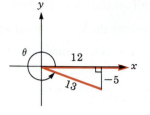

Figure 7–3

Use the double-angle identities to find the functions of 2θ.

a. $\sin 2\theta = 2 \sin \theta \cos \theta = 2\left(-\dfrac{5}{13}\right)\left(\dfrac{12}{13}\right) = -\dfrac{120}{169}$

b. $\cos 2\theta = \cos^2 \theta - \sin^2 \theta = \left(\dfrac{12}{13}\right)^2 - \left(-\dfrac{5}{13}\right)^2 = \dfrac{119}{169}$

c. $\tan 2\theta = \dfrac{\sin 2\theta}{\cos 2\theta} = \dfrac{-\dfrac{120}{169}}{\dfrac{119}{169}} = -\dfrac{120}{119}$

Example 4   Verify the identity $\cos 2x = \cos^4 x - \sin^4 x$.

Solution   Because the right side factors, work on that side:

$$\cos^4 x - \sin^4 x = (\cos^2 x + \sin^2 x)(\cos^2 x - \sin^2 x)$$
$$= 1 \cdot (\cos^2 x - \sin^2 x)$$
$$= \cos 2x$$

Example 5   Verify the identity $\cos u \sin 2u = 2 \sin u - 2 \sin^3 u$.

Solution   Although the right side looks more complicated, work on the left side because it involves a double angle.

$$\cos u \sin 2u = \cos u(2 \sin u \cos u)$$
$$= 2 \cos^2 u \sin u$$
$$= 2(1 - \sin^2 u) \sin u$$
$$= 2 \sin u - 2 \sin^3 u$$

Example 6   Verify the identity $\cos 2u \sec^2 u = 2 - \sec^2 u$.

Solution   Work on the left side to eliminate the double angle:

$$\cos 2u \sec^2 u = (2 \cos^2 u - 1) \sec^2 u$$
$$= 2 \cos^2 u \sec^2 u - \sec^2 u$$
$$= 2 - \sec^2 u$$

Example 7   Verify that $\sin(u + v) + \sin(u - v) = 2 \sin u \cos v$.

Solution   Work on the left side of the equation:

$$\sin(u + v) + \sin(u - v)$$
$$= \sin u \cos v + \cos u \sin v + \sin u \cos v - \cos u \sin v$$
$$= \sin u \cos v + \sin u \cos v$$
$$= 2 \sin u \cos v$$

∎

Example 8   Verify the identity $2 \cot 2x = \cot x - \tan x$.

Solution   Work on the left side of the equation:

$$2 \cot 2x = 2 \left(\frac{1}{\tan 2x} \right)$$
$$= 2 \left(\frac{1 - \tan^2 x}{2 \tan x} \right)$$
$$= \frac{1 - \tan^2 x}{\tan x}$$
$$= \frac{1}{\tan x} - \frac{\tan^2 x}{\tan x}$$
$$= \cot x - \tan x$$

∎

▬ Exercise 7.3 ▬▬▬▬▬▬▬▬▬▬▬▬▬▬▬▬▬▬▬▬▬▬▬▬▬

In Exercises 1–26 write the given expression in terms of a single trigonometric function of twice the given angle.

1. $2 \sin \alpha \cos \alpha$ 2. $2 \cos^2 \alpha - 1$ 3. $2 \sin 3\theta \cos 3\theta$ 4. $2 \cos^2 2A - 1$

5. $\cos^2 \beta - \sin^2 \beta$ 6. $1 - 2 \sin^2 \beta$ 7. $2 \cos^2 \dfrac{\beta}{2} - 1$ 8. $\sin 5\theta \cos 5\theta$

9. $4 \sin \theta \cos \theta$ 10. $\cos^2 \dfrac{\theta}{2} - \sin^2 \dfrac{\theta}{2}$ 11. $4 \sin^2 2\theta \cos^2 2\theta$ 12. $2 - 4 \sin^2 6B$

13. $\cos^2 \alpha - \dfrac{1}{2}$ 14. $2 - 4 \sin^2 \dfrac{\alpha}{4}$ 15. $\cos^2 9\theta - \sin^2 9\theta$ 16. $4 \sin 4B \cos 4B$

17. $4 \sin^2 5\theta \cos^2 5\theta$ 18. $\dfrac{2 \tan A}{1 - \tan^2 A}$ 19. $\dfrac{2 \tan 4C}{1 - \tan^2 4C}$ 20. $3 - 6 \cos^2 6x$

21. $\dfrac{\tan \dfrac{A}{2}}{\dfrac{1}{2} - \dfrac{1}{2} \tan^2 \dfrac{A}{2}}$ 22. $\dfrac{2 \tan \alpha}{2 - \sec^2 \alpha}$ 23. $\cos^4 4x - \sin^4 4x$ 24. $\dfrac{1}{2} \sec \theta \csc \theta$

25. $1 - 2 \cos^2 5x$ 26. $\dfrac{1 - \tan^2 3x}{2 \tan 3x}$

In Exercises 27–38 use a double-angle identity to find the value of the given expression. Then check your work by evaluating the expression directly.

27. $\sin 2(30°)$ 28. $\cos 2(30°)$ 29. $\tan 2(45°)$ 30. $\cot 2(135°)$

31. $\cos 2(240°)$ 32. $\sin 2(150°)$ 33. $\cot 2(225°)$ 34. $\tan 2(315°)$

35. $\sin 2\left(\dfrac{\pi}{3}\right)$ **36.** $\cos 2\left(\dfrac{7\pi}{6}\right)$ **37.** $\cos 2\left(\dfrac{11\pi}{6}\right)$ **38.** $\sin 2\left(\dfrac{5\pi}{3}\right)$

In Exercises 39–50 find the exact value of the sine, cosine, and tangent of 2θ using the given information.

39. $\sin \theta = \dfrac{12}{13}$; θ in QI

40. $\cos \theta = \dfrac{5}{13}$; θ in QIV

41. $\tan \theta = \dfrac{12}{5}$; θ in QIII

42. $\sin \theta = \dfrac{3}{5}$; θ in QII

43. $\cos \theta = -\dfrac{4}{5}$; θ in QIII

44. $\cot \theta = -\dfrac{15}{8}$; θ in QIV

45. $\sin \theta = -\dfrac{24}{25}$; θ in QIV

46. $\cos \theta = -\dfrac{7}{25}$; θ in QIII

47. $\sin \theta = \dfrac{40}{41}$; $\cos \theta$ is positive

48. $\sin \theta = -\dfrac{40}{41}$; $\tan \theta$ is negative

49. $\cos \theta = \dfrac{40}{41}$; $\tan \theta$ is negative

50. $\cos \theta = 0$

In Exercises 51–77 verify the given identity.

51. $\dfrac{\tan 2x}{\sin 2x} = \sec 2x$

52. $\dfrac{\cot 2x}{\cos 2x} = \csc 2x$

53. $2 \csc 2A = \sec A \csc A$

54. $\sin 2A + 2 \sin A = \dfrac{-2 \sin^3 A}{\cos A - 1}$

55. $\cos 2x - \dfrac{\sin 2x}{\cos x} + 2 \sin^2 x = 1 - 2 \sin x$

56. $2 \cos^4 \theta + 2 \sin^2 \theta \cos^2 \theta - 1 = \cos 2\theta$

57. $\sin 2A - 2 \sin A = -\dfrac{2 \sin^3 A}{\cos A + 1}$

58. $2 \cos A - 1 = \dfrac{\cos 2A + \cos A}{\cos A + 1}$

59. $2 \sin^4 \theta + 2 \sin^2 \theta \cos^2 \theta - 1 = -\cos 2\theta$

60. $\sin 2\theta - \tan \theta = \cos 2\theta \tan \theta$

61. $2 \cos A + 1 = \dfrac{\cos 2A - \cos A}{\cos A - 1}$

62. $(\sin A + \cos A)^2 = 1 + \sin 2A$

63. $\sec 2\theta = \dfrac{\tan \theta + \cot \theta}{\cot \theta - \tan \theta}$

64. $\sin 4x = 4 \cos x \sin x - 8 \cos x \sin^3 x$

65. $4 \csc^2 2A = \sec^2 A + \csc^2 A$

66. $1 - \dfrac{1}{2} \sin 2x = \dfrac{\sin^3 x + \cos^3 x}{\sin x + \cos x}$

67. $\cos 4x = 8 \cos^4 x - 8 \cos^2 x + 1$

68. $-\tan 2x = \dfrac{2}{\tan x - \cot x}$

69. $2 \tan 2x = \dfrac{\sin x - \cos x}{\sin x + \cos x} - \dfrac{\sin x + \cos x}{\sin x - \cos x}$

70. $\sin^4 \theta = \dfrac{\cos 4\theta}{8} - \dfrac{\cos 2\theta}{2} + \dfrac{3}{8}$

71. $\dfrac{\sin x + \sin 2x}{\cos x - \cos 2x} = \dfrac{\sec x + 2}{\csc x - \cot x + \tan x}$

72. $\dfrac{\cos x - \sin x}{\cos 2x} = \dfrac{\cos x + \sin x}{1 + \sin 2x}$

73. $\dfrac{1 - \tan^2 x}{1 + \tan^2 x} = \cos 2x$

74. $\dfrac{\sec^2 x}{\sec 2x} = 2 - \sec^2 x$

75. $\dfrac{1 + \tan x}{1 - \tan x} = \dfrac{\cos 2x}{1 - \sin 2x}$

76. $\cos 2x = \dfrac{1}{\tan 2x \tan x + 1}$

77. $\tan A \sin 2A + \cos 2A = 1$

7.4   THE HALF-ANGLE IDENTITIES

To develop the **half-angle identities**, we begin by solving the identity $\cos 2x = 2 \cos^2 x - 1$ for $\cos x$.

$$\cos 2x = 2 \cos^2 x - 1$$

$$2 \cos^2 x = 1 + \cos 2x$$

$$\cos^2 x = \frac{1 + \cos 2x}{2}$$

$$\cos x = \pm \sqrt{\frac{1 + \cos 2x}{2}}$$

With an appropriate choice of the sign preceding the radical, the previous equation is true for all values of x. To derive the first half-angle identity, we let $x = \frac{u}{2}$:

$$\cos \frac{u}{2} = \pm \sqrt{\frac{1 + \cos u}{2}}$$

The sign preceding the radical sign in the previous equation is determined by the quadrant in which $\frac{u}{2}$ lies.

We solve $\cos 2x = 1 - 2 \sin^2 x$ for $\sin x$ and let $x = \frac{u}{2}$ to obtain the half-angle identity for the sine function:

$$\cos 2x = 1 - 2 \sin^2 x$$

$$2 \sin^2 x = 1 - \cos 2x$$

$$\sin^2 x = \frac{1 - \cos 2x}{2}$$

$$\sin x = \pm \sqrt{\frac{1 - \cos 2x}{2}}$$

$$\sin \frac{u}{2} = \pm \sqrt{\frac{1 - \cos u}{2}}$$

Again, the $+$ or $-$ sign is chosen by the quadrant in which $\frac{u}{2}$ lies.

Example 1   Use a half-angle identity to find $\sin 15°$.

Solution   Because $15°$ is $\frac{1}{2}(30°)$, it follows that

$$\sin 15° = \sin \frac{30°}{2} = + \sqrt{\frac{1 - \cos 30°}{2}}$$

$$= \sqrt{\frac{1 - \frac{\sqrt{3}}{2}}{2}}$$

$$= \sqrt{\frac{2 - \sqrt{3}}{4}}$$

$$= \frac{\sqrt{2 - \sqrt{3}}}{2}$$

The + sign is chosen because 15° is a first-quadrant angle and the sine of a first-quadrant angle is positive. ■

Example 2   Use a half-angle identity to find $\cos \frac{7\pi}{12}$.

Solution   Because $\frac{7\pi}{12}$ is $\frac{1}{2}(\frac{7\pi}{6})$, it follows that

$$\cos \frac{7\pi}{12} = \cos\left(\frac{1}{2} \cdot \frac{7\pi}{6}\right) = -\sqrt{\frac{1 + \cos \frac{7\pi}{6}}{2}}$$

$$= -\sqrt{\frac{1 - \frac{\sqrt{3}}{2}}{2}}$$

$$= -\frac{\sqrt{2 - \sqrt{3}}}{2}$$

Here, the − sign is chosen because $\frac{7\pi}{12}$ is a second-quadrant angle and the cosine of a second-quadrant angle is negative. ■

To develop an identity involving $\tan \frac{u}{2}$, we substitute $\frac{u}{2}$ for x in the identity

$$\tan x = \frac{\sin x}{\cos x}$$

and proceed as follows:

$$\tan \frac{u}{2} = \frac{\sin \frac{u}{2}}{\cos \frac{u}{2}} = \frac{\pm \sqrt{\frac{1 - \cos u}{2}}}{\pm \sqrt{\frac{1 + \cos u}{2}}}$$

$$= \pm \sqrt{\frac{1 - \cos u}{1 + \cos u}}$$

If the fraction within the radical is multiplied by

$$\frac{1 - \cos u}{1 - \cos u}$$

we obtain

$$\tan \frac{u}{2} = \pm \sqrt{\frac{(1 - \cos u)(1 - \cos u)}{(1 + \cos u)(1 - \cos u)}}$$

$$= \pm \sqrt{\frac{(1 - \cos u)^2}{1 - \cos^2 u}}$$

$$= \pm \sqrt{\frac{(1 - \cos u)^2}{\sin^2 u}}$$

1. $\tan \dfrac{u}{2} = \dfrac{1 - \cos u}{\sin u}$

If we use the previous identity, it is not necessary to choose the appropriate $+$ or $-$ sign. The selection is automatic because $1 - \cos u$ is never negative and $\sin u$ and $\tan \frac{u}{2}$ always agree in sign. You will be asked to prove this fact in an exercise.

If we multiply the right side of Equation 1 by

$$\frac{1 + \cos u}{1 + \cos u}$$

and simplify, we obtain a different version of the half-angle formula for $\tan \frac{u}{2}$:

$$\tan \frac{u}{2} = \frac{\sin u}{1 + \cos u}$$

Example 3   Use the identity $\tan \dfrac{u}{2} = \dfrac{1 - \cos u}{\sin u}$ to find $\tan \dfrac{\pi}{8}$.

Solution   Because $\frac{\pi}{8}$ is $\frac{1}{2}\left(\frac{\pi}{4}\right)$, it follows that

$$\tan \frac{\pi}{8} = \tan \frac{\frac{\pi}{4}}{2} = \frac{1 - \cos \frac{\pi}{4}}{\sin \frac{\pi}{4}}$$

$$= \frac{1 - \dfrac{\sqrt{2}}{2}}{\dfrac{\sqrt{2}}{2}}$$

$$= \frac{2 - \sqrt{2}}{\sqrt{2}} \qquad \text{Multiply numerator and denominator by 2.}$$

$$= \sqrt{2} - 1 \qquad \text{Multiply numerator and denominator by } \sqrt{2}. \qquad ■$$

Example 4   Use the identity $\tan \dfrac{u}{2} = \dfrac{\sin u}{1 + \cos u}$ to find $\tan 157.5°$.

Solution Because $157.5°$ is $\frac{315°}{2}$,

$$\tan 157.5° = \tan \frac{315°}{2} = \frac{\sin 315°}{1 + \cos 315°}$$

$$= \frac{-\dfrac{\sqrt{2}}{2}}{1 + \dfrac{\sqrt{2}}{2}}$$

$$= \frac{-\sqrt{2}}{2 + \sqrt{2}}$$ Multiply numerator and denominator by 2.

$$= \frac{-\sqrt{2}(2 - \sqrt{2})}{(2 + \sqrt{2})(2 - \sqrt{2})}$$

$$= 1 - \sqrt{2}$$

Thus, $\tan 157.5° = 1 - \sqrt{2}$. ∎

Example 5 Find **a.** $\sin \dfrac{u}{2}$ and **b.** $\cos \dfrac{u}{2}$ if $\sin u = \dfrac{3}{5}$ and u is in the interval $\left[\dfrac{\pi}{2}, \pi \right]$.

Solution Use the information that $\sin u = \frac{3}{5}$ and that u is in QII to draw the angle in standard position, as in Figure 7–4. The value of $\cos u$ can be read from the figure:

$$\cos u = -\frac{4}{5}$$

Then use the half-angle identities to find the functions of $\frac{u}{2}$:

a. $\sin \dfrac{u}{2} = \sqrt{\dfrac{1 - \left(\dfrac{-4}{5} \right)}{2}} = \sqrt{\dfrac{9}{10}} = \dfrac{3\sqrt{10}}{10}$

b. $\cos \dfrac{u}{2} = \sqrt{\dfrac{1 + \left(\dfrac{-4}{5} \right)}{2}} = \sqrt{\dfrac{1}{10}} = \dfrac{\sqrt{10}}{10}$

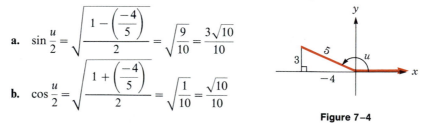

Figure 7–4

Choose the radicals to be positive because, if u is in the interval $[\frac{\pi}{2}, \pi]$, then $\frac{u}{2}$ is in QI. ∎

Example 6 Write $\dfrac{\sin(-20A)}{-\cos 20A - 1}$ as a trigonometric function of $10A$.

Solution Because $\sin(-20A) = -\sin 20A$, it follows that

$$\frac{\sin(-20A)}{-\cos 20A - 1} = \frac{-\sin 20A}{-(1 + \cos 20A)} = \frac{\sin 20A}{1 + \cos 20A}$$

One of the identities for $\tan \frac{u}{2}$ is

$$\tan \frac{u}{2} = \frac{\sin u}{1 + \cos u}$$

Thus, if $u = 20A$, you have

$$\frac{\sin 20A}{1 + \cos 20A} = \tan \frac{20A}{2} = \tan 10A$$

■

Example 7   Verify the identity $\tan \dfrac{\theta}{2} = \csc \theta - \cot \theta$.

Solution   Work on the left side to eliminate the half-angle:

$$\tan \frac{\theta}{2} = \frac{1 - \cos \theta}{\sin \theta}$$

$$= \frac{1}{\sin \theta} - \frac{\cos \theta}{\sin \theta}$$

$$= \csc \theta - \cot \theta$$

■

Example 8   Verify the identity $2 \sin^2 \dfrac{x}{2} \tan x = \tan x - \sin x$.

Solution   Work on the left side:

$$2 \sin^2 \frac{x}{2} \tan x = 2\left(\frac{1 - \cos x}{2}\right)\left(\frac{\sin x}{\cos x}\right)$$

$$= \frac{\sin x - \sin x \cos x}{\cos x}$$

$$= \frac{\sin x}{\cos x} - \sin x$$

$$= \tan x - \sin x$$

■

Example 9   Verify the identity $\sec^2 \dfrac{x}{2} = \dfrac{2 \tan \dfrac{x}{2}}{\sin x}$.

Solution   Work on the right side:

$$\frac{2 \tan \dfrac{x}{2}}{\sin x} = \frac{2\left(\dfrac{\sin x}{1 + \cos x}\right)}{\sin x}$$

$$= \frac{2}{1 + \cos x}$$

$$= \frac{1}{\dfrac{1 + \cos x}{2}}$$

$$= \frac{1}{\cos^2 \dfrac{x}{2}}$$

$$= \sec^2 \frac{x}{2}$$

■

Exercise 7.4

In Exercises 1–12 use half-angle identities to find the required values. ***Do not use a calculator.***

1. $\cos 15°$

2. $\tan 15°$

3. $\tan 105°$

4. $\sin 105°$

5. $\sin \dfrac{\pi}{8}$

6. $\cos \dfrac{\pi}{8}$

7. $\cos \dfrac{\pi}{12}$

8. $\sin \dfrac{\pi}{12}$

9. $\tan 165°$

10. $\cos 165°$

11. $\cot \dfrac{5\pi}{4}$

12. $\tan \dfrac{7\pi}{4}$

In Exercises 13–24 use the given information to find the exact value of the sine, cosine, and tangent of $\frac{\theta}{2}$. Assume that $0° \le \theta < 360°$.

13. $\sin \theta = \dfrac{3}{5}$; θ in QI

14. $\cos \theta = \dfrac{12}{13}$; θ in QI

15. $\tan \theta = \dfrac{4}{3}$; θ in QIII

16. $\tan \theta = \dfrac{3}{4}$; θ in QIII

17. $\cos \theta = \dfrac{8}{17}$; θ in QIV

18. $\sin \theta = -\dfrac{7}{25}$; θ in QIII

19. $\cot \theta = \dfrac{40}{9}$; θ in QI

20. $\sec \theta = -\dfrac{41}{40}$; θ in QII

21. $\csc \theta = \dfrac{17}{8}$; θ in QII

22. $\csc \theta = -\dfrac{5}{3}$; θ in QIV

23. $\sec \theta = \dfrac{3}{2}$; θ in QIV

24. $\cos \theta = -0.1$; θ in QIII

In Exercises 25–36 write the given expression as a single trigonometric function of half the given angle.

25. $\sqrt{\dfrac{1 + \cos 30°}{2}}$

26. $\sqrt{\dfrac{1 - \cos 30°}{2}}$

27. $\dfrac{1 - \cos 200°}{\sin 200°}$

28. $\dfrac{\sin 50°}{1 + \cos 50°}$

29. $\csc 80° - \cot 80°$

30. $\dfrac{2 \tan 140°}{\sin 280°}$ (280° is the given angle)

31. $\sqrt{\dfrac{1 - \cos 2\pi}{1 + \cos 2\pi}}$

32. $\dfrac{1 - \cos 4\theta}{\sin 4\theta}$

33. $\dfrac{1 - \cos \dfrac{x}{2}}{\sin \dfrac{x}{2}}$

34. $\dfrac{1 - \cos 2x}{1 + \cos 2x}$

35. $\dfrac{\sin 10A}{1 + \cos 10A}$

36. $\dfrac{1 + \cos 4x}{1 - \cos 4x}$

In Exercises 37–50 verify the given identity.

37. $\sin^2 \dfrac{\theta}{2} = \dfrac{1}{2}(1 - \cos \theta)$

38. $\cos^2 \dfrac{\theta}{2} = \dfrac{1}{2}(1 + \cos \theta)$

39. $\sec^2 \dfrac{\theta}{2} = \dfrac{2}{1 + \cos \theta}$

40. $\csc^2 \dfrac{\theta}{2} = \dfrac{2}{1 - \cos \theta}$

41. $\cot \dfrac{\theta}{2} = \dfrac{1 + \cos \theta}{\sin \theta}$

42. $\cot \dfrac{\theta}{2} = \dfrac{\sin \theta}{1 - \cos \theta}$

43. $\csc^2 \dfrac{\theta}{2} = 2 \csc^2 \theta + 2 \cot \theta \csc \theta$

44. $\sin^2 \dfrac{\theta}{2} = \dfrac{\sec \theta - 1}{2 \sec \theta}$

45. $-\sec B = \dfrac{\sec^2 \dfrac{B}{2}}{\sec^2 \dfrac{B}{2} - 2}$

46. $\tan\left(\dfrac{\pi}{4} + \dfrac{\theta}{2}\right) = \dfrac{1 + \cos \theta + \sin \theta}{1 + \cos \theta - \sin \theta}$

47. $\csc \theta = \dfrac{1}{2} \csc \dfrac{\theta}{2} \sec \dfrac{\theta}{2}$

48. $\dfrac{1}{2} \sin x \tan \dfrac{x}{2} \csc^2 \dfrac{x}{2} = 1$

49. $\tan \dfrac{B}{2} \cos B + \tan \dfrac{B}{2} = \sin B$

50. $\left(\cos \dfrac{\alpha}{2} - \sin \dfrac{\alpha}{2} \right)^2 = 1 - \sin \alpha$

51. Show that $\sin A$ and $\tan \dfrac{A}{2}$ always agree in sign.

7.5   MORE IDENTITIES

In this section we shall develop identities that can be used to change the form of certain expressions. We begin by discussing the *product-to-sum identities*.

The Product-to-Sum Identities.

1. $\sin u \cos v = \dfrac{1}{2} \left[\sin(u + v) + \sin(u - v) \right]$

2. $\cos u \sin v = \dfrac{1}{2} \left[\sin(u + v) - \sin(u - v) \right]$

3. $\cos u \cos v = \dfrac{1}{2} \left[\cos(u + v) + \cos(u - v) \right]$

4. $\sin u \sin v = \dfrac{1}{2} \left[\cos(u - v) - \cos(u + v) \right]$

Equation 1 can be verified by adding the left and right sides of the identities

$$\sin(u + v) = \sin u \cos v + \cos u \sin v$$

and

$$\sin(u - v) = \sin u \cos v - \cos u \sin v$$

to obtain

$$\sin(u + v) + \sin(u - v) = 2 \sin u \cos v$$

We then divide both sides of the previous equation by 2 to obtain

$$\sin u \cos v = \dfrac{1}{2} \left[\sin(u + v) + \sin(u - v) \right]$$

Equation 2 can be verified by subtracting the left and right sides of the identities for $\sin(u + v)$ and $\sin(u - v)$.

Equation 3 can be verified by adding the left and right sides of the identities

$$\cos(u + v) = \cos u \cos v - \sin u \sin v$$

and

$$\cos(u - v) = \cos u \cos v + \sin u \sin v$$

to obtain

$$\cos(u + v) + \cos(u - v) = 2 \cos u \cos v$$

We then divide both sides of the previous equation by 2 to obtain

$$\cos u \cos v = \frac{1}{2}[\cos(u + v) + \cos(u - v)]$$

Equation 4 can be verified by subtracting the left and right sides of the identities for $\cos(u + v)$ and $\cos(u - v)$.

Example 1 Use a product-to-sum identity to calculate the value of $\sin 67.5° \cos 22.5°$.

Solution Let $u = 67.5°$ and $v = 22.5°$, substitute these values into Equation 1, and simplify:

$$\sin u \cos v = \frac{1}{2}[\sin(u + v) + \sin(u - v)]$$

$$\sin 67.5° \cos 22.5° = \frac{1}{2}[\sin(67.5° + 22.5°) + \sin(67.5° - 22.5°)]$$

$$= \frac{1}{2}(\sin 90° + \sin 45°)$$

$$= \frac{1}{2}\left(1 + \frac{\sqrt{2}}{2}\right)$$

$$= \frac{1}{2} + \frac{\sqrt{2}}{4} \qquad \blacksquare$$

Another set of identities is referred to as the *sum-to-product* or the *factoring identities*.

The Sum-to-Product Identities.

5. $\sin u + \sin v = 2 \sin \dfrac{u + v}{2} \cos \dfrac{u - v}{2}$

6. $\sin u - \sin v = 2 \cos \dfrac{u + v}{2} \sin \dfrac{u - v}{2}$

7. $\cos u + \cos v = 2 \cos \dfrac{u + v}{2} \cos \dfrac{u - v}{2}$

8. $\cos u - \cos v = -2 \sin \dfrac{u + v}{2} \sin \dfrac{u - v}{2}$

To verify Equation 5, we let $u = x + y$ and $v = x - y$ and solve the system

$$\begin{cases} u = x + y \\ v = x - y \end{cases}$$

for x and y to obtain

$$x = \frac{u+v}{2} \quad \text{and} \quad y = \frac{u-v}{2}$$

We then substitute $\frac{u+v}{2}$ for x, $\frac{u-v}{2}$ for y, u for $x+y$, and v for $x-y$ in the equation

$$\sin x \cos y = \frac{1}{2}\left[\sin(x+y) + \sin(x-y)\right] \qquad \text{See Equation 1.}$$

to obtain

$$\sin \frac{u+v}{2} \cos \frac{u-v}{2} = \frac{1}{2}\left[\sin u + \sin v\right]$$

or

$$\sin u + \sin v = 2 \sin \frac{u+v}{2} \cos \frac{u-v}{2}$$

Identities 6–8 can be verified by making similar substitutions in Identities 2–4, respectively.

Example 2   Verify the identity $\sin 3\theta + \sin \theta = 2 \sin 2\theta \cos \theta$.

Solution   Use Equation 5 with $u = 3\theta$ and $v = \theta$.

$$\sin u + \sin v = 2 \sin \frac{u+v}{2} \cos \frac{u-v}{2}$$

$$\sin 3\theta + \sin \theta = 2 \sin \frac{3\theta + \theta}{2} \cos \frac{3\theta - \theta}{2}$$

$$= 2 \sin \frac{4\theta}{2} \cos \frac{2\theta}{2}$$

$$= 2 \sin 2\theta \cos \theta$$

Example 3   Find the value of $\cos \frac{5\pi}{12} - \cos \frac{\pi}{12}$.

Solution   Use Equation 8 with $u = \frac{5\pi}{12}$ and $v = \frac{\pi}{12}$.

$$\cos u - \cos v = -2 \sin \frac{u+v}{2} \sin \frac{u-v}{2}$$

$$\cos \frac{5\pi}{12} - \cos \frac{\pi}{12} = -2 \sin \frac{\frac{5\pi}{12} + \frac{\pi}{12}}{2} \sin \frac{\frac{5\pi}{12} - \frac{\pi}{12}}{2}$$

$$= -2 \sin \frac{\pi}{4} \sin \frac{\pi}{6}$$

$$= -2\left(\frac{\sqrt{2}}{2}\right)\left(\frac{1}{2}\right)$$

$$= -\frac{\sqrt{2}}{2}$$

Example 4   Write $\cos 2\theta + \cos 6\theta$ as a product of two functions.

Solution   Substitute 2θ for u and 6θ for v in Equation 7 and simplify:

$$\cos 2\theta + \cos 6\theta = 2 \cos \frac{2\theta + 6\theta}{2} \cos \frac{2\theta - 6\theta}{2}$$

$$= 2 \cos 4\theta \cos(-2\theta)$$

$$= 2 \cos 4\theta \cos 2\theta \qquad\qquad \cos(-2\theta) = \cos 2\theta. \qquad\blacksquare$$

Example 5   Verify the identity $\tan 3x = \dfrac{\sin 4x + \sin 2x}{\cos 4x + \cos 2x}$.

Solution   Work on the right side:

$$\frac{\sin 4x + \sin 2x}{\cos 4x + \cos 2x} = \frac{2 \sin \dfrac{4x + 2x}{2} \cos \dfrac{4x - 2x}{2}}{2 \cos \dfrac{4x + 2x}{2} \cos \dfrac{4x - 2x}{2}}$$

$$= \frac{2 \sin 3x \cos x}{2 \cos 3x \cos x}$$

$$= \frac{\sin 3x}{\cos 3x}$$

$$= \tan 3x \qquad\qquad\qquad\qquad\qquad\blacksquare$$

Sums of the Form $A \sin x + B \cos x$

The graph of $y = A \sin x + B \cos x$ is a sine curve of the form $y = k \sin(x + \phi)$. The amplitude k and the phase shift angle ϕ are determined by the values of A and B. To show that this is true, we proceed as follows.

From the terms of the expression $A \sin x + B \cos x$, factor out a common factor of $\sqrt{A^2 + B^2}$.

$$A \sin x + B \cos x = \sqrt{A^2 + B^2} \left(\frac{A}{\sqrt{A^2 + B^2}} \sin x + \frac{B}{\sqrt{A^2 + B^2}} \cos x \right)$$

Because the sum of the squares of the coefficients

$$\frac{A}{\sqrt{A^2 + B^2}} \qquad \text{and} \qquad \frac{B}{\sqrt{A^2 + B^2}}$$

is 1, one of these coefficients is $\sin \phi$ and one is $\cos \phi$, for some angle ϕ. We let ϕ be an angle such that

$$\sin \phi = \frac{B}{\sqrt{A^2 + B^2}} \qquad \text{and} \qquad \cos \phi = \frac{A}{\sqrt{A^2 + B^2}}$$

Then the previous equation can be written as

$$A \sin x + B \cos x = \sqrt{A^2 + B^2} \, (\cos \phi \sin x + \sin \phi \cos x)$$

$$= \sqrt{A^2 + B^2} \, \sin(x + \phi)$$

and we have the following theorem:

Theorem. $A \sin x + B \cos x = k \sin(x + \phi)$ where $k = \sqrt{A^2 + B^2}$ and ϕ is an angle for which

$$\sin \phi = \frac{B}{\sqrt{A^2 + B^2}} \qquad \text{and} \qquad \cos \phi = \frac{A}{\sqrt{A^2 + B^2}}$$

Example 6   Express $3 \sin x + 4 \cos x$ in the form $k \sin(x + \phi)$.

Solution   Begin by evaluating k:

$$k = \sqrt{A^2 + B^2} = \sqrt{3^2 + 4^2} = \sqrt{25} = 5$$

Angle ϕ is an angle such that $\sin \phi = \frac{4}{5}$ and $\cos \phi = \frac{3}{5}$. Use a calculator to determine that $\phi \approx 53.1°$. Thus,

$$3 \sin x + 4 \cos x \approx 5 \sin(x + 53.1°)$$

∎

Example 7   Write $5 \sin 3x - 12 \cos 3x$ as a single expression involving the sine function only.

Solution   The expression $5 \sin 3x - 12 \cos 3x$ can be written as $k \sin(3x + \phi)$, where

$$k = \sqrt{5^2 + (-12)^2} = \sqrt{25 + 144} = \sqrt{169} = 13$$

The angle ϕ is such that $\sin \phi = -\frac{12}{13} \approx -0.9231$ and $\cos \phi = \frac{5}{13} \approx 0.3846$. Because $\sin \phi$ is negative and $\cos \phi$ is positive, ϕ must lie in the fourth quadrant. Thus, $\phi \approx -67.4°$ and

$$5 \sin 3x - 12 \cos 3x \approx 13 \sin(3x - 67.4°)$$

∎

Example 8   Graph the function $y = \cos x - \sin x$.

Solution   Note that $\cos x - \sin x$ (or $-\sin x + \cos x$) can be changed to the form $k = \sin(x + \phi)$ where

$$k = \sqrt{(-1)^2 + 1^2} = \sqrt{2}, \qquad \sin \phi = \frac{\sqrt{2}}{2}, \qquad \cos \phi = -\frac{\sqrt{2}}{2}$$

Because $\sin \phi$ is positive and $\cos \phi$ is negative, angle ϕ is a second-quadrant angle. The second-quadrant angle with a sine of $\frac{\sqrt{2}}{2}$ is $\frac{3\pi}{4}$ radians. Thus, $y = \cos x - \sin x$ is equivalent to

$$y = \sqrt{2} \sin\left(x + \frac{3\pi}{4}\right)$$

The graph of this equation is a sine curve with an amplitude of $\sqrt{2}$ and a phase shift of $\frac{3\pi}{4}$ to the left. The graph appears in Figure 7–5. This equation could have been graphed by using the method of addition of ordinates.

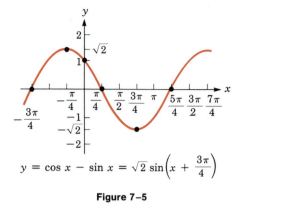

$$y = \cos x - \sin x = \sqrt{2}\,\sin\left(x + \frac{3\pi}{4}\right)$$

Figure 7–5

Exercise 7.5

In Exercises 1–12 express each product as a sum or difference and find its value. **Do not use a calculator or tables.**

1. $\cos 75° \cos 15°$ **2.** $\sin 15° \cos 75°$ **3.** $\sin 165° \sin 105°$

4. $\sin 15° \cos 15°$ **5.** $\cos 22.5° \cos 67.5°$ **6.** $\cos 105° \sin 15°$

7. $\sin \dfrac{\pi}{12} \sin \dfrac{5\pi}{12}$ **8.** $\sin \dfrac{5\pi}{12} \cos \dfrac{13\pi}{12}$ **9.** $\cos \dfrac{7\pi}{12} \cos \dfrac{5\pi}{12}$

10. $\cos \dfrac{7\pi}{12} \sin \dfrac{13\pi}{12}$ **11.** $\cos \dfrac{\pi}{12} \sin \dfrac{5\pi}{12}$ **12.** $\sin \dfrac{\pi}{12} \cos \dfrac{5\pi}{12}$

In Exercises 13–24 express each quantity as a product and find its value. **Do not use a calculator or tables.**

13. $\cos 75° + \cos 15°$ **14.** $\sin 15° + \sin 75°$ **15.** $\sin 165° - \sin 105°$

16. $\sin 15° - \sin 75°$ **17.** $\cos 165° - \cos 105°$ **18.** $\cos 105° - \cos 15°$

19. $\sin \dfrac{\pi}{12} + \sin \dfrac{5\pi}{12}$ **20.** $\sin \dfrac{5\pi}{12} - \sin \dfrac{13\pi}{12}$ **21.** $\cos \dfrac{7\pi}{12} + \cos \dfrac{5\pi}{12}$

22. $\sin \dfrac{7\pi}{12} + \sin \dfrac{13\pi}{12}$ **23.** $\sin \dfrac{5\pi}{12} - \sin \dfrac{\pi}{12}$ **24.** $\cos \dfrac{11\pi}{12} - \cos \dfrac{7\pi}{12}$

In Exercises 25–36 write the given expression in the form $k \sin(x + \phi)$.

25. $6 \sin x + 8 \cos x$ **26.** $12 \sin x + 5 \cos x$ **27.** $6 \sin x - 8 \cos x$

28. $12 \sin x - 5 \cos x$ **29.** $2 \sin x + \cos x$ **30.** $\sin x - \cos x$

31. $\sin x + \cos x$ **32.** $2 \sin x - \cos x$ **33.** $-\sin x + 5 \cos x$

34. $\sqrt{3} \sin x + 3 \cos x$ **35.** $\sqrt{3} \sin x - 3 \cos x$ **36.** $-\sin x - 5 \cos x$

In Exercises 37–42 use the method of Example 8 to graph each function.

37. $y = \sin x + \cos x$ **38.** $y = \sin x - \cos x$ **39.** $y = \sin x - \sqrt{3} \cos x$

40. $y = \sqrt{3} \sin x + \cos x$ **41.** $y = \sin 2x + \sqrt{3} \cos 2x$ **42.** $y = \sin 2x - \sqrt{3} \cos 2x$

In Exercises 43–52 verify each identity.

43. $\dfrac{\sin A + \sin B}{\sin A - \sin B} = \tan \dfrac{1}{2}(A + B) \cot \dfrac{1}{2}(A - B)$ **44.** $\dfrac{\sin A + \sin B}{\cos A + \cos B} = \tan \dfrac{1}{2}(A + B)$

45. $\dfrac{\sin A + \sin B}{\cos A - \cos B} = -\cot \dfrac{1}{2}(A - B)$

46. $\dfrac{\cos A + \cos B}{\sin A - \sin B} = \cot \dfrac{1}{2}(A - B)$

47. $\dfrac{\cos A + \cos 5A}{\cos A - \cos 5A} = \cot 3A \cot 2A$

48. $\sin^2 A - \sin^2 B = \sin(A + B)\sin(A - B)$

49. $\cos^2 A - \cos^2 B = \sin(B + A)\sin(B - A)$

50. $\cos 2A(1 + 2\cos A) = \cos A + \cos 2A + \cos 3A$

51. $2\cos 5A \sin 2A = \sin 7A - \sin 3A$

52. $\cot 7A \cot 5A = \dfrac{\cos 12A + \cos 2A}{\cos 2A - \cos 12A}$

In Exercises 53–56 assume that $A + B + C = 180°$. Verify each identity.

53. $\sin A + \sin B + \sin C = 4\cos \dfrac{A}{2}\cos \dfrac{B}{2}\cos \dfrac{C}{2}$

54. $\cos A + \cos B - \cos C = 4\cos \dfrac{A}{2}\cos \dfrac{B}{2}\sin \dfrac{C}{2} - 1$

55. $\tan A + \tan B + \tan C = \tan A \tan B \tan C$

56. $\cos A + \cos B + \cos C = 1 + 4\sin \dfrac{A}{2}\sin \dfrac{B}{2}\sin \dfrac{C}{2}$

7.6   TRIGONOMETRIC EQUATIONS

The equation $\sin 2x = 2\sin x$ is not an identity because it is not true for all values of x. It is true, however, for some values of x. For example, the equation is true if $x = 0$, as the following check shows:

$$\sin 2x \overset{?}{=} 2\sin x$$
$$\sin 2(0) \overset{?}{=} 2\sin 0$$
$$\sin 0 \overset{?}{=} 2(0)$$
$$0 = 0$$

Other values of x, such as π, 2π, 3π, and so on, also satisfy this equation. The process of finding all such values that satisfy the equation is called **solving the equation**. Example 1 provides a formal solution of the equation $\sin 2x = 2\sin x$.

Example 1   Solve $\sin 2x = 2\sin x$ for x, where x is a real number.

Solution   Use the identity for $\sin 2x$ to rewrite the left side of the equation as $2\sin x \cos x$. Then proceed as follows:

$$\sin 2x = 2\sin x$$
$$2\sin x \cos x = 2\sin x$$

$2\sin x \cos x - 2\sin x = 0$ Add $-2\sin x$ to both sides.

$2\sin x(\cos x - 1) = 0$ Factor out $2\sin x$.

$2\sin x = 0$ or $\cos x - 1 = 0$ Set each factor equal to 0.

$\sin x = 0$ | $\cos x = 1$

Thus, the solutions of the equation are those real numbers x whose sine is 0 or whose cosine is 1. Numbers with a sine of 0 are

$$\ldots, -4\pi, \quad -3\pi, \quad -2\pi, \quad -\pi, \quad 0, \quad \pi, \quad 2\pi, \quad 3\pi, \quad 4\pi, \ldots$$

Numbers with a cosine of 1 are

$$\ldots, \; -4\pi, \;\; -2\pi, \;\; 0, \;\; 2\pi, \;\; 4\pi, \ldots$$

The solutions of the given equation are the real numbers

$$\ldots, \; -4\pi, \;\; -3\pi, \;\; -2\pi, \;\; -\pi, \;\; 0, \;\; \pi, \;\; 2\pi, \;\; 3\pi, \;\; 4\pi, \ldots$$

These solutions can be denoted concisely as $n\pi$, where n is an integer.

Example 2   Solve $2 \sin^2 \theta + \sin \theta = 1$ for all values of θ where $0° \leq \theta < 360°$.

Solution   Subtract 1 from both sides of the equation, factor the left side, set each factor equal to 0, and solve each resulting equation for θ:

$$2 \sin^2 \theta + \sin \theta = 1$$
$$2 \sin^2 \theta + \sin \theta - 1 = 0$$
$$(2 \sin \theta - 1)(\sin \theta + 1) = 0$$

$2 \sin \theta - 1 = 0$	or	$\sin \theta + 1 = \;\;\; 0$
$2 \sin \theta = 1$		$\sin \theta = -1$
$\sin \theta = \dfrac{1}{2}$		$\theta = \;\;\; 270°$
$\theta = 30°, 150°$		

These are the solutions that lie in the designated interval from $0°$ to $360°$. Verify that each solution satisfies the given equation.

Example 3   Solve $\sin 3x = \dfrac{1}{2}$ for all x in the interval $[0, 2\pi)$.

Solution   There are infinitely many values of $3x$ that will satisfy the equation $\sin 3x = \frac{1}{2}$. Because you are given that $0 \leq x < 2\pi$, you must find all such values of $3x$ where

$$3(0) \leq 3x < 3(2\pi)$$

or

$$0 \leq 3x < 6\pi$$

There are six such values:

$$3x = \frac{\pi}{6}, \; \frac{5\pi}{6}, \; \frac{13\pi}{6}, \; \frac{17\pi}{6}, \; \frac{25\pi}{6}, \; \frac{29\pi}{6}$$

The values of x are then found by dividing each of the values of $3x$ by 3:

$$x = \frac{\pi}{18}, \; \frac{5\pi}{18}, \; \frac{13\pi}{18}, \; \frac{17\pi}{18}, \; \frac{25\pi}{18}, \; \frac{29\pi}{18}$$

Note that the largest of these, $\frac{29\pi}{18}$, is still less than 2π. Each solution satisfies the given equation.

Example 4   Solve $\sin \theta = \cos \theta$ for θ, where $0° \leq \theta < 360°$.

Solution   If $\cos\theta \neq 0$, you can divide both sides of the equation by $\cos\theta$:

$$\sin\theta = \cos\theta$$

$$\frac{\sin\theta}{\cos\theta} = 1$$

$$\tan\theta = 1$$

$$\theta = 45°, 225°$$

Both values satisfy the given equation. If $\cos\theta = 0$, then $\theta = 90°$ or $\theta = 270°$, and neither of these values satisfies the equation. Thus, the only solutions are 45° and 225°.   ■

Example 5   Solve the equation $2\cos^3\theta = \cos\theta$ for θ, where $0° \leq \theta < 360°$.

Solution   Do not divide both sides by $\cos\theta$. Instead, subtract $\cos\theta$ from both sides of the equation, factor out $\cos\theta$, and proceed as follows:

$$2\cos^3\theta = \cos\theta$$

$$2\cos^3\theta - \cos\theta = 0$$

$$\cos\theta(2\cos^2\theta - 1) = 0$$

$$\cos\theta = 0 \qquad \text{or} \qquad 2\cos^2\theta - 1 = 0$$

$$\theta = 90°, 270° \qquad\qquad \cos^2\theta = \frac{1}{2}$$

$$\cos\theta = \pm\frac{1}{\sqrt{2}} = \pm\frac{\sqrt{2}}{2}$$

$$\theta = 45°, 135°, 225°, 315°$$

All six values satisfy the given equation. Note that, if both sides of the original equation had been divided by $\cos\theta$, the solutions 90° and 270° would have been lost.   ■

Example 6   Solve $4\sin^2\frac{x}{2} = 1$ for x, where x is in the interval $[0, 2\pi)$.

Solution   Proceed as follows:

$$4\sin^2\frac{x}{2} = 1$$

$$4\sin^2\frac{x}{2} - 1 = 0 \qquad\qquad \text{Add } -1 \text{ to both sides.}$$

$$\left(2\sin\frac{x}{2} - 1\right)\left(2\sin\frac{x}{2} + 1\right) = 0 \qquad\qquad \text{Factor the left side.}$$

$$2\sin\frac{x}{2} - 1 = 0 \qquad \text{or} \qquad 2\sin\frac{x}{2} + 1 = 0 \qquad\qquad \text{Set each factor equal to 0.}$$

$$\sin\frac{x}{2} = \frac{1}{2} \qquad\qquad\qquad \sin\frac{x}{2} = -\frac{1}{2}$$

$$\frac{x}{2} = \frac{\pi}{6}, \frac{5\pi}{6} \qquad\qquad\qquad \frac{x}{2} = \frac{7\pi}{6}, \frac{11\pi}{6}$$

$$x = \frac{\pi}{3}, \frac{5\pi}{3} \qquad\qquad\qquad x = \frac{7\pi}{3}, \frac{11\pi}{3}$$

Because $\frac{7\pi}{3}$ and $\frac{11\pi}{3}$ are not in $[0, 2\pi)$, they must be excluded. Thus, the only solutions are

$$x = \frac{\pi}{3} \quad \text{and} \quad x = \frac{5\pi}{3}$$

Verify that each solution satisfies the given equation.   ∎

Example 7   Use a half-angle identity to solve $4 \sin^2 \frac{x}{2} = 1$, where x is in the interval $[0, 2\pi)$.

Solution   From the half-angle identity for $\sin \frac{x}{2}$, it follows that

$$\sin^2 \frac{x}{2} = \frac{1 - \cos x}{2}$$

Substituting $\dfrac{1 - \cos x}{2}$ for $\sin^2 \dfrac{x}{2}$ in the original equation we obtain

$$4 \sin^2 \frac{x}{2} = 1$$

$$4\left(\frac{1 - \cos x}{2}\right) = 1$$

$$1 - \cos x = \frac{1}{2}$$

$$\cos x = \frac{1}{2}$$

$$x = \frac{\pi}{3}, \frac{5\pi}{3}$$

Verify that each solution satisfies the given equation.   ∎

Example 8   Solve $2 \sin x \cos x + \cos x - 2 \sin x - 1 = 0$ for x, where x is in the interval $[0, 2\pi)$.

Solution   You can use the technique of *factoring by grouping*.

$$2 \sin x \cos x + \cos x - 2 \sin x - 1 = 0$$

$$\cos x(2 \sin x + 1) - 1(2 \sin x + 1) = 0$$

$$(2 \sin x + 1)(\cos x - 1) = 0$$

$2 \sin x + 1 = 0$ or $\cos x - 1 = 0$

$$\sin x = -\frac{1}{2} \qquad\qquad \cos x = 1$$

$$\qquad\qquad\qquad\qquad x = 0$$

$$x = \frac{7\pi}{6}, \frac{11\pi}{6}$$

Verify that each solution satisfies the given equation.   ∎

Example 9   Solve $\sin \theta + \cos \theta = 1$ for θ, where $0° \leq \theta < 360°$.

Solution   Use the identity

$$A \sin \theta + B \cos \theta = \sqrt{A^2 + B^2} \sin(\theta + \phi)$$

to write $\sin \theta + \cos \theta$ in the form $k \sin(\theta + \phi)$:

$$\sin \theta + \cos \theta = 1$$
$$\sqrt{1^2 + 1^2} \sin(\theta + \phi) = 1$$
$$\sqrt{2} \sin(\theta + \phi) = 1$$

Remember that ϕ is an angle for which $\sin \phi = \frac{1}{\sqrt{2}}$ and $\cos \phi = \frac{1}{\sqrt{2}}$. Thus, $\phi = 45°$ and you have

$$\sqrt{2} \sin(\theta + 45°) = 1$$
$$\sin(\theta + 45°) = \frac{1}{\sqrt{2}}$$

$$\theta + 45° = 45° \quad \text{or} \quad \theta + 45° = 135°$$
$$\theta = 0° \qquad\qquad\qquad \theta = 90°$$

Verify that each solution satisfies the given equation.   ■

Exercise 7.6

In Exercises 1–4 solve for all values of the variable. Assume that x is a real number and that θ is an angle in degrees. **Do not use a calculator or tables.**

1. $\sin \theta = \dfrac{\sqrt{3}}{2}$ **2.** $\cos x = \dfrac{-\sqrt{2}}{2}$ **3.** $\sin 2x = 0$ **4.** $\cos 2\theta = 1$

In Exercises 5–46 solve for all values of the variable between 0° and 360°, including 0°. **Do not use a calculator or tables.**

5. $\cos \dfrac{\theta}{2} = \dfrac{1}{2}$ **6.** $\sin \dfrac{\theta}{2} = \dfrac{\sqrt{3}}{2}$ **7.** $\sin \theta \cos \theta = 0$

8. $\sin \theta \cos \theta - \sin \theta = 0$ **9.** $\cos \theta \sin \theta - \cos \theta = 0$ **10.** $\cos \theta \sin \theta + \cos \theta = 0$

11. $\sin \theta \cos \theta + \sin \theta = 0$ **12.** $\cos^2 \theta = 1$ **13.** $\sin^2 \theta = \dfrac{1}{2}$

14. $\sin^2 \theta - \dfrac{1}{2} \sin \theta - \dfrac{1}{2} = 0$ **15.** $\cos^2 \theta - \dfrac{1}{2} \cos \theta - \dfrac{1}{2} = 0$ **16.** $\sin^2 \theta - 3 \sin \theta + 2 = 0$

17. $4 \sin^2 \theta + 4 \sin \theta + 1 = 0$ **18.** $\sin 2\theta - \cos \theta = 0$ **19.** $\sin 2A + \cos A = 0$

20. $\cos^2 A + 4 \cos A + 3 = 0$ **21.** $\cos B = \sin B$ **22.** $\cos B = \sqrt{3} \sin B$

23. $\cos \theta = \cos \dfrac{\theta}{2}$ **24.** $\cos \theta = \sin \dfrac{\theta}{2}$ **25.** $\cos 2\theta = \cos \theta$

26. $\cos 2\theta = \sin \theta$ **27.** $\tan A = -\sin A$ **28.** $\cot A = -\cos A$

29. $\cos^2 C - \sin^2 C = \dfrac{1}{2}$ **30.** $\cos^2 C + \sin^2 C = \dfrac{1}{2}$ **31.** $\sin \theta + \cos \theta = \sqrt{2}$
    (*Hint:* Square both sides.)

32. $\sin \theta - \cos \theta = \sqrt{2}$ **33.** $\cos 2A = 1 - \sin A$ **34.** $\cos 2A = \cos A - 1$

35. $\cos 2B + \cos B + 1 = 0$ **36.** $\cos^2 B + \sin B - 1 = 0$ **37.** $\sin 4\theta = \sin 2\theta$

38. $\cos 4\theta = \cos 2\theta$

39. $\cos B = 1 + \sqrt{3} \sin B$

40. $\tan 2B + \sec 2B = 1$

41. $9 \cos^4 A = \sin^4 A$

42. $6 \cos^2 A = -9 \sin A$

43. $\sec C = \tan C + \cos C$

44. $1 - \tan C = \sqrt{2} \sec C$

45. $2 \cos x \sin x - \sqrt{2} \cos x = \sqrt{2} \sin x - 1$

46. $4 \sin x \cos x - 2 \sin x = 2 \cos x - 1$

In Exercises 47–52 solve for θ, where $0° \le \theta < 360°$. *Recall that the expression* $a \sin x + b \cos x$ *can be written as* $k \sin(x + \phi)$.

47. $\dfrac{\sqrt{3}}{2} \sin \theta + \dfrac{1}{2} \cos \theta = \dfrac{1}{2}$

48. $\sqrt{2} \sin \theta + \sqrt{2} \cos \theta = \sqrt{3}$

49. $\dfrac{1}{2} \sin \theta + \dfrac{\sqrt{3}}{2} \cos \theta = 1$

50. $\cos \theta - \sin \theta = \sqrt{2}$

51. $\cos \theta - \sqrt{3} \sin \theta = 1$

52. $\sin \theta + \cos \theta = -\sqrt{2}$

In Exercises 53–62 solve for x, where $0 \le x < 2\pi$. *Consider using a sum-to-product or a product-to-sum identity.*

53. $\sin x \cos x = \dfrac{1}{2}$

54. $\cos x \sin x = \dfrac{\sqrt{3}}{4}$

55. $2 \sin \dfrac{3x}{2} \cos \dfrac{x}{2} = \sin x$

56. $2 \cos \dfrac{3x}{2} \cos \dfrac{x}{2} = \cos 2x$

57. $\sin 3x \cos x = \dfrac{1}{2} \sin 2x$

58. $\cos 3x \cos x = \cos^2 2x$

59. $\sin 4x = -\sin 2x$

60. $\cos 4x = -\cos 2x$

61. $\cos 9x = -\cos 3x$

62. $\sin 12x = -\sin 4x$

In Exercises 63–76 solve for x, where $0 \le x < 2\pi$.

63. $2 \sin^4 x - 9 \sin^2 x + 4 = 0$

64. $2 \sin^3 x + \sin^2 x = \sin x$

65. $4 \cos x \sin^2 x - \cos x = 0$

66. $\tan^2 x = 1 + \sec x$

67. $\csc^4 x = 2 \csc^2 x - 1$

68. $\tan^2 x - 5 \tan x + 6 = 0$

69. $4 \sin^2 x + 4 \sin x + 1 = 0$

70. $2 \sin 5x = 1$

71. $2 \sin^2 5x = 1$

72. $2 \sin^2 2x + \sin 2x - 5 = 0$

73. $\tan^2 x - \tan x = 0$

74. $2 \sin x - \sqrt{2} \tan x = \sqrt{2} \sec x - 2$

75. $\cot^2 x \cos x + \cot^2 x - 3 \cos x - 3 = 0$

76. $\tan x + \cot x = -2$

77. Suppose that $\sin \sqrt{2}\, \pi t = 0$. Show that the values of $\cos \sqrt{2}\, \pi t$ are 1 or -1.

78. Suppose that $\cos \dfrac{t}{2} = 0$. Show that the values of $\sin \dfrac{t}{2}$ are 1 or -1.

79. Suppose that $\cos t - \sin t = 0$. Show that the values of $\cos t + \sin t$ are $\sqrt{2}$ and $-\sqrt{2}$.

80. Suppose that $\cos 2t = 1$. Show that the value of $2 \sin t \cos t$ is 0.

7.7   INVERSES OF THE TRIGONOMETRIC FUNCTIONS

The set of ordered pairs of real numbers (x, y) that satisfy the equation $y = \sin x$ determines a function because every real number x gives a single value y. The graph of $y = \sin x$ appears in Figure 7–6.

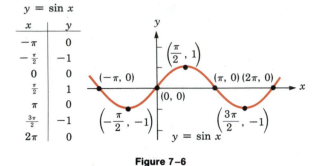

$y = \sin x$

x	y
$-\pi$	0
$-\frac{\pi}{2}$	-1
0	0
$\frac{\pi}{2}$	1
π	0
$\frac{3\pi}{2}$	-1
2π	0

Figure 7–6

Interchanging the x and y in the equation $y = \sin x$ determines the inverse relation $x = \sin y$, whose graph appears in Figure 7–7.

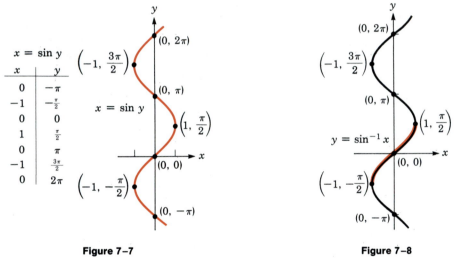

$x = \sin y$

x	y
0	$-\pi$
-1	$-\frac{\pi}{2}$
0	0
1	$\frac{\pi}{2}$
0	π
-1	$\frac{3\pi}{2}$
0	2π

Figure 7–7

Figure 7–8

Because the graph of $x = \sin y$ does not pass the vertical line test, it does not represent a function. However, the portion of the graph lying between $y = -\frac{\pi}{2}$ and $y = \frac{\pi}{2}$ does pass the vertical line test and does determine a function. The colored portion of the graph in Figure 7–8 is the graph of a function, denoted by $y = \sin^{-1} x$, called the **inverse sine function**, sometimes called the **arcsine function**.

> **Definition.** The **inverse sine** (or arcsine) **function**, $y = \sin^{-1} x$ (or $y =$ arcsin x), has a domain of $\{x: -1 \le x \le 1\}$ and a range of $\{y: -\frac{\pi}{2} \le y \le \frac{\pi}{2}\}$.
>
> $$y = \sin^{-1} x \text{ if and only if } x = \sin y \text{ and } -\frac{\pi}{2} \le y \le \frac{\pi}{2}$$

The solid black curve in Figure 7–9 is the portion of the graph of $y = \sin x$ for which x is in the interval $\left[-\frac{\pi}{2}, \frac{\pi}{2}\right]$. That portion represents an increasing

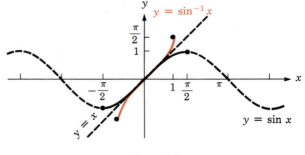

Figure 7-9

function, which has an inverse and, therefore, is one-to-one. Its inverse is the colored curve, which is the graph of $y = \sin^{-1} x$. Note that the solid black curve and the colored curve are symmetric about the line $y = x$.

Although the domain and range of the inverse sine function are defined to be sets of real numbers, it is often convenient to think of the expression $\sin^{-1} x$ as meaning the "angle whose sine is x." In this situation, the range of the inverse sine function can be thought of as a set of angles measured in radians.

Example 1   Find **a.** $\sin^{-1} \dfrac{1}{2}$, **b.** $\sin^{-1}\left(-\dfrac{\sqrt{2}}{2}\right)$, **c.** $\sin^{-1} \pi$, **d.** $\sin^{-1} \dfrac{\pi}{4}$, and **e.** arcsin 0.8330.

Solution   **a.**   The expression $\sin^{-1} \frac{1}{2}$ represents the number in the interval $\left[-\frac{\pi}{2}, \frac{\pi}{2}\right]$ whose sine is $\frac{1}{2}$. Because $\frac{\pi}{6}$ is the only number in this interval whose sine is $\frac{1}{2}$,

$$\sin^{-1} \frac{1}{2} = \frac{\pi}{6}$$

b.   The expression $\sin^{-1}\left(-\frac{\sqrt{2}}{2}\right)$ represents the number in the interval $\left[-\frac{\pi}{2}, \frac{\pi}{2}\right]$ that has a sine of $-\frac{\sqrt{2}}{2}$. Because $-\frac{\pi}{4}$ is the only number in this interval whose sine is $-\frac{\sqrt{2}}{2}$,

$$\sin^{-1}\left(-\frac{\sqrt{2}}{2}\right) = -\frac{\pi}{4}$$

c.   The expression $\sin^{-1} \pi$ represents the number in the interval $\left[-\frac{\pi}{2}, \frac{\pi}{2}\right]$ with a sine of π. Because no number has a sine greater than 1, $\sin^{-1} \pi$ is undefined.

d.   To find $\sin^{-1} \frac{\pi}{4}$, use a calculator set for radians. Divide π by 4 and then press the $\boxed{\text{INV}}$ and $\boxed{\text{SIN}}$ keys in that order. The number whose sine is $\frac{\pi}{4}$ is approximately 0.9033, as displayed on the calculator.

e.   To find arcsin 0.8330, use a calculator set for radians. Enter .833 and press the $\boxed{\text{INV}}$ and $\boxed{\text{SIN}}$ keys in that order. The number whose sine is 0.8330 is approximately 0.9845, as displayed on the calculator.   ■

Because $y = \sin x$ (where x is restricted to the interval $\left[-\frac{\pi}{2}, \frac{\pi}{2}\right]$) and $y = \sin^{-1} x$ are inverse functions, we consider the effect of performing them in succession. Because $\sin^{-1} x$ means "the number whose sine is x," the expression $\sin(\sin^{-1} x)$ means "the sine of the number whose sine is x." The obvious answer to this question is x. Thus,

If $-1 \le x \le 1$, then $\sin(\sin^{-1} x) = x$.

Also,

> **If** $-\dfrac{\pi}{2} \le x \le \dfrac{\pi}{2}$, **then** $\sin^{-1}(\sin x) = x.$

However, if x is not restricted to the proper interval, then $\sin^{-1}(\sin x)$ might not be x. For example,

$$\sin^{-1}\left(\sin \frac{5\pi}{6}\right) = \sin^{-1}\frac{1}{2} = \frac{\pi}{6}$$

Similar considerations produce the inverse functions of the remaining trigonometric functions.

> **Definition.** The **inverse cosine** (or **arccosine**) **function**, denoted by $y = \cos^{-1} x$ (or $y = \arccos x$), has a domain of $\{x : -1 \le x \le 1\}$ and a range of $\{y : 0 \le y \le \pi\}$.
>
> $y = \cos^{-1} x$ if and only if $x = \cos y$ and $0 \le y \le \pi$

The graph of $y = \cos^{-1} x$, shown in Figure 7–10, is the reflection of a portion of the cosine curve with the line $y = x$ as an axis of symmetry.

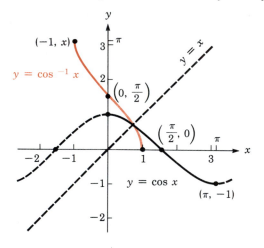

Figure 7–10

> **Definition.** The **inverse tangent** (or **arctangent**) **function**, denoted by $y = \tan^{-1} x$ (or $y = \arctan x$), has a domain of the real numbers and a range of $\{y : -\frac{\pi}{2} < y < \frac{\pi}{2}\}$.
>
> $y = \tan^{-1} x$ if and only if $x = \tan y$ and $-\dfrac{\pi}{2} < y < \dfrac{\pi}{2}$

The graph of $y = \tan^{-1} x$, shown in Figure 7–11, is the reflection of a portion of the tangent curve with the line $y = x$ as an axis of symmetry.

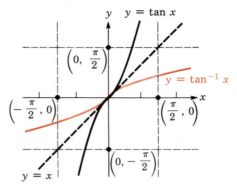

Figure 7–11

Example 2   Find **a.** $\tan^{-1}(-1)$, **b.** $\cos^{-1} \dfrac{\sqrt{3}}{2}$, and **c.** $\tan^{-1} \dfrac{\pi}{4}$.

Solution   **a.**   Because $-\frac{\pi}{4}$ is the only number in the interval $(-\frac{\pi}{2}, \frac{\pi}{2})$ whose tangent is -1,

$$\tan^{-1}(-1) = -\frac{\pi}{4}$$

b.   Because $\frac{\pi}{6}$ is the only number in the interval $[0, \pi]$ whose cosine is $\frac{\sqrt{3}}{2}$,

$$\cos^{-1} \frac{\sqrt{3}}{2} = \frac{\pi}{6}$$

c.   To find $\tan^{-1} \frac{\pi}{4}$, use a calculator set in radian mode. Divide π by 4, and then press the $\boxed{\text{INV}}$ and $\boxed{\text{TAN}}$ keys in succession to obtain

$$\tan^{-1} \frac{\pi}{4} \approx 0.6658$$

■

Example 3   Find $\cos(\sin^{-1} 1)$.

Solution   Because $\frac{\pi}{2}$ is the number whose sine is 1, it follows that

$$\cos(\sin^{-1} 1) = \cos \frac{\pi}{2} = 0$$

■

Example 4   Find $\cos(\sin^{-1} x)$ given that x is in the interval $(0, 1]$.

Solution   Because the value of x is unknown, you don't know what number has a sine of x. However, because $-\frac{\pi}{2} \le \sin^{-1} x \le \frac{\pi}{2}$, the expression $\sin^{-1} x$ could represent a first- or fourth-quadrant angle. In Figure 7–12, the angle denoted by $\sin^{-1} x$ has a sine of x ($x > 0$), because the side opposite $\sin^{-1} x$ is labeled x and the hypotenuse is 1. By using

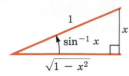

Figure 7–12

the Pythagorean theorem, you can determine that the length of the third side of the triangle is $\sqrt{1 - x^2}$. You can then read the value of the cosine of $\sin^{-1} x$ from the figure:

$$\cos(\sin^{-1} x) = \frac{\text{adjacent side}}{\text{hypotenuse}}$$

$$= \frac{\sqrt{1 - x^2}}{1}$$

$$= \sqrt{1 - x^2}$$

Because $-\frac{\pi}{2} \le \sin^{-1} x \le \frac{\pi}{2}$, the expression $\cos(\sin^{-1} x)$ is never negative and the positive value of the radical is correct. ■

Example 5   Write $\tan(\cos^{-1} x)$ as an algebraic expression in x, where x is in the interval $[-1, 1]$.

Solution   In this example, there are two cases: when x is in the interval $[0, 1]$ and when x is in the interval $[-1, 0)$.

If $x \ge 0$, then $\cos^{-1} x$ can represent an acute angle drawn in standard position, as in Figure 7–13. Because $\cos^{-1} x$ represents the angle whose cosine is x, the side adjacent to the angle can be labeled x and the hypotenuse labeled 1. The length of the remaining side of the right triangle is determined by the Pythagorean theorem to be $\sqrt{1 - x^2}$. Then,

$$\tan(\cos^{-1} x) = \frac{\sqrt{1 - x^2}}{x}$$

The tangent of an acute angle is positive. Because x is also positive, the radical was chosen to be positive.

Now consider the case when $x < 0$. Then, $\cos^{-1} x$ represents a second-quadrant angle drawn in standard position, as in Figure 7–14. Because $\cos^{-1} x$ is the angle whose cosine is x, the adjacent side can be labeled x and the hypotenuse labeled 1. The length of the remaining side is $\sqrt{1 - x^2}$. Thus,

$$\tan(\cos^{-1} x) = \frac{\sqrt{1 - x^2}}{x}$$

The tangent of a second-quadrant angle is negative. Because x is also negative, the radical must again be chosen positive. Taken together, the results of these two cases determine that, for all x in the interval $[-1, 1]$,

$$\tan(\cos^{-1} x) = \frac{\sqrt{1 - x^2}}{x}$$

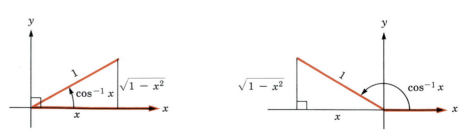

Figure 7–13   **Figure 7–14**   ■

Example 6   Write $\sin(\sin^{-1} x + \sin^{-1} y)$ as an algebraic expression in the variables x and y.

Solution   Use the identity $\sin(u + v) = \sin u \cos v + \cos u \sin v$ to remove parentheses.

$$\sin(\sin^{-1} x + \sin^{-1} y) = \sin(\sin^{-1} x)\cos(\sin^{-1} y) + \cos(\sin^{-1} x)\sin(\sin^{-1} y)$$

Because $\sin(\sin^{-1} x) = x$ and $\cos(\sin^{-1} x) = \sqrt{1 - x^2}$ (see Example 4), it follows that

$$\sin(\sin^{-1} x + \sin^{-1} y) = x\sqrt{1 - y^2} + \sqrt{1 - x^2}(y)$$
$$= x\sqrt{1 - y^2} + y\sqrt{1 - x^2} \qquad ■$$

The domain, range, and graph of each of the remaining inverse trigonometric functions are shown in Table 7–1. There is no agreed-upon definition of the ranges of the inverse secant and inverse cosecant functions. We have given definitions that correspond to the ranges of the inverse cosine and the inverse sine functions. Because the inverse cotangent, inverse secant, and inverse cosecant functions are seldom used, we shall not discuss them in detail.

Table 7–1

Function	Domain	Range	Graph
$y = \cot^{-1} x$	all real numbers x	$0 < y < \pi$	
$y = \sec^{-1} x$	$x \le -1$ or $x \ge 1$	$0 \le y \le \pi*$ and $y \ne \dfrac{\pi}{2}$	
$y = \csc^{-1} x$	$x \le -1$ or $x \ge 1$	$-\dfrac{\pi}{2} \le y \le \dfrac{\pi}{2}^{\dagger}$ and $y \ne 0$	

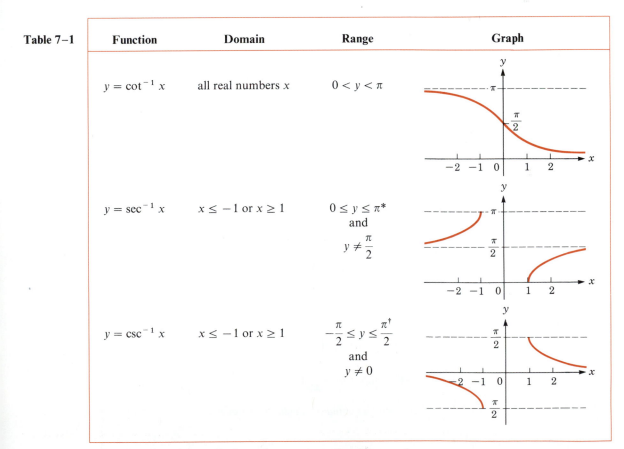

* Some books restrict y to the intervals $-\pi \le y < -\frac{\pi}{2}$ and $0 \le y < \frac{\pi}{2}$.
† Some books restrict y to the intervals $-\pi < y \le -\frac{\pi}{2}$ and $0 < y \le \frac{\pi}{2}$.

▬ Exercise 7.7 ▬▬▬▬▬▬▬▬▬▬▬▬▬▬▬▬▬▬▬▬▬▬▬▬▬▬▬▬

In Exercises 1–12 find the value of x, if any. **Do not use a calculator or tables.** *Note that each answer should be a real number.*

1. $\sin^{-1}\dfrac{1}{2} = x$

2. $\cos^{-1}\dfrac{\sqrt{3}}{2} = x$

3. $\cos^{-1} 0 = x$

4. $\sin^{-1}\sqrt{3} = x$

5. $\tan^{-1} 1 = x$

6. $\tan^{-1} 0 = x$

7. $\sin^{-1} 3 = x$

8. $\tan^{-1}(-\sqrt{3}) = x$

9. $\cos^{-1}\left(-\dfrac{\sqrt{2}}{2}\right) = x$

10. $\sin^{-1}\dfrac{\sqrt{2}}{2} = x$

11. $\arcsin\dfrac{\sqrt{3}}{2} = x$

12. $\arccos\dfrac{1}{2} = x$

In Exercises 13–24 find the sine, cosine, and tangent of θ. **Do not use a calculator or tables.**

13. $\sin^{-1}\dfrac{1}{2} = \theta$

14. $\cos^{-1}\dfrac{\sqrt{3}}{2} = \theta$

15. $\tan^{-1} 0 = \theta$

16. $\tan^{-1} 1 = \theta$

17. $\cos^{-1}\left(-\dfrac{\sqrt{3}}{2}\right) = \theta$

18. $\sin^{-1}\left(-\dfrac{\sqrt{3}}{2}\right) = \theta$

19. $\arcsin 1 = \theta$

20. $\arctan 1 = \theta$

21. $\cos^{-1}(-1) = \theta$

22. $\sin^{-1}\dfrac{\sqrt{2}}{2} = \theta$

23. $\arccos\dfrac{\sqrt{2}}{2} = \theta$

24. $\arcsin 0 = \theta$

In Exercises 25–36 find the value of x. **Do not use a calculator or tables.**

25. $\sin\left(\sin^{-1}\dfrac{1}{2}\right) = x$

26. $\cos\left(\cos^{-1}\dfrac{1}{2}\right) = x$

27. $\tan(\tan^{-1} 1) = x$

28. $\sin(\sin^{-1} 0) = x$

29. $\cos(\cos^{-1} 1) = x$

30. $\tan(\tan^{-1} 0) = x$

31. $\sin\left(\cos^{-1}\dfrac{\sqrt{3}}{2}\right) = x$

32. $\cos\left(\sin^{-1}\dfrac{1}{2}\right) = x$

33. $\tan\left[\sin^{-1}\left(-\dfrac{\sqrt{3}}{2}\right)\right] = x$

34. $\cot\left(\cos^{-1}\dfrac{\sqrt{2}}{2}\right) = x$

35. $\cos(\arctan 1) = x$

36. $\sin(\arctan 0) = x$

In Exercises 37–48 evaluate each expression without using a calculator or tables.

37. $\cos\left(\sin^{-1}\dfrac{4}{5}\right)$

38. $\sin\left(\cos^{-1}\dfrac{3}{5}\right)$

39. $\sin\left(\cos^{-1}\dfrac{5}{13}\right)$

40. $\cos\left[\sin^{-1}\left(-\dfrac{5}{13}\right)\right]$

41. $\tan\left[\sin^{-1}\left(-\dfrac{4}{5}\right)\right]$

42. $\tan\left(\cos^{-1}\dfrac{3}{5}\right)$

43. $\tan\left(\cos^{-1}\dfrac{5}{13}\right)$

44. $\tan\left[\sin^{-1}\left(-\dfrac{12}{13}\right)\right]$

45. $\cos\left[\tan^{-1}\left(-\dfrac{5}{12}\right)\right]$

46. $\sin\left(\tan^{-1}\dfrac{12}{5}\right)$

47. $\sin\left(\arccos\dfrac{9}{41}\right)$

48. $\cos\left(\arcsin\dfrac{40}{41}\right)$

In Exercises 49–60 find the required value without using a calculator or tables.

49. $\sin\left(\sin^{-1}\dfrac{1}{2} + \cos^{-1}\dfrac{1}{2}\right)$

50. $\sin\left(\sin^{-1}\dfrac{1}{2} - \cos^{-1}\dfrac{1}{2}\right)$

51. $\cos\left(\sin^{-1}\dfrac{4}{5} - \cos^{-1}\dfrac{4}{5}\right)$

52. $\cos\left(\sin^{-1}\dfrac{5}{13} + \cos^{-1}\dfrac{5}{13}\right)$

53. $\sin 2\left(\sin^{-1}\dfrac{\sqrt{2}}{2}\right)$

54. $\cos 2\left(\sin^{-1}\dfrac{\sqrt{2}}{2}\right)$

55. $\tan 2\left(\sin^{-1}\dfrac{\sqrt{2}}{2}\right)$

56. $\cot 2\left(\sin^{-1}\dfrac{\sqrt{2}}{2}\right)$

57. $\sin\dfrac{1}{2}\left(\cos^{-1}\dfrac{1}{2}\right)$

58. $\cos\dfrac{1}{2}\left(\cos^{-1}\dfrac{1}{2}\right)$

59. $\tan\dfrac{1}{2}\left(\arccos\dfrac{15}{17}\right)$

60. $\cot\dfrac{1}{2}\left(\arcsin\dfrac{8}{17}\right)$

In Exercises 61–72 rewrite each value as an algebraic expression in the variable x.

61. $\sin(\tan^{-1} x)$

62. $\cos(\tan^{-1} x)$

63. $\tan(\sin^{-1} x)$

64. $\tan(\cos^{-1} x)$

65. $\sin(\cos^{-1} x)$

66. $\cos(\sin^{-1} x)$

67. $\sin(2\sin^{-1} x)$

68. $\cos(2\cos^{-1} x)$

69. $\tan(2\arctan x)$

70. $\sin(2\arccos x)$

71. $\cos(2\sin^{-1} x)$

72. $\sin\left(\dfrac{1}{2}\sin^{-1} x\right)$

In Exercises 73–82 evaluate each of the expressions.

73. $\sin^{-1}\left(\sin\dfrac{11\pi}{6}\right)$

74. $\cos^{-1}\left(\cos\dfrac{7\pi}{6}\right)$

75. $\cos^{-1}\left(\sin\dfrac{3\pi}{4}\right)$

76. $\sin^{-1}\left(\cos\dfrac{5\pi}{4}\right)$

77. $\cot^{-1}\sqrt{3}$

78. $\cot^{-1}\left(-\dfrac{\sqrt{3}}{3}\right)$

79. $\sec^{-1}(-1)$

80. $\sec^{-1}(-2)$

81. $\csc^{-1}(-1)$

82. $\csc^{-1} 2$

7.8   TRIGONOMETRIC FORM OF A COMPLEX NUMBER

The complex numbers share many of the properties of real numbers. However, there is no way of ordering the complex numbers that is consistent with the ordering established for the real numbers. For example, it makes no sense to say that $3i$ is larger than 5 or that $5 + 5i$ is greater than $-5 + i$.

Because the complex numbers are not linearly ordered, it is impossible to graph them on a number line in any way that preserves the ordering of the real numbers. However, there *is* a method for graphing the complex numbers. To graph complex numbers, we construct two perpendicular axes and consider one axis to be the axis of real numbers and the other to be the axis of imaginary numbers. (See Figure 7–15.) These axes determine what is called the **complex**

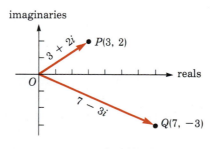

Figure 7–15

plane. Although these axes resemble the x-axis and y-axis encountered in previous chapters, instead of plotting x- and y-values, we plot ordered pairs of real numbers (a, b), where a and b are the real and imaginary parts of the complex number $a + bi$. The axis of reals is used for plotting a and the axis of imaginaries is used for plotting b.

Example 1   Graph the complex number $3 + 2i$.

Solution   To graph the complex number $3 + 2i$, plot the point $P(3, 2)$ as in Figure 7–15. The vector drawn from the origin, point O, to point P is the graph of the complex number $3 + 2i$.   ■

Example 2   Graph the complex number $7 - 3i$.

Solution   To graph the complex number $7 - 3i$, plot the point $Q(7, -3)$ as in Figure 7–15. The vector **OQ** is the graph of the complex number $7 - 3i$.   ■

The length of the vector that represents a complex number is considered to be the **absolute value** of that complex number.

Absolute Value of a Complex Number. If $a + bi$ is a complex number, then

$$|a + bi| = \sqrt{a^2 + b^2}$$

See Figure 7–16 and note that $\sqrt{a^2 + b^2}$ is the length of the hypotenuse of a right triangle with sides of length $|a|$ and $|b|$. On the complex plane, $|a + bi|$ is the distance between the point (a, b) and the origin $(0, 0)$. This is consistent with the definition of the absolute value of a real number. The expression $|x|$ represents the distance on the number line between the point with coordinate x and the origin O.

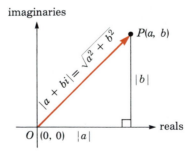

Figure 7–16

Example 3   Find the value of $|3 + 4i|$.

Solution $|3 + 4i| = \sqrt{3^2 + 4^2} = \sqrt{25} = 5$ ■

Example 4   Evaluate $|2 - 5i|$.

Solution $|2 - 5i| = \sqrt{2^2 + (-5)^2} = \sqrt{29}$ ■

Trigonometric Form of a Complex Number

If a complex number is written in the form $a + bi$, it is said to be written in **rectangular** or **algebraic form**. To write a complex number in a different form, we refer to Figure 7–17, in which $a = r \cos \theta$ and $b = r \sin \theta$, and proceed as follows:

$$a + bi = r \cos \theta + (r \sin \theta)i \quad \text{Substitute } r \cos \theta \text{ for } a \text{ and } r \sin \theta \text{ for } b.$$
$$= r(\cos \theta + i \sin \theta) \quad \text{Factor out } r.$$

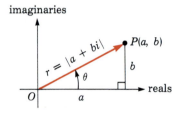

Figure 7–17

The expression $r(\cos \theta + i \sin \theta)$ is called the **trigonometric form** of the complex number $a + bi$.

If $r = |a + bi|$ and θ is any angle for which $\sin \theta = \frac{b}{r}$ and $\cos \theta = \frac{a}{r}$, then

$$a + bi = r(\cos \theta + i \sin \theta)$$

If a complex number is written in $r(\cos \theta + i \sin \theta)$ form, r is called the **modulus** and θ is called the **argument**. The notation $r(\cos \theta + i \sin \theta)$ is often abbreviated as r cis θ. Thus

$$r \text{ cis } \theta = r(\cos \theta + i \sin \theta)$$

Example 5   Write the complex number $3 + 4i$ in trigonometric form.

Solution   To write $3 + 4i$ in the form $r(\cos \theta + i \sin \theta)$, you must find values for r and θ. To do so, graph the complex number as in Figure 7–18. Because triangle OAP is a right triangle, $r^2 = 3^2 + 4^2$ or $r = 5$. From the figure, it follows that $\tan \theta = \frac{4}{3}$, so $\theta \approx 53.1°$.

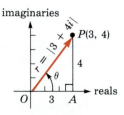

Figure 7–18

Substituting these values for r and θ gives

$$3 + 4i \approx 5(\cos 53.1° + i \sin 53.1°)$$

Note that $5(\cos 53.1° + i \sin 53.1°)$ can be written as 5 cis 53.1°.

Example 6 Write the complex number $-2 - 7i$ in trigonometric form.

Solution To write $-2 - 7i$ in the form $r(\cos \theta + i \sin \theta)$, you must find r and θ. To do so, graph $-2 - 7i$ as in Figure 7–19 and calculate r:

$$r = |-2 - 7i| = \sqrt{(-2)^2 + (-7)^2} = \sqrt{4 + 49} = \sqrt{53}$$

Because $\tan \alpha = \frac{7}{2}$, it follows that $\alpha \approx 74.1°$. Find θ by noting that θ is a third-quadrant angle, so that

$$\theta = 180° + \alpha \approx 180° + 74.1° = 254.1°$$

Thus,

$$-2 - 7i \approx \sqrt{53}(\cos 254.1° + i \sin 254.1°) = \sqrt{53} \text{ cis } 254.1°$$

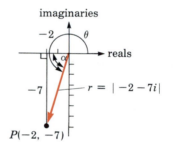

Figure 7–19

Example 7 Change the complex number 10 cis 60° to $a + bi$ form.

Solution
$$10 \text{ cis } 60° = 10(\cos 60° + i \sin 60°)$$
$$= 10\left(\frac{1}{2} + i\frac{\sqrt{3}}{2}\right)$$
$$= 5 + 5\sqrt{3}i$$

It is easy to add complex numbers if they are written in rectangular form—simply add their real parts and add their imaginary parts. However, it is difficult to multiply and divide complex numbers written in $a + bi$ form. Fortunately, it is easy to multiply and divide complex numbers if they are written in trigonometric form.

Suppose that $N_1 = r_1(\cos \theta_1 + i \sin \theta_1)$ and $N_2 = r_2(\cos \theta_2 + i \sin \theta_2)$. The product $N_1 N_2$ can be found as follows.

$$N_1 N_2 = r_1(\cos \theta_1 + i \sin \theta_1) \cdot r_2(\cos \theta_2 + i \sin \theta_2)$$
$$= r_1 r_2[\cos \theta_1 \cos \theta_2 + i \cos \theta_1 \sin \theta_2 + i \sin \theta_1 \cos \theta_2 - \sin \theta_1 \sin \theta_2]$$
$$= r_1 r_2[(\cos \theta_1 \cos \theta_2 - \sin \theta_1 \sin \theta_2) + i(\cos \theta_1 \sin \theta_2 + \sin \theta_1 \cos \theta_2)]$$
$$= r_1 r_2[\cos(\theta_1 + \theta_2) + i \sin(\theta_1 + \theta_2)]$$

This result can be stated more compactly by using r cis θ notation:

$$r_1 \text{ cis } \theta_1 \cdot r_2 \text{ cis } \theta_2 = r_1 r_2 \text{ cis}(\theta_1 + \theta_2)$$

Thus, to form a product of two complex numbers written in trigonometric form, multiply the moduli r_1 and r_2 and add the arguments θ_1 and θ_2. This property generalizes to products containing any number of factors.

$$r_1(\cos \theta_1 + i \sin \theta_1) \cdot r_2(\cos \theta_2 + i \sin \theta_2) \cdots \cdots r_n(\cos \theta_n + i \sin \theta_n)$$
$$= r_1 r_2 \cdots \cdots r_n[\cos(\theta_1 + \theta_2 + \cdots + \theta_n) + i \sin(\theta_1 + \theta_2 + \cdots + \theta_n)]$$

The previous equation can be written more compactly as

$$r_1 \text{ cis } \theta_1 \cdot r_2 \text{ cis } \theta_2 \cdots \cdots r_n \text{ cis } \theta_n$$
$$= r_1 r_2 \cdots \cdots r_n \text{ cis}(\theta_1 + \theta_2 + \cdots + \theta_n)$$

Example 8   Find the product of $2(\cos 40° + i \sin 40°)$ and $3(\cos 30° + i \sin 30°)$.

Solution   To find the product, simply multiply the moduli and add the arguments.

$$2(\cos 40° + i \sin 40°) \cdot 3(\cos 30° + i \sin 30°)$$
$$= 2 \cdot 3[\cos(40° + 30°) + i \sin(40° + 30°)]$$
$$= 6(\cos 70° + i \sin 70°) \qquad \blacksquare$$

Example 9   Find the product of 2 cis 10°, 4 cis 20°, and 9 cis 50°.

Solution   To find the product, multiply the moduli and add the arguments.

$$(2 \text{ cis } 10°)(4 \text{ cis } 20°)(9 \text{ cis } 50°) = 2 \cdot 4 \cdot 9 \text{ cis}(10° + 20° + 50°)$$
$$= 72 \text{ cis } 80° \qquad \blacksquare$$

If two complex numbers are written in trigonometric form, their quotient can be found by dividing their moduli and subtracting their arguments. To show that this is true, we let $N_1 = r_1(\cos \theta_1 + i \sin \theta_1)$ and $N_2 = r_2(\cos \theta_2 + i \sin \theta_2)$. Then we have

$$\frac{N_1}{N_2} = \frac{r_1(\cos \theta_1 + i \sin \theta_1)}{r_2(\cos \theta_2 + i \sin \theta_2)}$$
$$= \frac{r_1}{r_2} \cdot \frac{\cos \theta_1 + i \sin \theta_1}{\cos \theta_2 + i \sin \theta_2} \cdot \frac{\cos \theta_2 - i \sin \theta_2}{\cos \theta_2 - i \sin \theta_2}$$
$$= \frac{r_1}{r_2} \cdot \frac{\cos \theta_1 \cos \theta_2 - i \cos \theta_1 \sin \theta_2 + i \sin \theta_1 \cos \theta_2 - i^2 \sin \theta_1 \sin \theta_2}{\cos^2 \theta_2 - i^2 \sin^2 \theta_2}$$
$$= \frac{r_1}{r_2} \cdot \frac{\cos \theta_1 \cos \theta_2 + \sin \theta_1 \sin \theta_2 + i(\sin \theta_1 \cos \theta_2 - \cos \theta_1 \sin \theta_2)}{\cos^2 \theta_2 + \sin^2 \theta_2}$$
$$= \frac{r_1}{r_2} \cdot \frac{\cos(\theta_1 - \theta_2) + i \sin(\theta_1 - \theta_2)}{1}$$
$$= \frac{r_1}{r_2}[\cos(\theta_1 - \theta_2) + i \sin(\theta_1 - \theta_2)]$$

Thus, to divide complex numbers written in trigonometric form, we use the following rule:

$$\frac{r_1(\cos \theta_1 + i \sin \theta_1)}{r_2(\cos \theta_2 + i \sin \theta_2)} = \frac{r_1}{r_2}\left[\cos(\theta_1 - \theta_2) + i \sin(\theta_1 - \theta_2)\right]$$

or in r cis θ notation

$$\frac{r_1 \text{ cis } \theta_1}{r_2 \text{ cis } \theta_2} = \frac{r_1}{r_2} \text{cis}(\theta_1 - \theta_2)$$

Example 10   Divide 8 cis 110° by 4 cis 50°.

Solution   Use the division rule.

$$\frac{8 \text{ cis } 110°}{4 \text{ cis } 50°} = \frac{8}{4} \text{cis}(110° - 50°)$$

$$= 2 \text{ cis } 60°$$

Note that 2 cis 60° can be written in the form 2(cos 60° + i sin 60°).   ∎

Exercise 7.8

In Exercises 1–12 graph the given complex number on the complex plane.

1. $2 + 3i$
2. $2 - 3i$
3. $-2 + 3i$
4. $-2 - 3i$

5. $(4 + i)^2$
6. $(4 - i)^3$
7. $\dfrac{2 + i}{3 - i}$
8. $\dfrac{-2 + 2i}{-3 - i}$

9. 6
10. -8
11. $7i$
12. $-5i$

In Exercises 13–24 compute each absolute value.

13. $|2 + 3i|$
14. $|-5 - i|$
15. $|-7 + 7i|$
16. $\left|\dfrac{1}{2} - \dfrac{1}{4}i\right|$

17. $|6|$
18. $|-6|$
19. $|5i|$
20. $|-4i|$

21. $\left|\dfrac{-3i}{2 + i}\right|$
22. $\left|\dfrac{5i}{i - 2}\right|$
23. $\left|\dfrac{4 - i}{4 + i}\right|$
24. $|i^{365}|$

In Exercises 25–36 write the given complex number in trigonometric form.

25. $6 + 0i$
26. $-7 + 0i$
27. $0 - 3i$
28. $0 + 4i$

29. $-1 - i$
30. $-1 + i$
31. $3 + 3i\sqrt{3}$
32. $7 - 7i$

33. $-1 - \sqrt{3}i$
34. $-3\sqrt{3} + 3i$
35. $-\sqrt{3} - i$
36. $1 + i\sqrt{3}$

In Exercises 37–48 write the given complex number in a + bi form.

37. $2(\cos 30° + i \sin 30°)$
38. $5(\cos 45° + i \sin 45°)$
39. $7(\cos 90° + i \sin 90°)$

40. $12(\cos 0° + i \sin 0°)$
41. $-2\left(\cos \dfrac{2\pi}{3} + i \sin \dfrac{2\pi}{3}\right)$
42. $3\left(\cos \dfrac{4\pi}{3} + i \sin \dfrac{4\pi}{3}\right)$

43. $\dfrac{1}{2}(\cos \pi + i \sin \pi)$
44. $-\dfrac{2}{3}(\cos 2\pi + i \sin 2\pi)$
45. $-3 \text{ cis } 225°$

46. $3 \text{ cis } 300°$
47. $11 \text{ cis } \dfrac{11\pi}{6}$
48. $9 \text{ cis } 3$

In Exercises 49–60 find each product.

49. $[4(\cos 30° + i \sin 30°)][2(\cos 60° + i \sin 60°)]$

50. $[3(\cos 45° + i \sin 45°)][2(\cos 120° + i \sin 120°)]$

51. $(\cos 300° + i \sin 300°)(\cos 0° + i \sin 0°)$

52. $[5(\cos 85° + i \sin 85°)][2(\cos 65° + i \sin 65°)]$

53. $[2(\cos \pi + i \sin \pi)][3(\cos \pi + i \sin \pi)]$

54. $\left(\cos \dfrac{\pi}{2} + i \sin \dfrac{\pi}{2}\right)\left(\cos \dfrac{3\pi}{2} + i \sin \dfrac{3\pi}{2}\right)$

55. $\left[2\left(\cos \dfrac{\pi}{3} + i \sin \dfrac{\pi}{3}\right)\right]\left[3\left(\cos \dfrac{\pi}{6} + i \sin \dfrac{\pi}{6}\right)\right]$

56. $\left[3\left(\cos \dfrac{5\pi}{6} + i \sin \dfrac{5\pi}{6}\right)\right]\left[4\left(\cos \dfrac{7\pi}{6} + i \sin \dfrac{7\pi}{6}\right)\right]$

57. $(3 \text{ cis } 12°)(2 \text{ cis } 22°)(5 \text{ cis } 82°)$

58. $(2 \text{ cis } 50°)(3 \text{ cis } 100°)(6 \text{ cis } 2°)$

59. $\left(3 \text{ cis } \dfrac{\pi}{2}\right)\left(4 \text{ cis } \dfrac{\pi}{3}\right)\left(3 \text{ cis } \dfrac{\pi}{4}\right)$

60. $\left(4 \text{ cis } \dfrac{\pi}{6}\right)\left(2 \text{ cis } \dfrac{2\pi}{3}\right)\left(\text{cis } \dfrac{\pi}{4}\right)$

In Exercises 61–68 find each quotient.

61. $\dfrac{12(\cos 60° + i \sin 60°)}{2(\cos 30° + i \sin 30°)}$

62. $\dfrac{24(\cos 150° + i \sin 150°)}{48(\cos 50° + i \sin 50°)}$

63. $\dfrac{18(\cos \pi + i \sin \pi)}{12\left(\cos \dfrac{\pi}{2} + i \sin \dfrac{\pi}{2}\right)}$

64. $\dfrac{15(\cos 2\pi + i \sin 2\pi)}{45(\cos \pi + i \sin \pi)}$

65. $\dfrac{12 \text{ cis } 250°}{5 \text{ cis } 120°}$

66. $\dfrac{365 \text{ cis } 370°}{20 \text{ cis } 255°}$

67. $\dfrac{\text{cis } \dfrac{2\pi}{3}}{2 \text{ cis } \dfrac{\pi}{6}}$

68. $\dfrac{250 \text{ cis } \dfrac{7\pi}{16}}{50 \text{ cis } \dfrac{\pi}{3}}$

In Exercises 69–72 simplify each expression.

69. $\dfrac{(2 \text{ cis } 60°)(3 \text{ cis } 20°)}{6 \text{ cis } 40°}$

70. $\dfrac{36 \text{ cis } 200°}{(2 \text{ cis } 40°)(9 \text{ cis } 10°)}$

71. $\dfrac{48 \text{ cis } \dfrac{11\pi}{6}}{\left(3 \text{ cis } \dfrac{\pi}{3}\right)\left(4 \text{ cis } \dfrac{2\pi}{3}\right)}$

72. $\dfrac{(96 \text{ cis } \pi)(12 \text{ cis } 2\pi)}{\left(48 \text{ cis } \dfrac{\pi}{2}\right)\left(3 \text{ cis } \dfrac{3\pi}{2}\right)}$

7.9 DE MOIVRE'S THEOREM

If two complex numbers are written in trigonometric form, their product is found by multiplying their moduli and adding their arguments. This fact makes it easy to find powers of complex numbers that are expressed in trigonometric form. For example, to find the cube of $3(\cos 40° + i \sin 40°)$, we proceed as follows:

$$[3(\cos 40° + i \sin 40°)]^3$$
$$= [3(\cos 40° + i \sin 40°)][3(\cos 40° + i \sin 40°)][3(\cos 40° + i \sin 40°)]$$
$$= 3^3[\cos(40° + 40° + 40°) + i \sin(40° + 40° + 40°)]$$
$$= 27[\cos 3(40°) + i \sin 3(40°)]$$
$$= 27(\cos 120° + i \sin 120°)$$

The generalization of the previous example is called **De Moivre's theorem**.

Abraham De Moivre
(1667–1754) De Moivre
also made several
contributions to the field
of probability.

> **De Moivre's Theorem.**   If n is a real number and $r(\cos \theta + i \sin \theta)$ is a complex number in trigonometric form, then
>
> $$[r(\cos \theta + i \sin \theta)]^n = r^n[\cos n\theta + i \sin n\theta]$$
>
> or
>
> $$(r \operatorname{cis} \theta)^n = r^n \operatorname{cis} n\theta$$

This theorem was first developed about 1730 by the French mathematician Abraham De Moivre.

Example 1   Find $[2(\cos 15° + i \sin 15°)]^4$.

Solution   Use De Moivre's theorem.

$$[2(\cos 15° + i \sin 15°)]^4 = 2^4[\cos 4 \cdot 15° + i \sin 4 \cdot 15°]$$
$$= 16[\cos 60° + i \sin 60°]$$

This result could be changed easily to $a + bi$ form if desired:

$$16(\cos 60° + i \sin 60°) = 16\left(\frac{1}{2} + i \frac{\sqrt{3}}{2}\right) = 8 + 8\sqrt{3}i$$

■

Example 2   Find $[\sqrt{2}(\cos 10° + i \sin 10°)]^{10}$.

Solution   Use De Moivre's theorem.

$$[\sqrt{2}(\cos 10° + i \sin 10°)]^{10} = (\sqrt{2})^{10}[(\cos 10 \cdot 10° + i \sin 10 \cdot 10°)]$$
$$= 32(\cos 100° + i \sin 100°)$$

■

De Moivre's theorem can be used to find all of the nth roots of any number. Because both real and complex numbers can be written in the form $a + bi$, they can be written in trigonometric form as well. Thus, one nth root of $a + bi$ is

$$\sqrt[n]{a + bi} = (a + bi)^{1/n}$$
$$= [r(\cos \theta + i \sin \theta)]^{1/n}$$
$$= \sqrt[n]{r}\left(\cos \frac{\theta}{n} + i \sin \frac{\theta}{n}\right)$$

Recall that the equation $x^2 = 9$ has two distinct roots, 3 and -3, and each qualifies as a square root of 9. In like manner, the equation $x^n = a + bi$ has n distinct roots, and each qualifies as an nth root of the complex number $a + bi$. It follows that there are n distinct roots of any complex number.

Because $\sin \theta = \sin(\theta + k \cdot 360°)$ and $\cos \theta = \cos(\theta + k \cdot 360°)$ for all integers k, De Moivre's theorem implies that

$$[r(\cos \theta + i \sin \theta)]^{1/n} = \{r[\cos(\theta + k \cdot 360°) + i \sin(\theta + k \cdot 360°)]\}^{1/n}$$

$$= r^{1/n}\left(\cos \frac{\theta + k \cdot 360°}{n} + i \sin \frac{\theta + k \cdot 360°}{n}\right)$$

Substituting the numbers $0, 1, 2, \ldots, (n - 1)$ for k yields the n nth roots of the given complex number.

Example 3   Find the three cube roots of 8.

Solution   Because 8 can be expressed as $8 + 0i$, graph the complex number $8 + 0i$ as in Figure 7–20 to see that $r = 8$ and $\theta = 0°$. Write $8 + 0i$ in trigonometric form, and use the equation

$$[r(\cos \theta + i \sin \theta)]^{1/n} = \sqrt[n]{r}\left[\cos \frac{\theta + k \cdot 360°}{n} + i \sin \frac{\theta + k \cdot 360°}{n}\right]$$

Substituting the values for n, r, and θ gives

$$[8(\cos 0° + i \sin 0°)]^{1/3} = 8^{1/3}\left[\cos \frac{0° + k \cdot 360°}{3} + i \sin \frac{0° + k \cdot 360°}{3}\right]$$

Substituting 0 for k and replacing $8^{1/3}$ with 2 gives

$$2\left(\cos \frac{0°}{3} + i \sin \frac{0°}{3}\right) = 2(\cos 0° + i \sin 0°)$$

$$= 2(1 + 0i)$$

$$= 2$$

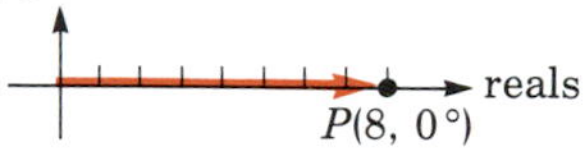

Figure 7–20

Now substitute 1 for k.

$$2\left(\cos \frac{0° + 360°}{3} + i \sin \frac{0° + 360°}{3}\right) = 2(\cos 120° + i \sin 120°)$$

$$= 2\left(-\frac{1}{2} + i \frac{\sqrt{3}}{2}\right)$$

$$= -1 + i\sqrt{3}$$

Finally, substitute 2 for k.

$$2\left(\cos \frac{0° + 720°}{3} + i \sin \frac{0° + 720°}{3}\right) = 2(\cos 240° + i \sin 240°)$$

$$= 2\left(-\frac{1}{2} + i \frac{-\sqrt{3}}{2}\right)$$

$$= -1 - i\sqrt{3}$$

The numbers 2, $-1 + i\sqrt{3}$, and $-1 - i\sqrt{3}$ are the three cube roots of 8. If these three cube roots of 8 are graphed in the complex plane, they are equally spaced around a circle of radius 2. If the endpoints of these vectors are joined by straight line segments, an equilateral triangle is formed as in Figure 7–21.

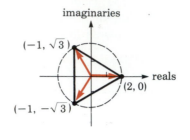

Figure 7–21

Example 4   Find the four fourth roots of $-16i$.

Solution   Express $-16i$ as $0 - 16i$ and graph it to determine that $r = 16$ and $\theta = 270°$. Hence,

$$\sqrt[4]{-16i} = [16(\cos 270° + i \sin 270°)]^{1/4}$$

$$= 16^{1/4}\left(\cos \frac{270° + k \cdot 360°}{4} + i \sin \frac{270° + k \cdot 360°}{4}\right)$$

Substitute 0, 1, 2, and 3 for k to get the four fourth roots of $-16i$.

$$2(\cos 67.5° + i \sin 67.5°) \approx \quad 0.77 + 1.85i$$
$$2(\cos 157.5° + i \sin 157.5°) \approx -1.85 + 0.77i$$
$$2(\cos 247.5° + i \sin 247.5°) \approx -0.77 - 1.85i$$
$$2(\cos 337.5° + i \sin 337.5°) \approx \quad 1.85 - 0.77i$$

Graphs of these four fourth roots are equally spaced around a circle of radius 2, and the endpoints of these vectors are the vertices of a square. See Figure 7–22.

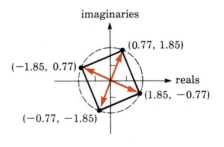

Figure 7–22

Example 5   Find the five fifth roots of $-4 - 4i$.

Solution   Express $-4 - 4i$ in trigonometric form by determining that $r = 4\sqrt{2}$ and $\theta = 225°$. Then, you have

$$[4\sqrt{2}(\cos 225° + i \sin 225°)]^{1/5}$$

$$= (4\sqrt{2})^{1/5}\left(\cos \frac{225° + k \cdot 360°}{5} + i \sin \frac{225° + k \cdot 360°}{5}\right)$$

$$= \sqrt{2}\left(\cos \frac{225° + k \cdot 360°}{5} + i \sin \frac{225° + k \cdot 360°}{5}\right)$$

Substituting 0, 1, 2, 3, and 4 for k generates the five fifth roots of $-4 - 4i$.

$$\sqrt{2}(\cos 45° + i \sin 45°) = \quad 1 + i$$
$$\sqrt{2}(\cos 117° + i \sin 117°) \approx -0.64 + 1.26i$$
$$\sqrt{2}(\cos 189° + i \sin 189°) \approx -1.40 - 0.22i$$
$$\sqrt{2}(\cos 261° + i \sin 261°) \approx -0.22 - 1.40i$$
$$\sqrt{2}(\cos 333° + i \sin 333°) \approx \quad 1.26 - 0.64i$$

If these five fifth roots are graphed, they are equally spaced around a circle with radius $\sqrt{2}$, and the endpoints of these vectors are the vertices of a regular pentagon. See Figure 7–23.

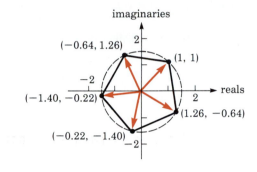

Figure 7–23

If $n > 2$ and the n nth roots of a complex number are graphed on the complex plane, the endpoints of the vectors that represent each root will always be at the vertices of a regular polygon. ■

Exercise 7.9

In Exercises 1–12 find the indicated power. Leave all answers in trigonometric form.

1. $[3(\cos 30° + i \sin 30°)]^3$ **2.** $[4(\cos 15° + i \sin 15°)]^6$ **3.** $(\cos 15° + i \sin 15°)^{12}$

4. $[2(\cos 120° + i \sin 120°)]^6$ **5.** $[5 \text{ cis } 2°]^5$ **6.** $[0.5 \text{ cis } 100°]^3$

7. $\left[3\left(\cos \dfrac{\pi}{4} + i \sin \dfrac{\pi}{4}\right)\right]^4$ **8.** $\left[2\left(\cos \dfrac{3\pi}{2} + i \sin \dfrac{3\pi}{2}\right)\right]^6$ **9.** $[4(\cos 3 + i \sin 3)]^4$

10. $[2(\cos 5 + i \sin 5)]^{20}$ **11.** $\left[\dfrac{1}{3} \text{ cis } \dfrac{\pi}{2}\right]^3$ **12.** $\left[\dfrac{1}{2} \text{ cis } \dfrac{\pi}{6}\right]^5$

In Exercises 13–18 find the indicated nth root of each expression in $a + bi$ form.

13. A cube root of $8(\cos 180° + i \sin 180°)$ **14.** A fifth root of $32(\cos 150° + i \sin 150°)$

15. A fifth root of $(\cos 300° + i \sin 300°)$ **16.** A fourth root of $64(\cos \pi + i \sin \pi)$

17. A sixth root of $64(\cos 2\pi + i \sin 2\pi)$ **18.** A sixth root of $3^6(\cos \pi + i \sin \pi)$

In Exercises 19–26 find and graph the indicated roots of each complex number. Change your answers to $a + bi$ form only if that answer would be exact. Otherwise, leave the answers in trigonometric form.

19. The three cube roots of -8 **20.** The four fourth roots of 16

21. The two square roots of i **22.** The three cube roots of i

23. The five fifth roots of $-i$

24. The three cube roots of $\dfrac{\sqrt{2}}{2} + \dfrac{\sqrt{2}}{2}i$

25. The four fourth roots of $-8 + 8\sqrt{3}i$

26. The six sixth roots of $-i$

In Exercises 27–30 substitute the given value of n into De Moivre's theorem with $r = 1$, and raise the binomial on the left to the nth power. Follow the additional directions.

27. $n = 2$. Set the real parts of the complex numbers equal to each other and thereby show that $\cos 2\theta = \cos^2 \theta - \sin^2 \theta$.

28. $n = 2$. Set the imaginary parts of the complex numbers equal to each other and thereby show that $\sin 2\theta = 2 \cos \theta \sin \theta$.

29. $n = 3$. Set the imaginary parts of the complex numbers equal to each other and thereby show that $\sin 3\theta = 3 \cos^2 \theta \sin \theta - \sin^3 \theta$.

30. $n = 3$. Set the real parts of the complex numbers equal to each other and thereby show that $\cos 3\theta = \cos^2 \theta - 3 \cos \theta \sin^2 \theta$.

31. Use the right side of the identity in Exercise 29 to show that $\sin 3\theta = 3 \sin \theta - 4 \sin^3 \theta$.

32. Use the right side of the identity in Exercise 30 to show that $\cos 3\theta = 4 \cos^3 \theta - 3 \cos \theta$.

7.10   MORE ON VECTORS

Recall that a *vector* is a directed line segment. Vectors are usually denoted by boldface letters such as **a** or **V**. However, in handwritten work we often express vectors with letters having arrows above them, such as $\vec{V}$. If a vector starts at point A and ends at point B, the vector can be denoted as **AB** or $\overrightarrow{AB}$.

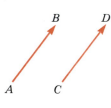

Figure 7–24

> **Definition.** Two **vectors** are equal if and only if they have the same length and the same direction.

Recall that the length of a vector **V** is called its **norm** and is denoted by $|\mathbf{V}|$. Because the norm of a vector is a real number that expresses magnitude only, it is a **scalar quantity**.

Figure 7–25

In Figure 7–24, vectors **AB** and **CD** have the same length and direction. Thus, $|\mathbf{AB}| = |\mathbf{CD}|$ and $\mathbf{AB} = \mathbf{CD}$. In Figure 7–25, $|\mathbf{OA}| = |\mathbf{OB}|$, but because **OA** and **OB** have different directions, $\mathbf{OA} \neq \mathbf{OB}$.

In Figure 7–26, vector **V** is placed on a coordinate system originating at the origin O and ending at the point $A(3, 2)$. If a vector, such as **V**, originates at the origin, it is completely determined by its endpoint. Thus, we can denote the vector **V** with the ordered pair $\langle 3, 2 \rangle$. Corner brackets $\langle\ \rangle$ are used to distinguish the vector from the point with coordinates $(3, 2)$. The distance formula can be used to find the norm of **V**:

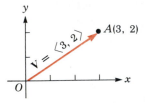

Figure 7–26

$$|\mathbf{V}| = \sqrt{3^2 + 2^2}$$
$$= \sqrt{13}$$

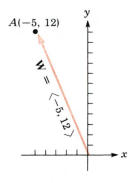

Figure 7–27

Similarly, in Figure 7–27, $\mathbf{W} = \langle -5, 12 \rangle$, and

$$|\mathbf{W}| = \sqrt{(-5)^2 + 12^2}$$
$$= \sqrt{169}$$
$$= 13$$

In general,

Theorem. If a vector $\mathbf{V}$ is placed on a coordinate system and is represented by $\langle a, b \rangle$, then the norm of $\mathbf{V}$ is given by

$$|\mathbf{V}| = \sqrt{a^2 + b^2}$$

Example 1 If vector $\mathbf{V} = \langle 3, -5 \rangle$, find $|\mathbf{V}|$.

Solution
$$|\mathbf{V}| = \sqrt{3^2 + (-5)^2}$$
$$= \sqrt{9 + 25}$$
$$= \sqrt{34}$$

Adding vectors is easy if we use ordered-pair notation. To find the coordinates of point C in the parallelogram in Figure 7–28, we add the corresponding coordinates of points A and B. Thus, if $\mathbf{OA} = \langle -3, 1 \rangle$ and $\mathbf{OB} = \langle 6, 1 \rangle$, then

$$\mathbf{OA} + \mathbf{OB} = \langle -3, 1 \rangle + \langle 6, 1 \rangle$$
$$= \langle -3 + 6, 1 + 1 \rangle$$
$$= \langle 3, 2 \rangle$$
$$= \mathbf{OC}$$

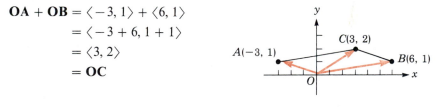

Figure 7–28

The previous discussion suggests the following definition.

Definition. If $\mathbf{V} = \langle a, b \rangle$ and $\mathbf{W} = \langle c, d \rangle$, then
$$\mathbf{V} + \mathbf{W} = \langle a + c, b + d \rangle$$

Example 2 If $\mathbf{V} = \langle 3, -5 \rangle$ and $\mathbf{W} = \langle 1, 2 \rangle$, find **a.** $\mathbf{V} + \mathbf{W}$ and **b.** $\mathbf{W} + \mathbf{V}$.

Solution

a. $\mathbf{V} + \mathbf{W} = \langle 3, -5 \rangle + \langle 1, 2 \rangle$
$\qquad = \langle 3 + 1, -5 + 2 \rangle$
$\qquad = \langle 4, -3 \rangle$

b. $\mathbf{W} + \mathbf{V} = \langle 1, 2 \rangle + \langle 3, -5 \rangle$
$\qquad = \langle 1 + 3, 2 + (-5) \rangle$
$\qquad = \langle 4, -3 \rangle$

Example 2 illustrates that the addition of two vectors is commutative.

If k is a real number and $\mathbf{V}$ is a vector, we can define a type of multiplication called **scalar multiplication**.

Definition. If k is a scalar and $\mathbf{V}$ is the vector $\langle a, b\rangle$, then

$$k\mathbf{V} = k\langle a, b\rangle = \langle ka, kb\rangle$$

Example 3 If $\mathbf{V} = \langle 3, -5\rangle$, find **a.** $2\mathbf{V}$ and **b.** $-8\mathbf{V}$.

Solution

a. $$\begin{aligned} 2\mathbf{V} &= 2\langle 3, -5\rangle \\ &= \langle 6, -10\rangle \end{aligned}$$

b. $$\begin{aligned} -8\mathbf{V} &= -8\langle 3, -5\rangle \\ &= \langle -24, 40\rangle \end{aligned}$$

■

Note that, if k is a real number and $\mathbf{V}$ is a vector, then the product $k\mathbf{V}$ is also a vector. Its norm is $|k|$ times the norm of $\mathbf{V}$ itself. If k is a positive real number, then $k\mathbf{V}$ has the same direction as $\mathbf{V}$. If k is a negative real number, then $k\mathbf{V}$ has the opposite direction of $\mathbf{V}$. See Figure 7–29. The vector $k\mathbf{V}$ is called a **scalar multiple** of $\mathbf{V}$.

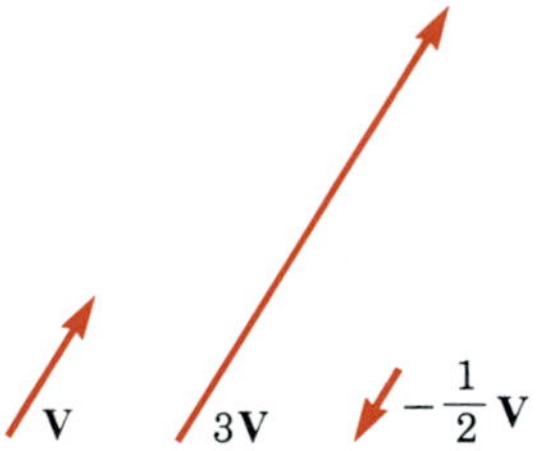

Figure 7–29

Subtraction of vectors is defined as follows:

Definition. If $\mathbf{V}$ and $\mathbf{W}$ are two vectors, then

$$\mathbf{V} - \mathbf{W} = \mathbf{V} + (-\mathbf{W})$$

Example 4 If $\mathbf{V} = \langle -2, 8\rangle$ and $\mathbf{W} = \langle -5, -8\rangle$, find **a.** $\mathbf{V} - \mathbf{W}$ and **b.** $\mathbf{W} - \mathbf{V}$.

Solution

a. $$\begin{aligned} \mathbf{V} - \mathbf{W} &= \mathbf{V} + (-\mathbf{W}) \\ &= \langle -2, 8\rangle + (-\langle -5, -8\rangle) \\ &= \langle -2, 8\rangle + \langle 5, 8\rangle \\ &= \langle -2 + 5, 8 + 8\rangle \\ &= \langle 3, 16\rangle \end{aligned}$$

b. $$\begin{aligned} \mathbf{W} - \mathbf{V} &= \mathbf{W} + (-\mathbf{V}) \\ &= \langle -5, -8\rangle + (-\langle -2, 8\rangle) \\ &= \langle -5, -8\rangle + \langle 2, -8\rangle \\ &= \langle -5 + 2, -8 + (-8)\rangle \\ &= \langle -3, -16\rangle \end{aligned}$$

■

Example 4 illustrates that the subtraction of two vectors is not commutative. However, $\mathbf{V} - \mathbf{W} = -(\mathbf{W} - \mathbf{V})$ for any two vectors $\mathbf{V}$ and $\mathbf{W}$.

If $\mathbf{V} = \langle a, b \rangle$, then $\mathbf{V} - \mathbf{V} = \langle a, b \rangle - \langle a, b \rangle = \langle 0, 0 \rangle$. The vector $\langle 0, 0 \rangle$, called the **zero vector**, is denoted as $\mathbf{0}$.

We summarize many properties of vectors as follows:

$$
\begin{array}{ll}
\mathbf{A} + \mathbf{B} = \mathbf{B} + \mathbf{A} & (\mathbf{A} + \mathbf{B}) + \mathbf{C} = \mathbf{A} + (\mathbf{B} + \mathbf{C}) \\
\mathbf{A} + \mathbf{0} = \mathbf{0} + \mathbf{A} = \mathbf{A} & \mathbf{A} + (-\mathbf{A}) = -\mathbf{A} + \mathbf{A} = \mathbf{0} \\
c(\mathbf{A} + \mathbf{B}) = c\mathbf{A} + c\mathbf{B} & (\mathbf{A} + \mathbf{B})c = \mathbf{A}c + \mathbf{B}c \\
(c + d)\mathbf{A} = c\mathbf{A} + d\mathbf{A} & \mathbf{A}(c + d) = \mathbf{A}c + \mathbf{A}d \\
1\mathbf{A} = \mathbf{A}1 = \mathbf{A} & 0\mathbf{A} = \mathbf{A}0 = \mathbf{0} \\
(cd)\mathbf{A} = c(d\mathbf{A}) = d(c\mathbf{A}) &
\end{array}
$$

Proof   The proof of each property follows from the previous definitions. For example, to prove that $(c + d)\mathbf{A} = c\mathbf{A} + d\mathbf{A}$, we let $\mathbf{A} = \langle a, b \rangle$ and proceed as follows:

$$
\begin{aligned}
(c + d)\mathbf{A} &= (c + d)\langle a, b \rangle \\
&= \langle (c + d)a, (c + d)b \rangle \\
&= \langle ca + da, cb + db \rangle \\
&= \langle ca, cb \rangle + \langle da, db \rangle \\
&= c\langle a, b \rangle + d\langle a, b \rangle \\
&= c\mathbf{A} + d\mathbf{A}
\end{aligned}
$$

$\square$

If $\mathbf{i}$ is the vector $\langle 1, 0 \rangle$, then $\mathbf{i}$ is a vector originating at the origin and pointing in the positive direction on the x-axis. Because $|\mathbf{i}| = 1$, $\mathbf{i}$ is called a **unit vector**. If $\mathbf{j}$ is the unit vector $\langle 0, 1 \rangle$, then $\mathbf{j}$ is a vector originating at the origin and pointing in the positive direction on the y-axis. Any vector can be written as the sum of scalar multiples of the vectors $\mathbf{i}$ and $\mathbf{j}$. For example, the vector $\langle 5, 2 \rangle$ can be written in this form by proceeding as follows:

$$
\begin{aligned}
\langle 5, 2 \rangle &= \langle 5, 0 \rangle + \langle 0, 2 \rangle \\
&= 5\langle 1, 0 \rangle + 2\langle 0, 1 \rangle \\
&= 5\mathbf{i} + 2\mathbf{j}
\end{aligned}
$$

The two vectors $5\mathbf{i}$ and $2\mathbf{j}$ are called the ***x-*** and ***y-*components** of the vector $\langle 5, 2 \rangle$. See Figure 7–30.

Definition.   The vectors $\mathbf{i} = \langle 1, 0 \rangle$ and $\mathbf{j} = \langle 0, 1 \rangle$ is called **unit-coordinate vectors**.

Any vector $\langle x, y \rangle$ is said to be **resolved** into its ***x-*** or **horizontal component** $x\mathbf{i}$ and its ***y-*** or **vertical component** $y\mathbf{j}$ when it is written in the form $x\mathbf{i} + y\mathbf{j}$.

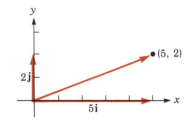

Figure 7–30

Example 5   Let $V = 2i + 3j$ and $W = 4i - j$. Calculate **a.** $5V + 3W$ and **b.** $V - W$.

Solution   **a.** $5V + 3W = 5(2i + 3j) + 3(4i - 1j)$

$= (10i + 15j) + (12i - 3j)$

$= 22i + 12j$

b. $V - W = V + (-1)W$

$= (2i + 3j) + (-1)(4i - 1j)$

$= (2i + 3j) + (-4i + 1j)$

$= -2i + 4j$ ∎

The definition of scalar multiplication provides the way to multiply a vector by a real number. We now define a way, called the **dot product**, to multiply one vector by another.

Definition.   The **dot product** of vectors **V** and **W** is the scalar

$$V \cdot W = |V| \, |W| \cos \theta$$

where θ is the angle between the vectors.

It is not convenient to calculate the dot product of two vectors by using the previous definition. However, we can calculate a dot product by using a theorem, which is stated without proof.

Theorem.   Let $V = ai + bj$ and $W = ci + dj$. Then

$$V \cdot W = ac + bd$$

Example 6   If $A = 2i + 3j$ and $B = 5i - 4j$, calculate $A \cdot B$.

Solution $A \cdot B = (2i + 3j) \cdot (5i - 4j)$

$= 2 \cdot 5 + 3 \cdot (-4)$

$= -2$ ∎

Example 7   Find the angle between $\mathbf{A} = 3\mathbf{i} + 4\mathbf{j}$ and $\mathbf{B} = 5\mathbf{i} - 12\mathbf{j}$.

Solution   See Figure 7–31. By definition, $\mathbf{A} \cdot \mathbf{B} = |\mathbf{A}|\,|\mathbf{B}| \cos \theta$, where θ is the angle between the vectors. Solve for $\cos \theta$, and proceed as follows:

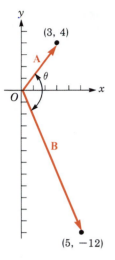

$$\mathbf{A} \cdot \mathbf{B} = |\mathbf{A}|\,|\mathbf{B}| \cos \theta$$

$$\cos \theta = \frac{\mathbf{A} \cdot \mathbf{B}}{|\mathbf{A}|\,|\mathbf{B}|}$$

$$= \frac{(3\mathbf{i} + 4\mathbf{j}) \cdot (5\mathbf{i} - 12\mathbf{j})}{\sqrt{3^2 + 4^2}\,\sqrt{5^2 + (-12)^2}}$$

$$= \frac{15 - 48}{5 \cdot 13}$$

$$\cos \theta = \frac{-33}{65}$$

$$\theta = \cos^{-1}\left(\frac{-33}{65}\right)$$

$$\theta \approx 120.5°$$

$\blacksquare$

Figure 7–31

If the dot product, $|\mathbf{A}|\,|\mathbf{B}| \cos \theta$, of two nonzero vectors $\mathbf{A}$ and $\mathbf{B}$ is 0, then $\cos \theta$ must be 0. If $\cos \theta = 0$, then $\theta = 90°$ and the two vectors must be perpendicular. Thus, the dot product provides a test for the perpendicularity of two vectors.

> **Theorem.**   Two nonzero vectors are perpendicular if and only if their dot product is zero.

Example 8   Are the vectors $\mathbf{A} = 6\mathbf{i} - 2\mathbf{j}$ and $\mathbf{B} = \mathbf{i} + 3\mathbf{j}$ perpendicular?

Solution   Calculate $\mathbf{A} \cdot \mathbf{B}$. If the product is zero, the vectors are perpendicular. Otherwise, they are not.

$$\mathbf{A} \cdot \mathbf{B} = (6\mathbf{i} - 2\mathbf{j}) \cdot (\mathbf{i} + 3\mathbf{j})$$

$$= 6 \cdot 1 + (-2)(+3)$$

$$= 6 - 6$$

$$= 0$$

Because the dot product is 0, the vectors $\mathbf{A}$ and $\mathbf{B}$ are perpendicular.

$\blacksquare$

Example 9   What are the horizontal and vertical components of a 2.0-pound force that makes an angle of 30° with the x-axis? Express the 2.0-pound force in $a\mathbf{i} + b\mathbf{j}$ form.

Solution   The horizontal component of the given force is vector **OA**, which is one leg of the right triangle *OAC*. See Figure 7–32. The norm of **OA** is found as follows.

$$|\mathbf{OA}| = |\mathbf{OC}| \cos \theta$$
$$= 2.0(\cos 30°)$$
$$= 2.0 \frac{\sqrt{3}}{2}$$
$$= \sqrt{3}$$

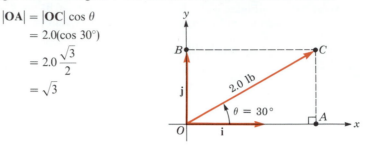

Figure 7–32

The horizontal component is $\sqrt{3}$ pounds.
    Similarly, the vertical component **OB** is found as follows.

$$|\mathbf{OB}| = |\mathbf{OC}| \sin \theta$$
$$= 2.0(\sin 30°)$$
$$= 1.0$$

The vertical component is 1.0 pound.
    Thus, you have $\mathbf{OC} = \sqrt{3}\mathbf{i} + \mathbf{j}$. ■

Example 10   A force of 2.0 pounds makes an angle of 30° with the horizontal. What is the component of this force in the direction $\mathbf{OB} = 12\mathbf{i} + 5\mathbf{j}$?

Solution   You must find the component of the given force in a direction other than that of an axis. See Figure 7–33.
    You must find the norm of vector **OA**. In right triangle *OAC*, the following relation holds.

$$|\mathbf{OA}| = |\mathbf{OC}| \cos \theta$$

Use the dot product to make this calculation. Proceed as follows.

$$|\mathbf{OA}| = |\mathbf{OC}| \cos \theta$$

$$= \frac{|\mathbf{OC}||\mathbf{OB}| \cos \theta}{|\mathbf{OB}|} \qquad \text{Multiply and divide by } |\mathbf{OB}|.$$

$$= \frac{\mathbf{OC} \cdot \mathbf{OB}}{|\mathbf{OB}|} \qquad \text{Use the definition of dot product.}$$

$$= \frac{(\sqrt{3}\mathbf{i} + \mathbf{j}) \cdot (12\mathbf{i} + 5\mathbf{j})}{\sqrt{12^2 + 5^2}} \qquad \text{Use the result from Example 9, and } \mathbf{OB} = 12\mathbf{i} + 5\mathbf{j}.$$

$$|\mathbf{OA}| = \frac{12\sqrt{3} + 5}{13}$$

The component of the 2.0-pound force in the direction of **OB** is $\dfrac{12\sqrt{3} + 5}{13}$ pounds.

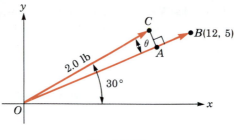

Figure 7–33

Exercise 7.10

In Exercises 1–12 let $U = \langle 2, -3 \rangle$, $V = \langle 5, -2 \rangle$, *and* $W = \langle -1, 1 \rangle$. *Calculate each quantity.*

1. $U + V$ 2. $V + W$ 3. $3U$ 4. $5V$
5. $2U + V$ 6. $3U - W$ 7. $|U|$ 8. $|3U|$
9. $|U + W|$ 10. $|V - W|$ 11. $|U| + |W|$ 12. $|3V - 5W|$

In Exercises 13–18 resolve each vector into its horizontal and its vertical components by writing each vector in $a\mathbf{i} + b\mathbf{j}$ *form.*

13. $\langle 3, 5 \rangle + \langle 5, 3 \rangle$ 14. $\langle -2, 7 \rangle + \langle 2, 3 \rangle$
15. A vector of length 10, making an angle of 30° with the x-axis
16. A vector of length 10, making an angle of 45° with the x-axis
17. A vector of length 23.3, making an angle of 37.2° with the x-axis
18. A vector of length 19.1, making an angle of 183.7° with the x-axis

In Exercises 19–24 find the dot product of the two given vectors.

19. $\langle 2, -3 \rangle$ and $\langle 3, -1 \rangle$ 20. $\langle 1, -5 \rangle$ and $\langle 5, 1 \rangle$ 21. $2\mathbf{i} + 5\mathbf{j}$ and $\mathbf{i} + \mathbf{j}$
22. $3\mathbf{i} - 3\mathbf{j}$ and $2\mathbf{i} + \mathbf{j}$ 23. $\mathbf{i}$ and $\mathbf{j}$ 24. $2\mathbf{i}$ and $3\mathbf{i}$

In Exercises 25–30 find the angle between the given vectors.

25. $\langle 2, 2 \rangle$ and $\langle 5, 0 \rangle$ 26. $\langle \sqrt{3}, 1 \rangle$ and $\langle 3, 3 \rangle$
27. $\langle \sqrt{3}, -1 \rangle$ and $\langle -1, \sqrt{3} \rangle$ 28. $2\mathbf{i} + 3\mathbf{j}$ and $-3\mathbf{i} + 2\mathbf{j}$
29. $3\mathbf{i} - \mathbf{j}$ and $3\mathbf{i} + \mathbf{j}$ 30. $3\mathbf{i} - 4\mathbf{j}$ and $5\mathbf{i} + 12\mathbf{j}$

In Exercises 31–36 indicate whether the given vectors are perpendicular.

31. $\langle 2, 3 \rangle$ and $\langle -3, 2 \rangle$ 32. $\langle 2, 3 \rangle$ and $\langle 3, 2 \rangle$ 33. $5\mathbf{i} + \mathbf{j}$ and $\mathbf{i} + \mathbf{j}$
34. $6\mathbf{i} - 2\mathbf{j}$ and $\mathbf{i} + 3\mathbf{j}$ 35. $\mathbf{i}$ and $\mathbf{j}$ 36. $-\mathbf{i}$ and $3\mathbf{i}$

In Exercises 37–40 find the component of the first vector in the direction of the second vector.

37. $\langle 3, 4 \rangle, \langle 5, 12 \rangle$ 38. $\langle 1, 1 \rangle, \langle 3, 2 \rangle$ 39. $6\mathbf{i} + 8\mathbf{j}, 4\mathbf{i} - 3\mathbf{j}$ 40. $\mathbf{i}, \mathbf{i} + \mathbf{j}$

41. Find an example to illustrate that $|U + V| \neq |U| + |V|$.
42. Find an example to support the distributive law $(a + b)V = aV + bV$.
43. Find an example to support the distributive law $a(V + W) = aV + aW$.
44. Find an example to support the associative law $a(V \cdot W) = (aV) \cdot W$.

45. Let $\mathbf{V} = a\mathbf{i} + b\mathbf{j}$. Prove that $\mathbf{V} \cdot \mathbf{V} = |\mathbf{V}|^2$.

46. Prove that $\mathbf{A} + \mathbf{0} = \mathbf{A}$.

47. Prove that $c(\mathbf{A} + \mathbf{B}) = c\mathbf{A} + c\mathbf{B}$.

48. Prove that $\mathbf{A}(c + d) = \mathbf{A}c + \mathbf{A}d$.

REVIEW EXERCISES

In Review Exercises 1–4 verify each identity.

1. $\dfrac{\sin \theta + \cos \theta \tan \theta}{\tan \theta} = 2 \cos \theta$

2. $\dfrac{\sin \theta}{\sec \theta} = \dfrac{1}{\tan \theta + \cot \theta}$

3. $2 \tan \theta = \dfrac{\sin \theta}{\cos \theta} + \dfrac{\sec \theta}{\csc \theta}$

4. $\sec \theta - \tan \theta = \dfrac{1}{\sec \theta + \tan \theta}$

In Review Exercises 5–8 express each quantity as a single function of one angle.

5. $\sin 20° \cos 51° + \cos 20° \sin 51°$

6. $\dfrac{\tan 20° - \tan 51°}{1 + \tan 20° \tan 51°}$

7. $\cos \dfrac{3\pi}{11} \cos \dfrac{\pi}{11} - \sin \dfrac{3\pi}{11} \sin \dfrac{\pi}{11}$

8. $1 - 2 \sin^2 \dfrac{\pi}{3}$

In Review Exercises 9–12 verify each identity.

9. $\cos(60° + \theta) = \dfrac{1}{2}(\cos \theta - \sqrt{3} \sin \theta)$

10. $\tan(180° - \theta) = -\tan \theta$

11. $\sin x = \dfrac{\sin 2x}{2 \cos x}$

12. $\cos \theta = \pm\sqrt{\sin^2 \theta + \cos 2\theta}$

In Review Exercises 13–18 use the given information to find the sine, cosine, and tangent of 2θ.

13. $\cos \theta = -\dfrac{12}{13}$; θ in QII

14. $\sin \theta = \dfrac{-5}{13}$; θ in QIII

15. $\tan \theta = \dfrac{-20}{21}$; θ in QIV

16. $\cos \theta = \dfrac{21}{29}$; θ in QI

17. $\sin \theta = -\dfrac{4}{5}$; θ in QIII

18. $\cot \theta = -\dfrac{3}{4}$; θ in QII

In Review Exercises 19–24 use the given information to find the sine, cosine, and tangent of $\dfrac{\theta}{2}$. Assume that $0° \leq \theta < 360°$.

19. $\cos \theta = \dfrac{12}{13}$; θ in QI

20. $\sin \theta = -\dfrac{5}{13}$; θ in QIV

21. $\tan \theta = -\dfrac{5}{12}$; θ in QIV

22. $\cos \theta = -\dfrac{3}{5}$; θ in QII

23. $\sin \theta = \dfrac{4}{5}$; θ in QII

24. $\cot \theta = \dfrac{3}{4}$; θ in QIII

In Review Exercises 25–28 evaluate each of the products. **Do not use a calculator or tables.**

25. $\sin 285° \cos 15°$

26. $\cos 285° \sin 15°$

27. $\sin \dfrac{5\pi}{4} \sin \dfrac{\pi}{12}$

28. $\cos \dfrac{\pi}{12} \cos \dfrac{5\pi}{4}$

In Review Exercises 29–32 express each quantity as a product.

29. $\sin 5° + \sin 7°$

30. $\sin 312° - \sin 140°$

31. $\cos \dfrac{3\pi}{5} - \cos \dfrac{\pi}{5}$

32. $\cos \dfrac{2\pi}{7} + \cos \dfrac{3\pi}{7}$

In Review Exercises 33–36 express each quantity as a product and evaluate without using a calculator or tables.

33. $\sin 285° + \sin 15°$ **34.** $\sin 15° - \sin 285°$ **35.** $\cos \dfrac{5\pi}{12} - \cos \dfrac{\pi}{12}$ **36.** $\cos \dfrac{\pi}{12} + \cos \dfrac{5\pi}{12}$

In Review Exercises 37–38 write each expression in the form $k \sin(x + \phi)$.

37. $y = \sin x + 2 \cos x$ **38.** $y = -\sin x + \cos x$

In Review Exercises 39–42 solve each equation for all values of the variable between 0 and 2π, including 0. **Do not use a calculator or tables.**

39. $\sin 2x = 1$ **40.** $\sin x + 1 = \tan x + \cos x$

41. $\cos^2 x - \sin^2 x = 0$ **42.** $\csc x \sec x = \sec x + \cot x$

In Review Exercises 43–48 find the number x, if it exists. **Do not use a calculator or tables.**

43. $\sin^{-1} \dfrac{\sqrt{3}}{2} = x$ **44.** $\cos^{-1}\left(-\dfrac{1}{2}\right) = x$ **45.** $\tan^{-1}(-\sqrt{3}) = x$

46. $\cos^{-1} \sqrt{3} = x$ **47.** $\cot^{-1} 0 = x$ **48.** $\csc^{-1}(0.5) = x$

In Review Exercises 49–54 find each value of θ, if any. Express all answers in radians. **Do not use a calculator or tables.**

49. $\sin^{-1}\left(-\dfrac{1}{2}\right) = \theta$ **50.** $\tan^{-1} \dfrac{\sqrt{3}}{3} = \theta$ **51.** $\cos^{-1} 0 = \theta$

52. $\operatorname{arcsec}(-1) = \theta$ **53.** $\arcsin 0 = \theta$ **54.** $\arccos 1 = \theta$

In Review Exercises 55–58 find the required value without using a calculator or tables.

55. $\sin\left[\sin^{-1}\left(-\dfrac{1}{2}\right) + \cos^{-1}\left(-\dfrac{1}{2}\right)\right]$ **56.** $\sin 2\left(\sin^{-1} \dfrac{1}{2}\right)$

57. $\sin \dfrac{1}{2}\left[\arccos\left(-\dfrac{1}{2}\right)\right]$ **58.** $\tan 2(\arcsin 1)$

In Review Exercises 59–60 rewrite each value as an algebraic expression in the variable u.

59. $\sin(\cos^{-1} u)$ **60.** $\tan(2 \tan^{-1} u)$

In Review Exercises 61–64 graph each complex number on the complex plane.

61. $4 - 5i$ **62.** $-7 + 2i$ **63.** 6 **64.** $3i$

In Review Exercises 65–68 compute each absolute value.

65. $|8 + 3i|$ **66.** $|10 - 10i|$ **67.** $\left|\dfrac{3i}{i + 3}\right|$ **68.** $\left|\dfrac{4 - 3i}{4 + 3i}\right|$

In Review Exercises 69–72 write the given complex number in trigonometric form.

69. $-2 + 2i$ **70.** $5 - 5i$ **71.** $3 + 3i\sqrt{3}$ **72.** 4

In Review Exercises 73–76 write the given complex number in $a + bi$ form.

73. $3(\cos 60° + i \sin 60°)$ **74.** $2(\cos 330° + i \sin 330°)$

75. $3\left(\cos \dfrac{4\pi}{3} + i \sin \dfrac{4\pi}{3}\right)$ **76.** $7\left(\cos \dfrac{5}{6}\pi + i \sin \dfrac{5}{6}\pi\right)$

In Review Exercises 77–82 perform the indicated operation. Simplify, but leave your answer in trigonometric form.

77. $(\text{cis } 60°)(\text{cis } 50°)$

78. $(2 \text{ cis } 330°)(3 \text{ cis } 240°)$

79. $\left[3\left(\cos \dfrac{\pi}{12} + i \sin \dfrac{\pi}{12}\right)\right]\left[2\left(\cos \dfrac{\pi}{6} + i \sin \dfrac{\pi}{6}\right)\right]$

80. $\left[7\left(\cos \dfrac{\pi}{5} + i \sin \dfrac{\pi}{5}\right)\right]\left[3\left(\cos \dfrac{4\pi}{5} + i \sin \dfrac{4\pi}{5}\right)\right]$

81. $\dfrac{10 \text{ cis } 60°}{5 \text{ cis } 10°}$

82. $\dfrac{20(\cos 50° + i \sin 50°)}{30(\cos 40° + i \sin 40°)}$

83. Find one cube root of $\cos 60° + i \sin 60°$.

84. Find one fourth root of $7\sqrt{2} + 7\sqrt{2}i$.

85. Find the three cube roots of 125.

86. Find the four fourth roots of 81.

In Review Exercises 87–90 assume that $\mathbf{V} = \langle 3, 7 \rangle$ *and* $\mathbf{W} = \langle -2, 5 \rangle$. *Calculate each quantity.*

87. $2\mathbf{V} + 3\mathbf{W}$

88. $|3\mathbf{V}| - |\mathbf{V}|$

89. $5(\mathbf{V} - \mathbf{W})$

90. $|3\mathbf{V} - \mathbf{W}|$

In Review Exercises 91–94 find the angle between the two given vectors.

91. $\langle 0, 5 \rangle$ and $\langle 2, 0 \rangle$

92. $\langle 8, 2 \rangle$ and $\langle 4, 1 \rangle$

93. $\sqrt{3}\mathbf{i} + \mathbf{j}$ and $\mathbf{i} - \mathbf{j}$

94. $2\mathbf{i} - 2\sqrt{3}\mathbf{j}$ and $\mathbf{i} + \sqrt{3}\mathbf{j}$

8 SYSTEMS OF EQUATIONS AND INEQUALITIES

Applications of mathematics often require solving several equations involving several variables. In this chapter, we discuss techniques for solving such systems of equations. We consider also the related topics of determinants, systems of inequalities, and linear programming.

8.1 SYSTEMS OF EQUATIONS

Any collection of two or more equations is called a **system of equations**. For example, the equations

$$\begin{cases} 5x - 2y = 1 \\ 2x + 3y = 8 \end{cases}$$

form a system of two linear equations in two variables. The brace indicates that the two equations are to be considered together. To understand what is meant by a **solution** of this system of equations, we consider their graphs in Figure 8–1.

Because the coordinates of any point on the graph of an equation satisfy the equation, there are infinitely many ordered pairs (x, y) that satisfy the first equation of the system. Similarly, the second equation of the system has infinitely many solutions.

There is, however, only one ordered pair of real numbers (x, y) that satisfies *both* equations simultaneously. It is the pair $(1, 2)$, the coordinates of the point of intersection of the two lines. The solution of the system is the pair $(1, 2)$ or, alternatively, $x = 1$ and $y = 2$.

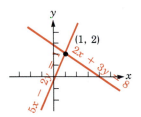

Figure 8–1

Definition. A **system of equations** in the variables x and y is a collection of two or more equations in those variables.

Any ordered pair of real numbers (x, y) that satisfies each of the equations of the system is a **simultaneous solution**, or just a **solution**, of the system.

Two systems of equations that have exactly the same solutions are called **equivalent systems**.

Example 1   Solve each system by estimating the coordinates of the point of intersection of the graphs of the equations of the system.

a. $\begin{cases} 3x + y = 1 \\ -x + 2y = 9 \end{cases}$ b. $\begin{cases} y = 4 - x \\ 2x + 2y = 8 \end{cases}$ c. $\begin{cases} 2x - 3y = 4 \\ 4x = -4 + 6y \end{cases}$

Solution   a.   The graph of each equation is a straight line. See Figure 8–2a. The solution of the system is the ordered pair $(-1, 4)$, given by the coordinates of the point where the lines intersect. We also say that the solution of the system is $x = -1$ and $y = 4$. Verify that these values satisfy each equation.

b.   The graph of each equation is a straight line, and the two lines coincide. See Figure 8–2b. Thus this system has infinitely many solutions. The coordinates of any point on the common line satisfy both equations.

c.   The graph of each equation is a straight line, and the lines are parallel. See Figure 8–2c. Because parallel lines do not intersect, the system has no solution.

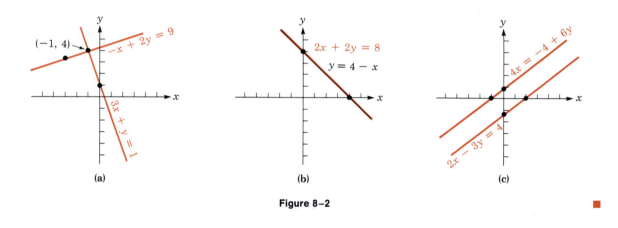

(a)   (b)   (c)

Figure 8–2

If the solutions of a system of equations exist, as in parts **a** and **b** of Example 1, the system is said to be **consistent**. If a system has no solutions, as in part **c**, it is called an **inconsistent** system of equations.

If a system of two linear equations in two variables has exactly one solution as in part **a**, or no solutions as in part **c**, the equations of the system are called **independent**. The equations of part **b** are **dependent**.

Graphing is not an effective method for solving most systems of equations, because it is difficult to read exact values from a graph and because graphing equations in several variables may be impossible. We now consider two algebraic methods that can be used to solve systems of equations exactly.

The Substitution Method

One algebraic method, called the **method of substitution,** can be used to solve a system of two equations in two variables. To do so we proceed as follows:

1. Solve one of the equations for one of the variables, say y.
2. Substitute the expression obtained for y for all occurrences of y in the other equation. This results in a single equation in the other variable, say x.
3. Solve the equation.
4. Substitute the solution found in Step 3 into the equation found in Step 1, and determine y.

Example 2   Use the substitution method to solve the system

$$\begin{cases} 3x + y = 1 \\ -x + 2y = 9 \end{cases}$$

Solution   Solve the first equation for y to obtain $y = 1 - 3x$. Then substitute $1 - 3x$ for y in the second equation:

$$\begin{cases} 3x + y = 1 \\ -x + 2y = 9 \end{cases} \longrightarrow y = \boxed{1 - 3x}$$

This substitution determines one linear equation in one variable. Solve it for x:

$$-x + 2(1 - 3x) = 9$$
$$-x + 2 - 6x = 9$$
$$-7x = 7$$
$$x = -1$$

To find y, substitute -1 for x in the equation $y = 1 - 3x$ and simplify:

$$y = 1 - 3x$$
$$= 1 - 3(-1)$$
$$= 1 + 3$$
$$= 4$$

The solution is $(-1, 4)$. ∎

The Addition Method

The addition method is based on the following three algebraic manipulations that will transform a system of equations into an **equivalent system**, one that has the same solutions as the original system.

1. The positions of any two equations of the system can be interchanged.
2. Both sides of any equation of the system can be multiplied or divided by any nonzero constant.
3. Any equation of the system can be altered by adding to its sides a constant multiple of the corresponding sides of another equation of the system.

Example 3   Use the addition method to solve the system $\begin{cases} 3x + y = 1 \\ -x + 2y = 9 \end{cases}$.

Solution   Multiply both sides of the second equation by 3, and add the result to the first equation to eliminate the variable x:

$$
\begin{array}{r}
3x + y = 1 \\
-3x + 6y = 27 \\
\hline
7y = 28 \\
y = 4
\end{array}
$$

Substitute **4** for y into either of the original equations and solve for x. For example, use the first equation:

$$
\begin{aligned}
3x + \mathbf{y} &= 1 \\
3x + \mathbf{4} &= 1 \\
3x &= -3 \\
x &= -1
\end{aligned}
$$

Thus, the solution of the given system is $(-1, 4)$.   ■

Example 4   Use the addition method to solve the system $\begin{cases} x + 2y = 3 \\ 2x + 4y = 6 \end{cases}$.

Solution   Multiply both sides of the first equation by -2, and add the results to the second equation:

$$
\begin{array}{r}
-2x - 4y = -6 \\
2x + 4y = 6 \\
\hline
0 = 0
\end{array}
$$

Although the result $0 = 0$ is true, it does not lead directly to a solution for y. This result is a clue, however, that the equations of the system are dependent. Note that the second equation is twice the first, so the two equations in this system are essentially the same: they are equivalent. If each equation were graphed, the resulting lines would coincide. The pairs of coordinates of all points on the *one* line described by the system make up the infinite set of solutions to the system. The system is consistent, and the equations are dependent.

To find some of the solutions, solve either equation for y to obtain

$$
y = -\frac{1}{2}x + \frac{3}{2}
$$

Then calculate solutions (x, y) by replacing x with any number in its domain. For example, let $x = -1$. Then

$$
\begin{aligned}
y &= -\frac{1}{2}(-1) + \frac{3}{2} \\
y &= 2
\end{aligned}
$$

Thus, one solution is the pair $(-1, 2)$. As another example, let $x = 2$:

$$y = -\frac{1}{2}(2) + \frac{3}{2}$$

$$= -1 + \frac{3}{2}$$

$$y = \frac{1}{2}$$

Another solution is the pair $(2, \frac{1}{2})$. *All* solutions of the system are of the form $(x, -\frac{1}{2}x + \frac{3}{2})$, where x is any real number.   ∎

Example 5   Solve the system $\begin{cases} x + y = 3 \\ x + y = 2 \end{cases}$.

Solution   Multiply both sides of the second equation by -1, and add the results to the first equation:

$$\begin{array}{r} x + y = 3 \\ -x - y = -2 \\ \hline 0 = 1 \end{array}$$

Because this result is impossible, the system has no solutions. The graphs of the equations in this system are parallel lines. The system is inconsistent, and the equations are independent.   ∎

The following example illustrates that the method of addition can be used to solve some systems of nonlinear equations.

Example 6   Solve the system $\begin{cases} x^2 - \log(y + 1) = 0 \\ x^2 + \log y = \log 12 \end{cases}$.

Solution   To eliminate the term involving x, multiply both sides of the first equation by -1, and add the results to the corresponding sides of the second equation:

$$\begin{array}{rl} -x^2 + \log(y + 1) = 0 & \\ x^2 + \log y = \log 12 & \\ \hline \log y + \log(y + 1) = \log 12 & \\ \log y(y + 1) = \log 12 & \quad \log a + \log b = \log ab. \\ y(y + 1) = 12 & \quad \text{If } \log a = \log b, \text{ then } a = b. \\ y^2 + y - 12 = 0 & \\ (y + 4)(y - 3) = 0 & \\ y + 4 = 0 \quad \text{or} \quad y - 3 = 0 & \\ y = -4 \quad \quad y = 3 & \end{array}$$

To be a solution of the system, the values determined for y must satisfy *both* equations of the system. Because the logarithm of a negative number does not exist, the value $y = -4$ does not satisfy the second equation; it is extraneous and must be discarded. Substitute the other value, $y = 3$, into the first equation and solve for x:

$$x^2 - \log(3 + 1) = 0$$
$$x^2 - \log 4 = 0$$
$$x^2 = \log 4$$
$$x = \sqrt{\log 4} \quad \text{or} \quad x = -\sqrt{\log 4}$$
$$x \approx 0.7759 \quad \bigg| \quad x \approx -0.7759 \qquad \text{Use a calculator.}$$

The system of equations has two solutions, the pairs $(0.7759, 3)$ and $(-0.7759, 3)$. ∎

The substitution and addition methods can be used to solve a system of three equations in three variables.

Example 7   Solve the system

$$
\begin{array}{ll}
\textbf{1.} & \left\{
\begin{aligned}
x + 2y + z &= 8 \\
2x + y - z &= 1 \\
x + y - 2z &= -3
\end{aligned}
\right.
\end{array}
$$

Solution   Use Equations 1 and 2 to eliminate the variable z. Then use Equations 1 and 3 to eliminate the variable z. Because you are eliminating the *same* variable, z, both times, you can then solve the resulting system of two equations in the two variables x and y.

Add Equations 1 and 2 to eliminate the variable z:

$$
\begin{array}{l}
\textbf{1.} \quad \left\{
\begin{aligned}
x + 2y + z &= 8 \\
2x + y - z &= 1
\end{aligned}
\right. \\
\hline
\phantom{\textbf{1.}\,\,} 3x + 3y = 9
\end{array}
$$

Divide both sides of the resulting equation by 3 to obtain

4. $x + y = 3$

The variable z can be eliminated again by using Equations 1 and 3. Multiply both sides of Equation 1 by the constant 2, and add the result to Equation 3:

$$
\begin{array}{l}
\phantom{\textbf{3.}\,\,}\left\{
\begin{aligned}
2x + 4y + 2z &= 16 \\
x + y - 2z &= -3
\end{aligned}
\right. \\
\hline
\textbf{5.} \quad\, 3x + 5y = 13
\end{array}
$$

Equations 4 and 5 form a system of two equations in two variables. Solve this system by substitution as follows:

$$
\begin{array}{ll}
\textbf{4.} & \left\{
\begin{aligned}
x + y &= 3 \\
3x + 5y &= 13
\end{aligned}
\right.
\end{array}
\quad \longrightarrow \quad y = \boxed{3 - x}
$$

$$3x + 5(3 - x) = 13$$
$$3x + 15 - 5x = 13$$
$$-2x = -2$$
$$x = 1$$

Substitute **1** for x in the equation $y = 3 - x$ to find y:

$y = 3 - 1$
$y = 2$

To find z, substitute 1 for x, and 2 for y in any of the original equations. You will find that $z = 3$. The solution to the given system is the triple

$(1, 2, 3)$

Because this solution is unique, the given system is consistent and the equations in the system are independent. Verify that $x = 1$, $y = 2$, and $z = 3$ satisfy each equation in the original system. ■

Example 8   An airplane flies 600 miles with the wind for 2 hours. The return trip against the wind takes 3 hours. Find the speed of the wind and the airspeed of the plane.

Solution   Let a represent the airspeed of the plane, and w represent the speed of the wind. The ground speed of the plane on the outbound trip is the combined speed $a + w$. On the return trip, against a head wind, the ground speed is $a - w$. The information of this problem, organized in the chart in Figure 8–3, gives a system of two equations in the two variables a and w.

	d	=	r	$\cdot$	t
outbound trip	600		$a + w$		2
return trip	600		$a - w$		3

Figure 8–3

Because $d = rt$, you have

$$\begin{cases} 600 = 2(a + w) \\ 600 = 3(a - w) \end{cases}$$

or

$$\begin{cases} 300 = a + w \\ 200 = a - w \end{cases}$$

Add the equations together to get

$500 = 2a$

or

$a = 250$

To calculate w, substitute **250** for a in a previous equation such as $300 = a + w$ and solve for w:

$$300 = \boldsymbol{a} + w$$
$$300 = \boldsymbol{250} + w$$
$$w = 50$$

The plane could do 250 miles per hour in still air. With a 50-mile-per-hour tail wind, the ground speed would be 300 miles per hour, and the 600-mile trip would take 2 hours. With a 50-mile-per-hour head wind, the ground speed would be 200 miles per hour, and the 600-mile trip would take 3 hours. The answers check. ■

Exercise 8.1

In Exercises 1–4 solve each system of equations by the graphing method.

1. $\begin{cases} 3x + y = 5 \\ x - 2y = -3 \end{cases}$

2. $\begin{cases} x - 2y = -3 \\ 3x + y = -9 \end{cases}$

3. $\begin{cases} \dfrac{3x}{2} + y = 1 \\ -\dfrac{x}{8} + \dfrac{3y}{16} = 1 \end{cases}$

4. $\begin{cases} x + y = 0 \\ 7x + 8y = 1 \end{cases}$

In Exercises 5–12 solve each system of equations by the substitution method, if possible.

5. $\begin{cases} y = x \\ y = 2x \end{cases}$

6. $\begin{cases} 2y = x \\ 3y = 2x \end{cases}$

7. $\begin{cases} 2x + 3y = 0 \\ y = 3x - 11 \end{cases}$

8. $\begin{cases} 2x + y = 3 \\ y = 5x - 11 \end{cases}$

9. $\begin{cases} \dfrac{5}{2}x + y = 8 \\ 2x - \dfrac{2}{3}y = 2 \end{cases}$

10. $\begin{cases} \dfrac{x + y}{3} - \dfrac{x - y}{2} = 1 \\ x = 2y \end{cases}$

11. $\begin{cases} x + 3y = 1 \\ 2x + 6y = 3 \end{cases}$

12. $\begin{cases} x - 3y = 14 \\ 3(x - 12) = 9y \end{cases}$

In Exercises 13–24 solve each system of equations by the addition method, if possible.

13. $\begin{cases} 5x - 3y = 12 \\ 2x - 3y = 3 \end{cases}$

14. $\begin{cases} 2x + 3y = 8 \\ -5x + y = -3 \end{cases}$

15. $\begin{cases} x - 7y = -11 \\ 8x + 2y = 28 \end{cases}$

16. $\begin{cases} 3x + 9y = 9 \\ -x + 5y = -3 \end{cases}$

17. $\begin{cases} 3(x - y) = y - 9 \\ 5(x + y) = -15 \end{cases}$

18. $\begin{cases} 2(x + y) = y + 1 \\ 3(x + 1) = y - 3 \end{cases}$

19. $\begin{cases} 2 = \dfrac{1}{x + y} \\ 2 = \dfrac{3}{x - y} \end{cases}$

20. $\begin{cases} \dfrac{1}{x + y} = 12 \\ \dfrac{3x}{y} = -4 \end{cases}$

21. $\begin{cases} 5x = 10y \\ -x + 2y = 0 \end{cases}$

22. $\begin{cases} -0.3x + 0.1y = -0.1 \\ 6x - 2y = 2 \end{cases}$

23. $\begin{cases} \dfrac{x + y}{2} + \dfrac{x - y}{5} = 2 \\ 2x = y + 2 \end{cases}$

24. $\begin{cases} \dfrac{3}{2}x + \dfrac{1}{3}y = 2 \\ \dfrac{2}{3}x + \dfrac{1}{9}y = 1 \end{cases}$

In Exercises 25–44 solve each system of equations, if possible.

25. $\begin{cases} x + y + z = 3 \\ 2x + y + z = 4 \\ 3x + y - z = 5 \end{cases}$

26. $\begin{cases} x - y - z = 0 \\ x + y - z = 0 \\ x - y + z = 2 \end{cases}$

27. $\begin{cases} x - y + z = 0 \\ x + y + 2z = -1 \\ -x - y + z = 0 \end{cases}$

28. $\begin{cases} 2x + y - z = 7 \\ x - y + z = 2 \\ x + y - 3z = 2 \end{cases}$

29. $\begin{cases} 2x + y = 4 \\ x - z = 2 \\ y + z = 1 \end{cases}$

30. $\begin{cases} 3x + y + z = 0 \\ 2x - y + z = 0 \\ 2x + y + z = 0 \end{cases}$

31. $\begin{cases} x + y + z = 6 \\ 2x + y + 3z = 17 \\ x + y + 2z = 11 \end{cases}$

32. $\begin{cases} x + y + z = 3 \\ 2x + y + z = 6 \\ x + 2y + 3z = 2 \end{cases}$

33. $\begin{cases} x + y + z = 3 \\ x + z = 2 \\ 2x + 2y + 2z = 3 \end{cases}$

34. $\begin{cases} x + y + z = 3 \\ x + z = 2 \\ 2x + y + 2z = 5 \end{cases}$

35. $\begin{cases} x + y = 2 \\ y + z = 2 \\ x - z = 0 \end{cases}$

36. $\begin{cases} x + y = 2 \\ y + z = 2 \\ 3x + 3y = 2 \end{cases}$

37. $\begin{cases} x + y + z = 4 \\ 2x + y + z = 5 \\ 3x + 2y + z = 7 \end{cases}$

38. $\begin{cases} x + 2y - z = 2 \\ 2x - y = -1 \\ 3x + y + z = 1 \end{cases}$

39. $\begin{cases} (x + y) + (y + z) + (z + x) = 6 \\ (x - y) + (y - z) + (z - x) = 0 \\ x + y + 2z = 4 \end{cases}$

40. $\begin{cases} (x + y) + (y + z) = 1 \\ (x + z) + (x + y) = 3 \\ (x - y) - (x - z) = -1 \end{cases}$

41. $\begin{cases} x + 2 \log y = 5 \\ 3x - \log y = 8 \end{cases}$

42. $\begin{cases} 2 \sin x - y = 0 \\ 4 \sin x + y = 6 \end{cases}$

43. $\begin{cases} y = 2 - 3 \cos x \\ 2 \cos x = 4y - 1 \end{cases}$

44. $\begin{cases} 3 \tan x + 2 \ln y = 5 \\ \tan x - \ln y = 0 \end{cases}$

45. Suppose that a retailer obtains a product from three sources, A, B, and C. The retailer buys as many units from source A as from the other two combined. She must pay source A at the rate of $4 per unit and sources B and C at the rate of $5 per unit. If she requires 100 units per month to satisfy customer demand, and if her cost for one month's supply is $450, how many units does she buy from each source?

46. Flutter Hi-Fi, Wow Stereo, and Rumble Electronics buy a total of 175 cassette tape decks from High Hiss Distributors each month. Because Rumble Electronics buys 25 more tape decks than the other two stores combined, Rumble's cost is only $160 per unit. The decks cost Wow Stereo $165 each and Flutter Hi-Fi $170 each. How many decks does each retailer buy each month if High Hiss receives $28,500 each month from the sale of tape decks to these companies?

47. A college student earns $142.50 per week working three part-time jobs. Half of the student's 30-hour work week is spent cooking hamburgers at a fast food chain, earning $4.50 per hour. In addition, the student earns $4.75 per hour working at a gas station and $5.50 per hour doing janitorial work. How many hours per week does the student work at each job?

48. A factory manufactures widgets, gidgets, and gadgets at a monthly cost of $6850 for 2150 units. It costs $2 to make a widget, $3 to make a gidget, and $4 to make a gadget. A widget sells for $3, a gidget for $4.50, and a gadget for $5.50. The monthly profit is $2975. How many of each item are manufactured?

49. A collection of nickels, dimes, and quarters has a value of $3.40. There are twice as many dimes as quarters, and there are 32 coins in all. How many of each kind are there?

50. The sum of the angles of a triangle is $180°$. The largest angle is $20°$ greater than the sum of the other two, and is $10°$ greater than 3 times the smallest. How large is each angle?

8.2 GAUSSIAN ELIMINATION AND MATRIX METHODS

A method called **Gaussian elimination** can be used to solve a system of linear equations. In this method, the system of equations is transformed into an equivalent system that can be solved by a process called **back substitution**.

In Gaussian elimination, it is convenient to solve the system by working with only the coefficients of the variables. All of the information needed to find a solution of

$$\begin{cases} x + 2y + z = 8 \\ 2x + y - z = 1 \\ x + y - 2z = -3 \end{cases}$$

Carl Friedrich Gauss
(1777–1855) Many people consider Gauss to be the greatest mathematician of all time. He often is called the "prince of the mathematicians."

for example, is contained in the following rectangular array of numbers, called a **matrix**.

$$\begin{bmatrix} 1 & 2 & 1 & \vdots & 8 \\ 2 & 1 & -1 & \vdots & 1 \\ 1 & 1 & -2 & \vdots & -3 \end{bmatrix}$$

Each row in this matrix represents one of the equations of the system. The first row, for example, represents the first equation, $1x + 2y + 1z = 8$.

Because the matrix above has three rows and four columns, it is called a 3×4 (read as "3 by 4") matrix. The 3×3 matrix to the left of the broken line is called the **coefficient matrix**. The entire matrix is called the **augmented matrix**.

Example 1   Use Gaussian elimination to solve the system

1. $\begin{cases} x + 2y + z = 8 \\ \textbf{2.} \quad 2x + y - z = 1 \\ \textbf{3.} \quad x + y - 2z = -3 \end{cases}$

Solution   First multiply each term in Equation 1 by -2, and add the result to Equation 2 to obtain Equation 4 below. Then multiply each term of Equation 1 by -1, and add the result to Equation 3 to obtain Equation 5. This gives an equivalent system, one with the same solution as the original system. The process is shown both in equation form and in matrix form. Note that the variable x does not appear in Equations 4 or 5:

1. $\begin{cases} x + 2y + z = 8 \\ \textbf{4.} \quad -3y - 3z = -15 \\ \textbf{5.} \quad -y - 3z = -11 \end{cases}$ $\begin{bmatrix} 1 & 2 & 1 & \vdots & 8 \\ 0 & -3 & -3 & \vdots & -15 \\ 0 & -1 & -3 & \vdots & -11 \end{bmatrix}$

Divide both sides of Equation 4 by -3 to obtain Equation 6:

1. $\begin{cases} x + 2y + z = 8 \\ \textbf{6.} \quad y + z = 5 \\ \textbf{5.} \quad -y - 3z = -11 \end{cases}$ $\begin{bmatrix} 1 & 2 & 1 & \vdots & 8 \\ 0 & 1 & 1 & \vdots & 5 \\ 0 & -1 & -3 & \vdots & -11 \end{bmatrix}$

Add Equation 6 to Equation 5 to obtain Equation 7:

1. $\begin{cases} x + 2y + z = 8 \\ \textbf{6.} \quad y + z = 5 \\ \textbf{7.} \quad -2z = -6 \end{cases}$ $\begin{bmatrix} 1 & 2 & 1 & \vdots & 8 \\ 0 & 1 & 1 & \vdots & 5 \\ 0 & 0 & -2 & \vdots & -6 \end{bmatrix}$

Divide both sides of Equation 7 by -2 to obtain the system

1. $\begin{cases} x + 2y + z = 8 \\ y + z = 5 \\ z = 3 \end{cases}$ $\begin{bmatrix} 1 & 2 & 1 & | & 8 \\ 0 & 1 & 1 & | & 5 \\ 0 & 0 & 1 & | & 3 \end{bmatrix}$
6.

The system can now be solved by back substitution. Because $z = 3$, you can substitute 3 for z in Equation 6, and solve for y:

6. $y + z = 5$
$y + 3 = 5$
$y = 2$

Substitute 2 for y and 3 for z in Equation 1, and solve for x:

1. $x + 2y + z = 8$
$x + 2(2) + 3 = 8$
$x = 1$

The solution of the given system of equations is the ordered triple

$(1, 2, 3)$

Verify that $x = 1$, $y = 2$, and $z = 3$ satisfy each of the original three equations.   ■

Note the triangular formation of 0s in the final matrix in Example 1. A matrix such as

$$\begin{bmatrix} 1 & 2 & 1 & | & 8 \\ 0 & 1 & 1 & | & 5 \\ 0 & 0 & 1 & | & 3 \end{bmatrix}$$

is said to be in **triangular form** if all entries below the diagonal running from the upper left to the lower right are 0.

Example 2   Use matrices to solve the system

$$\begin{cases} x + 2y + 3z = 4 \\ 2x - y - 2z = 0 \\ x - 3y - 3z = -2 \end{cases}$$

Solution   This system is represented by the augmented matrix

$$\begin{bmatrix} 1 & 2 & 3 & | & 4 \\ 2 & -1 & -2 & | & 0 \\ 1 & -3 & -3 & | & -2 \end{bmatrix}$$

Change the individual rows of this matrix by adding multiples of one row to another. When combining rows in this way, you are adding multiples of the coefficients of one equation to the corresponding coefficients of another, always obtaining coefficients that form an equivalent set of equations. The goal is to produce a matrix in triangular form that represents an equivalent system of equations.

To begin, use the 1 in the upper left-hand corner of the augmented matrix to "zero out" the rest of the first column. The notation "$(-2)R1 + R2 \rightarrow R2$" means "multiply row one by -2 and add the result to row two to get a new row two."

$$(-2)R1 + R2 \rightarrow R2$$

$$\begin{bmatrix} 1 & 2 & 3 & | & 4 \\ 2 & -1 & -2 & | & 0 \\ 1 & -3 & -3 & | & -2 \end{bmatrix} \Leftrightarrow \begin{bmatrix} 1 & 2 & 3 & | & 4 \\ 0 & -5 & -8 & | & -8 \\ 1 & -3 & -3 & | & -2 \end{bmatrix}$$

Read $\Leftrightarrow$ as "is equivalent to."

Next, multiply row one by -1 and add the result to row three to get a new row three:

$$(-1)R1 + R3 \rightarrow R3$$

$$\begin{bmatrix} 1 & 2 & 3 & | & 4 \\ 0 & -5 & -8 & | & -8 \\ 1 & -3 & -3 & | & -2 \end{bmatrix} \Leftrightarrow \begin{bmatrix} 1 & 2 & 3 & | & 4 \\ 0 & -5 & -8 & | & -8 \\ 0 & -5 & -6 & | & -6 \end{bmatrix}$$

Because each row of a matrix represents an equation, multiplying an entire row by a constant is equivalent to multiplying both sides of an equation by a constant. Hence, multiply row two by -1 to eliminate some minus signs and add the result to row three to get

$$(-1)R2 \rightarrow R2 \qquad\qquad R2 + R3 \rightarrow R3$$

$$\begin{bmatrix} 1 & 2 & 3 & | & 4 \\ 0 & -5 & -8 & | & -8 \\ 0 & -5 & -6 & | & -6 \end{bmatrix} \Leftrightarrow \begin{bmatrix} 1 & 2 & 3 & | & 4 \\ 0 & 5 & 8 & | & 8 \\ 0 & -5 & -6 & | & -6 \end{bmatrix} \Leftrightarrow \begin{bmatrix} 1 & 2 & 3 & | & 4 \\ 0 & 5 & 8 & | & 8 \\ 0 & 0 & 2 & | & 2 \end{bmatrix}$$

Finally, multiply the third row by $\frac{1}{2}$:

$$(\tfrac{1}{2})R3 \rightarrow R3$$

$$\begin{bmatrix} 1 & 2 & 3 & | & 4 \\ 0 & 5 & 8 & | & 8 \\ 0 & 0 & 2 & | & 2 \end{bmatrix} \Leftrightarrow \begin{bmatrix} 1 & 2 & 3 & | & 4 \\ 0 & 5 & 8 & | & 8 \\ 0 & 0 & 1 & | & 1 \end{bmatrix}$$

This final matrix is in triangular form and represents the system

$$\begin{cases} \textbf{1.} & x + 2y + 3z = 4 \\ \textbf{2.} & 5y + 8z = 8 \\ \textbf{3.} & z = 1 \end{cases}$$

To solve this system by back substitution, substitute **1** for z in Equation 2, and solve for y:

2. $\quad 5y + 8z = 8$

$\qquad 5y + 8(\mathbf{1}) = 8$

$\qquad\qquad y = 0$

Then, substitute **0** for y and **1** for z in Equation 1, and solve for x:

1. $\qquad x + 2y + 3z = 4$

$\qquad x + 2(\mathbf{0}) + 3(\mathbf{1}) = 4$

$\qquad\qquad x = 1$

The solution of the original system is the ordered triple

$(1, 0, 1)$

Verify that $x = 1$, $y = 0$, and $z = 1$ satisfy each of the original equations. ∎

In the preceding example, one matrix was transformed into another by certain manipulations called **elementary row operations**.

Elementary Row Operations.

Type 1 row operation.   Two rows of a matrix can be interchanged.

Type 2 row operation.   The elements of a row of a matrix can be multiplied by a nonzero constant.

Type 3 row operation.   A row of a matrix can be altered by adding to it a multiple of any other row.

Each type of elementary row operation yields a new matrix representing a system of equations with the same solution as the original system. A type 1 row operation is equivalent to writing the equations of a system in a different order. A type 2 row operation is equivalent to multiplying both sides of an equation by a nonzero constant, and a type 3 row operation is equivalent to adding a multiple of one equation to another.

In a type 3 row operation, it is important to remember which row is being changed and which row is causing that change. For example, if twice row three is added to row one, it is row one that is being altered. Row three, which was used to accomplish the change, stays the same.

Any matrix that can be obtained from another matrix by a sequence of elementary row operations is called **row equivalent** to the original matrix. Thus, the symbol ⇔ can be read as "is row equivalent to."

If a matrix has the following properties, it is said to be in **echelon form**:

1. The lead entry (the first nonzero entry) of each row is 1.
2. Lead entries appear farther to the right as you move down the rows of the matrix.
3. Rows containing only 0s are at the bottom of the matrix.

Example 3   Use matrix methods to solve the following system of four equations in three variables:

$$\begin{cases} x + 2y - z = 6 \\ x - y + z = -2 \\ 2x \quad\quad + z = 1 \\ x + y + z = 2 \end{cases}$$

Solution   Note that this system has more equations than it has variables. To solve it, form the augmented matrix and use elementary row operations to write the augmented matrix in echelon form.

$(-1)R1 + R2 \rightarrow R2$

$$\begin{bmatrix} 1 & 2 & -1 & | & 6 \\ 1 & -1 & 1 & | & -2 \\ 2 & 0 & 1 & | & 1 \\ 1 & 1 & 1 & | & 2 \end{bmatrix} \Leftrightarrow \begin{bmatrix} 1 & 2 & -1 & | & 6 \\ 0 & -3 & 2 & | & -8 \\ 2 & 0 & 1 & | & 1 \\ 1 & 1 & 1 & | & 2 \end{bmatrix}$$

$(-2)R1 + R3 \rightarrow R3$
$(-1)R1 + R4 \rightarrow R4$

$$\Leftrightarrow \begin{bmatrix} 1 & 2 & -1 & | & 6 \\ 0 & -3 & 2 & | & -8 \\ 0 & -4 & 3 & | & -11 \\ 0 & -1 & 2 & | & -4 \end{bmatrix}$$

$(-1)R4 \rightarrow R4$

$$\Leftrightarrow \begin{bmatrix} 1 & 2 & -1 & | & 6 \\ 0 & -3 & 2 & | & -8 \\ 0 & -4 & 3 & | & -11 \\ 0 & 1 & -2 & | & 4 \end{bmatrix}$$

$(4)R4 + R3 \rightarrow R3$
$(3)R4 + R2 \rightarrow R2$

$$\Leftrightarrow \begin{bmatrix} 1 & 2 & -1 & | & 6 \\ 0 & 0 & -4 & | & 4 \\ 0 & 0 & -5 & | & 5 \\ 0 & 1 & -2 & | & 4 \end{bmatrix}$$

$(-\frac{1}{4})R2 \rightarrow R2$
$(-\frac{1}{5})R3 \rightarrow R3$

$$\Leftrightarrow \begin{bmatrix} 1 & 2 & -1 & | & 6 \\ 0 & 0 & 1 & | & -1 \\ 0 & 0 & 1 & | & -1 \\ 0 & 1 & -2 & | & 4 \end{bmatrix}$$

$R2 \leftrightarrow R4$ (exchange $R2$ and $R4$)

$$\Leftrightarrow \begin{bmatrix} 1 & 2 & -1 & | & 6 \\ 0 & 1 & -2 & | & 4 \\ 0 & 0 & 1 & | & -1 \\ 0 & 0 & 1 & | & -1 \end{bmatrix}$$

$(-1)R3 + R4 \rightarrow R4$

$$\Leftrightarrow \begin{bmatrix} 1 & 2 & -1 & | & 6 \\ 0 & 1 & -2 & | & 4 \\ 0 & 0 & 1 & | & -1 \\ 0 & 0 & 0 & | & 0 \end{bmatrix}$$

This final matrix is in echelon form. All lead entries are 1, the lead entries appear farther to the right as you move down the rows of the matrix, and the row of 0s is last. This matrix represents the following system of equations, which can be solved by back substitution:

$$\begin{array}{rl} \textbf{1.} & \\ \textbf{2.} & \\ \textbf{3.} & \\ \textbf{4.} & \end{array} \left\{ \begin{array}{rcrcr} x & + & 2y & - & z & = & 6 \\ & & y & - & 2z & = & 4 \\ & & & & z & = & -1 \\ 0x & + & 0y & - & 0z & = & 0 \end{array} \right.$$

Because Equation 4 is satisfied by *all* numbers x, y, and z, it is unnecessary and can be ignored. From Equation 3, you know that $z = -1$. So, substitute -1 for z in Equation 2 and solve for y:

$$\begin{aligned} \textbf{2.} \qquad y - 2z &= 4 \\ y - 2(-1) &= 4 \\ y + 2 &= 4 \\ y &= 2 \end{aligned}$$

Then, substitute -1 for z and 2 for y in Equation 1 and solve for x:

$$\begin{aligned} \textbf{1.} \qquad x + 2y - z &= 6 \\ x + 2(2) - (-1) &= 6 \\ x + 4 + 1 &= 6 \\ x &= 1 \end{aligned}$$

Thus, the solution of the original system of equations is the ordered triple

$$(x, y, z) = (1, 2, -1)$$

The system is consistent because it has a solution. Check this solution to verify that it satisfies each of the four original equations. ■

Example 4 Use matrices to solve the system

$$\left\{ \begin{array}{rcrcrcr} x & + & 2y & + & z & = & 8 \\ 2x & + & y & - & z & = & 1 \\ x & - & y & - & 2z & = & -7 \end{array} \right.$$

Solution Set up the augmented matrix for the system, and use row operations to reduce it to echelon form:

$$(-2)R1 + R2 \rightarrow R2$$
$$(-1)R1 + R3 \rightarrow R3$$

$$\left[\begin{array}{ccc|c} 1 & 2 & 1 & 8 \\ 2 & 1 & -1 & 1 \\ 1 & -1 & -2 & -7 \end{array} \right] \Leftrightarrow \left[\begin{array}{ccc|c} 1 & 2 & 1 & 8 \\ 0 & -3 & -3 & -15 \\ 0 & -3 & -3 & -15 \end{array} \right]$$

$$(-\tfrac{1}{3})R2 \rightarrow R2$$
$$(-\tfrac{1}{3})R3 \rightarrow R3$$

$$\Leftrightarrow \left[\begin{array}{ccc|c} 1 & 2 & 1 & 8 \\ 0 & 1 & 1 & 5 \\ 0 & 1 & 1 & 5 \end{array} \right]$$

$$(-1)R2 + R3 \rightarrow R3$$

$$\Leftrightarrow \begin{bmatrix} 1 & 2 & 1 & | & 8 \\ 0 & 1 & 1 & | & 5 \\ 0 & 0 & 0 & | & 0 \end{bmatrix}$$

This matrix is in echelon form and represents the system

$$\begin{array}{ll} \textbf{1.} \\ \textbf{2.} \\ \textbf{3.} \end{array} \left\{ \begin{array}{l} x + 2y + z = 8 \\ y + z = 5 \\ 0x + 0y + 0z = 0 \end{array} \right.$$

As before, the bottom equation can be ignored. Solve this system by back substitution. First, solve Equation 2 for the variable y:

$$y = 5 - z$$

Then, substitute $\mathbf{5 - z}$ for y in Equation 1 and solve for x:

$$\begin{array}{ll} \textbf{1.} & x + 2y + z = 8 \\ & x + 2(\mathbf{5 - z}) + z = 8 \\ & x + 10 - z = 8 \\ & x = -2 + z \end{array}$$

The solution of this system is

$$(x, y, z) = (-2 + z, 5 - z, z)$$

There is no unique solution for this system. The variable z can be any real number, but once you pick a number z, the numbers x and y are determined. For example, if z were equal to 3, then x would equal $-2 + 3$, or 1, and y would equal $5 - 3$, or 2. Thus, one possible solution to this system is $x = 1$, $y = 2$, and $z = 3$. Picking $z = 2$ gives another solution: (0, 3, 2). Because this system has infinitely many solutions, it is consistent, but the equations are dependent. ∎

Example 5   Use matrices to solve the system

$$\left\{ \begin{array}{l} w + x + y + z = 3 \\ x + 2y - z = -2 \\ w - y - z = 2 \\ 2w + x = 2 \end{array} \right.$$

Solution   Row reduce the augmented matrix as follows:

$$(-1)R1 + R3 \rightarrow R3$$

$$\begin{bmatrix} 1 & 1 & 1 & 1 & | & 3 \\ 0 & 1 & 2 & -1 & | & -2 \\ 1 & 0 & -1 & -1 & | & 2 \\ 2 & 1 & 0 & 0 & | & 2 \end{bmatrix} \Leftrightarrow \begin{bmatrix} 1 & 1 & 1 & 1 & | & 3 \\ 0 & 1 & 2 & -1 & | & -2 \\ 0 & -1 & -2 & -2 & | & -1 \\ 2 & 1 & 0 & 0 & | & 2 \end{bmatrix}$$

$$(-1)R3 \rightarrow R3$$
$$(-2)R1 + R4 \rightarrow R4$$

$$\Leftrightarrow \begin{bmatrix} 1 & 1 & 1 & 1 & | & 3 \\ 0 & 1 & 2 & -1 & | & -2 \\ 0 & 1 & 2 & 2 & | & 1 \\ 0 & -1 & -2 & -2 & | & -4 \end{bmatrix}$$

$$R3 + R4 \rightarrow R4$$

$$\Leftrightarrow \begin{bmatrix} 1 & 1 & 1 & 1 & | & 3 \\ 0 & 1 & 2 & -1 & | & -2 \\ 0 & 1 & 2 & 2 & | & 1 \\ 0 & 0 & 0 & 0 & | & -3 \end{bmatrix}$$

The last row of the final matrix represents the equation

$$0w + 0x + 0y + 0z = -3$$

Obviously, *no* values of w, x, y, and z could make $0 = -3$. Because a solution must satisfy *each* equation, the given system has no solution. Hence, it is inconsistent. ■

Exercise 8.2

In Exercises 1–28 solve each system of equations using matrix methods.

1. $\begin{cases} 2x + y = 3 \\ x - 3y = 5 \end{cases}$

2. $\begin{cases} x + 2y = -1 \\ 3x - 5y = 19 \end{cases}$

3. $\begin{cases} x - 7y = -2 \\ 5x - 2y = -10 \end{cases}$

4. $\begin{cases} 3x - y = 3 \\ 2x + y = -3 \end{cases}$

5. $\begin{cases} 2x - y = 5 \\ x + 3y = 6 \end{cases}$

6. $\begin{cases} 3x - 5y = -25 \\ 2x + y = 5 \end{cases}$

7. $\begin{cases} x - 2y = 3 \\ -2x + 4y = 6 \end{cases}$

8. $\begin{cases} 3x - y = 7 \\ -x + \dfrac{1}{3}y = -\dfrac{7}{3} \end{cases}$

9. $\begin{cases} x - y + z = 3 \\ 2x - y + z = 4 \\ x + 2y - z = -1 \end{cases}$

10. $\begin{cases} 2x + y - z = 1 \\ x + y - z = 0 \\ 3x + y + 2z = 2 \end{cases}$

11. $\begin{cases} x + y - z = -1 \\ 3x + y = 4 \\ y - 2z = -4 \end{cases}$

12. $\begin{cases} 3x + y = 7 \\ x - z = 0 \\ y - 2z = -8 \end{cases}$

13. $\begin{cases} x - y + z = 2 \\ 2x + y + z = 5 \\ 3x - 4z = -5 \end{cases}$

14. $\begin{cases} x + z = -1 \\ 3x + y = 2 \\ 2x + y + 5z = 3 \end{cases}$

15. $\begin{cases} x + y + 2z = 4 \\ -x - y - 3z = -5 \\ 2x + y + z = 2 \end{cases}$

16. $\begin{cases} 2x - y + z = 6 \\ 3x + y - z = 2 \\ -x + 3y - 3z = 8 \end{cases}$

17. $\begin{cases} x + y = -2 \\ 3x - y = 6 \\ 2x + 2y = -4 \\ x - y = 4 \end{cases}$

18. $\begin{cases} x - y = -3 \\ 2x + y = -3 \\ 3x - y = -7 \\ 4x + y = -7 \end{cases}$

19. $\begin{cases} x + 2y + z = 4 \\ 3x - y - z = 2 \end{cases}$

20. $\begin{cases} x + 2y - 3z = -5 \\ 5x + y - z = -11 \end{cases}$

21. $\begin{cases} w + x - y + z = 2 \\ 2w - x - 2y + z = 0 \\ w - 2x - y + z = -1 \end{cases}$

22. $\begin{cases} w + x \qquad\quad = 1 \\ w \qquad + y \qquad = 0 \\ \quad x \qquad + z = 0 \end{cases}$

23. $\begin{cases} x + 2y + \ z = 4 \\ x - \ y + \ z = 1 \\ 2x + \ y + 2z = 2 \end{cases}$

24. $\begin{cases} x + \ y = 3 \\ 2x + \ y = 1 \\ 3x + 2y = 2 \end{cases}$

25. $\begin{cases} 2x - 2y + 3z + \ t = 2 \\ x + \ y + \ z + \ t = 5 \\ -x + 2y - 3z + 2t = 2 \\ x + \ y + 2z - \ t = 4 \end{cases}$

26. $\begin{cases} x + \ y + 2z + \ t = 1 \\ x + 2y + \ z + \ t = 2 \\ 2x + \ y + \ z + \ t = 4 \\ x + \ y + \ z + 2t = 3 \end{cases}$

27. $\begin{cases} x + \ y \qquad + \ t = \ 4 \\ x \qquad + z + \ t = \ 2 \\ 2x + 2y + z + 2t = \ 8 \\ x - \ y + z - \ t = -2 \end{cases}$

28. $\begin{cases} x - \ y + 2z + \ t = \ 3 \\ 3x - 2y - \ z - \ t = \ 4 \\ 2x + \ y + 2z - \ t = 10 \\ x + 2y + \ z - 3t = \ 8 \end{cases}$

In Exercises 29–32 solve each system of equations using matrix methods.

29. $\begin{cases} x^2 + \ y^2 + \ z^2 = \ 14 \\ 2x^2 + 3y^2 - 2z^2 = -7 \\ x^2 - 5y^2 + \ z^2 = \ 8 \end{cases}$

(*Hint:* Solve first as a linear system in x^2, y^2, and z^2.)

30. $\begin{cases} \dfrac{3}{x} + \dfrac{1}{y} + \dfrac{1}{z} = \ 4 \\ \dfrac{1}{x} - \dfrac{3}{y} - \dfrac{2}{z} = -3 \\ \dfrac{7}{x} - \dfrac{9}{y} + \dfrac{3}{z} = -14 \end{cases}$

31. $\begin{cases} 5\sqrt{x} + 2\sqrt{y} + \ \sqrt{z} = 22 \\ \sqrt{x} + \ \sqrt{y} - \ \sqrt{z} = \ 5 \\ 3\sqrt{x} - 2\sqrt{y} - 3\sqrt{z} = 10 \end{cases}$

32. $\begin{cases} \sqrt{x} + \dfrac{2}{y} - \ z^2 = 0 \\ 2\sqrt{x} - \dfrac{5}{y} + \ z^2 = 3 \\ -\sqrt{x} + \dfrac{1}{y} + 2z^2 = 7 \end{cases}$

8.3 MATRIX ALGEBRA

It is possible to treat matrices as mathematical entities that, like numbers, can be added, subtracted, and multiplied. We now develop rules for doing this arithmetic, as we begin the study of **matrix algebra**.

> **Definition.** An $m \times n$ **matrix** is a rectangular array of mn numbers arranged in m rows and n columns.

Arthur Cayley (1821–1895)
Cayley was a major founder
of the theory of matrices.

A matrix with n rows and n columns is called a **square matrix of order n**.

To indicate a particular element of a matrix, we use **double subscript notation**. The symbol a_{23} represents the entry in row two, column three, of the matrix A. Similarly, the symbol a_{ij} represents the entry in the ith row and the jth column of A.

We will use any of the following notations to denote the matrix A:

$$A_{m \times n}, \quad A, \quad [a_{ij}], \quad \begin{bmatrix} a_{11} & a_{12} & a_{13} & \cdots & a_{1n} \\ a_{21} & a_{22} & a_{23} & \cdots & a_{2n} \\ \vdots & & & & \vdots \\ a_{m1} & a_{m2} & a_{m3} & \cdots & a_{mn} \end{bmatrix}$$

Definition of Equality of Matrices. If $A = [a_{ij}]$ and $B = [b_{ij}]$ are both $m \times n$ matrices, then

$$A = B \quad \text{if and only if} \quad a_{ij} = b_{ij}$$

for all i and j, where $i = 1, 2, 3, \ldots, m$ and $j = 1, 2, 3, \ldots, n$.

The previous definition points out that two matrices must be identical to be equal. They must be the same size and have the same corresponding entries.

The Sum of Two Matrices. Let $A = [a_{ij}]$ and $B = [b_{ij}]$ be two $m \times n$ matrices. Their sum, $A + B$, is the $m \times n$ matrix C, found by adding the corresponding entries of matrices A and B:

$$A + B = C = [c_{ij}]$$

where $c_{ij} = a_{ij} + b_{ij}$, for $i = 1, 2, 3, \ldots, m$ and $j = 1, 2, 3, \ldots, n$.

Example 1 Add the matrices

$$\begin{bmatrix} 2 & 1 & 3 \\ 1 & -1 & 0 \end{bmatrix} \quad \text{and} \quad \begin{bmatrix} 1 & -1 & 2 \\ -1 & 1 & 5 \end{bmatrix}$$

Solution Because each matrix is 2×3, their sum is defined and can be calculated by adding their corresponding elements:

$$\begin{bmatrix} 2 & 1 & 3 \\ 1 & -1 & 0 \end{bmatrix} + \begin{bmatrix} 1 & -1 & 2 \\ -1 & 1 & 5 \end{bmatrix} = \begin{bmatrix} 2+1 & 1-1 & 3+2 \\ 1-1 & -1+1 & 0+5 \end{bmatrix}$$

$$= \begin{bmatrix} 3 & 0 & 5 \\ 0 & 0 & 5 \end{bmatrix} \quad \blacksquare$$

Example 2 If possible, add the matrices

$$\begin{bmatrix} 2 & 4 & 3 \\ 1 & 1 & 1 \end{bmatrix} \quad \text{and} \quad \begin{bmatrix} 1 & 2 \\ 2 & 3 \end{bmatrix}$$

Solution   The first matrix is 2×3 and the second is 2×2. Because these matrices are of different sizes, they cannot be added.   ∎

Several of the properties of real numbers discussed in Chapter 1 apply to matrices also.

Theorem.   The addition of two $m \times n$ matrices is commutative.

Proof   Let $A = [a_{ij}]$ and $B = [b_{ij}]$ be $m \times n$ matrices. Then $A + B = C = [c_{ij}]$, where $c_{ij} = a_{ij} + b_{ij}$ for each $i = 1, 2, 3, \ldots, m$ and $j = 1, 2, 3, \ldots, n$. On the other hand, $B + A = D = [d_{ij}]$, where $d_{ij} = b_{ij} + a_{ij}$ for each i and j.

Because each entry in each matrix is a real number and the addition of real numbers is commutative, it follows that

$$c_{ij} = a_{ij} + b_{ij} = b_{ij} + a_{ij} = d_{ij}$$

for all i and j, $i = 1, 2, 3, \ldots, m$ and $j = 1, 2, 3, \ldots, n$.

By the definition of equality of matrices, $C = D$, and therefore $A + B = B + A$.   □

Theorem.   The addition of three $m \times n$ matrices is associative.

The proof of the previous theorem is left as an exercise.

In the collection of all $m \times n$ matrices, there is a matrix called the **zero matrix**.

Definition.   Let A be any $m \times n$ matrix. There is an $m \times n$ matrix **0**, called the **zero matrix**, or the **additive identity matrix**, for which

$$A + 0 = 0 + A = A$$

The matrix **0** consists of m rows and n columns of 0s.

To illustrate the above definition, we note that the matrix

$$\begin{bmatrix} 0 & 0 & 0 \\ 0 & 0 & 0 \\ 0 & 0 & 0 \end{bmatrix}$$

is the 3×3 zero matrix, and that

$$\begin{bmatrix} 0 & 0 & 0 \\ 0 & 0 & 0 \\ 0 & 0 & 0 \end{bmatrix} + \begin{bmatrix} 1 & 2 & 3 \\ 4 & 5 & 6 \\ 7 & 8 & 9 \end{bmatrix} = \begin{bmatrix} 1 & 2 & 3 \\ 4 & 5 & 6 \\ 7 & 8 & 9 \end{bmatrix}$$

Matrices are similar to real numbers in another respect: every matrix has an additive inverse.

Definition. Any $m \times n$ matrix A has an **additive inverse**, an $m \times n$ matrix $-A$ with the property that the sum of A and $-A$ is the zero matrix:

$$A + (-A) = (-A) + A = \mathbf{0}$$

The entries of $-A$ are the negatives of the corresponding entries of A.

The additive inverse of the 2×3 matrix $A = \begin{bmatrix} 1 & -3 & 2 \\ 0 & 1 & -5 \end{bmatrix}$ is the matrix

$$-A = \begin{bmatrix} -1 & 3 & -2 \\ 0 & -1 & 5 \end{bmatrix}$$

because their sum is the zero matrix:

$$A + (-A) = \begin{bmatrix} 1 & -3 & 2 \\ 0 & 1 & -5 \end{bmatrix} + \begin{bmatrix} -1 & 3 & -2 \\ 0 & -1 & 5 \end{bmatrix}$$

$$= \begin{bmatrix} 1-1 & -3+3 & 2-2 \\ 0+0 & 1-1 & -5+5 \end{bmatrix}$$

$$= \begin{bmatrix} 0 & 0 & 0 \\ 0 & 0 & 0 \end{bmatrix}$$

The Difference of Two Matrices. If A and B are $m \times n$ matrices, their difference, $A - B$, is the sum of A and the additive inverse of B:

$$A - B = A + (-B)$$

Example 3

$$\begin{bmatrix} 2 & -5 \\ 3 & 1 \end{bmatrix} - \begin{bmatrix} 4 & -5 \\ -3 & 9 \end{bmatrix} = \begin{bmatrix} 2 & -5 \\ 3 & 1 \end{bmatrix} + \begin{bmatrix} -4 & 5 \\ 3 & -9 \end{bmatrix}$$

$$= \begin{bmatrix} -2 & 0 \\ 6 & -8 \end{bmatrix}$$ ■

We illustrate how to find the product of two matrices by computing the product of a 2×3 matrix A and a 3×3 matrix B. The result is the 2×3 matrix C.

$$AB = \begin{bmatrix} 1 & 2 & 3 \\ 4 & 5 & 6 \end{bmatrix} \begin{bmatrix} a & b & c \\ d & e & f \\ g & h & i \end{bmatrix} = C$$

Each entry of matrix C is the result of a calculation that involves a row of A and a column of B. For example, the first-row, third-column entry of matrix

C is found by keeping a running total of the products of corresponding entries of the first row of A and the third column of B:

$$\begin{bmatrix} 1 & 2 & 3 \\ 4 & 5 & 6 \end{bmatrix} \begin{bmatrix} a & b & c \\ d & e & f \\ g & h & i \end{bmatrix} = \begin{bmatrix} ? & ? & 1c + 2f + 3i \\ ? & ? & ? \end{bmatrix}$$

Similarly, the second-row, second-column entry of matrix C is formed by a calculation involving the second row of A and the second column of B.

$$\begin{bmatrix} 1 & 2 & 3 \\ 4 & 5 & 6 \end{bmatrix} \begin{bmatrix} a & b & c \\ d & e & f \\ g & h & i \end{bmatrix} = \begin{bmatrix} ? & ? & 1c + 2f + 3i \\ ? & 4b + 5e + 6h & ? \end{bmatrix}$$

To calculate the first-row, first-column entry of matrix C, we use the first row of A and the first column of B.

$$\begin{bmatrix} 1 & 2 & 3 \\ 4 & 5 & 6 \end{bmatrix} \begin{bmatrix} a & b & c \\ d & e & f \\ g & h & i \end{bmatrix} = \begin{bmatrix} 1a + 2d + 3g & ? & 1c + 2f + 3i \\ ? & 4b + 5e + 6h & ? \end{bmatrix}$$

The complete product C is

$$\begin{bmatrix} 1 & 2 & 3 \\ 4 & 5 & 6 \end{bmatrix} \begin{bmatrix} a & b & c \\ d & e & f \\ g & h & i \end{bmatrix} = \begin{bmatrix} 1a + 2d + 3g & 1b + 2e + 3h & 1c + 2f + 3i \\ 4a + 5d + 6g & 4b + 5e + 6h & 4c + 5f + 6i \end{bmatrix}$$

For the product $A \cdot B$ to exist, the number of columns of A must equal the number of rows of B. If the product exists, it will have as many rows as A and as many columns as B:

$$\begin{array}{ccccc} A & \cdot & B & = & C \\ m \times n & & n \times p & & m \times p \end{array}$$

These must agree.

The product is $m \times p$.

More formally, we have the following definition.

The Product of Two Matrices. Let $A = [a_{ij}]$ be an $m \times n$ matrix and $B = [b_{ij}]$ be an $n \times p$ matrix. The product, AB, is the $m \times p$ matrix C found as follows: $AB = C = [c_{ij}]$, where c_{ij} is the sum of the products of the corresponding entries in the ith row of A and the jth column of B:

$$c_{ij} = a_{i1}b_{1j} + a_{i2}b_{2j} + a_{i3}b_{3j} + \cdots + a_{in}b_{nj}$$

for $i = 1, 2, 3, \ldots, m$ and $j = 1, 2, 3, \ldots, p$.

Example 4 Find C if $AB = \begin{bmatrix} 1 & 2 \\ 3 & 4 \\ 5 & 6 \end{bmatrix} \begin{bmatrix} a & b \\ c & d \end{bmatrix} = C.$

Solution Because the first matrix is 3×2 and the second matrix is 2×2, the product C exists, and it is a 3×2 matrix. The first-row, first-column entry of C is the total of the products of corresponding entries in the first row of A and the first column of B: $c_{11} = 1a + 2c$. Similarly, c_{12} is computed by using the first row of A and the second column of B: $c_{12} = 1b + 2d$. The entire product is

$$\begin{bmatrix} 1 & 2 \\ 3 & 4 \\ 5 & 6 \end{bmatrix} \begin{bmatrix} a & b \\ c & d \end{bmatrix} = \begin{bmatrix} 1a + 2c & 1b + 2d \\ 3a + 4c & 3b + 4d \\ 5a + 6c & 5b + 6d \end{bmatrix}$$ ■

Example 5 Find the product $\begin{bmatrix} 1 & -1 & 2 \\ 1 & 3 & 0 \\ 0 & 1 & 1 \end{bmatrix} \begin{bmatrix} 2 & 1 \\ 1 & 3 \\ 0 & 1 \end{bmatrix}.$

Solution Because the matrices are 3×3 and 3×2, the product is a 3×2 matrix.

$$\begin{bmatrix} 1 & -1 & 2 \\ 1 & 3 & 0 \\ 0 & 1 & 1 \end{bmatrix} \begin{bmatrix} 2 & 1 \\ 1 & 3 \\ 0 & 1 \end{bmatrix}$$

$$= \begin{bmatrix} 1 \cdot 2 + (-1) \cdot 1 + 2 \cdot 0 & 1 \cdot 1 + (-1) \cdot 3 + 2 \cdot 1 \\ 1 \cdot 2 + 3 \cdot 1 + 0 \cdot 0 & 1 \cdot 1 + 3 \cdot 3 + 0 \cdot 1 \\ 0 \cdot 2 + 1 \cdot 1 + 1 \cdot 0 & 0 \cdot 1 + 1 \cdot 3 + 1 \cdot 1 \end{bmatrix}$$

$$= \begin{bmatrix} 1 & 0 \\ 5 & 10 \\ 1 & 4 \end{bmatrix}$$ ■

Example 6 Find the product $\begin{bmatrix} 1 & 2 & 3 \end{bmatrix} \begin{bmatrix} 4 \\ 5 \\ 6 \end{bmatrix}.$

Solution Because the first matrix is 1×3 and the second matrix is 3×1, the product is a 1×1 matrix:

$$\begin{bmatrix} 1 & 2 & 3 \end{bmatrix} \begin{bmatrix} 4 \\ 5 \\ 6 \end{bmatrix} = \begin{bmatrix} 1 \cdot 4 + 2 \cdot 5 + 3 \cdot 6 \end{bmatrix} = \begin{bmatrix} 32 \end{bmatrix}$$ ■

Example 7 If $A = \begin{bmatrix} 1 & 1 \\ 0 & 0 \end{bmatrix}$ and $B = \begin{bmatrix} 0 & 1 \\ 0 & 1 \end{bmatrix}$, calculate AB and BA and thereby show that multiplication of matrices is not commutative.

Solution

$$AB = \begin{bmatrix} 1 & 1 \\ 0 & 0 \end{bmatrix} \begin{bmatrix} 0 & 1 \\ 0 & 1 \end{bmatrix} = \begin{bmatrix} 0 & 2 \\ 0 & 0 \end{bmatrix}$$

$$BA = \begin{bmatrix} 0 & 1 \\ 0 & 1 \end{bmatrix} \begin{bmatrix} 1 & 1 \\ 0 & 0 \end{bmatrix} = \begin{bmatrix} 0 & 0 \\ 0 & 0 \end{bmatrix}$$

Because the products are not equal, matrix multiplication is not commutative.   ■

Example 8   Find values for x, y, and z such that

$$\begin{bmatrix} 1 & 2 & 3 \\ 2 & -1 & -2 \\ 1 & -3 & -3 \end{bmatrix} \begin{bmatrix} x \\ y \\ z \end{bmatrix} = \begin{bmatrix} 4 \\ 0 \\ -2 \end{bmatrix}$$

Solution   Find the product of the first two matrices and set it equal to the third matrix.

$$\begin{bmatrix} 1 & 2 & 3 \\ 2 & -1 & -2 \\ 1 & -3 & -3 \end{bmatrix} \begin{bmatrix} x \\ y \\ z \end{bmatrix} = \begin{bmatrix} 1x + 2y + 3z \\ 2x - 1y - 2z \\ 1x - 3y - 3z \end{bmatrix} = \begin{bmatrix} 4 \\ 0 \\ -2 \end{bmatrix}$$

The product will be equal to the third matrix if and only if their corresponding components are equal. Set the corresponding components equal to get

$$\begin{cases} x + 2y + 3z = 4 \\ 2x - y - 2z = 0 \\ x - 3y - 3z = -2 \end{cases}$$

This system was solved in Example 2 of Section 8.2. Its solution is $x = 1$, $y = 0$, and $z = 1$.   ■

The number 1 is called the identity for multiplication because multiplying a number by 1 does not change that number. There is a **multiplicative identity matrix** with a similar property.

Definition.   Let A be an $n \times n$ matrix. There is an $n \times n$ **identity matrix** I for which

$$AI = IA = A$$

The matrix I consists of 1s on its diagonal and 0s elsewhere.

$$I = \begin{bmatrix} 1 & 0 & 0 & \cdots & 0 \\ 0 & 1 & 0 & \cdots & 0 \\ 0 & 0 & 1 & \cdots & 0 \\ \vdots & \vdots & \vdots & & \vdots \\ 0 & 0 & 0 & \cdots & 1 \end{bmatrix}$$

Note that an identity matrix is a square matrix—it has the same number of rows and columns.

Example 9 illustrates the previous definition for the 3×3 identity matrix.

Example 9 Find **a.** $\begin{bmatrix} 1 & 0 & 0 \\ 0 & 1 & 0 \\ 0 & 0 & 1 \end{bmatrix}\begin{bmatrix} 1 & 2 & 3 \\ 4 & 5 & 6 \\ 7 & 8 & 9 \end{bmatrix}$ and **b.** $\begin{bmatrix} 1 & 2 & 3 \\ 4 & 5 & 6 \\ 7 & 8 & 9 \end{bmatrix}\begin{bmatrix} 1 & 0 & 0 \\ 0 & 1 & 0 \\ 0 & 0 & 1 \end{bmatrix}$.

Solution **a.** $\begin{bmatrix} 1 & 0 & 0 \\ 0 & 1 & 0 \\ 0 & 0 & 1 \end{bmatrix}\begin{bmatrix} 1 & 2 & 3 \\ 4 & 5 & 6 \\ 7 & 8 & 9 \end{bmatrix}$

$$= \begin{bmatrix} 1\cdot1+0\cdot4+0\cdot7 & 1\cdot2+0\cdot5+0\cdot8 & 1\cdot3+0\cdot6+0\cdot9 \\ 0\cdot1+1\cdot4+0\cdot7 & 0\cdot2+1\cdot5+0\cdot8 & 0\cdot3+1\cdot6+0\cdot9 \\ 0\cdot1+0\cdot4+1\cdot7 & 0\cdot2+0\cdot5+1\cdot8 & 0\cdot3+0\cdot6+1\cdot9 \end{bmatrix}$$

$$= \begin{bmatrix} 1 & 2 & 3 \\ 4 & 5 & 6 \\ 7 & 8 & 9 \end{bmatrix}$$

b. $\begin{bmatrix} 1 & 2 & 3 \\ 4 & 5 & 6 \\ 7 & 8 & 9 \end{bmatrix}\begin{bmatrix} 1 & 0 & 0 \\ 0 & 1 & 0 \\ 0 & 0 & 1 \end{bmatrix} = \begin{bmatrix} 1 & 2 & 3 \\ 4 & 5 & 6 \\ 7 & 8 & 9 \end{bmatrix}$ ∎

Exercise 8.3

In Exercises 1–8 find values of x and y, if any, that will make the two matrices equal.

1. $\begin{bmatrix} x & y \\ 1 & 3 \end{bmatrix} = \begin{bmatrix} 2 & 5 \\ 1 & 3 \end{bmatrix}$

2. $\begin{bmatrix} x & 5 \\ 3 & y \end{bmatrix} = \begin{bmatrix} 0 & 5 \\ 3 & 2 \end{bmatrix}$

3. $\begin{bmatrix} x & y \\ 1 & 3 \end{bmatrix} = \begin{bmatrix} 2 & 5 \\ 1 & 4 \end{bmatrix}$

4. $\begin{bmatrix} x & y \\ 1 & x+y \end{bmatrix} = \begin{bmatrix} 2 & 1 \\ 1 & 2 \end{bmatrix}$

5. $\begin{bmatrix} x+y & 3+x \\ -2 & 5y \end{bmatrix} = \begin{bmatrix} 3 & 4 \\ -2 & 10 \end{bmatrix}$

6. $\begin{bmatrix} x+y & x-y \\ 2x & 3y \end{bmatrix} = \begin{bmatrix} -x & x-2 \\ -y & 8-y \end{bmatrix}$

7. $\begin{bmatrix} x & 3x \\ y & x+1 \end{bmatrix} = \begin{bmatrix} y & 6 \\ 2 & 3 \end{bmatrix}$

8. $\begin{bmatrix} x \\ y \end{bmatrix} = \begin{bmatrix} 1 & 2 \\ 3 & 4 \end{bmatrix}$

In Exercises 9–18 perform the indicated operation, if possible.

9. $\begin{bmatrix} 2 & 1 & -1 \\ -3 & 2 & 5 \end{bmatrix} + \begin{bmatrix} -3 & 1 & 2 \\ -3 & -2 & -5 \end{bmatrix}$

10. $\begin{bmatrix} 3 & 1 \\ 2 & 2 \end{bmatrix} + \begin{bmatrix} 2 & 1 \\ -1 & 0 \end{bmatrix} + \begin{bmatrix} -5 & -2 \\ -1 & -2 \end{bmatrix}$

11. $\begin{bmatrix} 3 & 2 & 1 \\ -2 & 3 & -3 \\ -4 & -2 & -1 \end{bmatrix} - \begin{bmatrix} -2 & 6 & -2 \\ 5 & 7 & -1 \\ -4 & -6 & 7 \end{bmatrix}$

12. $\begin{bmatrix} -2 & 7 & -3 \\ 3 & 6 & -7 \\ -9 & -2 & -5 \end{bmatrix} + \begin{bmatrix} -5 & -4 & -3 \\ -1 & 2 & 10 \\ -1 & -3 & -4 \end{bmatrix}$

13. $\begin{bmatrix} 1 & 3 & -1 \\ 2 & 1 & 5 \\ 1 & 3 & 0 \end{bmatrix} + \begin{bmatrix} 2 \\ 0 \\ -3 \end{bmatrix}$

14. $\begin{bmatrix} 3 \\ 2 \\ 3 \end{bmatrix} - \begin{bmatrix} -3 & 5 & -6 \\ -3 & -5 & -6 \\ 4 & 6 & -6 \end{bmatrix}$

15. $[1 \quad 2 \quad 3] + [4 \quad 5 \quad 6]$

16. $\begin{bmatrix} 1 \\ 2 \\ 3 \end{bmatrix} + [4 \quad -5 \quad -6]$

17. $\begin{bmatrix} 1 & 3 & -4 \\ 2 & -1 & 3 \\ 1 & 5 & 7 \end{bmatrix} + \begin{bmatrix} 3 & 2 & -8 \\ 9 & 11 & 17 \\ 2 & 1 & 3 \end{bmatrix} - \begin{bmatrix} 1 & 3 & -5 \\ 2 & -9 & 5 \\ 3 & 10 & 11 \end{bmatrix}$

18. $\begin{bmatrix} -3 & -2 & 15 \\ 2 & -5 & 9 \end{bmatrix} - \begin{bmatrix} 3 & 2 & -15 \\ -2 & 5 & -9 \end{bmatrix} + \begin{bmatrix} 6 & 4 & -30 \\ -3 & 12 & -15 \end{bmatrix}$

In Exercises 19–30 find each product, if possible.

19. $\begin{bmatrix} 2 & 3 \\ 3 & -2 \end{bmatrix}\begin{bmatrix} 1 & 2 \\ 0 & -2 \end{bmatrix}$

20. $\begin{bmatrix} -2 & 3 \\ 3 & -2 \end{bmatrix}\begin{bmatrix} 2 & 4 \\ -5 & 7 \end{bmatrix}$

21. $\begin{bmatrix} -4 & -2 \\ 21 & 0 \end{bmatrix}\begin{bmatrix} -5 & 6 \\ 21 & -1 \end{bmatrix}$

22. $\begin{bmatrix} -5 & 4 \\ 4 & -5 \end{bmatrix}\begin{bmatrix} 6 & -2 \\ 1 & 3 \end{bmatrix}$

23. $\begin{bmatrix} 2 & 1 & 3 \\ 1 & 2 & -1 \\ 0 & 1 & 0 \end{bmatrix}\begin{bmatrix} 1 & 2 & 3 \\ 2 & -2 & 1 \\ 0 & 0 & 1 \end{bmatrix}$

24. $\begin{bmatrix} 2 & 1 & 1 \\ 1 & 1 & 2 \\ 1 & -2 & -1 \end{bmatrix}\begin{bmatrix} 1 & 2 & 3 \\ 1 & 2 & -3 \\ -1 & -1 & 3 \end{bmatrix}$

25. $[1 \quad -2 \quad -3]\begin{bmatrix} 4 \\ -5 \\ -6 \end{bmatrix}$

26. $\begin{bmatrix} 1 \\ -2 \\ -3 \end{bmatrix}[4 \quad -5 \quad -6]$

27. $[1 \quad 2 \quad 3][4 \quad 5 \quad 6]$

28. $\begin{bmatrix} 2 & 3 & 4 \\ 1 & 2 & 3 \\ -2 & 2 & 2 \end{bmatrix}\begin{bmatrix} -1 \\ 2 \\ 3 \end{bmatrix}$

29. $[1 \quad 2 \quad 3]\begin{bmatrix} 1 & 2 & 3 \\ 4 & 5 & 6 \\ 7 & 8 & 9 \end{bmatrix}$

30. $\begin{bmatrix} 1 & 2 & 3 \\ 1 & 2 & 1 \\ 1 & -1 & -1 \end{bmatrix}\begin{bmatrix} 1 & 2 \\ 2 & 1 \\ 1 & 1 \end{bmatrix}$

In Exercises 31–36 perform the indicated operations.

31. $\begin{bmatrix} 1 & 2 \\ 2 & 3 \end{bmatrix}\left(\begin{bmatrix} 2 & 1 & -5 \\ 1 & 1 & 2 \end{bmatrix} + \begin{bmatrix} -2 & -1 & 6 \\ 0 & -1 & -1 \end{bmatrix}\right)$

32. $\begin{bmatrix} 1 & 2 \\ 2 & 3 \end{bmatrix}\begin{bmatrix} 2 & 1 & -5 \\ 1 & 1 & 2 \end{bmatrix} + \begin{bmatrix} 1 & 2 \\ 2 & 3 \end{bmatrix}\begin{bmatrix} -2 & -1 & 6 \\ 0 & -1 & -1 \end{bmatrix}$

33. $\begin{bmatrix} 1 & 2 & 3 \\ 2 & 3 & 1 \\ 1 & 2 & 1 \end{bmatrix}\begin{bmatrix} 2 & 1 & 1 \\ 3 & -1 & -1 \\ 2 & -2 & 2 \end{bmatrix} + \begin{bmatrix} -2 & 3 & 4 \\ 1 & 1 & 1 \\ 0 & 1 & 0 \end{bmatrix}$

34. $\begin{bmatrix} 2 & 1 & 0 \\ 1 & -2 & -1 \\ 1 & 1 & -1 \end{bmatrix}\left(\begin{bmatrix} 1 & 0 & 1 \\ 1 & 1 & 2 \\ 1 & 2 & -1 \end{bmatrix} + \begin{bmatrix} -1 & -1 & 2 \\ 0 & 0 & 1 \\ 1 & 0 & -1 \end{bmatrix}\right)$

35. $\left(\begin{bmatrix} 1 & 2 \\ 2 & 3 \end{bmatrix}\begin{bmatrix} 1 \\ -3 \end{bmatrix} + \begin{bmatrix} -2 \\ 1 \end{bmatrix}\right)\left([1 \quad 2]\begin{bmatrix} 1 \\ -3 \end{bmatrix} + [4]\right)$

36. $\begin{bmatrix} 1 \\ 2 \end{bmatrix}[-3 \quad -4] - \begin{bmatrix} 0 & 3 \\ 2 & 1 \end{bmatrix}\begin{bmatrix} 2 & 0 \\ 1 & -1 \end{bmatrix}$

In Exercises 37–42 let $A = \begin{bmatrix} 1 & 3 \\ 2 & 5 \end{bmatrix}$, $B = \begin{bmatrix} -1 \\ 3 \end{bmatrix}$, and $C = \begin{bmatrix} 3 & 2 \end{bmatrix}$. *Perform, if possible, the indicated operations.*

37. $A - BC$ **38.** $AB + B$ **39.** $CB - AB$ **40.** CAB

41. ABC **42.** $CA + C$

43. Let $A = \begin{bmatrix} 1 & 1 \\ 1 & 1 \end{bmatrix}$. Find A^7. (*Hint:* Calculate A^2 and A^3. Do you see a pattern?)

44. Let a, b, and c be real numbers. If $ab = ac$ and $a \neq 0$, then $b = c$. Find 2×2 matrices A, B, and C, where $A \neq 0$, to show that such a law does not hold for all matrices.

45. In the real number system, the numbers 0 and 1 are the only numbers that equal their own squares: if $a^2 = a$, then $a = 0$ or $a = 1$. Find a 2×2 matrix A that is neither the zero matrix nor the identity matrix such that $A^2 = AA = A$.

46. Another property of the real numbers is that, if $ab = 0$, then either $a = 0$ or $b = 0$. To show that this property is not true for matrices, find two nonzero 2×2 matrices, A and B, such that $AB = 0$.

47. Multiplication of three $n \times n$ matrices is associative: $(AB)C = A(BC)$. Verify this by an example chosen from the set of 2×2 matrices.

48. If A is an $m \times n$ matrix and B and C are each $n \times p$ matrices, then $A(B + C) = AB + AC$. Illustrate this with an example showing that matrix multiplication distributes over matrix addition.

49. Prove that the addition of matrices is associative.

8.4 MATRIX INVERSION

Two real numbers are called **multiplicative inverses** if their product is the multiplicative identity 1. Some square matrices have multiplicative inverses also.

> **Definition.** If A and B are $n \times n$ matrices, I is the $n \times n$ identity matrix, and
>
> $$AB = BA = I$$
>
> then A and B are called **multiplicative inverses**. Matrix A is the **inverse** of B, and B is the **inverse** of A.

It can be shown that the inverse of a matrix A, if it exists, is unique. The inverse of A is denoted by A^{-1}.

Example 1 If $A = \begin{bmatrix} 1 & 1 & 0 \\ 4 & 3 & 0 \\ 2 & 1 & -1 \end{bmatrix}$ and $B = \begin{bmatrix} -3 & 1 & 0 \\ 4 & -1 & 0 \\ -2 & 1 & -1 \end{bmatrix}$, show that A and B are inverses.

Solution   Multiply the matrices in each order to show that the product is the identity matrix.

$$AB = \begin{bmatrix} 1 & 1 & 0 \\ 4 & 3 & 0 \\ 2 & 1 & -1 \end{bmatrix} \begin{bmatrix} -3 & 1 & 0 \\ 4 & -1 & 0 \\ -2 & 1 & -1 \end{bmatrix}$$

$$= \begin{bmatrix} -3+4 & 1-1 & 0 \\ -12+12 & 4-3 & 0 \\ -6+4+2 & 2-1-1 & 1 \end{bmatrix} = \begin{bmatrix} 1 & 0 & 0 \\ 0 & 1 & 0 \\ 0 & 0 & 1 \end{bmatrix}$$

$$BA = \begin{bmatrix} -3 & 1 & 0 \\ 4 & -1 & 0 \\ -2 & 1 & -1 \end{bmatrix} \begin{bmatrix} 1 & 1 & 0 \\ 4 & 3 & 0 \\ 2 & 1 & -1 \end{bmatrix} = \begin{bmatrix} 1 & 0 & 0 \\ 0 & 1 & 0 \\ 0 & 0 & 1 \end{bmatrix} \quad\blacksquare$$

If a matrix has an inverse, it is called a **nonsingular matrix**. Otherwise, it is called a **singular matrix**. The following theorem, stated without proof, provides a way of calculating the inverse of a nonsingular matrix.

> **Theorem.** If a sequence of elementary row operations performed on the $n \times n$ matrix A reduces A to the $n \times n$ identity matrix I, then those same row operations, performed in the same order on the identity matrix I, will transform I into A^{-1}. Furthermore, if *no* sequence of row operations will reduce A to I, then A is singular.

To use the previous theorem, we perform elementary row operations on matrix A to change it to the identity matrix I. At the same time, we perform these elementary row operations on the identity matrix I. This changes I into A^{-1}.

A notation for this process uses an n-row by $2n$-column matrix, with matrix A as the left half and matrix I as the right half. If A is nonsingular, the proper row operations performed on $[A \mid I]$ will transform it into $[I \mid A^{-1}]$.

Example 2   Find the inverse of matrix A if $A = \begin{bmatrix} 2 & -4 \\ 4 & -7 \end{bmatrix}$.

Solution   Set up a 2×4 matrix with A on the left and I on the right of the broken line:

$$[A \mid I] = \begin{bmatrix} 2 & -4 & \mid & 1 & 0 \\ 4 & -7 & \mid & 0 & 1 \end{bmatrix}$$

Perform row operations on the entire matrix to transform the left half into I:

$$(\tfrac{1}{2})R1 \rightarrow R1$$
$$(-2)R1 + R2 \rightarrow R2$$

$$\begin{bmatrix} 2 & -4 & \mid & 1 & 0 \\ 4 & -7 & \mid & 0 & 1 \end{bmatrix} \Leftrightarrow \begin{bmatrix} 1 & -2 & \mid & \tfrac{1}{2} & 0 \\ 0 & 1 & \mid & -2 & 1 \end{bmatrix}$$

$$(2)R2 + R1 \rightarrow R1$$

$$\Leftrightarrow \begin{bmatrix} 1 & 0 & | & -\frac{7}{2} & 2 \\ 0 & 1 & | & -2 & 1 \end{bmatrix}$$

Matrix A has been transformed into I. Thus, the right side of the previous matrix is A^{-1}. Verify this by finding AA^{-1} and $A^{-1}A$ and showing that each product is I:

$$AA^{-1} = \begin{bmatrix} 2 & -4 \\ 4 & -7 \end{bmatrix} \begin{bmatrix} -\frac{7}{2} & 2 \\ -2 & 1 \end{bmatrix} = \begin{bmatrix} 1 & 0 \\ 0 & 1 \end{bmatrix}$$

$$A^{-1}A = \begin{bmatrix} -\frac{7}{2} & 2 \\ -2 & 1 \end{bmatrix} \begin{bmatrix} 2 & -4 \\ 4 & -7 \end{bmatrix} = \begin{bmatrix} 1 & 0 \\ 0 & 1 \end{bmatrix}$$

Example 3   Find the inverse of matrix A if $A = \begin{bmatrix} 1 & 1 & 0 \\ 1 & 2 & 1 \\ 2 & 3 & 2 \end{bmatrix}$.

Solution   Set up a 3×6 matrix with A on the left and I on the right of the broken line:

$$[A \mid I] = \begin{bmatrix} 1 & 1 & 0 & | & 1 & 0 & 0 \\ 1 & 2 & 1 & | & 0 & 1 & 0 \\ 2 & 3 & 2 & | & 0 & 0 & 1 \end{bmatrix}$$

Perform row operations on the entire matrix to transform the left half into I.

$$(-1)R1 + R2 \rightarrow R2$$
$$(-2)R1 + R3 \rightarrow R3$$

$$\begin{bmatrix} 1 & 1 & 0 & | & 1 & 0 & 0 \\ 1 & 2 & 1 & | & 0 & 1 & 0 \\ 2 & 3 & 2 & | & 0 & 0 & 1 \end{bmatrix} \Leftrightarrow \begin{bmatrix} 1 & 1 & 0 & | & 1 & 0 & 0 \\ 0 & 1 & 1 & | & -1 & 1 & 0 \\ 0 & 1 & 2 & | & -2 & 0 & 1 \end{bmatrix}$$

$$(-1)R2 + R1 \rightarrow R1$$
$$(-1)R2 + R3 \rightarrow R3$$

$$\Leftrightarrow \begin{bmatrix} 1 & 0 & -1 & | & 2 & -1 & 0 \\ 0 & 1 & 1 & | & -1 & 1 & 0 \\ 0 & 0 & 1 & | & -1 & -1 & 1 \end{bmatrix}$$

$$R3 + R1 \rightarrow R1$$
$$(-1)R3 + R2 \rightarrow R2$$

$$\Leftrightarrow \begin{bmatrix} 1 & 0 & 0 & | & 1 & -2 & 1 \\ 0 & 1 & 0 & | & 0 & 2 & -1 \\ 0 & 0 & 1 & | & -1 & -1 & 1 \end{bmatrix}$$

The left half has been transformed into the identity matrix, and the right half has become A^{-1}. Thus,

$$A^{-1} = \begin{bmatrix} 1 & -2 & 1 \\ 0 & 2 & -1 \\ -1 & -1 & 1 \end{bmatrix}$$

Example 4   Find the inverse of $A = \begin{bmatrix} 1 & 2 \\ 2 & 4 \end{bmatrix}$, if possible.

Solution   Form the 2×4 matrix

$$[A \mid I] = \begin{bmatrix} 1 & 2 & \vdots & 1 & 0 \\ 2 & 4 & \vdots & 0 & 1 \end{bmatrix}$$

and begin to transform the left side of the matrix into the identity matrix I:

$$(-2)R1 + R2 \rightarrow R2$$

$$\begin{bmatrix} 1 & 2 & \vdots & 1 & 0 \\ 2 & 4 & \vdots & 0 & 1 \end{bmatrix} \Leftrightarrow \begin{bmatrix} 1 & 2 & \vdots & 1 & 0 \\ 0 & 0 & \vdots & -2 & 1 \end{bmatrix}$$

In obtaining the second-row, first-column position of A, the entire second row of A is "zeroed out." Because it is impossible to transform A to the identity, matrix A is singular and has no inverse.   ∎

The next example shows how the inverse of a nonsingular matrix can be used to solve a system of equations.

Example 5   Solve the system

$$\begin{cases} x + y & = 3 \\ x + 2y + z = -2 \\ 2x + 3y + 2z = 1 \end{cases}$$

Solution   This system can be written as a single equation involving three matrices.

1. $\begin{bmatrix} 1 & 1 & 0 \\ 1 & 2 & 1 \\ 2 & 3 & 2 \end{bmatrix} \begin{bmatrix} x \\ y \\ z \end{bmatrix} = \begin{bmatrix} 3 \\ -2 \\ 1 \end{bmatrix}$

The 3×3 matrix on the left is the matrix whose inverse was found in Example 3. Multiply each side of Equation 1 on the left by this inverse to obtain an equivalent system of equations. The solution of this system can be read directly from the matrix to the right of the equals sign:

$$\begin{bmatrix} 1 & -2 & 1 \\ 0 & 2 & -1 \\ -1 & -1 & 1 \end{bmatrix} \begin{bmatrix} 1 & 1 & 0 \\ 1 & 2 & 1 \\ 2 & 3 & 2 \end{bmatrix} \begin{bmatrix} x \\ y \\ z \end{bmatrix} = \begin{bmatrix} 1 & -2 & 1 \\ 0 & 2 & -1 \\ -1 & -1 & 1 \end{bmatrix} \begin{bmatrix} 3 \\ -2 \\ 1 \end{bmatrix}$$

$$\begin{bmatrix} 1 & 0 & 0 \\ 0 & 1 & 0 \\ 0 & 0 & 1 \end{bmatrix} \begin{bmatrix} x \\ y \\ z \end{bmatrix} = \begin{bmatrix} 8 \\ -5 \\ 0 \end{bmatrix}$$

$$\begin{bmatrix} x \\ y \\ z \end{bmatrix} = \begin{bmatrix} 8 \\ -5 \\ 0 \end{bmatrix}$$

The solution of this system of equations is $x = 8$, $y = -5$, $z = 0$. Verify that these results satisfy all three of the original equations.   ■

The equations of Example 5 can be thought of as the matrix equation $AX = B$, where A is the coefficient matrix,

$$A = \begin{bmatrix} 1 & 1 & 0 \\ 1 & 2 & 1 \\ 2 & 3 & 2 \end{bmatrix}$$

X is a column matrix of the variables,

$$X = \begin{bmatrix} x \\ y \\ z \end{bmatrix}$$

and B is a column matrix of the constants from the right sides of the equations,

$$B = \begin{bmatrix} 3 \\ -2 \\ 1 \end{bmatrix}$$

When each side of $AX = B$ is multiplied on the left by A^{-1}, the solution of the system appears as the column of numbers in the matrix that is the product of A^{-1} and B:

$$AX = B$$
$$A^{-1}AX = A^{-1}B$$
$$IX = A^{-1}B$$
$$X = A^{-1}B$$

This method is especially useful for finding solutions of several systems of equations that differ from each other *only* in the column matrix B. If the coefficient matrix A remains unchanged from one system of equations to the next, then A^{-1} needs to be found only once. The solution of each system is found by a single matrix multiplication, $A^{-1}B$.

▬▬ Exercise 8.4 ▬▬

In Exercises 1–16 find the inverse of each given matrix, if possible.

1. $\begin{bmatrix} 3 & -4 \\ -2 & 3 \end{bmatrix}$

2. $\begin{bmatrix} 2 & 3 \\ 3 & 5 \end{bmatrix}$

3. $\begin{bmatrix} 3 & 7 \\ 2 & 5 \end{bmatrix}$

4. $\begin{bmatrix} 1 & -2 \\ 2 & -5 \end{bmatrix}$

5. $\begin{bmatrix} 1 & 2 & 3 \\ 2 & 5 & 3 \\ 1 & 0 & 8 \end{bmatrix}$

6. $\begin{bmatrix} 2 & 1 & -1 \\ 2 & 2 & -1 \\ -1 & -1 & 1 \end{bmatrix}$

7. $\begin{bmatrix} 3 & 2 & 1 \\ 1 & 1 & -1 \\ 4 & 3 & 1 \end{bmatrix}$

8. $\begin{bmatrix} -2 & 1 & -3 \\ 2 & 3 & 0 \\ 1 & 0 & 1 \end{bmatrix}$ **9.** $\begin{bmatrix} 1 & 3 & 5 \\ 0 & 1 & 6 \\ 1 & 4 & 11 \end{bmatrix}$ **10.** $\begin{bmatrix} 1 & 1 & 1 \\ 2 & 2 & 2 \\ 3 & 3 & 3 \end{bmatrix}$

11. $\begin{bmatrix} 1 & 2 & 3 \\ 0 & 1 & 2 \\ 0 & 0 & 1 \end{bmatrix}$ **12.** $\begin{bmatrix} 1 & 2 & 3 \\ 0 & 1 & 1 \\ 0 & -1 & 0 \end{bmatrix}$ **13.** $\begin{bmatrix} 1 & 6 & 4 \\ 1 & -2 & -5 \\ 2 & 4 & -1 \end{bmatrix}$

14. $\begin{bmatrix} 1 & 1 & 1 \\ 1 & 0 & -1 \\ 1 & 2 & 3 \end{bmatrix}$ **15.** $\begin{bmatrix} 1 & 2 & 3 & 4 \\ 0 & 1 & 2 & 3 \\ 0 & 0 & 1 & 2 \\ 0 & 0 & 0 & 1 \end{bmatrix}$ **16.** $\begin{bmatrix} 1 & 0 & 0 & 0 \\ 1 & 1 & 0 & 0 \\ 1 & 1 & 1 & 0 \\ 1 & 2 & 2 & 1 \end{bmatrix}$

In Exercises 17–26 use the method of Example 5 to solve each system of equations.

17. $\begin{cases} 3x - 4y = 1 \\ -2x + 3y = 5 \end{cases}$
18. $\begin{cases} 2x + 3y = 7 \\ 3x + 5y = -5 \end{cases}$
19. $\begin{cases} 3x + 7y = 0 \\ 2x + 5y = -10 \end{cases}$
20. $\begin{cases} x - 2y = 12 \\ 2x - 5y = 13 \end{cases}$

21. $\begin{cases} x + 2y + 3z = 1 \\ 2x + 5y + 3z = 3 \\ x \quad\;\; + 8z = -2 \end{cases}$
22. $\begin{cases} 2x + y - z = 3 \\ 2x + 2y - z = -1 \\ -x - y + z = 4 \end{cases}$
23. $\begin{cases} 3x + 2y + z = 2 \\ x + y - z = -1 \\ 4x + 3y + z = 0 \end{cases}$

24. $\begin{cases} -2x + y - 3z = 5 \\ 2x + 3y \quad\;\; = 1 \\ x \quad\;\; + z = -2 \end{cases}$
25. $\begin{cases} x + 2y + 3z = 1 \\ 2x + 2y + 2z = 2x + y \\ z = 3 \end{cases}$
26. $\begin{cases} x + 2y + 3z = 2 \\ x + y + z = x \\ x - y \quad\;\; = x \end{cases}$

27. If the $n \times n$ matrix A is nonsingular, and if B and C are $n \times n$ matrices such that $AB = AC$, prove that $B = C$.

28. If B is an $n \times n$ matrix that behaves as an identity ($AB = BA = A$, for any $n \times n$ matrix A), prove that $B = I$.

29. If $A = \begin{bmatrix} 0 & 1 \\ 1 & 0 \end{bmatrix}$, compute $A^2, A^3, A^4, \ldots$. Give a general rule for A^n, where n is a natural number.

30. If A and B are 2×2 matrices, is $(AB)^2 = A^2B^2$? Support your answer.

31. Prove that $\begin{bmatrix} a & b \\ c & d \end{bmatrix}$ has an inverse if and only if $ad - bc \neq 0$.

32. If $A = \begin{bmatrix} 1 & 1 \\ 0 & 1 \end{bmatrix}$, compute A^n for various values of n ($n = 2, 3, 4, \ldots$). What do you notice?

33. If $A = \begin{bmatrix} 1 & 0 \\ 1 & 1 \end{bmatrix}$, compute A^n for various values of n ($n = 2, 3, 4, \ldots$). What do you notice?

34. For what value of x will $\begin{bmatrix} 3 & 8 \\ 6 & x \end{bmatrix}$ not have a multiplicative inverse? (*Hint:* See Exercise 31.)

35. For what values of x will $\begin{bmatrix} x & 8 \\ 2 & x \end{bmatrix}$ not have a multiplicative inverse?

36. Does $(AB)^{-1} = A^{-1}B^{-1}$? Support your answer with an example chosen from 2×2 matrices.

37. Use an example chosen from 2×2 matrices to illustrate that $(AB)^{-1} = B^{-1}A^{-1}$.

38. Let A be any 3×3 matrix. Find a 3×3 matrix E such that the product EA is the result of performing the row operation $R1 \leftrightarrow R2$ on matrix A (the row operation that exchanges rows one and two of matrix A). What is E^{-1}?

39. Let A be any 3×3 matrix. Find another 3×3 matrix E such that the product EA is the result of performing the row operation $(3)R1 + R3 \rightarrow R3$ on matrix A. What is E^{-1}?

8.5 SOLUTION OF SYSTEMS OF EQUATIONS BY DETERMINANTS

There is a function, called the **determinant function**, that associates a numerical value with every square matrix. For any square matrix A, the symbol $\det(A)$ or the symbol $|A|$ represents the determinant of matrix A. We begin by defining the determinant of a 2×2 matrix.

Gabriel Cramer
(1704–1752) Although other mathematicians had worked with determinants, it was the work of Cramer that popularized them.

Definition. If a, b, c, and d are numbers, then the determinant of $A = \begin{bmatrix} a & b \\ c & d \end{bmatrix}$ is

$$\det(A) = \begin{vmatrix} a & b \\ c & d \end{vmatrix} = ad - bc$$

The determinant of a 2×2 matrix A is the number that is equal to the product of the entries on the major diagonal

$$\begin{vmatrix} a & b \\ c & d \end{vmatrix}$$

minus the product of the entries on the other diagonal

$$\begin{vmatrix} a & b \\ c & d \end{vmatrix}$$

Example 1

a. $\begin{vmatrix} 1 & 2 \\ 3 & 4 \end{vmatrix} = 1(4) - 2(3) = 4 - 6 = -2$

b. $\begin{vmatrix} -1 & -3 \\ 2 & 0 \end{vmatrix} = -1(0) - (-3)(2) = 0 - (-6) = 6$

c. $\begin{vmatrix} -2 & 3 \\ -\pi & \frac{1}{2} \end{vmatrix} = -2\left(\frac{1}{2}\right) - (3)(-\pi) = -1 + 3\pi$ ∎

To see how determinants can be used to solve a system of equations, we consider the system

$$\begin{cases} ax + by = e \\ cx + dy = f \end{cases}$$

By multiplying the first equation by d, the second equation by $-b$, and adding, the terms that involve y drop out:

$$\begin{aligned} adx + bdy &= ed \\ -bcx - bdy &= -bf \\ \hline adx - bcx &= ed - bf \end{aligned}$$

We can solve the resulting equation for x:

$$adx - bcx = ed - bf$$
$$(ad - bc)x = ed - bf$$

1. $$x = \frac{ed - bf}{ad - bc} \quad \text{provided } ad - bc \neq 0$$

Solving this system for y gives

2. $$y = \frac{af - ec}{ad - bc} \quad \text{provided } ad - bc \neq 0$$

The numerators and denominators of Equations 1 and 2 can be expressed as determinants:

$$x = \frac{\begin{vmatrix} e & b \\ f & d \end{vmatrix}}{\begin{vmatrix} a & b \\ c & d \end{vmatrix}} = \frac{ed - bf}{ad - bc} \qquad y = \frac{\begin{vmatrix} a & e \\ c & f \end{vmatrix}}{\begin{vmatrix} a & b \\ c & d \end{vmatrix}} = \frac{af - ec}{ad - bc}$$

If these formulas are compared with the original system of equations,

$$\begin{cases} ax + by = e \\ cx + dy = f \end{cases}$$

it is apparent that the denominator determinant consists of the coefficients of the variables of each equation:

$$\text{denominator determinant} = \begin{vmatrix} a & b \\ c & d \end{vmatrix}$$

Each numerator determinant is a modified copy of the denominator determinant. The column of coefficients of the variable for which we are solving is replaced with the column of constants that appears to the right of the equal signs. Thus, when solving for x, the coefficients of x (a and c) are replaced in the numerator determinant by the constants e and f:

$$x = \frac{\begin{vmatrix} e & b \\ f & d \end{vmatrix}}{\begin{vmatrix} a & b \\ c & d \end{vmatrix}} \qquad \begin{cases} ax + by = e \\ cx + dy = f \end{cases}$$

Similarly, when solving for y, the coefficients of y (b and d) are replaced in the numerator determinant by the constants e and f.

$$y = \frac{\begin{vmatrix} a & e \\ c & f \end{vmatrix}}{\begin{vmatrix} a & b \\ c & d \end{vmatrix}} \qquad \begin{cases} ax + by = e \\ cx + dy = f \end{cases}$$

The method of using determinants to solve systems of equations is called **Cramer's rule**.

Cramer's Rule for Two Equations in Two Variables. The solution of the system of equations

$$\begin{cases} ax + by = e \\ cx + dy = f \end{cases}$$

is the pair

$$x = \frac{\begin{vmatrix} e & b \\ f & d \end{vmatrix}}{\begin{vmatrix} a & b \\ c & d \end{vmatrix}} \quad \text{and} \quad y = \frac{\begin{vmatrix} a & e \\ c & f \end{vmatrix}}{\begin{vmatrix} a & b \\ c & d \end{vmatrix}} \quad \text{provided} \begin{vmatrix} a & b \\ c & d \end{vmatrix} \neq 0$$

If the denominators and the numerators of these fractions are all 0, the system is consistent but the equations are dependent.

If the denominators are 0 and one numerator is not 0, the system is inconsistent.

Example 2   Use determinants to solve the system

$$\begin{cases} 3x + 2y = 7 \\ -x + 5y = 9 \end{cases}$$

Solution   Use Cramer's rule:

$$x = \frac{\begin{vmatrix} 7 & 2 \\ 9 & 5 \end{vmatrix}}{\begin{vmatrix} 3 & 2 \\ -1 & 5 \end{vmatrix}} = \frac{7 \cdot 5 - 2 \cdot 9}{3 \cdot 5 - 2(-1)} = \frac{35 - 18}{15 + 2} = \frac{17}{17} = 1$$

$$y = \frac{\begin{vmatrix} 3 & 7 \\ -1 & 9 \end{vmatrix}}{\begin{vmatrix} 3 & 2 \\ -1 & 5 \end{vmatrix}} = \frac{3 \cdot 9 - 7(-1)}{3 \cdot 5 - 2(-1)} = \frac{27 + 7}{15 + 2} = \frac{34}{17} = 2$$

The solution of the given system is $x = 1$ and $y = 2$. The pair $(1, 2)$ satisfies both of the equations in the given system.   ∎

Cramer's rule can be used to solve many systems of n linear equations in n variables where $n > 2$. But to do so, we must develop a method of evaluating determinants larger than 2×2 that will give the correct solutions. Such a method is called **expansion by minors**.

Definition.   Let A be a square matrix of order n, and let a_{ij} be the element in the ith row and the jth column of A. The **minor** M_{ij} of a_{ij} is the determinant of the $n - 1 \times n - 1$ matrix formed by those elements of A that do not lie in row i or in column j.

To find the minor of any chosen element of a determinant, we cross out the row and the column of that element. The minor is the determinant of the square array of numbers that remains.

Definition.   The **cofactor** A_{ij} of the element a_{ij} in an $n \times n$ matrix A is

$$A_{ij} = (-1)^{i+j} M_{ij}$$

Thus, the cofactor of a_{ij} is the minor of a_{ij} if $i + j$ is even, or the negative of the minor of a_{ij} if $i + j$ is odd.

Example 3   Find the cofactor of **a.** 7 and **b.** 2 in the matrix $\begin{bmatrix} 1 & 2 & 3 \\ 4 & 5 & 6 \\ 7 & 8 & 9 \end{bmatrix}$.

Solution   **a.**   In the determinant of the 3×3 matrix A

$$\begin{vmatrix} 1 & 2 & 3 \\ 4 & 5 & 6 \\ 7 & 8 & 9 \end{vmatrix}$$

the minor of the 7 that appears in the third row, first column, is the determinant of the 2×2 matrix

$$\begin{vmatrix} 2 & 3 \\ 5 & 6 \end{vmatrix}$$

Because 7 is in the third row, first column, its row number plus the column number is even $(3 + 1 = 4)$. Thus, the cofactor of 7 is its minor:

$$\begin{vmatrix} 2 & 3 \\ 5 & 6 \end{vmatrix} = 2 \cdot 6 - 3 \cdot 5 = 12 - 15 = -3$$

b.   The minor of 2, which appears in the first row, second column of A is

$$\begin{vmatrix} 4 & 6 \\ 7 & 9 \end{vmatrix} = 4 \cdot 9 - 6 \cdot 7 = 36 - 42 = -6$$

This is found by ignoring the row and column in which the 2 is located. Because the sum of the row number and column number is odd $(1 + 2 = 3)$, the cofactor of the element 2 is the *negative* of its minor. Thus, the cofactor of 2 is $-(-6) = +6$.   ■

Definition. The **determinant of the square matrix** A is the sum of the products of the elements in the first row of the matrix and the cofactors of those elements.

$$|A| = \begin{vmatrix} a_{11} & a_{12} & \cdots & a_{1n} \\ a_{21} & a_{22} & \cdots & a_{2n} \\ \vdots & \vdots & & \vdots \\ a_{n1} & a_{n2} & \cdots & a_{nn} \end{vmatrix} = a_{11}A_{11} + a_{12}A_{12} + \cdots + a_{1n}A_{1n}$$

A determinant need not be expanded along its *first* row. It can be proved that the determinant of any square matrix is the sum of the products of the elements of *any* row (or column) of the matrix and the cofactors of those elements.

Example 4 Evaluate the determinant $\begin{vmatrix} 1 & 2 & -3 \\ -1 & 0 & 1 \\ -2 & 2 & 1 \end{vmatrix}$ along **a.** the first row, **b.** the third row, and **c.** the second column.

Solution **a.** Multiply each element in the first row by its cofactor and form the sum of these three products:

$$\begin{vmatrix} \mathbf{1} & \mathbf{2} & \mathbf{-3} \\ -1 & 0 & 1 \\ -2 & 2 & 1 \end{vmatrix} = \mathbf{1}\begin{vmatrix} 0 & 1 \\ 2 & 1 \end{vmatrix} \mathbf{-2}\begin{vmatrix} -1 & 1 \\ -2 & 1 \end{vmatrix} + (\mathbf{-3})\begin{vmatrix} -1 & 0 \\ -2 & 2 \end{vmatrix}$$

Evaluate the three 2×2 determinants and simplify:

$$\begin{vmatrix} 1 & 2 & -3 \\ -1 & 0 & 1 \\ -2 & 2 & 1 \end{vmatrix} = 1(-2) - 2(1) - 3(-2)$$

$$= -2 - 2 + 6$$

$$= \mathbf{2}$$

b. Multiply each element in the third row by its cofactor and form the sum of these three products:

$$\begin{vmatrix} 1 & 2 & -3 \\ -1 & 0 & 1 \\ \mathbf{-2} & \mathbf{2} & \mathbf{1} \end{vmatrix} = \mathbf{-2}\begin{vmatrix} 2 & -3 \\ 0 & 1 \end{vmatrix} \mathbf{-2}\begin{vmatrix} 1 & -3 \\ -1 & 1 \end{vmatrix} + \mathbf{1}\begin{vmatrix} 1 & 2 \\ -1 & 0 \end{vmatrix}$$

$$= -2(2) - 2(-2) + 1(2)$$

$$= -4 + 4 + 2$$

$$= \mathbf{2}$$

c. Multiply each element of the second column by its cofactor and form the sum of these three products:

$$\begin{vmatrix} 1 & 2 & -3 \\ -1 & 0 & 1 \\ -2 & 2 & 1 \end{vmatrix} = -2\begin{vmatrix} -1 & 1 \\ -2 & 1 \end{vmatrix} + 0\begin{vmatrix} 1 & -3 \\ -2 & 1 \end{vmatrix} - 2\begin{vmatrix} 1 & -3 \\ -1 & 1 \end{vmatrix}$$

$$= -2(1) + 0(-5) - 2(-2)$$
$$= -2 + 4$$
$$= 2$$

Note that the same result is obtained in each part of this example. In fact, the same result will be obtained if you expand the determinant along *any* row or column. ∎

Example 5 Evaluate the determinant $\begin{vmatrix} 0 & 0 & 2 & 0 \\ 1 & 2 & 17 & -3 \\ -1 & 0 & 28 & 1 \\ -2 & 2 & -37 & 1 \end{vmatrix}$.

Solution Because the determinant can be evaluated along *any* row or column, choose the row or column with the most 0s. So, expand this determinant along its first row. Note that three of the four minors need not be evaluated because each will be multiplied by 0.

$$\begin{vmatrix} 0 & 0 & 2 & 0 \\ 1 & 2 & 17 & -3 \\ -1 & 0 & 28 & 1 \\ -2 & 2 & -37 & 1 \end{vmatrix} = 0\begin{vmatrix} \text{Who cares?} \end{vmatrix} - 0\begin{vmatrix} \text{Who cares?} \end{vmatrix} + 2\begin{vmatrix} 1 & 2 & -3 \\ -1 & 0 & 1 \\ -2 & 2 & 1 \end{vmatrix} - 0\begin{vmatrix} \text{Who cares?} \end{vmatrix}$$

$$= 2(2) \qquad \text{See Example 4.}$$
$$= 4 \qquad\qquad\qquad\qquad ∎$$

Example 5 suggests the following theorem.

Theorem. If any row or column of a square matrix consists entirely of 0s, the value of the determinant of the matrix is 0.

Proof A determinant can be evaluated along any of its rows or columns. Evaluate the determinant along its row or column of 0s. Each entry in that row or column is 0, and the product of each entry with its cofactor is also 0. The value of the determinant is the sum of these products. Hence, the value of the determinant is 0. □

The next example illustrates how Cramer's rule can be used to solve a system of three equations in three variables.

Example 6 Use Cramer's rule to solve the following system of equations:

$$\begin{cases} 2x - y + 2z = 3 \\ x - y + z = 2 \\ x + y + 2z = 3 \end{cases}$$

Solution   Each of the values x, y, and z is the ratio of two 3×3 determinants. The denominator of each quotient is the determinant consisting of the nine coefficients of the variables. The numerators for x, y, and z are modified copies of this denominator determinant. The column of constants is substituted for the coefficients of the variable for which you are solving.

$$\begin{cases} 2x - y + 2z = 3 \\ x - y + z = 2 \\ x + y + 2z = 3 \end{cases}$$

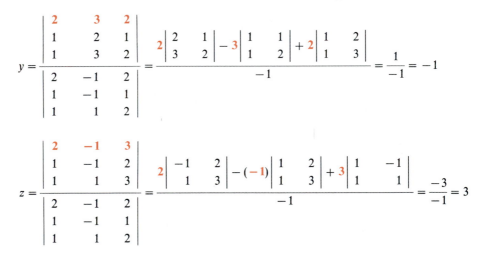

$$x = \frac{\begin{vmatrix} 3 & -1 & 2 \\ 2 & -1 & 1 \\ 3 & 1 & 2 \end{vmatrix}}{\begin{vmatrix} 2 & -1 & 2 \\ 1 & -1 & 1 \\ 1 & 1 & 2 \end{vmatrix}} = \frac{3\begin{vmatrix} -1 & 1 \\ 1 & 2 \end{vmatrix} - (-1)\begin{vmatrix} 2 & 1 \\ 3 & 2 \end{vmatrix} + 2\begin{vmatrix} 2 & -1 \\ 3 & 1 \end{vmatrix}}{2\begin{vmatrix} -1 & 1 \\ 1 & 2 \end{vmatrix} - (-1)\begin{vmatrix} 1 & 1 \\ 1 & 2 \end{vmatrix} + 2\begin{vmatrix} 1 & -1 \\ 1 & 1 \end{vmatrix}} = \frac{2}{-1} = -2$$

$$y = \frac{\begin{vmatrix} 2 & 3 & 2 \\ 1 & 2 & 1 \\ 1 & 3 & 2 \end{vmatrix}}{\begin{vmatrix} 2 & -1 & 2 \\ 1 & -1 & 1 \\ 1 & 1 & 2 \end{vmatrix}} = \frac{2\begin{vmatrix} 2 & 1 \\ 3 & 2 \end{vmatrix} - 3\begin{vmatrix} 1 & 1 \\ 1 & 2 \end{vmatrix} + 2\begin{vmatrix} 1 & 2 \\ 1 & 3 \end{vmatrix}}{-1} = \frac{1}{-1} = -1$$

$$z = \frac{\begin{vmatrix} 2 & -1 & 3 \\ 1 & -1 & 2 \\ 1 & 1 & 3 \end{vmatrix}}{\begin{vmatrix} 2 & -1 & 2 \\ 1 & -1 & 1 \\ 1 & 1 & 2 \end{vmatrix}} = \frac{2\begin{vmatrix} -1 & 2 \\ 1 & 3 \end{vmatrix} - (-1)\begin{vmatrix} 1 & 2 \\ 1 & 3 \end{vmatrix} + 3\begin{vmatrix} 1 & -1 \\ 1 & 1 \end{vmatrix}}{-1} = \frac{-3}{-1} = 3$$

The triple $(-2, -1, 3)$ satisfies each of the equations in the given system.   ∎

Type 3 row operations can be used to simplify the calculations involved in determinant expansion.

> **Theorem.**   A type 3 elementary row operation performed on a square matrix does not alter the value of its determinant.

Justification.   The following discussion will illustrate the previous theorem for one particular row operation on a 3×3 matrix.

Evaluating $|D|$ by expanding along the first row, we get

$$|D| = \begin{vmatrix} a & b & c \\ d & e & f \\ g & h & i \end{vmatrix} = a\begin{vmatrix} e & f \\ h & i \end{vmatrix} - b\begin{vmatrix} d & f \\ g & i \end{vmatrix} + c\begin{vmatrix} d & e \\ g & h \end{vmatrix}$$

If we perform the type 3 row operation $kR3 + R1 \to R1$ on matrix D, the resulting modified determinant, $|D'|$, has a new first row. We expand that determinant on its first row:

$$|D'| = \begin{vmatrix} a+kg & b+kh & c+ki \\ d & e & f \\ g & h & i \end{vmatrix} = (a+kg)\begin{vmatrix} e & f \\ h & i \end{vmatrix} - (b+kh)\begin{vmatrix} d & f \\ g & i \end{vmatrix} + (c+ki)\begin{vmatrix} d & e \\ g & h \end{vmatrix}$$

Using the distributive law to remove the parentheses and rearranging the terms gives

$$|D'| = a\begin{vmatrix} e & f \\ h & i \end{vmatrix} - b\begin{vmatrix} d & f \\ g & i \end{vmatrix} + c\begin{vmatrix} d & e \\ g & h \end{vmatrix} + kg\begin{vmatrix} e & f \\ h & i \end{vmatrix} - kh\begin{vmatrix} d & f \\ g & i \end{vmatrix} + ki\begin{vmatrix} d & e \\ g & h \end{vmatrix}$$

The first three terms of the above expansion are identical to the original determinant $|D|$. The original determinant and the modified determinant differ by the amount that is represented by the last three terms of the expansion of $|D'|$. Hence,

$$|D'| = |D| + k\left(g\begin{vmatrix} e & f \\ h & i \end{vmatrix} - h\begin{vmatrix} d & f \\ g & i \end{vmatrix} + i\begin{vmatrix} d & e \\ g & h \end{vmatrix} \right)$$

$$|D'| = |D| + k(gei - ghf - hdi + hgf + idh - ige)$$

All terms within the parentheses subtract out, and we have

$$|D'| = |D| + k \cdot 0$$
$$|D'| = |D|$$

A similar result holds for type 3 column operations.

Theorem.   If any column of a square matrix is altered by adding to it any multiple of another column, the value of the determinant of the matrix is unchanged.

These two theorems provide a way to reduce the work involved in evaluating a large determinant.

Example 7   Evaluate $|A|$ if $|A| = \begin{vmatrix} 1 & 2 & -1 & 2 \\ 2 & 1 & 1 & 1 \\ 1 & 2 & -3 & 2 \\ 2 & -1 & -1 & 1 \end{vmatrix}$.

Solution   Expanding the given determinant along any row or column leads to four determinants involving 3×3 matrices. The row operation $(-1)R3 + R1 \to R1$ gives three 0s in the first row. The introduction of these 0s simplifies the work because only one 3×3 determinant needs to be evaluated. Expand the new determinant along the first row.

$$|A| = \begin{vmatrix} 0 & 0 & 2 & 0 \\ 2 & 1 & 1 & 1 \\ 1 & 2 & -3 & 2 \\ 2 & -1 & -1 & 1 \end{vmatrix} = 2\begin{vmatrix} 2 & 1 & 1 \\ 1 & 2 & 2 \\ 2 & -1 & 1 \end{vmatrix}$$

To introduce more 0s, perform the column operation $(-2)C3 + C1 \rightarrow C1$ on the 3×3

determinant $\begin{vmatrix} 2 & 1 & 1 \\ 1 & 2 & 2 \\ 2 & -1 & 1 \end{vmatrix}$ and expand that result on its first column:

$$|A| = 2\begin{vmatrix} 0 & 1 & 1 \\ -3 & 2 & 2 \\ 0 & -1 & 1 \end{vmatrix} = 2\left[-(-3)\begin{vmatrix} 1 & 1 \\ -1 & 1 \end{vmatrix} \right]$$

$$= 2 \cdot 3[1 - (-1)]$$
$$= 2 \cdot 3 \cdot 2$$
$$= 12$$ ■

Two more theorems describe the effect of type 2 and type 1 row operations on a determinant.

Theorem.   If any row or column of a square matrix is multiplied by a constant k, the value of the determinant of the matrix is multiplied by k.

Theorem.   If two rows or columns of a square matrix are interchanged, the value of the determinant of the matrix is multiplied by -1.

Exercise 8.5

In Exercises 1–16 evaluate each determinant.

1. $\begin{vmatrix} 2 & 1 \\ -2 & 3 \end{vmatrix}$

2. $\begin{vmatrix} -3 & -6 \\ 2 & -5 \end{vmatrix}$

3. $\begin{vmatrix} 2 & -3 \\ -3 & 5 \end{vmatrix}$

4. $\begin{vmatrix} 5 & 8 \\ -6 & -2 \end{vmatrix}$

5. $\begin{vmatrix} 2 & -3 & 5 \\ -2 & 1 & 3 \\ 1 & 3 & -2 \end{vmatrix}$

6. $\begin{vmatrix} 1 & 3 & 1 \\ -2 & 5 & 3 \\ 3 & -2 & -2 \end{vmatrix}$

7. $\begin{vmatrix} 1 & -1 & 2 \\ 2 & 1 & 3 \\ 1 & 1 & -1 \end{vmatrix}$

8. $\begin{vmatrix} 1 & 3 & 1 \\ 2 & 1 & -1 \\ 2 & -1 & 1 \end{vmatrix}$

9. $\begin{vmatrix} 2 & 1 & -1 \\ 1 & 3 & 5 \\ 2 & -5 & 3 \end{vmatrix}$

10. $\begin{vmatrix} 3 & 1 & -2 \\ -3 & 2 & 1 \\ 1 & 3 & 0 \end{vmatrix}$

11. $\begin{vmatrix} 0 & 1 & -3 \\ -3 & 5 & 2 \\ 2 & -5 & 3 \end{vmatrix}$

12. $\begin{vmatrix} 1 & -7 & -2 \\ -2 & 0 & 3 \\ -1 & 7 & 1 \end{vmatrix}$

13. $\begin{vmatrix} 1 & 2 & 1 & 3 \\ -2 & 1 & -3 & 1 \\ -1 & 0 & 1 & -2 \\ 2 & -1 & -1 & 3 \end{vmatrix}$

14. $\begin{vmatrix} -1 & 3 & -2 & 5 \\ 2 & 1 & 0 & 1 \\ 1 & 3 & -2 & 5 \\ 2 & -1 & 0 & -1 \end{vmatrix}$

15. $\begin{vmatrix} 1 & 2 & 3 & 4 & 5 \\ 0 & 1 & 2 & 3 & 4 \\ 0 & 0 & 1 & 2 & 3 \\ 0 & 0 & 0 & 1 & 2 \\ 0 & 0 & 0 & 0 & 1 \end{vmatrix}$

16. $\begin{vmatrix} 1 & 1 & 1 & 1 & 1 \\ 1 & 1 & 1 & 1 & 2 \\ 1 & 1 & 1 & 2 & 2 \\ 1 & 1 & 2 & 2 & 2 \\ 1 & 2 & 2 & 2 & 2 \end{vmatrix}$

In Exercises 17–28 use Cramer's rule to find the solution to each system of equations, if possible.

17. $\begin{cases} 3x + 2y = 7 \\ 2x - 3y = -4 \end{cases}$

18. $\begin{cases} x - 5y = -6 \\ 3x + 2y = -1 \end{cases}$

19. $\begin{cases} x - y = 3 \\ 3x - 7y = 9 \end{cases}$

20. $\begin{cases} 2x - y = -6 \\ x + y = 0 \end{cases}$

21. $\begin{cases} x + 2y + z = 2 \\ x - y + z = 2 \\ x + y + 3z = 4 \end{cases}$

22. $\begin{cases} x + 2y - z = -1 \\ 2x + y - z = 1 \\ x - 3y - 5z = 17 \end{cases}$

23. $\begin{cases} 2x - y + z = 5 \\ 3x - 3y + 2z = 10 \\ x + 3y + z = 0 \end{cases}$

24. $\begin{cases} x - y - z = 2 \\ x + y + z = 2 \\ -x - y + z = -4 \end{cases}$

25. $\begin{cases} \dfrac{x}{2} + \dfrac{y}{3} + \dfrac{z}{2} = 11 \\ \dfrac{x}{3} + y - \dfrac{z}{6} = 6 \\ \dfrac{x}{2} + \dfrac{y}{6} + z = 16 \end{cases}$

26. $\begin{cases} \dfrac{x}{2} + \dfrac{y}{5} + \dfrac{z}{3} = 17 \\ \dfrac{x}{5} + \dfrac{y}{2} + \dfrac{z}{5} = 32 \\ x + \dfrac{y}{3} + \dfrac{z}{2} = 30 \end{cases}$

27. $\begin{cases} 2p - q + 3r - s = 3 \\ p + q - 2s = 0 \\ 3p - r = 2 \\ p - q + 3s = 3 \end{cases}$

28. $\begin{cases} a + b + c + d = -1 \\ a + b + c + 2d = 0 \\ a + b + 2c + 3d = 1 \\ a + 2b + 3c + 4d = 0 \end{cases}$

29. Use the method of addition to solve the system $\begin{cases} ax + by = e \\ cx + dy = f \end{cases}$ for y, and thereby show that

$$y = \frac{af - ec}{ad - bc}$$

30. Use an example chosen from 2×2 matrices to show that the determinant of the product of two matrices is the product of the determinants of those two matrices.

31. Find an example among 2×2 determinants to show that the determinant of a sum of two matrices is not equal to the sum of the determinants of those matrices.

32. Find a 2×2 matrix A for which $|A| \neq 0$. Then find A^{-1}, and verify that $|A^{-1}| = \dfrac{1}{|A|}$.

33. Show that $\begin{vmatrix} a & b & c \\ 0 & d & e \\ 0 & 0 & f \end{vmatrix} = adf.$

34. Show that $\begin{vmatrix} a & b & c & d \\ 0 & e & f & g \\ 0 & 0 & h & i \\ 0 & 0 & 0 & j \end{vmatrix} = aehj.$

35. Show that multiplying the first row of $\begin{vmatrix} 2 & -1 & 3 \\ 1 & 2 & -1 \\ 3 & 2 & -1 \end{vmatrix}$ by 3 multiplies the value of the determinant by 3.

36. Interchange two rows of the determinant in Exercise 35 and show that this has the effect of multiplying the value of the determinant by -1.

37. Interchange two columns of the determinant in Exercise 35 and show that this has the effect of multiplying the value of the determinant by -1.

38. Find the value of k if $\begin{vmatrix} 2a & 2b & 2c \\ 3d & 3e & 3f \\ 5g & 5h & 5i \end{vmatrix} = k \begin{vmatrix} a & b & c \\ d & e & f \\ g & h & i \end{vmatrix}.$

In Exercises 39–42 expand the determinants and solve for x.

39. $\begin{vmatrix} 3 & x \\ 1 & 2 \end{vmatrix} = \begin{vmatrix} 2 & -1 \\ x & -5 \end{vmatrix}$

40. $\begin{vmatrix} 4 & x^2 \\ 1 & -1 \end{vmatrix} = \begin{vmatrix} x & 4 \\ 2 & 3 \end{vmatrix}$

41. $\begin{vmatrix} 3 & x & 1 \\ x & 0 & -2 \\ 4 & 0 & 1 \end{vmatrix} = \begin{vmatrix} 2 & x \\ x & 4 \end{vmatrix}$

42. $\begin{vmatrix} x & -1 & 2 \\ -2 & x & 3 \\ 4 & -3 & -1 \end{vmatrix} = \begin{vmatrix} 2 & 2 \\ 5 & x \end{vmatrix}$

43. A determinant is a function that associates a number with every square matrix. Give the domain and the range of that function.

44. Select 2×2 matrices A and B and show that $|AB| = |BA|$.

45. If A and B are matrices and if $|AB| = 0$, must $|A| = 0$ or $|B| = 0$? Support your answer.

46. If A and B are matrices and if $|AB| = 0$, must $A = 0$ or $B = 0$? Support your answer.

8.6 PARTIAL FRACTIONS

In this section we will solve systems of equations to express a complicated fraction as the sum of several simpler fractions. This process, called **decomposing a fraction into partial fractions**, is important in calculus. We begin by reviewing addition of fractions.

Example 1 Find the sum: $\dfrac{2}{x} + \dfrac{6}{x+1} + \dfrac{-1}{(x+1)^2}.$

Solution Write each fraction in a form with the least common denominator, $x(x + 1)^2$, and add:

$$\frac{2}{x} + \frac{6}{x + 1} + \frac{-1}{(x + 1)^2} = \frac{2(x + 1)^2}{x(x + 1)^2} + \frac{6x(x + 1)}{(x + 1)x(x + 1)} + \frac{-1x}{(x + 1)^2 x}$$

$$= \frac{2x^2 + 4x + 2 + 6x^2 + 6x - x}{x(x + 1)^2}$$

$$= \frac{8x^2 + 9x + 2}{x(x + 1)^2}$$ ∎

Example 2 Express the fraction $\dfrac{3x^2 - x + 1}{x(x - 1)^2}$ as the sum of several fractions with denominators of the smallest degree possible.

Solution Example 1 leads you to suspect that constants A, B, and C can be found such that

$$\frac{3x^2 - x + 1}{x(x - 1)^2} = \frac{A}{x} + \frac{B}{x - 1} + \frac{C}{(x - 1)^2}$$

After you write the terms on the right side as fractions with the common denominator $x(x - 1)^2$, combine them:

$$\frac{3x^2 - x + 1}{x(x - 1)^2} = \frac{A(x - 1)^2}{x(x - 1)^2} + \frac{Bx(x - 1)}{x(x - 1)(x - 1)} + \frac{Cx}{(x - 1)^2 x}$$

$$= \frac{Ax^2 - 2Ax + A + Bx^2 - Bx + Cx}{x(x - 1)^2}$$

$$= \frac{(A + B)x^2 + (-2A - B + C)x + A}{x(x - 1)^2}$$

Johann Bernoulli
(1667–1748)
A Swiss mathematician
and teacher of Euler. His
method of decomposition by
partial fractions was a
major contribution
to the calculus.

Because the fractions are equal, the numerator $3x^2 - x + 1$ must equal the numerator $(A + B)x^2 + (-2A - B + C)x + A$. Thus,

$$\begin{cases} A + B & = \ \ 3 & \text{The coefficients of } x^2. \\ -2A - B + C = -1 & \text{The coefficients of } x. \\ A & = \ \ 1 & \text{The constants.} \end{cases}$$

This system of three equations in three variables can be solved by substitution. The solutions are $A = 1$, $B = 2$, and $C = 3$. Hence,

$$\frac{3x^2 - x + 1}{x(x - 1)^2} = \frac{A}{x} + \frac{B}{x - 1} + \frac{C}{(x - 1)^2} = \frac{1}{x} + \frac{2}{x - 1} + \frac{3}{(x - 1)^2}$$ ∎

Example 3 Express the fraction $\dfrac{2x^2 + x + 1}{x^3 + x}$ as the sum of fractions with denominators of the smallest possible degree.

Solution Factoring the denominator suggests that this fraction can be expressed as the sum of two fractions, one with a denominator of x and the other with a denominator of $x^2 + 1$.

$$\frac{2x^2 + x + 1}{x(x^2 + 1)} = \frac{}{x} + \frac{}{x^2 + 1}$$

Because the denominator x of the first fraction is of first degree, its numerator must be of degree 0—that is, a constant. Because the denominator $x^2 + 1$ of the second fraction is of second degree, the numerator might be a first-degree polynomial or a constant. You can allow for both possibilities by using a numerator of $Bx + C$. If $B = 0$, then $Bx + C$ is a constant. If $B \neq 0$, then $Bx + C$ is a first-degree polynomial. Thus,

$$\frac{2x^2 + x + 1}{x(x^2 + 1)} = \frac{A}{x} + \frac{Bx + C}{x^2 + 1} = \frac{A(x^2 + 1) + (Bx + C)x}{x(x^2 + 1)}$$

$$= \frac{Ax^2 + A + Bx^2 + Cx}{x(x^2 + 1)} = \frac{(A + B)x^2 + Cx + A}{x(x^2 + 1)}$$

Equate the corresponding coefficients of the polynomials $2x^2 + x + 1$ and $(A + B)x^2 + Cx + A$ to produce the following system of equations:

$$\begin{cases} A + B = 2 \\ \quad\ C = 1 \\ A = 1 \end{cases}$$

The solutions are $A = 1$, $B = 1$, $C = 1$. Therefore, the given fraction can be written as a sum:

$$\frac{2x^2 + x + 1}{x^3 + x} = \frac{1}{x} + \frac{x + 1}{x^2 + 1} \qquad\qquad \blacksquare$$

We summarize the process illustrated in these examples: Let $\frac{P(x)}{Q(x)}$ be the quotient of two polynomials with real coefficients, with the degree of $P(x)$ being less than the degree of $Q(x)$. Suppose also that the fraction $\frac{P(x)}{Q(x)}$ has been simplified.

The polynomial $Q(x)$ can always be factored as a product of first-degree and irreducible second-degree expressions. If all the identical factors of $Q(x)$ are collected into single factors of the form $(ax + b)^n$ or the form $(ax^2 + bx + c)^n$, then the partial fractions required for the decomposition of $\frac{P(x)}{Q(x)}$ can be found. Each factor of $Q(x)$ of the form $(ax + b)^n$ generates a sum of n partial fractions of the form

$$\frac{A_1}{ax + b} + \frac{A_2}{(ax + b)^2} + \cdots + \frac{A_n}{(ax + b)^n}$$

where each A_i represents a constant. Each factor of the form $(ax^2 + bx + c)^n$ generates the sum of n fractions of the form

$$\frac{B_1 x + C_1}{ax^2 + bx + c} + \frac{B_2 x + C_2}{(ax^2 + bx + c)^2} + \cdots + \frac{B_n x + C_n}{(ax^2 + bx + c)^n}$$

where each B_i and C_i is a constant. After finding a least common denominator and adding the fractions, we obtain a fractional expression that must be equivalent to $\frac{P(x)}{Q(x)}$. Equating the corresponding coefficients of the numerators gives a system of linear equations that can be solved for the constants A_i, B_i, and C_i.

Example 4   What fractions should be used in the decomposition of the rational expression

$$\frac{3x^7 - 5x^5 + 3x + 2}{x^3(x - 3)(x + 2)^2(2x^2 + x + 3)^2(x^2 + 1)^3}$$

Solution   The factor x^3 in the denominator requires three possible fractions in the decomposition:

$$\frac{A}{x} + \frac{B}{x^2} + \frac{C}{x^3}$$

The factor $x - 3$ adds one more to the list:

$$\frac{A}{x} + \frac{B}{x^2} + \frac{C}{x^3} + \frac{D}{x - 3}$$

The factor $(x + 2)^2$ generates two more fractions, each with a constant as numerator:

$$\frac{A}{x} + \frac{B}{x^2} + \frac{C}{x^3} + \frac{D}{x - 3} + \frac{E}{x + 2} + \frac{F}{(x + 2)^2}$$

The factor $(2x^2 + x + 3)^2$ produces two more fractions, each requiring first-degree numerators:

$$\frac{A}{x} + \frac{B}{x^2} + \frac{C}{x^3} + \frac{D}{x - 3} + \frac{E}{x + 2} + \frac{F}{(x + 2)^2} + \frac{Gx + H}{2x^2 + x + 3} + \frac{Jx + K}{(2x^2 + x + 3)^2}$$

Finally, the factor $(x^2 + 1)^3$ requires three more fractions, also with first-degree numerators:

$$\frac{A}{x} + \frac{B}{x^2} + \frac{C}{x^3} + \frac{D}{x - 3} + \frac{E}{x + 2} + \frac{F}{(x + 2)^2} + \frac{Gx + H}{2x^2 + x + 3}$$

$$+ \frac{Jx + K}{(2x^2 + x + 3)^2} + \frac{Lx + M}{x^2 + 1} + \frac{Nx + P}{(x^2 + 1)^2} + \frac{Rx + S}{(x^2 + 1)^3}$$

If you find a common denominator and combine the fractions, equating the corresponding coefficients of the numerators will give 16 equations in 16 variables. These can be solved for the variables $A, B, C, \ldots, S$.   ■

Example 5   Express the fraction $\dfrac{x^2 + 4x + 2}{x^2 + x}$ as the sum of several fractions with denominators of the smallest degree possible.

Solution   The method of partial fractions requires that the degree of the numerator be less than the degree of the denominator. Because the degree of both the numerator and denominator is the same in this example, you must perform a long division and express the given fraction in $quotient + \dfrac{remainder}{divisor}$ form:

$$
\begin{array}{r}
1 \\
x^2 + x\,\overline{)\,x^2 + 4x + 2} \\
\underline{x^2 + x} \\
3x + 2
\end{array}
$$

Hence,

$$\frac{x^2 + 4x + 2}{x^2 + x} = 1 + \frac{3x + 2}{x^2 + x}$$

Because the degree of the numerator of the fraction $\dfrac{3x + 2}{x^2 + x}$ is less than the degree of the denominator, you can find the partial fraction decomposition of this fraction:

$$\frac{3x + 2}{x^2 + x} = \frac{3x + 2}{x(x + 1)}$$

$$= \frac{A}{x} + \frac{B}{x + 1}$$

$$= \frac{A(x + 1) + Bx}{x(x + 1)}$$

$$= \frac{(A + B)x + A}{x(x + 1)}$$

Equate the corresponding coefficients in the numerator, and solve the resulting system of equations:

$$\begin{cases} A + B = 3 \\ \quad\ A = 2 \end{cases}$$

The solution is $A = 2$, $B = 1$, and the decomposition of the given fraction is

$$\frac{x^2 + 4x + 2}{x^2 + x} = 1 + \frac{2}{x} + \frac{1}{x + 1}$$

■

Exercise 8.6

In Exercises 1–24 decompose each expression into partial fractions.

1. $\dfrac{3x + 1}{(x + 1)(x - 1)}$

2. $\dfrac{-2x + 11}{x^2 - x - 6}$

3. $\dfrac{-3x^2 + x - 5}{(x + 1)(x^2 + 2)}$

4. $\dfrac{-x^2 - 3x - 5}{x^3 + x^2 + 2x + 2}$

5. $\dfrac{-2x^2 + x - 2}{x^3 - x^2}$

6. $\dfrac{2x^2 - 7x + 2}{x(x - 1)^2}$

7. $\dfrac{2x^2 + 1}{x^4 + x^2}$

8. $\dfrac{x^2 + x + 1}{x^3}$

9. $\dfrac{5x^2 + 2x + 2}{x^3 + x}$

10. $\dfrac{-2x^3 + 7x^2 + 6}{x^2(x^2 + 2)}$

11. $\dfrac{x^3 + 4x^2 + 2x + 1}{x^4 + x^3 + x^2}$

12. $\dfrac{x^3 + 4x^2 + 3x + 6}{(x^2 + 2)(x^2 + x + 2)}$

13. $\dfrac{x^3 + 3x^2 + 6x + 6}{(x^2 + x + 5)(x^2 + 1)}$

14. $\dfrac{x^2 - 2x - 3}{(x - 1)^3}$

15. $\dfrac{x^4 - x^3 + x^2 - x + 1}{x(x^2 + 1)^2}$

16. $\dfrac{x^2 + 2}{x^3 + 3x^2 + 3x + 1}$

17. $\dfrac{x^3 + 3x^2 + 2x + 4}{(x^2 + 1)(x^2 + x + 2)}$

18. $\dfrac{4x^3 + 5x^2 + 3x + 4}{x^2(x^2 + 1)}$

19. $\dfrac{2x^4 + 6x^3 + 20x^2 + 22x + 25}{x(x^2 + 2x + 5)^2}$

20. $\dfrac{3x^3 + 5x^2 + 3x + 1}{x^2(x^2 + x + 1)}$

21. $\dfrac{2x^3 + 6x^2 + 3x + 2}{x^3 + x^2}$

22. $\dfrac{x^3}{x^2 + 3x + 2}$

23. $\dfrac{x^4 + x^3 + x^2 + x + 1}{x^2}$

24. $\dfrac{x^3 + 2x^2 + 3x + 4}{x^3}$

8.7 SYSTEMS OF INEQUALITIES AND LINEAR PROGRAMMING

We now consider the graphs of systems of inequalities.

Example 1   Graph the solution set of the system

$$\begin{cases} x + y \le 1 \\ 2x - y > 2 \end{cases}$$

Solution   On the same set of coordinate axes, graph each inequality. See Figure 8–4. The graph of the inequality $x + y \le 1$ includes the line graph of the equation $x + y = 1$ and all points below it. Because the boundary line is included, it is drawn as a solid line. The graph of the inequality $2x - y > 2$ contains only those points below the line graph of the equation $2x - y = 2$. Because the boundary line is not included, it is drawn as a broken line. The area that is shaded twice represents the simultaneous solutions of the given system of inequalities. Any point in the doubly shaded region has coordinates that satisfy both inequalities in the system.

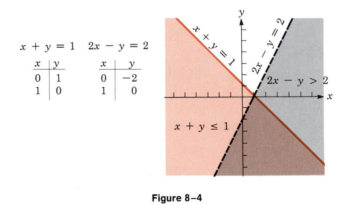

$x + y = 1$		$2x - y = 2$	
x	y	x	y
0	1	0	-2
1	0	1	0

Figure 8–4

Example 2   Graph the solution set of the system

$$\begin{cases} y < x^2 \\ y > \dfrac{x^2}{4} - 2 \end{cases}$$

Solution   The graph of the equation $y = x^2$ is a parabola opening upward with vertex at the origin. See Figure 8–5. The points with coordinates that satisfy the inequality $y < x^2$ are those points below the parabola.

The graph of $y = \frac{x^2}{4} - 2$ is also a parabola opening upward. However, this time the points with coordinates that satisfy the inequality are those points above the parabola. Thus, the graph of the solution set of this system is the shaded area between the two parabolas.

$$y = x^2 \qquad y = \frac{x^2}{4} - 2$$

x	y
0	0
1	1
-1	1
2	4
-2	4

x	y
0	-2
2	-1
-2	-1
4	2
-4	2

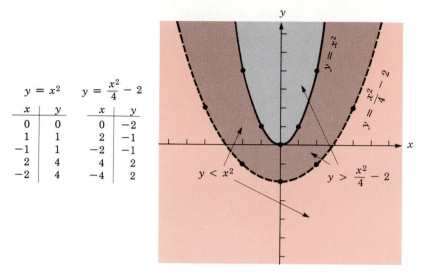

Figure 8–5

Example 3   Graph the solution set of the system

$$\begin{cases} x \geq 1 \\ y \geq x \\ 4x + 5y < 20 \end{cases}$$

Solution   The graph of the solution set of the inequality $x \geq 1$ includes those points on the graph of the equation $x = 1$ and to the right. See Figure 8–6a. The graph of the solution set of the inequality $y \geq x$ includes those points on the graph of the equation $y = x$ and above it. See Figure 8–6b. The graph of the solution set of the inequality $4x + 5y < 20$ includes those points below the line graph of the equation $4x + 5y = 20$. See Figure 8–6c.

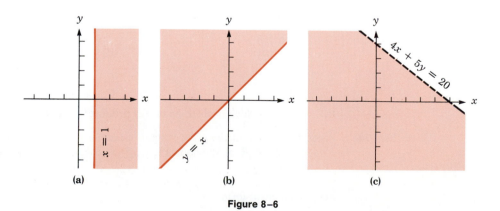

Figure 8–6

If these three graphs are merged onto a single coordinate system, the graph of the original system of inequalities includes those points within the shaded triangle together with the points on the sides of the triangle drawn as solid lines. See Figure 8–7.

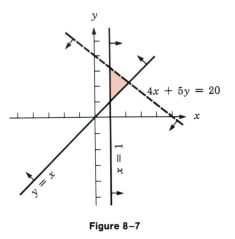

Figure 8–7

Linear Programming

Systems of inequalities provide the basis for an area of applied mathematics known as **linear programming**. Linear programming is used to help answer such questions as "How can a business make as much money as possible?" or "How can I plan a nutritious menu at the least cost?" In such problems, the solution depends on certain **constraints**: The business has limited resources, and the nutritious meal must contain sufficient vitamins, minerals, and so on. Any solution that satisfies the constraints is called a **feasible solution**. In linear programming, the constraints are expressed as a system of linear inequalities, and the quantity that is to be maximized (or minimized) is expressed as a linear function of several variables.

Example 4   Many ordered pairs (x, y) satisfy each inequality in the system

$$\begin{cases} x + y \geq 1 \\ x - y \leq 1 \\ x - y \geq 0 \\ x \leq 2 \end{cases}$$

If $Z = y - 2x$, which of these pairs will produce the greatest value of Z?

Solution   Find the solution of the given system of inequalities, and find the coordinates of each corner of region R, as shown in Figure 8–8. Then rewrite the equation

$$Z = y - 2x$$

in the equivalent form

$$y = 2x + Z$$

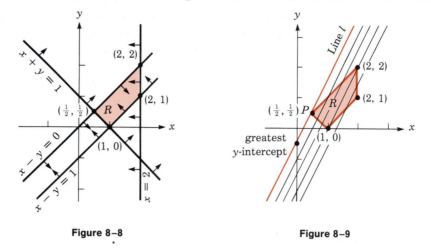

Figure 8–8

Figure 8–9

This is the equation of a straight line with slope of 2 and y-intercept of Z. Many such lines pass through the region R. To decide which of these provides the greatest value of Z, refer to Figure 8–9 and find the line with the greatest y-intercept. It is line l passing through point P, the left-most corner of R. The coordinates of P are $(\frac{1}{2}, \frac{1}{2})$. Thus, the greatest value of Z possible (subject to the given constraints) is

$$Z = \frac{1}{2} - 2\left(\frac{1}{2}\right) = -\frac{1}{2}$$

Example 4 illustrates this fact: *If a linear function that is subject to the constraints of a system of linear inequalities in two variables has a maximum (or minimum) value, it is always attained at a corner or along an entire edge of the region R that represents the solution of the system.*

Example 5   Fred and Donna are in a part-time business manufacturing clock cases. Fred must work 4 hours and Donna 2 hours to complete one case for a grandfather clock. To build one case for a wall clock, Fred must work 3 hours and Donna 4 hours. Neither partner wishes to work more than 20 hours per week. If they receive a profit of $80 for each grandfather clock and $64 for each wall clock, how many of each should they build each week to maximize their profit?

Solution   If the partners manufacture cases for x grandfather clocks and y wall clocks each week, their profit P (in dollars) is

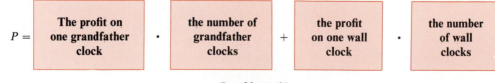

$$P = 80x + 64y$$

The time requirements are summarized in the following chart:

Partner	Time for one grandfather clock	Time for one wall clock
Fred	4 hours	3 hours
Donna	2 hours	4 hours

The profit function is subject to the following constraints:

$$\begin{cases} x \geq 0 \\ y \geq 0 \\ 4x + 3y \leq 20 \\ 2x + 4y \leq 20 \end{cases}$$

The inequalities $x \geq 0$ and $y \geq 0$ state that the number of clock cases to be built cannot be negative. The inequality $4x + 3y \leq 20$ is a constraint on Fred's time because he spends 4 hours on each of the x grandfather clocks and 3 hours on each of the y wall clocks, and his total time cannot exceed 20 hours. Similarly, the inequality $2x + 4y \leq 20$ is a constraint on Donna's time.

Graph each of the constraints to find the **feasibility region R**, as in Figure 8–10. The four corners of region R have coordinates of $(0, 0)$, $(0, 5)$, $(2, 4)$, and $(5, 0)$. Substitute each pair of numbers into the equation $P = 80x + 64y$ to find the maximum profit, P.

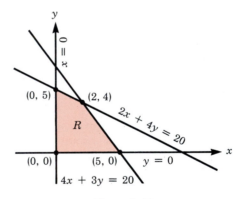

Figure 8–10

Corner	Profit
$(0, 0)$	$P = \$80(0) + \$64(0) = \quad \$0$
$(0, 5)$	$P = \$80(0) + \$64(5) = \$320$
$(2, 4)$	$P = \$80(2) + \$64(4) = \$416$
$(5, 0)$	$P = \$80(5) + \$64(0) = \$400$

Donna and Fred will maximize their profit if they build cases for 2 grandfather clocks and 4 wall clocks each week. If they do so, they will earn \$416. ■

Exercise 8.7

In Exercises 1–20 graph the solution set of each system of inequalities.

1. $\begin{cases} y < 3x + 2 \\ y < -2x + 3 \end{cases}$
2. $\begin{cases} y \leq x - 2 \\ y \geq 2x + 1 \end{cases}$
3. $\begin{cases} 3x + 2y > 6 \\ x + 3y \leq 2 \end{cases}$
4. $\begin{cases} x + y < 2 \\ x + y \leq 1 \end{cases}$

5. $\begin{cases} 3x + y \leq 1 \\ -x + 2y \geq 9 \end{cases}$
6. $\begin{cases} x + 2y < 3 \\ 2x - 4y < 8 \end{cases}$
7. $\begin{cases} 2x - y > 4 \\ y < -x^2 + 2 \end{cases}$
8. $\begin{cases} x \leq y^2 \\ y \geq x \end{cases}$

9. $\begin{cases} y > x^2 - 4 \\ y < -x^2 + 4 \end{cases}$
10. $\begin{cases} x \geq y^2 \\ y \geq x^2 \end{cases}$
11. $\begin{cases} 2x + 3y \leq 5 \\ 3x + y \leq 1 \\ x \leq 0 \end{cases}$
12. $\begin{cases} 2x + y \leq 2 \\ y \geq x \\ x \geq 0 \end{cases}$

13. $\begin{cases} x - y < 4 \\ y \geq 0 \\ xy = 12 \end{cases}$
14. $\begin{cases} xy \leq 1 \\ x \geq 0 \\ y \geq 0 \end{cases}$
15. $\begin{cases} x \geq 0 \\ y \geq 0 \\ 9x + 3y \leq 18 \\ 3x + 6y \leq 18 \end{cases}$
16. $\begin{cases} x + y \geq 1 \\ x - y \leq 1 \\ x - y \geq 0 \\ x \leq 2 \end{cases}$

17. $\begin{cases} y < \sqrt{x} \\ x \geq 0 \end{cases}$
18. $\begin{cases} y < -\sqrt{x} \\ x \geq 0 \end{cases}$
19. $\begin{cases} |x - 2| > 3 \\ |y| > 1 \end{cases}$
20. $\begin{cases} |x + 2| < 3 \\ |y| > 2 \end{cases}$

In Exercises 21–26 maximize P subject to the given constraints.

21. $P = 2x + 3y$
$\begin{cases} x \geq 0 \\ y \geq 0 \\ x + y \leq 4 \end{cases}$

22. $P = 3x + 2y$
$\begin{cases} x \geq 0 \\ y \geq 0 \\ x + y \leq 4 \end{cases}$

23. $P = y + \dfrac{1}{2}x$
$\begin{cases} 2y - x \leq 1 \\ y - 2x \geq -2 \\ x \geq 0 \\ y \geq 0 \end{cases}$

24. $P = 4y - x$
$\begin{cases} 2y - x \leq 1 \\ x \leq 2 \\ x + y \geq 1 \\ y \geq 0 \end{cases}$

25. $P = 2x + y$
$\begin{cases} y - x \leq 2 \\ 2x + 3y \leq 6 \\ 3x + y \leq 3 \\ y \geq 0 \end{cases}$

26. $P = 3x - 2y$
$\begin{cases} y - x \leq 1 \\ x - y \leq 1 \\ x \geq -1 \\ x \leq 1 \end{cases}$

27. Sally and Sandra each have 12 hours a week to make furniture. They obtain a \$50 profit from each table and a \$10 profit from each chair. Sally must work 3 hours and Sandra 2 to make a chair. Sally must work 2 hours and Sandra 6 to make a table. How many tables and how many chairs should they make each week to maximize profits?

28. Two machines, *A* and *B*, work 24 hours a day to manufacture chewing gum and bubble gum. There is a profit of \$150 per case of chewing gum and \$100 per case of bubble gum. To make a case of chewing gum, machine *A* must run 2 hours and machine *B* must run 8 hours. To make a case of bubble gum, machine *A* must run 4 hours and machine *B* must run 2 hours. How many cases of each should be produced each week to maximize profits?

29. Sarah and her sister Heidi have decided to plant a garden. Sarah wants to plant strawberries; Heidi wants to plant pumpkins. Neither crop may use more than $\frac{3}{4}$ of the 40 square meters of available space. Each square meter of strawberries will earn the children \$4, and each square meter of pumpkins will earn \$3. The children plan to eat half of the strawberries themselves. How should the planting be divided to maximize the total income?

30. A distributor of hybrid corn has two storehouses, A and B. At A is stored 110 tons of corn; 190 tons are stored at B. Farmer X ordered 60 tons to be delivered to his farm, and farmer Y ordered 80 tons. The shipping costs appear in the following table:

Storehouse	Farmer	Costs per ton
A	X	$5
A	Y	$7
B	X	$8
B	Y	$14

How should the orders be filled to minimize the shipping costs? (*Hint:* Let x represent the number of tons shipped from A to X, and let y represent the number of tons shipped from A to Y.)

31. Bill packs two foods for his camping trip. One ounce of food X costs 35¢ and provides 150 calories and 21 units of vitamins. One ounce of food Y costs 27¢ and provides 60 calories and 42 units of vitamins. Every day, Bill needs at least 3000 calories and at least 1260 units of vitamins, but he does not want to carry more than 60 ounces of food for each day of his trip. How much of each food should Bill pack to minimize the cost?

32. To manufacture two products X and Y, three workers are scheduled as follows:

Worker	Hours required for product X	Hours required for product Y	Time available per week
A	1	4	18
B	2	3	12
C	2	1	6

The profit on product x is $40, and on product Y, $60. How many of each product should be produced to maximize the profit?

REVIEW EXERCISES

In Review Exercises 1–4 solve each system of equations by the method of graphing.

1. $\begin{cases} 2x - y = -1 \\ x + y = 7 \end{cases}$

2. $\begin{cases} 5x + 2y = 1 \\ 2x - y = -5 \end{cases}$

3. $\begin{cases} y = 5x + 7 \\ x = y - 7 \end{cases}$

4. $\begin{cases} x = y + 5 \\ y = -4 + \dfrac{x}{2} \end{cases}$

In Review Exercises 5–8 solve each system of equations by substitution.

5. $\begin{cases} y = 3x + 2 \\ y = 5x \end{cases}$

6. $\begin{cases} 2y + x = 0 \\ x = y + 3 \end{cases}$

7. $\begin{cases} 2x + y = -3 \\ x - y = 3 \end{cases}$

8. $\begin{cases} \dfrac{x + y}{2} + \dfrac{x - y}{3} = 1 \\ y = 3x - 2 \end{cases}$

In Review Exercises 9–12 solve each system of equations by addition.

9. $\begin{cases} x + 5y = 7 \\ 3x + y = -7 \end{cases}$
10. $\begin{cases} 2x + 3y = 11 \\ 3x - 7y = -41 \end{cases}$
11. $\begin{cases} 2(x + y) - x = 0 \\ 3(x + y) + 2y = 1 \end{cases}$
12. $\begin{cases} \dfrac{x + y}{2} + \dfrac{x - y}{3} = \dfrac{7}{2} \\ \dfrac{x + y}{5} + \dfrac{x - y}{2} = \dfrac{5}{2} \end{cases}$

In Review Exercises 13–16 solve each system of equations by any method.

13. $\begin{cases} 3x + 2y - z = 2 \\ x + y - z = 0 \\ 2x + 3y - z = 1 \end{cases}$
14. $\begin{cases} 5x - y + z = 3 \\ 3x + y + 2z = 2 \\ x + y = 2 \end{cases}$
15. $\begin{cases} 2x - y + z = 1 \\ x - y + 2z = 3 \\ x - y + z = 1 \end{cases}$
16. $\begin{cases} x + 2y - z = -6 \\ x + y - z = -4 \\ 3y - 2z = -12 \end{cases}$

In Review Exercises 17–22 solve each system of equations by matrix methods, if possible.

17. $\begin{cases} 2x + 5y = 7 \\ 3x - y = 2 \end{cases}$
18. $\begin{cases} x + 3y - z = 8 \\ 2x + y - 2z = 11 \\ x - y + 5z = -8 \end{cases}$
19. $\begin{cases} x + 3y + z = 7 \\ 2x - y + z = 0 \\ 3x + 2y + 2z = 7 \end{cases}$

20. $\begin{cases} x + y + z = 4 \\ 3x - 2y - 2z = -3 \\ 4x - y - z = 0 \end{cases}$
21. $\begin{cases} w + x - 3y + z = 2 \\ x - y + 2z = 0 \\ w + y - z = 2 \\ w - x + 2z = -3 \end{cases}$
22. $\begin{cases} w + x + z = 3 \\ x - y + z = 3 \\ w + 2x - y + 2z = 6 \\ y - z = -2 \end{cases}$

In Review Exercises 23–32 perform the indicated matrix arithmetic.

23. $\begin{bmatrix} 3 & 2 & 1 \\ 3 & 2 & 1 \end{bmatrix} + \begin{bmatrix} -2 & 1 & 3 \\ 1 & -2 & 1 \end{bmatrix}$

24. $\begin{bmatrix} 2 & 3 & 5 \\ 1 & -2 & 4 \\ 2 & 1 & -2 \end{bmatrix} - \begin{bmatrix} 0 & -2 & 1 \\ 3 & 4 & -2 \\ 6 & -4 & 1 \end{bmatrix}$

25. $\begin{bmatrix} 2 & 3 \\ -1 & 2 \end{bmatrix} \begin{bmatrix} 1 & -2 \\ -3 & 1 \end{bmatrix}$

26. $\begin{bmatrix} -2 & 3 & 5 \\ 1 & -2 & -3 \end{bmatrix} \begin{bmatrix} 2 & 1 \\ -1 & 2 \\ -2 & 3 \end{bmatrix}$

27. $\begin{bmatrix} 1 & -3 & 2 \end{bmatrix} \begin{bmatrix} 2 \\ 1 \\ 3 \end{bmatrix}$

28. $\begin{bmatrix} 1 & -1 & -2 \\ 2 & -1 & 1 \end{bmatrix} \begin{bmatrix} 1 & 3 & -1 \\ 2 & -1 & 5 \\ 1 & -5 & 3 \end{bmatrix}$

29. $\begin{bmatrix} 1 \\ 2 \\ 1 \\ 5 \end{bmatrix} \begin{bmatrix} 2 & -1 & 1 & 3 \end{bmatrix}$

30. $\begin{bmatrix} 1 & -5 & 3 \\ 2 & 1 & -1 \end{bmatrix} \begin{bmatrix} 2 \\ -2 \\ 3 \end{bmatrix} + \begin{bmatrix} 1 & -1 \\ -1 & 3 \end{bmatrix} \begin{bmatrix} 1 \\ -2 \end{bmatrix}$

31. $\begin{bmatrix} 1 & -3 & 2 \end{bmatrix} \begin{bmatrix} 2 \\ 1 \\ -5 \end{bmatrix} + \begin{bmatrix} 1 & -3 \end{bmatrix} \begin{bmatrix} 2 \\ 5 \end{bmatrix} + \begin{bmatrix} 6 \end{bmatrix}$

32. $\left(\begin{bmatrix} 1 & -3 \\ 3 & 1 \end{bmatrix} + \begin{bmatrix} -1 & 3 \\ 1 & 1 \end{bmatrix} \right) \begin{bmatrix} 1 \\ -5 \end{bmatrix}$

In Review Exercises 33–38 find the inverse of each matrix, if possible.

33. $\begin{bmatrix} 1 & 3 \\ -3 & 5 \end{bmatrix}$

34. $\begin{bmatrix} 4 & 7 \\ 5 & 9 \end{bmatrix}$

35. $\begin{bmatrix} 1 & 3 & -5 \\ 0 & 1 & 9 \\ 0 & 0 & 1 \end{bmatrix}$

36. $\begin{bmatrix} 1 & 0 & 0 \\ 2 & 0 & -2 \\ 1 & 2 & 2 \end{bmatrix}$

37. $\begin{bmatrix} 1 & 0 & 8 \\ 3 & 7 & 6 \\ 1 & 2 & 3 \end{bmatrix}$

38. $\begin{bmatrix} -1 & 1 & 0 \\ -2 & 1 & 0 \\ 3 & -1 & -1 \end{bmatrix}$

In Review Exercises 39–40 use the inverse of the coefficient matrix to solve each system of equations.

39. $\begin{cases} 4x - y + 2z = 0 \\ x + y + 2z = 1 \\ x \quad\ + z = 0 \end{cases}$

40. $\begin{cases} w + 3x + y + 3z = 1 \\ w + 4x + y + 3z = 2 \\ x + y \quad\quad = 1 \\ w + 2x - y + 2z = 1 \end{cases}$

In Review Exercises 41–44 evaluate each determinant.

41. $\begin{vmatrix} 3 & -2 \\ 1 & -3 \end{vmatrix}$

42. $\begin{vmatrix} 1 & 3 & -1 \\ 1 & 2 & 1 \\ 1 & 0 & 2 \end{vmatrix}$

43. $\begin{vmatrix} 1 & -2 & 3 \\ 2 & -1 & 3 \\ 1 & -1 & 0 \end{vmatrix}$

44. $\begin{vmatrix} 1 & 2 & 3 & 4 \\ -1 & 3 & -3 & 2 \\ 0 & 0 & 0 & -1 \\ 3 & 3 & 4 & 3 \end{vmatrix}$

In Review Exercises 45–48 use Cramer's rule to solve each system of equations.

45. $\begin{cases} x + 3y = -5 \\ -2x + y = -4 \end{cases}$

46. $\begin{cases} x - y + z = -1 \\ 2x - y + 3z = -4 \\ x - 3y + z = -1 \end{cases}$

47. $\begin{cases} x - 3y + z = 7 \\ x + y - 3z = -9 \\ x + y + z = 3 \end{cases}$

48. $\begin{cases} w + x - y + z = 4 \\ 2w + x \quad\ + z = 4 \\ x + 2y + z = 0 \\ w \quad\ + y + z = 2 \end{cases}$

In Review Exercises 49–52 decompose each fraction into partial fractions.

49. $\dfrac{4x^2 + 4x + 1}{x^3 + x}$

50. $\dfrac{4x^3 + 3x + x^2 + 2}{x^4 + x^2}$

51. $\dfrac{x^2 + 5}{x^3 + x^2 + 5x}$

52. $\dfrac{x^2 + 1}{(x + 1)^3}$

In Review Exercises 53–56 maximize P subject to the given conditions.

53. $P = 2x + y$
$\begin{cases} x \geq 0 \\ y \geq 0 \\ x + y \leq 3 \end{cases}$

54. $P = 2x - 3y$
$\begin{cases} x \geq 0 \\ y \leq 3 \\ x - y \leq 4 \end{cases}$

55. $P = 3x - y$
$\begin{cases} y \geq 1 \\ y \leq 2 \\ y \leq 3x + 1 \\ x \leq 1 \end{cases}$

56. $P = y - 2x$
$\begin{cases} x + y \geq 1 \\ x \leq 1 \\ y \leq \dfrac{x}{2} + 2 \\ x + y \leq 2 \end{cases}$

57. A company manufactures two fertilizers, X and Y. Each 50-pound bag of fertilizer requires three ingredients, which are available in the limited quantities shown below:

Ingredient	Number of pounds in fertilizer X	Number of pounds in fertilizer Y	Total number of pounds available
Nitrogen	6	10	20,000
Phosphorus	8	6	16,400
Potash	6	4	12,000

The profit on each bag of fertilizer X is $6, and on each bag of Y, $5. How many bags of each product should be produced to maximize the profit?

9

ANALYTIC GEOMETRY

The graphs of second-degree equations in x and y represent figures that have interested mathematicians since the time of the ancient Greeks. However, the equations of those graphs were not carefully studied until the seventeenth century, when René Descartes (1596–1650) and Blaise Pascal (1623–1662) began investigating them.

> **Definition.** If A, B, C, D, E, and F are real numbers, and if at least one of A, B, and C is not 0, then
>
> $$Ax^2 + Bxy + Cy^2 + Dx + Ey + F = 0$$
>
> is called the **general form of a second-degree equation in x and y**.

Descartes discovered that the graphs of second-degree equations always fall into one of seven categories: a single point, a pair of straight lines, a circle, a parabola, an ellipse, a hyperbola, or no graph at all. These graphs are called **conic sections** because each is the intersection of a plane and a right-circular cone. See Figure 9–1.

The conic sections have many practical applications. For example, the properties of parabolas are used in building flashlights, satellite antennas, and solar

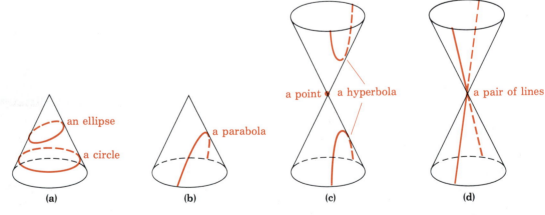

(a)   (b)   a point · a hyperbola (c)   a pair of lines (d)

Figure 9–1

furnaces. The orbits of the planets around the sun are ellipses. Hyperbolas are used in navigation and the design of gears.

9.1 THE CIRCLE

The most familiar of the conic sections is the circle.

> **Definition.** A **circle** is the set of all points in a plane that are a fixed distance from a point called its **center**. The fixed distance is called the **radius of the circle**.

To find the general equation of a circle with radius r and center at the point $C(h, k)$, we must find all points $P(x, y)$ such that the length of the line segment PC is r. See Figure 9–2. We can use the distance formula to find the length of CP, which is r:

$$r = \sqrt{(x - h)^2 + (y - k)^2}$$

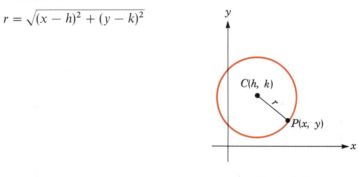

Figure 9–2

Pierre de Fermat
(1601–1665) Pierre de Fermat shares the honor with Descartes for discovering analytic geometry, and with Pascal for developing the theory of probability. But to Fermat alone goes credit for founding number theory.

We square both sides to get

$$r^2 = (x - h)^2 + (y - k)^2$$

This equation is called the **standard form of the equation of a circle**.

> **Theorem.** Any equation that can be written in the form
> $$(x - h)^2 + (y - k)^2 = r^2, \quad r > 0$$
> has a graph that is a circle with radius r and center at the point (h, k).

If $r = 1$, the circle is called a **unit circle**. If $r = 0$, the circle reduces to a single point called a **point circle**. If $r < 0$, a circle does not exist. If the center of the circle is the origin, then $(h, k) = (0, 0)$, and we have the following result.

> **Theorem.** Any equation that can be written in the form
>
> $$x^2 + y^2 = r^2, \quad r > 0$$
>
> has a graph that is a circle with radius r and with center at the origin.

We can use the previous theorems to write the equations of many circles.

Example 1   Find the equation of the circle with radius 5 and center (3, 2). Express the equation in general form.

Solution   Substitute **5** for r, **3** for h, and **2** for k in the standard form of the equation of the circle and simplify:

$$(x - \boldsymbol{h})^2 + (y - \boldsymbol{k})^2 = \boldsymbol{r}^2$$
$$(x - \boldsymbol{3})^2 + (y - \boldsymbol{2})^2 = \boldsymbol{5}^2$$
$$x^2 - 6x + 9 + y^2 - 4y + 4 = 25$$
$$x^2 + y^2 - 6x - 4y - 12 = 0$$

This final equation is a special case of the general form of a second-degree equation. The coefficient of the xy-term equals 0, and the coefficients of x^2 and y^2 are both 1.   ∎

Example 2   Find the equation of the circle with endpoints of its diameter at $(8, -3)$ and $(-4, 13)$.

Solution   First find the center (h, k) of the circle by finding the midpoint of its diameter. Use the midpoint formulas with $(x_1, y_1) = (8, -3)$ and $(x_2, y_2) = (-4, 13)$:

$$h = \frac{x_1 + x_2}{2} \qquad k = \frac{y_1 + y_2}{2}$$
$$h = \frac{8 + (-4)}{2} \qquad k = \frac{-3 + 13}{2}$$
$$= \frac{4}{2} \qquad\qquad = \frac{10}{2}$$
$$= 2 \qquad\qquad = 5$$

Thus, the center of the circle is the point $(h, k) = (2, 5)$.

To find the radius of the circle, use the distance formula to find the distance between the center and one endpoint of the diameter. Because one endpoint is $(8, -3)$, substitute **8** for x_1, **−3** for y_1, **2** for x_2, and **5** for y_2 in the distance formula and simplify:

$$r = \sqrt{(\boldsymbol{x_2} - \boldsymbol{x_1})^2 + (\boldsymbol{y_2} - \boldsymbol{y_1})^2}$$
$$r = \sqrt{(\boldsymbol{2} - \boldsymbol{8})^2 + [\boldsymbol{5} - (\boldsymbol{-3})]^2}$$
$$= \sqrt{(-6)^2 + (8)^2}$$
$$= \sqrt{36 + 64}$$
$$= \sqrt{100}$$
$$= 10$$

Thus, the radius of the circle is 10.

To find the equation of the circle with radius 10 and center at the point (2, 5), substitute **2** for h, **5** for k, and **10** for r in the standard form of the equation of the circle and simplify:

$$(x - \mathbf{h})^2 + (y - \mathbf{k})^2 = \quad r^2$$
$$(x - \mathbf{2})^2 + (y - \mathbf{5})^2 = \quad \mathbf{10}^2$$
$$x^2 - 4x + 4 + y^2 - 10y + 25 = 100$$
$$x^2 + y^2 - 4x - 10y - 71 = \quad 0$$

■

Example 3   Graph the circle $x^2 + y^2 - 4x + 2y = 20$.

Solution   To find the coordinates of the center and the radius, write the equation in standard form by completing the square on both x and y and then simplifying:

$$x^2 + y^2 - 4x + 2y = 20$$
$$x^2 - 4x + y^2 + 2y = 20$$
$$x^2 - 4x + \mathbf{4} + y^2 + 2y + \mathbf{1} = 20 + \mathbf{4} + \mathbf{1} \qquad \text{Add 4 and 1 to both sides to complete the square.}$$
$$(x - 2)^2 + (y + 1)^2 = 25 \qquad \text{Factor } x^2 - 4x + 4 \text{ and } y^2 + 2y + 1.$$
$$(x - 2)^2 + [y - (-1)]^2 = 5^2$$

Note that the radius of the circle is 5 and the coordinates of its center are $h = 2$ and $k = -1$. Plot the center of the circle and construct the circle with a radius of 5 units, as shown in Figure 9–3.

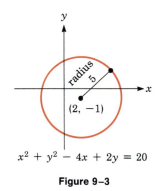

$$x^2 + y^2 - 4x + 2y = 20$$

Figure 9–3

■

Exercise 9.1

In Exercises 1–26 write an equation for the circle with the given properties.

1. Center at the origin; $r = 1$
2. Center at the origin; $r = 4$
3. Center at (6, 8); $r = 4$
4. Center at (5, 3); $r = 2$
5. Center at $(-5, 3)$ and tangent to the y-axis
6. Center at $(-7, -2)$ and tangent to the x-axis
7. Center at $(3, -4)$; $r = \sqrt{2}$
8. Center at $(-9, 8)$; $r = 2\sqrt{3}$
9. Ends of diameter at $(3, -2)$ and $(3, 8)$
10. Ends of diameter at $(5, 9)$ and $(-5, -9)$
11. Ends of diameter at $(-6, 9)$ and $(-4, -7)$
12. Ends of diameter at $(17, 0)$ and $(-3, -3)$
13. Center at $(-3, 4)$ and circle passing through the origin

14. Center at $(4, 0)$ and circle passing through the origin
15. Center at $(-2, -6)$ and circle passing through the origin
16. Center at $(-19, -13)$ and circle passing through the origin
17. Center at $(0, -3)$ and circle passing through $(6, 8)$
18. Center at $(2, 4)$ and circle passing through $(1, 1)$
19. Center at $(5, 8)$ and circle passing through $(-2, -9)$
20. Center at $(7, -5)$ and circle passing through $(-3, -7)$
21. Center at $(-4, -2)$ and circle passing through $(3, 5)$
22. Center at $(0, -7)$ and circle passing through $(0, 7)$
23. Radius of 6 and center at the intersection of $3x + y = 1$ and $-2x - 3y = 4$
24. Radius of 8 and center at the intersection of $x + 2y = 8$ and $2x - 3y = -5$
25. Radius of $\sqrt{10}$ and center at the intersection of $x - y = 12$ and $3x - y = 12$
26. Radius of $2\sqrt{2}$ and center at the intersection of $6x - 4y = 8$ and $2x + 3y = 7$
27. Can a circle with a radius of 10 have endpoints of its diameter at $(6, 8)$ and $(-2, -2)$?
28. Can a circle with radius 25 have endpoints of its diameter at $(0, 0)$ and $(6, 24)$?

In Exercises 29–38 graph the circle.

29. $x^2 + y^2 - 25 = 0$
30. $x^2 + y^2 - 8 = 0$
31. $(x - 1)^2 + (y + 2)^2 = 4$
32. $(x + 1)^2 + (y - 2)^2 = 9$
33. $x^2 + y^2 + 2x - 26 = 0$
34. $x^2 + y^2 - 4y = 12$
35. $9x^2 + 9y^2 - 12y = 5$
36. $4x^2 + 4y^2 + 4y = 15$
37. $4x^2 + 4y^2 - 4x + 8y + 1 = 0$
38. $9x^2 + 9y^2 - 6x + 18y + 1 = 0$

39. Write the equation of the circle passing through $(0, 8)$, $(5, 3)$, and $(4, 6)$.
40. Write the equation of the circle passing through $(-2, 0)$, $(2, 8)$, and $(5, -1)$.
41. Find the area of the circle $3x^2 + 3y^2 + 6x + 12y = 0$. (*Hint: $A = \pi r^2$.*)
42. Find the circumference of the circle $x^2 + y^2 + 4x - 10y - 20 = 0$. (*Hint: $C = 2\pi r$.*)
43. The rectangle in Illustration 1 is inscribed in the upper half of the circle $x^2 + y^2 = r^2$, with one vertex at $P(a, 0)$. Express the area of the rectangle as a function of a.
44. The triangle *ABP* in Illustration 2 is inscribed in the upper half of the circle $x^2 + y^2 = 1$. Find the area of the triangle as a function of a.

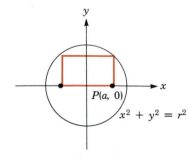

Illustration 1

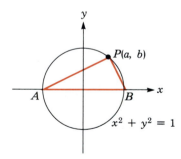

Illustration 2

9.2 THE PARABOLA

We have encountered parabolas in the discussion of quadratic functions. We now examine their equations in greater detail.

> **Definition.** A **parabola** is the set of all points in a plane such that each point in the set is equidistant from a line l, called the **directrix**, and a fixed point F, called the **focus**. The point on the parabola that is closest to the directrix is called the **vertex**. The line passing through the vertex and the focus is called the **axis**.

We will consider parabolas that open to the left, to the right, upward, and downward and that have a vertex at point (h, k). There is a standard form for the equation of each of these parabolas.

Consider the parabola in Figure 9–4, which opens to the right and has as its vertex the point $V(h, k)$. Let $P(x, y)$ be any point on the parabola. Because each point on the parabola is the same distance from the focus (point F) and from the directrix, we can let $d(DV) = d(VF) = p$, where p is some positive constant. Because of the geometry of the figure,

$$d(MP) = p - h + x$$

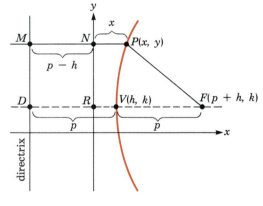

Figure 9–4

Using the distance formula,

$$d(PF) = \sqrt{[x - (p + h)]^2 + (y - k)^2}$$

By the definition of the parabola, $d(MP) = d(PF)$. Thus,

$$p - h + x = \sqrt{[x - (p + h)]^2 + (y - k)^2}$$
$$(p - h + x)^2 = [x - (p + h)]^2 + (y - k)^2 \qquad \text{Square both sides.}$$

Finally, we expand the expression on each side of the equation and simplify:

$$p^2 - ph + px - ph + h^2 - hx + px - hx + x^2$$
$$= x^2 - 2px - 2hx + p^2 + 2ph + h^2 + (y - k)^2$$
$$-2ph + 2px = -2px + 2ph + (y - k)^2$$
$$4px - 4ph = (y - k)^2$$
$$4p(x - h) = (y - k)^2$$

The above argument proves the following theorem.

> **Theorem.**   The standard form of the equation of a parabola with vertex at point (h, k) and opening to the right is
>
> $$(y - k)^2 = 4p(x - h)$$
>
> where p is the distance from the vertex to the focus.

If the parabola has its vertex at the origin, both h and k are equal to zero, and we have the following theorem.

> **Theorem.**   The standard form of the equation of a parabola with vertex at the origin and opening to the right is
>
> $$y^2 = 4px$$
>
> where p is the distance from the vertex to the focus.

Equations of parabolas that open to the right, left, upward, and downward are summarized in Table 9–1. If $p > 0$, then

Table 9–1	Parabola opening	Vertex at origin	Vertex at $V(h, k)$
	Right	$y^2 = 4px$	$(y - k)^2 = 4p(x - h)$
	Left	$y^2 = -4px$	$(y - k)^2 = -4p(x - h)$
	Upward	$x^2 = 4py$	$(x - h)^2 = 4p(y - k)$
	Downward	$x^2 = -4py$	$(x - h)^2 = -4p(y - k)$

Example 1   Find the equation of the parabola with vertex at the origin and focus at $(3, 0)$.

Solution   Sketch the parabola as in Figure 9–5. Because the focus is to the right of the vertex, the parabola opens to the right. Because the vertex is the origin, the standard form of the equation is $y^2 = 4px$. The distance between the focus and the vertex is 3, which is p. Therefore, the equation of the parabola is $y^2 = 4(3)x$, or

$$y^2 = 12x$$

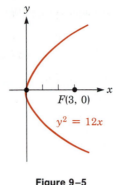

Figure 9–5

Example 2 Find the equation of the parabola that opens upward, has vertex at the point (4, 5), and passes through the point (0, 7).

Solution Use the standard form of a parabola that opens upward: $(x - h)^2 = 4p(y - k)$. Because the point (0, 7) is on the curve, substitute **0** for x and **7** for y in the equation. Because the vertex (h, k) is (4, 5), also substitute **4** for h and **5** for k. Then solve the equation to determine p:

$$(\textbf{x} - \textbf{h})^2 = 4p(\textbf{y} - \textbf{k})$$
$$(\textbf{0} - \textbf{4})^2 = 4p(\textbf{7} - \textbf{5})$$
$$16 = 8p$$
$$2 = p$$

To find the equation of the parabola, substitute **4** for h, **5** for k, and **2** for p in the standard form of the equation and simplify:

$$(x - \textbf{h})^2 = 4\textbf{p}(y - \textbf{k})$$
$$(x - \textbf{4})^2 = 4 \cdot \textbf{2}(y - \textbf{5})$$
$$(x - 4)^2 = 8(y - 5)$$

The graph of this equation appears in Figure 9–6.

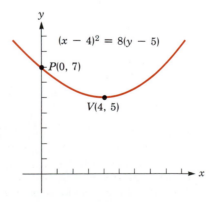

Figure 9–6

Example 3   Find the equations of the two parabolas each having its vertex at (2, 4) and passing through the point (0, 0).

Solution   Sketch the two possible parabolas as shown in Figure 9–7.

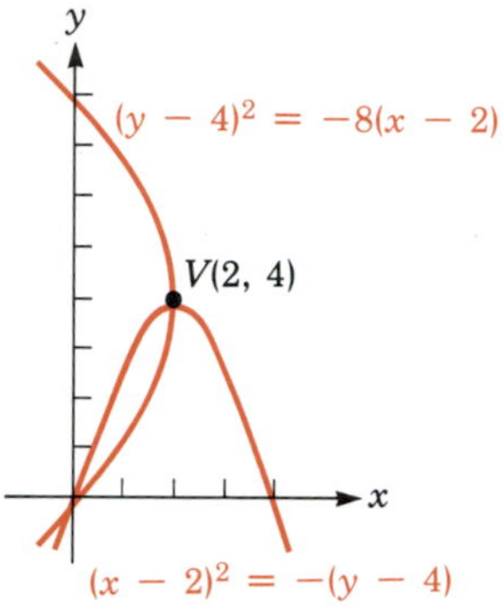

$(y - 4)^2 = -8(x - 2)$

$V(2, 4)$

x

$(x - 2)^2 = -(y - 4)$

Figure 9–7

Part 1.   To find the parabola that opens to the left, use the standard form of the equation $(y - k)^2 = -4p(x - h)$. Because the curve passes through the point $(x, y) = (0, 0)$ and the vertex is $(h, k) = (2, 4)$, substitute **0** for x, **0** for y, **2** for h, and **4** for k in the equation $(y - k)^2 = -4p(x - h)$ and solve for p:

$$(y = k)^2 = -4p(x = h)$$
$$(0 = 4)^2 = -4p(0 = 2)$$
$$16 = 8p$$
$$2 = p$$

Since $h = 2$, $k = 4$, $p = 2$, and the parabola opens to the left, its equation is

$$(y - k)^2 = -4p(x - h)$$
$$(y - 4)^2 = -4(2)(x - 2)$$
$$(y - 4)^2 = -8(x - 2)$$

Part 2.   To find the equation of the parabola that opens downward, use the standard form $(x - h)^2 = -4p(y - k)$. Substitute **2** for h, **4** for k, **0** for x, and **0** for y in the equation and solve for p:

$$(x - h)^2 = -4p(y - k)$$
$$(0 - 2)^2 = -4p(0 - 4)$$
$$4 = 16p$$
$$\frac{1}{4} = p$$

Since $h = 2$, $k = 4$, and $p = \frac{1}{4}$, and the parabola opens downward, its equation is

$$(x - h)^2 = -4p(y - k)$$
$$(x - 2)^2 = -4\left(\frac{1}{4}\right)(y - 4)$$
$$(x - 2)^2 = -(y - 4)$$

■

Example 4   Find the vertex and y-intercepts of the parabola $y^2 + 8x - 4y = 28$. Then graph the parabola.

Solution   Complete the square on y to write the equation in standard form:

$$y^2 + 8x - 4y = 28$$
$$y^2 - 4y = -8x + 28 \qquad \text{Add } -8x \text{ to both sides.}$$
$$y^2 - 4y + 4 = -8x + 28 + 4 \qquad \text{Add 4 to both sides.}$$
$$(y - 2)^2 = -8(x - 4) \qquad \text{Factor both sides.}$$

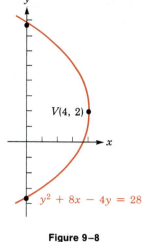

Observe that this equation represents a parabola opening to the left with vertex at (4, 2). To find the points where the graph intersects the y-axis, substitute **0** for x in the equation of the parabola.

$$(y - 2)^2 = -8(x - 4)$$
$$(y - 2)^2 = -8(0 - 4) \qquad \text{Substitute 0 for } x.$$
$$y^2 - 4y + 4 = 32 \qquad \text{Remove parentheses.}$$
$$y^2 - 4y - 28 = 0$$

Use the quadratic formula to determine that the roots of this quadratic equation are $y \approx 7.7$ and $y \approx -3.7$.

The points with coordinates of approximately (0, 7.7) and (0, −3.7) are on the graph of the parabola. Using this information and the knowledge that the graph opens to the left and has a vertex at (4, 2), draw the curve as shown in Figure 9–8.

Figure 9–8

Example 5 A stone is thrown straight up. The equation $s = 128t - 16t^2$ expresses the height of the stone in feet t seconds after it was thrown. Find the maximum height reached by the stone.

Solution The stone goes straight up and then straight down. The graph of $s = 128t - 16t^2$, expressing the height of the stone t seconds after it was thrown, is a parabola (see Figure 9–9). To find the maximum height reached by the stone, calculate the y-coordinate, k, of the vertex of the parabola. To find k, write the equation of the parabola, $s = 128t - 16t^2$, in standard form. To change this equation into standard form, complete the square on t:

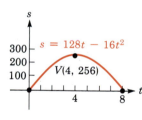

Figure 9–9

$$s = 128t - 16t^2$$
$$16t^2 - 128t = -s \qquad \text{Multiply both sides by } -1.$$
$$t^2 - 8t = \frac{-s}{16} \qquad \text{Divide both sides by 16.}$$
$$t^2 - 8t + 16 = \frac{-s}{16} + 16 \qquad \text{Add 16 to both sides.}$$
$$(t - 4)^2 = \frac{-s + 256}{16} \qquad \text{Factor } t^2 - 8t + 16 \text{ and combine terms.}$$
$$(t - 4)^2 = -\frac{1}{16}(s - 256) \qquad \text{Factor out } -\tfrac{1}{16}.$$

This equation indicates that the maximum height of 256 feet was reached in 4 seconds.

Exercise 9.2

In Exercises 1–16 find the equation of each parabola.

1. Vertex at (0, 0) and focus at (0, 3)

2. Vertex at (0, 0) and focus at (0, −3)

3. Vertex at (0, 0) and focus at (3, 0)

4. Vertex at (0, 0) and focus at (−3, 0)

5. Vertex at (3, 5) and focus at (3, 2)   **6.** Vertex at (3, 5) and focus at $(-3, 5)$

7. Vertex at (3, 5) and focus at $(3, -2)$ **8.** Vertex at (3, 5) and focus at (6, 5)

9. Vertex at (2, 2) and the parabola passing through (0, 0)

10. Vertex at $(-2, -2)$ and the parabola passing through (0, 0)

11. Vertex at $(-4, 6)$ and the parabola passing through (0, 3)

12. Vertex at $(-2, 3)$ and the parabola passing through $(0, -3)$

13. Vertex at (6, 8) and the parabola passing through (5, 10) and (5, 6)

14. Vertex at (2, 3) and the parabola passing through $(1, \frac{13}{4})$ and $(-1, \frac{21}{4})$

15. Vertex at (3, 1) and the parabola passing through (4, 3) and (2, 3)

16. Vertex at $(-4, -2)$ and the parabola passing through $(-3, 0)$ and $(\frac{9}{4}, 3)$

In Exercises 17–26 change each equation to standard form and graph each parabola.

17. $y = x^2 + 4x + 5$ **18.** $2x^2 - 12x - 7y = 10$

19. $y^2 + 4x - 6y = -1$ **20.** $x^2 - 2y - 2x = -7$

21. $y^2 + 2x - 2y = 5$ **22.** $y^2 - 4y = -8x + 20$

23. $x^2 - 6y + 22 = -4x$ **24.** $4y^2 - 4y + 16x = 7$

25. $4x^2 - 4x + 32y = 47$ **26.** $4y^2 - 16x + 17 = 20y$

27. A parabolic arch spans 30 meters and has a maximum height of 10 meters. Derive the equation of the arch using the vertex of the arch as the origin.

28. Find the maximum value of y in the parabola $x^2 + 8y - 8x = 8$.

29. The sum of two numbers is 20 and their product is maximum. What are the numbers?

30. The sum of two numbers is 10 and the sum of their squares is minimum. What are the numbers?

31. A resort owner plans to build and rent n cabins for d dollars per week. The price, d, that she can charge for each cabin depends on the number of cabins she builds, where $d = -45(\frac{n}{32} - \frac{1}{2})$. Find the number of cabins that she should build to maximize her weekly income.

32. A toy rocket is s meters above the earth at the end of t seconds, where $s = -16t^2 + 80\sqrt{3}t$. Find the maximum height of the rocket.

33. An engineer plans to build a tunnel whose arch is in the shape of a parabola. The tunnel will span a two-lane highway that is 8 meters wide. To allow safe passage for most vehicles, the tunnel must be 5 meters high at a distance of 1 meter from the tunnel's edge. What will be the maximum height of the tunnel?

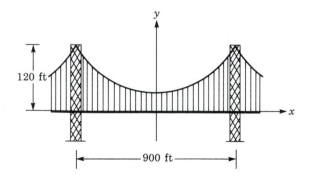

Illustration 1

34. The towers of a suspension bridge are 900 feet apart and rise 120 feet above the roadway. The cable between the towers has the shape of a parabola with a vertex 15 feet above the roadway. Find the equation of the parabola with respect to the indicated coordinate system. See Illustration 1.

35. A satellite antenna with a parabolic cross section is a dish 6 feet in diameter and 1 foot deep at its center. How far is the focus from the center of the dish?

36. A stone tossed upward is s meters above the earth after t seconds, where $s = -16t^2 + 128t$. Show that the stone's height x seconds *after* it is thrown is equal to its height x seconds *before* it returns to the ground.

37. Derive the standard form of the equation of a parabola that opens downward and has vertex at the origin.

38. Show that the result in Example 1 is a special case of the general form of the equation of second degree.

In Exercises 39–40 find the equation of the form $y = ax^2 + bx + c$ *that determines a parabola passing through the three given points.*

39. $(1, 8)$, $(-2, -1)$, and $(2, 15)$ 40. $(1, -3)$, $(-2, 12)$, and $(-1, 3)$

9.3 THE ELLIPSE

A third important conic is the ellipse.

> **Definition.**   An **ellipse** is the set of all points P in a plane such that the sum of the distances from P to two other fixed points F and F' is a positive constant.

In the ellipse shown in Figure 9–10, the two fixed points F and F' are called **foci** of the ellipse, the midpoint of the chord FF' is called the **center**, the chord VV' is called the **major axis**, and each endpoint of the major axis is called a **vertex**. The chord BB', perpendicular to the major axis and passing through the center C, is called the **minor axis**.

To simplify the algebra, we will derive the equation of the ellipse shown in Figure 9–11, which has its center at $(0, 0)$. Because the origin is the midpoint of the chord FF', we can let $d(OF) = d(OF') = c$, where $c > 0$. Then the coordinates of point F are $(c, 0)$, and the coordinates of F' are $(-c, 0)$. We also let $P(x, y)$ be any point on the ellipse.

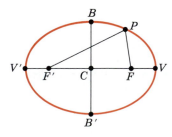

Figure 9–10

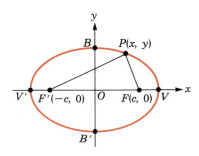

Figure 9–11

The definition of an ellipse requires that the sum of $d(F'P)$ and $d(PF)$ be a positive constant, which we will call $2a$. Thus,

1. $d(F'P) + d(PF) = 2a$

We use the distance formula to compute the lengths of $F'P$ and PF:

$$d(F'P) = \sqrt{[x - (-c)]^2 + y^2}$$
$$d(PF) = \sqrt{(x - c)^2 + y^2}$$

and substitute these values into Equation 1 to obtain

$$\sqrt{[x - (-c)]^2 + y^2} + \sqrt{(x - c)^2 + y^2} = 2a$$

or

$$\sqrt{[x + c]^2 + y^2} = 2a - \sqrt{(x - c)^2 + y^2}$$

We square both sides of this equation and simplify to get

$$(x + c)^2 + y^2 = 4a^2 - 4a\sqrt{(x - c)^2 + y^2} + [(x - c)^2 + y^2]$$
$$x^2 + 2cx + c^2 + y^2 = 4a^2 - 4a\sqrt{(x - c)^2 + y^2} + x^2 - 2cx + c^2 + y^2$$
$$4cx = 4a^2 - 4a\sqrt{(x - c)^2 + y^2}$$
$$cx = a^2 - a\sqrt{(x - c)^2 + y^2}$$
$$cx - a^2 = -a\sqrt{(x - c)^2 + y^2}$$

We square both sides again and simplify to obtain

$$c^2x^2 - 2a^2cx + a^4 = a^2[(x - c)^2 + y^2]$$
$$c^2x^2 - 2a^2cx + a^4 = a^2(x^2 - 2cx + c^2 + y^2)$$
$$c^2x^2 - 2a^2cx + a^4 = a^2x^2 - 2a^2cx + a^2c^2 + a^2y^2$$
$$c^2x^2 + a^4 = a^2x^2 + a^2c^2 + a^2y^2$$
$$a^4 - a^2c^2 = a^2x^2 - c^2x^2 + a^2y^2$$

2. $$a^2(a^2 - c^2) = (a^2 - c^2)x^2 + a^2y^2$$

Because the shortest path between two points is a line segment, $d(F'P) + d(PF) > d(F'F)$. Therefore, $2a > 2c$. This implies that $a > c$ and that $a^2 - c^2$ is a positive number, which we will call b^2. Letting $b^2 = a^2 - c^2$ and substituting into Equation 2, we have

$$a^2b^2 = b^2x^2 + a^2y^2$$

Dividing both sides of this equation by a^2b^2 gives the standard form of the equation for an ellipse with center at the origin and major axis on the x-axis:

$$\frac{x^2}{a^2} + \frac{y^2}{b^2} = 1 \quad \text{where } a > b > 0$$

To find the coordinates of the vertices V and V', we substitute $\mathbf{0}$ for y and solve for x:

$$\frac{x^2}{a^2} + \frac{y^2}{b^2} = 1$$

$$\frac{x^2}{a^2} + \frac{\mathbf{0}^2}{b^2} = 1$$

$$\frac{x^2}{a^2} = 1$$

$$x^2 = a^2$$

$$x = a \qquad \text{or} \qquad x = -a$$

Thus, the coordinates of V are $(a, 0)$, and the coordinates of V' are $(-a, 0)$. In other words, a is the distance between the center of the ellipse, $(0, 0)$, and either of its vertices, and the center of the ellipse is the midpoint of the major axis.

To find the coordinates of B and B', we substitute $\mathbf{0}$ for x and solve for y:

$$\frac{\mathbf{x}^2}{a^2} + \frac{y^2}{b^2} = 1$$

$$\frac{\mathbf{0}^2}{a^2} + \frac{y^2}{b^2} = 1$$

$$y^2 = b^2$$

$$y = b \qquad \text{or} \qquad y = -b$$

Thus, the coordinates of B are $(0, b)$, and the coordinates of B' are $(0, -b)$. The distance between the center of the ellipse and either endpoint of the minor axis is b.

Theorem.   The standard form of the equation of an ellipse with center at the origin and major axis on the x-axis is

$$\frac{x^2}{a^2} + \frac{y^2}{b^2} = 1 \quad \text{where } a > b > 0$$

If the major axis of an ellipse with center at $(0, 0)$ lies on the y-axis, the standard form of the equation of the ellipse is

$$\frac{y^2}{a^2} + \frac{x^2}{b^2} = 1 \quad \text{where } a > b > 0$$

In either case, the length of the major axis is $2a$, and the length of the minor axis is $2b$.

If we develop the equation of the ellipse with center at (h, k), we obtain the following results.

Theorem.   The standard form of the equation of an ellipse with center at (h, k) and major axis parallel to the x-axis is

$$\frac{(x-h)^2}{a^2} + \frac{(y-k)^2}{b^2} = 1 \quad \text{where } a > b > 0$$

If the major axis of an ellipse with center at (h, k) is parallel to the y-axis, the standard form of the equation of the ellipse is

$$\frac{(y-k)^2}{a^2} + \frac{(x-h)^2}{b^2} = 1 \quad \text{where } a > b > 0$$

In either case, the length of the major axis is $2a$, and the length of the minor axis is $2b$.

Example 1   Find the equation of the ellipse with center at the origin, major axis of length 6 units located on the x-axis, and minor axis of length 4 units.

Solution   Because the center of the ellipse is the origin and the length of the major axis is 6, $a = 3$ and the coordinates of the vertices of the ellipse are $(3, 0)$ and $(-3, 0)$, as shown in Figure 9–12.

Because the length of the minor axis is 4, the value of b is 2 and the coordinates of B and B' are $(0, 2)$ and $(0, -2)$. To find the desired equation, substitute **3** for a and **2** for b in the standard form of the equation of an ellipse with center at the origin and major axis on the x-axis. Then simplify the equation:

$$\frac{x^2}{a^2} + \frac{y^2}{b^2} = 1$$

$$\frac{x^2}{3^2} + \frac{y^2}{2^2} = 1$$

$$\frac{x^2}{9} + \frac{y^2}{4} = 1$$

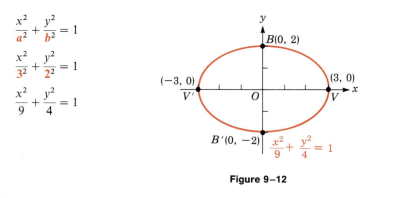

Figure 9–12

Example 2   Find the equation of the ellipse with focus $(0, 3)$ and vertices V and V' at $(3, 3)$ and $(-5, 3)$.

Solution   Because the midpoint of the major axis is the center of the ellipse, the coordinates of the center are $(-1, 3)$. Look at Figure 9–13 and note that the major axis is parallel to the x-axis. The standard form of the equation to use is

$$\frac{(x - h)^2}{a^2} + \frac{(y - k)^2}{b^2} = 1 \quad \text{where } a > b > 0$$

The distance between the center of the ellipse and a vertex is $a = 4$; the distance between the focus and the center is $c = 1$. In the ellipse, $b^2 = a^2 - c^2$. From this equation, compute b^2:

$$b^2 = a^2 - c^2$$
$$= 4^2 - 1^2$$
$$= 15$$

To find the equation of the ellipse, substitute -1 for h, 3 for k, 16 for a^2, and 15 for b^2 in the standard form of the equation for an ellipse and simplify:

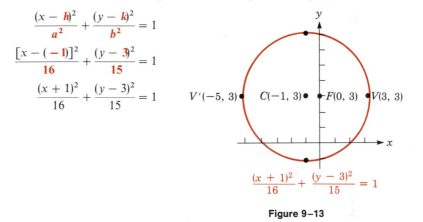

$$\frac{(x - h)^2}{a^2} + \frac{(y - k)^2}{b^2} = 1$$

$$\frac{[x - (-1)]^2}{16} + \frac{(y - 3)^2}{15} = 1$$

$$\frac{(x + 1)^2}{16} + \frac{(y - 3)^2}{15} = 1$$

$V'(-5, 3)$ $C(-1, 3)$ $F(0, 3)$ $V(3, 3)$

$$\frac{(x + 1)^2}{16} + \frac{(y - 3)^2}{15} = 1$$

Figure 9–13

Example 3 The orbit of the earth is approximately an ellipse, with the sun at one focus. The ratio of c to a (called the *eccentricity* of the ellipse) is about $\frac{1}{62}$, and the length of the major axis is approximately 186,000,000 miles. How close does the earth get to the sun?

Solution Assume that this ellipse has its center at the origin and vertices V' and V at $(-93,000,000, 0)$ and $(93,000,000, 0)$, as shown in Figure 9–14. This implies that $a = 93,000,000$.

$$\frac{c}{a} = \frac{1}{62} \quad \text{or} \quad c = \frac{1}{62} a$$

Because $a = 93,000,000$,

$$c = \frac{1}{62}(93,000,000)$$
$$= 1,500,000$$

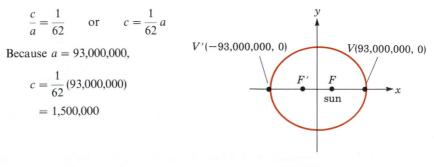

$V'(-93,000,000, 0)$ $V(93,000,000, 0)$

F' F
sun

Figure 9–14

$d(FV)$ represents the shortest possible distance between the earth and the sun. (You'll be asked to prove this in the exercises.) Thus,

$$d(FV) = a - c = 93{,}000{,}000 - 1{,}500{,}000 = 91{,}500{,}000 \text{ miles}$$

The earth's point of closest approach to the sun (called the *perigee*) is approximately 91.5 million miles. ■

Example 4   Graph the ellipse $\dfrac{(x + 2)^2}{4} + \dfrac{(y - 2)^2}{9} = 1$.

Solution   The center of the ellipse is at $(-2, 2)$, and the major axis is parallel to the y-axis. Because $a = 3$, the vertices are 3 units above and below the center at points $(-2, 5)$ and $(-2, -1)$. Because $b = 2$, the endpoints of the minor axis are 2 units to the right and left of the center at points $(0, 2)$ and $(-4, 2)$. Using these four points as guides, sketch the ellipse, as shown in Figure 9–15.

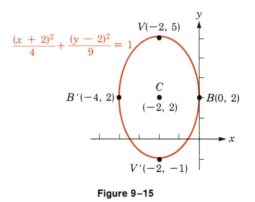

Figure 9–15

■

Example 5   Graph the equation $4x^2 + 9y^2 - 16x - 18y = 11$.

Solution   Write the equation in standard form by completing the square on x and y as follows:

$$4x^2 + 9y^2 - 16x - 18y = 11$$
$$4x^2 - 16x + 9y^2 - 18y = 11$$
$$4(x^2 - 4x) + 9(y^2 - 2y) = 11$$
$$4(x^2 - 4x + 4) + 9(y^2 - 2y + 1) = 11 + 16 + 9$$
$$4(x - 2)^2 + 9(y - 1)^2 = 36$$
$$\frac{(x - 2)^2}{9} + \frac{(y - 1)^2}{4} = 1$$

You can now see that the graph of the given equation is an ellipse with center at $(2, 1)$ and major axis parallel to the x-axis. Because $a = 3$, the vertices are at $(-1, 1)$ and $(5, 1)$. Because $b = 2$, the endpoints of the minor axis are at $(2, -1)$ and $(2, 3)$. Using these four points as guides, sketch the ellipse, as shown in Figure 9–16.

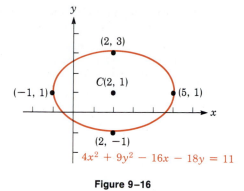

$(2, 3)$

$C(2, 1)$

$(-1, 1)$ $(5, 1)$

$(2, -1)$

$4x^2 + 9y^2 - 16x - 18y = 11$

Figure 9–16

Exercise 9.3

In Exercises 1–6 write the equation of the ellipse that has its center at the origin.

1. Focus at $(3, 0)$ and a vertex at $(5, 0)$

2. Focus at $(0, 4)$ and a vertex at $(0, 7)$

3. Focus at $(0, 1)$; $\frac{4}{3}$ is one-half the length of the minor axis

4. Focus at $(1, 0)$; $\frac{4}{3}$ is one-half the length of the minor axis

5. Focus at $(0, 3)$ and major axis equal to 8

6. Focus at $(5, 0)$ and major axis equal to 12

In Exercises 7–16 write the equation of each ellipse.

7. Center at $(3, 4)$; $a = 3$, $b = 2$; the major axis is parallel to the y-axis.

8. Center at $(3, 4)$; the curve passes through $(3, 10)$ and $(3, -2)$; $b = 2$.

9. Center at $(3, 4)$; $a = 3$, $b = 2$; the major axis is parallel to the x-axis.

10. Center at $(3, 4)$; the curve passes through $(8, 4)$ and $(-2, 4)$; $b = 2$.

11. Foci at $(-2, 4)$ and $(8, 4)$; $b = 4$

12. Foci at $(-8, 5)$ and $(4, 5)$; $b = 3$

13. Vertex at $(6, 4)$ and foci at $(-4, 4)$ and $(4, 4)$

14. Center at $(-4, 5)$; $\dfrac{c}{a} = \dfrac{1}{3}$; vertex at $(-4, -1)$

15. Foci at $(6, 0)$ and $(-6, 0)$; $\dfrac{c}{a} = \dfrac{3}{5}$

16. Vertices at $(2, 0)$ and $(-2, 0)$; $\dfrac{2b^2}{a} = 2$

In Exercises 17–24 graph each ellipse.

17. $\dfrac{x^2}{25} + \dfrac{y^2}{49} = 1$

18. $4x^2 + y^2 = 4$

19. $\dfrac{x^2}{16} + \dfrac{(y + 2)^2}{36} = 1$

20. $(x - 1)^2 + \dfrac{4y^2}{25} = 4$

21. $x^2 + 4y^2 - 4x + 8y + 4 = 0$

22. $x^2 + 4y^2 - 2x - 16y = -13$

23. $16x^2 + 25y^2 - 160x - 200y + 400 = 0$

24. $3x^2 + 2y^2 + 7x - 6y = -1$

25. The moon has an orbit that is an ellipse with the earth at one focus. If the major axis of the orbit is 378,000 miles and the ratio of c to a is approximately $\frac{11}{200}$, how far does the moon get from earth? (This farthest point in an orbit is called the *apogee*.)

26. An arch is a semiellipse 10 meters wide and 5 meters high. Write the equation of the ellipse if the ellipse is centered at the origin.

27. A track is built in the shape of an ellipse and has a maximum length of 100 meters and a maximum width of 60 meters. Write the equation of the ellipse and find its focal width; that is, find the length of a chord that is perpendicular to the major axis and that passes through either focus of the ellipse.

28. An arch has the shape of a semiellipse and has a maximum height of 5 meters. The foci are on the ground with a distance between them of 24 meters. Find the total distance from one focus to any point on the arch and back to the other focus.

29. Consider the ellipse in Illustration 1. If F is a focus of the ellipse and B is an endpoint of the minor axis, use the distance formula to prove that $d(FB)$ is a. (*Hint:* Remember that in an ellipse $a^2 - c^2 = b^2$.)

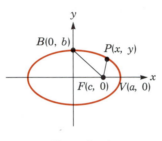

Illustration 1

30. Consider the ellipse in Illustration 1. If F is a focus of the ellipse and P is any point on the ellipse, use the distance formula to show that $d(FP)$ is $a - \frac{c}{a}x$. (*Hint:* Remember that in an ellipse $a^2 - c^2 = b^2$.)

31. Consider the ellipse in Illustration 2. Chord AA' passes through the focus F and is perpendicular to the major axis. Show that the length of AA' (called the **focal width** of the ellipse) is $\frac{2b^2}{a}$

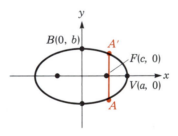

Illustration 2

32. Prove that $d(FV)$ in Example 3 does represent the shortest distance between the earth and the sun. (*Hint:* You might find the result of Exercise 30 helpful.)

33. The ends of a piece of string 6 meters long are attached to two thumbtacks that are 2 meters apart. A pencil catches the loop and draws it tight. As the pencil is moved about the thumbtacks (always keeping the tension), an ellipse is produced with the thumbtacks as foci. Write the equation of the ellipse. (*Hint:* You'll have to establish a coordinate system.)

34. Prove that $a > b$ in the development of the standard form of the equation of an ellipse.

35. Show that the expansion of the standard equation of an ellipse is a special case of the general second-degree equation.

36. The distance between point $P(x, y)$ and the point $(0, 2)$ is $\frac{1}{3}$ of the distance of point P from the line $y = 18$. Find the equation of the curve on which point P lies.

9.4 THE HYPERBOLA

The definition of the hyperbola is similar to the definition of the ellipse except that we demand a constant *difference* of $2a$ instead of a constant sum.

> **Definition.** A **hyperbola** is the set of all points P in a plane such that the difference of the distances from point P to two other points in the plane, F and F', is a positive constant.

Points F and F' (see Figure 9–17) are called the **foci** of the hyperbola, and the midpoint of chord FF' is called the **center** of the hyperbola. The points V and V', where the hyperbola intersects the line segment FF', are called the **vertices** of the hyperbola, and the line segment VV' is called the **transverse axis**.

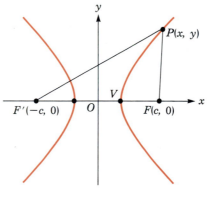

Figure 9–17

As with the ellipse, we will develop the equation of the hyperbola centered at the origin. Because the origin is the midpoint of chord FF', we can let $d(F'O) = d(OF) = c > 0$. Therefore, F is at $(c, 0)$ and F' is at $(-c, 0)$. The definition requires that $|d(F'P) - d(PF)| = 2a$, where $2a$ is a positive constant. Using the distance formula to compute the lengths of $F'P$ and PF gives

$$d(F'P) = \sqrt{[x - (-c)]^2 + y^2}$$
$$d(PF) = \sqrt{(x - c)^2 + y^2}$$

Substituting these values into the equation $d(F'P) - d(PF) = 2a$ gives

$$\sqrt{(x + c)^2 + y^2} - \sqrt{(x - c)^2 + y^2} = 2a$$

or

$$\sqrt{(x + c)^2 + y^2} = 2a + \sqrt{(x - c)^2 + y^2}$$

Squaring both sides of this equation and simplifying gives

$$(x + c)^2 + y^2 = 4a^2 + 4a\sqrt{(x - c)^2 + y^2} + (x - c)^2 + y^2$$

$$x^2 + 2cx + c^2 + y^2 = 4a^2 + 4a\sqrt{(x - c)^2 + y^2} + x^2 - 2cx + c^2 + y^2$$

$$4cx = 4a^2 + 4a\sqrt{(x - c)^2 + y^2}$$

$$cx - a^2 = a\sqrt{(x - c)^2 + y^2}$$

Squaring both sides again and simplifying gives

$$c^2x^2 - 2a^2cx + a^4 = a^2(x^2 - 2cx + c^2 + y^2)$$

$$c^2x^2 - 2a^2cx + a^4 = a^2x^2 - 2a^2cx + a^2c^2 + a^2y^2$$

$$c^2x^2 + a^4 = a^2x^2 + a^2c^2 + a^2y^2$$

1. $(c^2 - a^2)x^2 - a^2y^2 = a^2(c^2 - a^2)$

Because $c > a$ (you will be asked to prove this in the exercises), $c^2 - a^2$ is a positive number. Thus, we can let $b^2 = c^2 - a^2$ and substitute b^2 for $c^2 - a^2$ in Equation 1 to get

$$b^2x^2 - a^2y^2 = a^2b^2$$

Dividing both sides of the previous equation by a^2b^2 gives the standard form of the equation for a hyperbola with center at the origin and foci on the x-axis:

$$\frac{x^2}{a^2} - \frac{y^2}{b^2} = 1$$

If $y = 0$, the preceding equation becomes

$$\frac{x^2}{a^2} = 1 \quad \text{or} \quad x^2 = a^2$$

Solving this equation for x gives

$$x = a \quad \text{or} \quad x = -a$$

This implies that the coordinates of V and V' are $(a, 0)$ and $(-a, 0)$ and that the distance between the center of the hyperbola and either vertex is a. This, in turn, implies that the center of the hyperbola is the midpoint of the segment $V'V$ as well as of the segment FF'.

If $x = 0$, the equation becomes

$$\frac{-y^2}{b^2} = 1 \quad \text{or} \quad y^2 = -b^2$$

Because this equation has no real solutions, the hyperbola cannot intersect the y-axis. These results suggest the following theorem.

Theorem.   The standard form of the equation of a hyperbola with center at the origin and foci on the *x*-axis is

$$\frac{x^2}{a^2} - \frac{y^2}{b^2} = 1$$

The standard form of the equation of a hyperbola with center at the origin and foci on the *y*-axis is

$$\frac{y^2}{a^2} - \frac{x^2}{b^2} = 1$$

As with the ellipse, the standard equation of the hyperbola can be developed with center at (h, k). We state the results without proof.

Theorem.   The standard form of the equation of a hyperbola with center at (h, k) and foci on a line parallel to the *x*-axis is

$$\frac{(x - h)^2}{a^2} - \frac{(y - k)^2}{b^2} = 1$$

The standard form of the equation of a hyperbola with center at (h, k) and foci on a line parallel to the *y*-axis is

$$\frac{(y - k)^2}{a^2} - \frac{(x - h)^2}{b^2} = 1$$

Example 1   Write the equation of the hyperbola with vertices $(3, -3)$ and $(3, 3)$ and with a focus at $(3, 5)$.

Solution   First, plot the vertices and focus, as shown in Figure 9–18. Note that the foci lie on a vertical line. Therefore, the standard form to use is

$$\frac{(y - k)^2}{a^2} - \frac{(x - h)^2}{b^2} = 1$$

Because the center of the hyperbola is midway between the vertices V and V', the center is point $(3, 0)$, so $h = 3$, and $k = 0$. The distance between the vertex and the center of the hyperbola is $a = 3$, and the distance between the focus and the center is $c = 5$. Also, in a hyperbola, $b^2 = c^2 - a^2$. Therefore, $b^2 = 5^2 - 3^2 = 16$. Substituting the values for h, k, a^2, and b^2 into the standard form of the equation gives the desired result:

$$\frac{(y - 0)^2}{9} - \frac{(x - 3)^2}{16} = 1$$

$$\frac{y^2}{9} - \frac{(x - 3)^2}{16} = 1$$

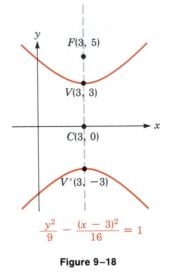

$$\frac{y^2}{9} - \frac{(x-3)^2}{16} = 1$$

Figure 9–18

Asymptotes of a Hyperbola

The values of a and b play an important role in graphing hyperbolas. To see their significance, we consider the hyperbola

$$\frac{x^2}{a^2} - \frac{y^2}{b^2} = 1$$

The center of this hyperbola is the origin, and the vertices are at $V(a, 0)$ and $V'(-a, 0)$. We plot points V, V', $B(0, b)$, and $B'(0, -b)$ and form rectangle $RSQP$, called the **fundamental rectangle**, as in Figure 9–19. The extended diagonals of this rectangle are asymptotes of the hyperbola. In the exercises, you will be asked to show that the equations of these two lines are

$$y = \frac{b}{a}x \qquad \text{and} \qquad y = -\frac{b}{a}x$$

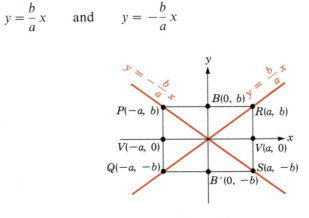

Figure 9–19

To show that the extended diagonals are asymptotes of the hyperbola, we solve the equation

$$\frac{x^2}{a^2} - \frac{y^2}{b^2} = 1$$

for y and modify its form:

$$\frac{x^2}{a^2} - \frac{y^2}{b^2} = 1$$

$$b^2x^2 - a^2y^2 = a^2b^2 \qquad \text{Multiply both sides by } a^2b^2.$$

$$y^2 = \frac{b^2x^2 - a^2b^2}{a^2} \qquad \begin{array}{l}\text{Add } -b^2x^2 \text{ to both sides and} \\ \text{divide both sides by } -a^2.\end{array}$$

$$y^2 = \frac{b^2x^2}{a^2}\left(1 - \frac{a^2}{x^2}\right) \qquad \text{Factor out a } b^2x^2 \text{ from the numerator.}$$

$$y = \pm\frac{bx}{a}\sqrt{1 - \frac{a^2}{x^2}} \qquad \text{Take the square root of both sides.}$$

If $|x|$ grows large without bound, the fraction $\frac{a^2}{x^2}$ in the previous equation approaches 0 and $\sqrt{1 - \frac{a^2}{x^2}}$ approaches 1. Hence, the hyperbola approaches the lines

$$y = \frac{b}{a}x \qquad \text{and} \qquad y = -\frac{b}{a}x$$

This fact makes it easy to sketch a hyperbola. We convert the equation into standard form, find the coordinates of its vertices, and plot them. Then, we construct the fundamental rectangle and its extended diagonals. Using the vertices and the asymptotes as guides, we make a quick and relatively accurate sketch, as in Figure 9–20. The segment BB' is called the **conjugate axis** of the hyperbola.

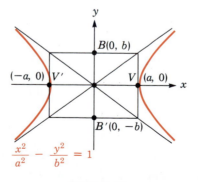

Figure 9–20

Example 2   Graph the hyperbola $x^2 - y^2 - 2x + 4y = 12$.

Solution   First complete the square on x and y to convert the equation into standard form:

$$x^2 - 2x - y^2 + 4y = 12$$
$$x^2 - 2x - (y^2 - 4y) = 12$$
$$x^2 - 2x + 1 - (y^2 - 4y + 4) = 12 + 1 - 4$$
$$(x - 1)^2 - (y - 2)^2 = 9$$
$$\frac{(x - 1)^2}{9} - \frac{(y - 2)^2}{9} = 1$$

From the standard form of the equation of a hyperbola, observe that the center is $(1, 2)$, that $a = 3$ and $b = 3$, and that the vertices are on a line segment parallel to the x-axis, as shown in Figure 9–21. Therefore, the vertices V and V' are 3 units to the right and left of the center and have coordinates of $(4, 2)$ and $(-2, 2)$. Points B and B', 3 units above and below the center, have coordinates $(1, 5)$ and $(1, -1)$. After plotting points V, V', B, and B', construct the fundamental rectangle and its extended diagonals. Using the vertices as points on the hyperbola and the extended diagonals as asymptotes, sketch the graph.

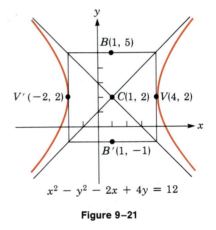

$$x^2 - y^2 - 2x + 4y = 12$$

Figure 9–21

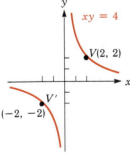

Figure 9–22

This discussion of the hyperbola has considered only cases where the segment that joins the foci is horizontal or vertical. However, there are hyperbolas where this is not true. For example, the graph of the equation $xy = 4$ is a hyperbola with vertices at $(2, 2)$ and $(-2, -2)$, as shown in Figure 9–22.

Exercise 9.4

In Exercises 1–12 write the equation of each hyperbola.

1. Vertices at $(5, 0)$ and $(-5, 0)$ and focus at $(7, 0)$
2. Focus at $(3, 0)$, vertex at $(2, 0)$, and center at $(0, 0)$
3. Center at $(2, 4)$; $a = 2$, $b = 3$; transverse axis is horizontal
4. Center at $(-1, 3)$, vertex at $(1, 3)$, and focus at $(2, 3)$

5. Center at $(5, 3)$, vertex at $(5, 6)$, hyperbola passes through $(1, 8)$

6. Foci at $(0, 10)$ and $(0, -10)$; $\dfrac{c}{a} = \dfrac{5}{4}$

7. Vertices at $(0, 3)$ and $(0, -3)$; $\dfrac{c}{a} = \dfrac{5}{3}$

8. Focus at $(4, 0)$, vertex at $(2, 0)$, and center at the origin

9. Center at $(1, -3)$; $a^2 = 4$, $b^2 = 16$

10. Center at $(1, 4)$, focus at $(7, 4)$, and vertex at $(3, 4)$

11. Center at the origin; hyperbola passes through points $(4, 2)$ and $(8, -6)$

12. Center at $(3, -1)$, y-intercept of -1, x-intercept of $3 + \dfrac{3\sqrt{5}}{2}$

In Exercises 13–16 find the area of the fundamental rectangle of each hyperbola.

13. $4(x - 1)^2 - 9(y + 2)^2 = 36$

14. $x^2 - y^2 - 4x - 6y = 6$

15. $x^2 + 6x - y^2 + 2y = -11$

16. $9x^2 - 4y^2 = 18x + 24y + 63$

In Exercises 17–20 write the equation of each hyperbola.

17. Center at $(-2, -4)$; $a = 2$; area of fundamental rectangle is 36 square units

18. Center at $(3, -5)$; $b = 6$; area of fundamental rectangle is 24 square units

19. One vertex at $(6, 0)$, one end of conjugate axis at $(0, \frac{5}{4})$

20. One vertex at $(3, 0)$, one focus at $(-5, 0)$, center at $(0, 0)$

In Exercises 21–29 graph the hyperbola.

21. $\dfrac{x^2}{9} - \dfrac{y^2}{4} = 1$

22. $\dfrac{y^2}{4} - \dfrac{x^2}{9} = 1$

23. $4x^2 - 3y^2 = 36$

24. $x^2 + 6x - y^2 + 2y = -11$

25. $y^2 - x^2 = 1$

26. $x^2 - y^2 - 4x - 6y = 6$

27. $4x^2 - 2y^2 + 8x - 8y = 8$

28. $9(y + 2)^2 - 4(x - 1)^2 = 36$

29. $y^2 - 4x^2 + 6y + 32x = 59$

In Exercises 30–32 graph each hyperbola by plotting points.

30. $xy = 9$

31. $-xy = 6$

32. $-xy = 20$

In Exercises 33–36 find the equation of each curve on which point P lies.

33. The difference of the distances between $P(x, y)$ and the points $(-2, 1)$ and $(8, 1)$ is 6.

34. The difference of the distances between $P(x, y)$ and the points $(3, -1)$ and $(3, 5)$ is 5.

35. The distance between point $P(x, y)$ and the point $(0, 3)$ is $\frac{3}{2}$ of the distance between P and the line $y = -2$.

36. The distance between point $P(x, y)$ and the point $(5, 4)$ is $\frac{5}{3}$ of the distance between P and the line $x = -3$.

37. Prove that $c > a$ for a hyperbola with center at the origin and line segment FF' on the x-axis.

38. Show that the equations of the extended diagonals of the fundamental rectangle of the hyperbola with equation $\frac{x^2}{a^2} - \frac{y^2}{b^2} = 1$ are $y = \frac{b}{a}x$ and $y = -\frac{b}{a}x$.

39. Show that the expansion of the standard form of the equation of a hyperbola is a special case of the general equation of second degree with $B = 0$.

40. The hyperbolas $\frac{x^2}{a^2} - \frac{y^2}{b^2} = 1$ and $-\frac{x^2}{a^2} + \frac{y^2}{b^2} = 1$ are **conjugate hyperbolas**. Show that they have the same asymptotes.

9.5 SYSTEMS OF SECOND-DEGREE EQUATIONS

We now discuss techniques for solving systems of two equations in two variables where at least one of the equations is of second degree.

Example 1   Solve this system of equations by graphing:

$$\begin{cases} x^2 + y^2 = 25 \\ 2x + y = 10 \end{cases}$$

Solution   The graph of the equation $x^2 + y^2 = 25$ is a circle with center at the origin and radius of 5. The graph of the equation $2x + y = 10$ is a straight line. Depending on whether the line is a secant (intersecting the circle at two points) or a tangent (intersecting the circle at one point) or does not intersect the circle at all, there are two, one, or no solutions to the system, respectively. After graphing the circle and the line, as shown in Figure 9–23, note that there are two intersection points, P and P', with the coordinates of (3, 4) and (5, 0). Thus, the solutions to the given system of equations are

$$\begin{cases} x = 3 \\ y = 4 \end{cases} \quad \text{and} \quad \begin{cases} x = 5 \\ y = 0 \end{cases}$$

Verify that these are *exact* solutions.

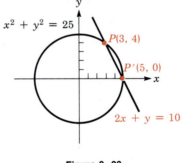

Figure 9–23

Graphical solutions of systems of equations usually give only approximate solutions. A second method, using algebra, can be used to find exact solutions.

Example 2   Solve the following system of equations algebraically:

$$\begin{cases} x^2 + y^2 = 25 \\ 2x + y = 10 \end{cases}$$

Solution   This system contains one equation of second degree and another of first degree. Solve systems of this kind by the substitution method. Solving the linear equation for y gives

$$2x + y = 10$$
$$y = -2x + 10$$

Substitute the expression $-2x + 10$ for y in the second-degree equation, and solve the resulting quadratic equation for x:

$$x^2 + y^2 = 25$$
$$x^2 + (-2x + 10)^2 = 25$$

$x^2 + 4x^2 - 40x + 100 = 25$	Remove parentheses.
$5x^2 - 40x + 75 = 0$	Combine terms.
$x^2 - 8x + 15 = 0$	Divide both sides by 5.
$(x - 5)(x - 3) = 0$	Factor $x^2 - 8x + 15 = 0$.

$$x = 5 \quad \text{or} \quad x = 3$$

Because $y = -2x + 10$, if $x = 5$ then $y = 0$, and if $x = 3$ then $y = 4$. The two solutions are

$$\begin{cases} x = 5 \\ y = 0 \end{cases} \quad \text{and} \quad \begin{cases} x = 3 \\ y = 4 \end{cases}$$

■

Example 3 Solve the following system of equations algebraically:

$$\begin{cases} 4x^2 + 9y^2 = 5 \\ \quad\quad y = x^2 \end{cases}$$

Solution Solve this system by substitution.

$4x^2 + 9y^2 = 5$	
$4y + 9y^2 = 5$	Substitute y for x^2.
$9y^2 + 4y - 5 = 0$	Add -5 to both sides.
$(9y - 5)(y + 1) = 0$	Factor $9y^2 + 4y^2 - 5$.

$$9y - 5 = 0 \quad \text{or} \quad y + 1 = 0$$
$$y = \frac{5}{9} \quad\quad\quad y = -1$$

Because $y = x^2$, the values of x are found by solving the equations

$$x^2 = \frac{5}{9} \quad \text{and} \quad x^2 = -1$$

Because the equation $x^2 = -1$ has no real solutions, this possibility is discarded. The solutions of the equation $x^2 = \frac{5}{9}$ are

$$x = \frac{\sqrt{5}}{3} \quad \text{and} \quad x = \frac{-\sqrt{5}}{3}$$

Thus, the solutions of the system are

$$\left(\frac{\sqrt{5}}{3}, \frac{5}{9}\right) \quad \text{and} \quad \left(\frac{-\sqrt{5}}{3}, \frac{5}{9}\right)$$

■

Example 4 Solve the following system of equations algebraically:

$$\begin{cases} 3x^2 + 2y^2 = 36 \\ 4x^2 - \ y^2 = \ 4 \end{cases}$$

Solution   In this system, both equations are of second degree and in the form $ax^2 + by^2 = c$. Solve systems like this by eliminating one of the variables by addition. Copy the first equation and multiply the second equation by 2 to obtain the equivalent system of equations

$$\begin{cases} 3x^2 + 2y^2 = 36 \\ 8x^2 - 2y^2 = 8 \end{cases}$$

Add the equations to eliminate the variable y, and solve the resulting equation for x:

$$11x^2 = 44$$
$$x^2 = 4$$
$$x = 2 \quad \text{or} \quad x = -2$$

To find y, substitute **2** for x and then **−2** for x in the first equation and proceed as follows:

For $x = 2$	**For $x = -2$**
$3x^2 + 2y^2 = 36$	$3x^2 + 2y^2 = 36$
$3(2)^2 + 2y^2 = 36$	$3(-2)^2 + 2y^2 = 36$
$12 + 2y^2 = 36$	$12 + 2y^2 = 36$
$2y^2 = 24$	$2y^2 = 24$
$y^2 = 12$	$y^2 = 12$
$y = +\sqrt{12}$ or $y = -\sqrt{12}$	$y = +\sqrt{12}$ or $y = -\sqrt{12}$
$y = 2\sqrt{3}$ $y = -2\sqrt{3}$	$y = 2\sqrt{3}$ $y = -2\sqrt{3}$

The four solutions of this system are

$$(2, 2\sqrt{3}), \quad (2, -2\sqrt{3}), \quad (-2, 2\sqrt{3}), \quad \text{and} \quad (-2, -2\sqrt{3}) \qquad \blacksquare$$

Exercise 9.5

In Exercises 1–10 solve each system of equations by graphing.

1. $\begin{cases} 8x^2 + 32y^2 = 256 \\ x = 2y \end{cases}$

2. $\begin{cases} x^2 + y^2 = 2 \\ x + y = 2 \end{cases}$

3. $\begin{cases} x^2 + y^2 = 90 \\ y = x^2 \end{cases}$

4. $\begin{cases} x^2 + y^2 = 5 \\ x + y = 3 \end{cases}$

5. $\begin{cases} x^2 + y^2 = 25 \\ 12x^2 + 64y^2 = 768 \end{cases}$

6. $\begin{cases} x^2 + y^2 = 13 \\ y = x^2 - 1 \end{cases}$

7. $\begin{cases} x^2 - 13 = -y^2 \\ y = 2x - 4 \end{cases}$

8. $\begin{cases} x^2 + y^2 = 20 \\ y = x^2 \end{cases}$

9. $\begin{cases} x^2 - 6x - y = -5 \\ x^2 - 6x + y = -5 \end{cases}$

10. $\begin{cases} x^2 - y^2 = -5 \\ 3x^2 + 2y^2 = 30 \end{cases}$

In Exercises 11–36 solve each system of equations algebraically for real values of x and y.

11. $\begin{cases} 25x^2 + 9y^2 = 225 \\ 5x + 3y = 15 \end{cases}$

12. $\begin{cases} x^2 + y^2 = 20 \\ y = x^2 \end{cases}$

13. $\begin{cases} x^2 + y^2 = 2 \\ x + y = 2 \end{cases}$

14. $\begin{cases} x^2 + y^2 = 36 \\ 49x^2 + 36y^2 = 1764 \end{cases}$

15. $\begin{cases} x^2 + y^2 = 5 \\ x + y = 3 \end{cases}$

16. $\begin{cases} x^2 - x - y = 2 \\ 4x - 3y = 0 \end{cases}$

17. $\begin{cases} x^2 + y^2 = 13 \\ y = x^2 - 1 \end{cases}$

18. $\begin{cases} x^2 + y^2 = 25 \\ 2x^2 - 3y^2 = 5 \end{cases}$

19. $\begin{cases} x^2 + y^2 = 30 \\ y = x^2 \end{cases}$

20. $\begin{cases} 9x^2 - 7y^2 = 81 \\ x^2 + y^2 = 9 \end{cases}$

21. $\begin{cases} x^2 + y^2 = 13 \\ x^2 - y^2 = 5 \end{cases}$

22. $\begin{cases} 2x^2 + y^2 = 6 \\ x^2 - y^2 = 3 \end{cases}$

23. $\begin{cases} x^2 + y^2 = 20 \\ x^2 - y^2 = -12 \end{cases}$

24. $\begin{cases} xy = -\dfrac{9}{2} \\ 3x + 2y = 6 \end{cases}$

25. $\begin{cases} y^2 = 40 - x^2 \\ y = x^2 - 10 \end{cases}$

26. $\begin{cases} x^2 - 6x - y = -5 \\ x^2 - 6x + y = -5 \end{cases}$

27. $\begin{cases} y = x^2 - 4 \\ x^2 - y^2 = -16 \end{cases}$

28. $\begin{cases} 6x^2 + 8y^2 = 182 \\ 8x^2 - 3y^2 = 24 \end{cases}$

29. $\begin{cases} x^2 - y^2 = -5 \\ 3x^2 + 2y^2 = 30 \end{cases}$

30. $\begin{cases} \dfrac{1}{x} + \dfrac{1}{y} = 5 \\ \dfrac{1}{x} - \dfrac{1}{y} = -3 \end{cases}$

31. $\begin{cases} \dfrac{1}{x} + \dfrac{2}{y} = 1 \\ \dfrac{2}{x} - \dfrac{1}{y} = \dfrac{1}{3} \end{cases}$

32. $\begin{cases} \dfrac{1}{x} + \dfrac{3}{y} = 4 \\ \dfrac{2}{x} - \dfrac{1}{y} = 7 \end{cases}$

33. $\begin{cases} 3y^2 = xy \\ 2x^2 + xy - 84 = 0 \end{cases}$

34. $\begin{cases} x^2 + y^2 = 10 \\ 2x^2 - 3y^2 = 5 \end{cases}$

35. $\begin{cases} xy = \dfrac{1}{6} \\ y + x = 5xy \end{cases}$

36. $\begin{cases} xy = \dfrac{1}{12} \\ y + x = 7xy \end{cases}$

37. The area of a rectangle is 63 square centimeters, and its perimeter is 32 centimeters. Find the dimensions of the rectangle.

38. The product of two integers is 32 and their sum is 12. Find the integers.

39. The sum of the squares of two numbers is 221, and the sum of the numbers is 212 less. Find the numbers.

40. Grant receives $225 annual income from one investment. Jeff invested $500 more than Grant, but at an annual rate of 1% less. Jeff's annual income is $240. What are the amount and rate of Grant's investment?

41. Carol receives $67.50 annual income from one investment. John invested $150 more than Carol at an annual rate of $1\frac{1}{2}\%$ more. John's annual income is $94.50. What are the amount and rate of Carol's investment? (*Hint:* There are two answers.)

42. Jim drove 306 miles. Jim's brother made the same trip at a speed 17 miles per hour slower than Jim did and required an extra $1\frac{1}{2}$ hours. What were Jim's rate and time?

9.6   TRANSLATION AND ROTATION OF AXES

The graph of the equation

$$(x - 3)^2 + (y - 1)^2 = 4$$

is a circle with radius of 2 and with center at the point (3, 1). See Figure 9–24. If we were to shift the black xy-coordinate system 3 units to the right and 1 unit up, we would establish the $x'y'$-coordinate system shown in color. With respect to this new $x'y'$-system, the center of the circle is the origin and its

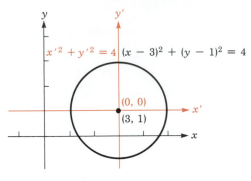

Figure 9–24

equation is

$$x'^2 + y'^2 = 4$$

In this section, we will discuss how to change the equation of a graph by shifting the position of the x- and y-axes. A shift to the left, right, up, or down is called a **translation of the coordinate axes**.

Figure 9–25 shows both an xy- and an $x'y'$-coordinate system. The color x'- and y'-axes are parallel to the black x- and y-axes, respectively, and the unit distance on each is the same. The origin of the $x'y'$-system is the point O' with $x'y'$-coordinates of $(0, 0)$ and with xy-coordinates of (h, k). The $x'y'$-system is called a **translated coordinate system**.

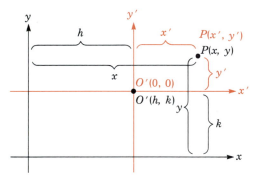

Figure 9–25

Relative to the xy-system in Figure 9–25, the coordinates of point P are (x, y). Relative to the $x'y'$-system, the coordinates of point P are (x', y'). By the geometry of the figure we can determine equations, called the **translation-of-axes formulas**, that enable us to find the coordinates of any point with respect to any translated coordinate system.

The Translation-of-Axes Formulas.

$$x' = x - h$$
$$y' = y - k$$

Example 1   The xy-coordinates of point O' in Figure 9–26 are $(-3, -5)$, and the xy-coordinates of point Q are $(1, -2)$. Find the $x'y'$-coordinates of point Q.

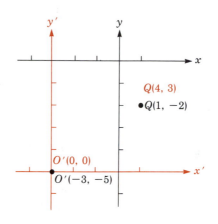

Figure 9–26

Solution   Use the translation-of-axes formulas with $(h, k) = (-3, -5)$ and $(x, y) = (1, -2)$:

$$x' = x - h \qquad y' = y - k$$
$$x' = 1 - (-3) \qquad y' = -2 - (-5)$$
$$= 4 \qquad\qquad = 3$$

The $x'y'$-coordinates of point Q are $(4, 3)$.   ■

Example 2   Find the equation of the parabola $y + 3 = (x - 2)^2$ with respect to a translated coordinate system with origin at $O'(2, -3)$. Graph the resulting equation with respect to the translated coordinate system.

Solution   Because the origin is translated to the point $(h, k) = (2, -3)$, substitute **2** for h and **-3** for k in the translation-of-axes formulas:

$$x' = x - \mathbf{h} \qquad y' = y - \mathbf{k}$$
$$x' = x - \mathbf{2} \qquad y' = y - (\mathbf{-3})$$
$$\qquad\qquad y' = y + 3$$

To obtain the equation of the same parabola with respect to the translated axes, substitute y' for $y + 3$ and x' for $x - 2$ in the given equation.

$$y + 3 = (x - 2)^2$$
$$y' = x'^2$$

The graph of the equation $y' = x'^2$ is a parabola with vertex at the origin of the $x'y'$-system. See Figure 9–27.

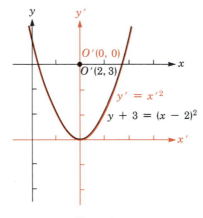

Figure 9–27 ■

Example 3   To what point should the origin of the xy-coordinate system be translated to remove the first-degree terms in the equation

$$4x^2 + y^2 + 8x - 6y + 9 = 0$$

Graph the resulting equation with respect to the translated coordinate system.

Solution   Begin by completing the square in both x and y and simplifying:

$$4x^2 + y^2 + 8x - 6y + 9 = 0$$

$\quad\quad 4x^2 + 8x + y^2 - 6y = -9$ Rearrange terms and add -9 to both sides.

$\quad\quad 4(x^2 + 2x) + y^2 - 6y = -9$ Factor 4 from $4x^2 + 8x$.

$\quad\quad 4(x^2 + 2x + 1) + y^2 - 6y + 9 = -9 + 4 + 9$ Add 4 and 9 to both sides.

$\quad\quad 4(x + 1)^2 + (y - 3)^2 = 4$ Factor both trinomials.

$\quad\quad \dfrac{(x + 1)^2}{1} + \dfrac{(y - 3)^2}{4} = 1$ Divide both sides by 4.

Let the origin of the $x'y'$-coordinate system be the point $O'(h, k) = O'(-1, 3)$. Then, by the translation-of-axes formulas, you have $x' = x + 1$ and $y' = y - 3$. Substitute x' for $x + 1$ and y' for $y - 3$ to obtain the equation

$$\frac{(x + 1)^2}{1} + \frac{(y - 3)^2}{4} = 1$$

$$\frac{x'^2}{1} + \frac{y'^2}{4} = 1$$

Note that this final equation contains no first-degree terms. Its graph is an ellipse centered at the origin of an $x'y'$-coordinate system whose origin has been translated to the point $(-1, 3)$ of the xy-system. See Figure 9–28.

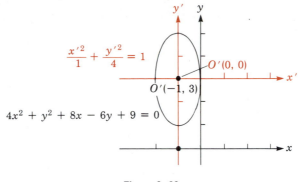

$$\frac{x'^2}{1} + \frac{y'^2}{4} = 1$$

$O'(0, 0)$

$O'(-1, 3)$

$$4x^2 + y^2 + 8x - 6y + 9 = 0$$

Figure 9–28

Example 4   Show that the slope of a nonvertical line is not affected by a translation of axes.

Solution   In the xy-system, the slope-intercept form of the equation of a nonvertical line with slope m is

$$y = mx + b$$

Use the translation-of-axes formulas to find the equation of the *same* line with respect to an $x'y'$-system whose origin has been translated to the point (h, k) of the xy-system. Substitute $x' + h$ for x and $y' + k$ for y and simplify. Proceed as follows:

$$y = mx + b$$
$$y' + k = m(x' + h) + b$$
$$y' + k = mx' + mh + b \qquad \text{Remove parentheses.}$$
$$y' = mx' + mh + b - k \qquad \text{Add } -k \text{ to both sides.}$$

This final equation is in slope-intercept form and represents a line with y'-intercept of $mh + b - k$ and a slope of m. Thus, the translation does not change the slope of the line.

Rotation of Axes

The equation of a graph can often be simplified by rotating the xy-axes through some angle θ. This process is called **rotation of the coordinate axes**.

Figure 9–29 shows both an xy- and an $x'y'$-coordinate system with a common origin O, and the same unit of measure on each axis. The color x'- and y'-axes are rotated through some angle θ with respect to the black x- and y-axes. The $x'y'$-system is called a **rotated coordinate system**.

Suppose the xy-coordinates of point P are (x, y) and its $x'y'$-coordinates are (x', y'). From point P, we draw PC perpendicular to the x'-axis and PA perpendicular to the x-axis. Segments CQ and CB are perpendicular to PA and

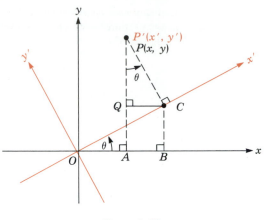

Figure 9–29

the x-axis, respectively, as indicated in the figure. This forms similar triangles OBC and PQC, in which

$$\sin \theta = \frac{BC}{OC} = \frac{BC}{x'} \qquad \text{or} \qquad BC = x' \sin \theta$$

$$\cos \theta = \frac{OB}{OC} = \frac{OB}{x'} \qquad \text{or} \qquad OB = x' \cos \theta$$

$$\sin \theta = \frac{QC}{PC} = \frac{AB}{y'} \qquad \text{or} \qquad AB = y' \sin \theta$$

and

$$\cos \theta = \frac{QP}{PC} = \frac{QP}{y'} \qquad \text{or} \qquad QP = y' \cos \theta$$

The variables x and y can be expressed in terms of x', y', and θ:

$$
\begin{aligned}
x &= OA & y &= AP \\
&= OB - AB & &= BC + QP \\
&= x' \cos \theta - y' \sin \theta & &= x' \sin \theta + y' \cos \theta
\end{aligned}
$$

The equations relating the xy-coordinates of point P to its coordinates in the $x'y'$-system are as follows:

The Equations of Rotation.

$$x = x' \cos \theta - y' \sin \theta$$
$$y = x' \sin \theta + y' \cos \theta$$

Example 5   The xy-coordinate system is rotated 30° counterclockwise to form the $x'y'$-coordinate system, as in Figure 9–30. The $x'y'$-coordinates of P are $(\sqrt{3} + 1, \sqrt{3} - 1)$. What are the xy-coordinates of P?

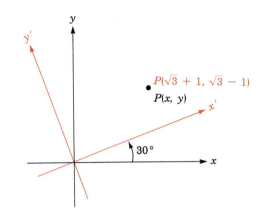

Figure 9–30

Solution   Use the equations

$$x = x' \cos \theta - y' \sin \theta$$
$$y = x' \sin \theta + y' \cos \theta$$

with $x' = \sqrt{3} + 1$, $y' = \sqrt{3} - 1$, and $\theta = 30°$.

$$x = x' \cos \theta - y' \sin \theta$$

$$= (\sqrt{3} + 1)\frac{\sqrt{3}}{2} - (\sqrt{3} - 1)\frac{1}{2} \qquad \text{Substitute } \sqrt{3} + 1 \text{ for } x', \sqrt{3} - 1 \text{ for } y',$$
$$\qquad\qquad\qquad\qquad\qquad\qquad \tfrac{\sqrt{3}}{2} \text{ for cos } 30°, \text{ and } \tfrac{1}{2} \text{ for sin } 30°.$$

$$= \frac{3}{2} + \frac{\sqrt{3}}{2} - \frac{\sqrt{3}}{2} + \frac{1}{2}$$

$$= 2$$

$$y = x' \sin \theta + y' \cos \theta$$

$$= (\sqrt{3} + 1)\frac{1}{2} + (\sqrt{3} - 1)\frac{\sqrt{3}}{2} \qquad \text{Substitute } \sqrt{3} + 1 \text{ for } x', \sqrt{3} - 1 \text{ for } y',$$
$$\qquad\qquad\qquad\qquad\qquad\qquad \tfrac{1}{2} \text{ for sin } 30°, \text{ and } \tfrac{\sqrt{3}}{2} \text{ for cos } 30°.$$

$$= \frac{\sqrt{3}}{2} + \frac{1}{2} + \frac{3}{2} - \frac{\sqrt{3}}{2}$$

$$= 2$$

With respect to the xy-system, the coordinates of point P are $(2, 2)$.   ■

A rotation of the coordinate axes can transform the general second-degree equation

$$Ax^2 + Bxy + Cy^2 + Dx + Ey + F = 0$$

into another equation with no xy term. To determine the appropriate angle θ, use the following result. The proof is omitted.

The Angle-of-Rotation Formula. If the coordinate axes are rotated through an angle θ, $0° < \theta < 90°$, determined by

$$\cot 2\theta = \frac{A - C}{B}$$

then the equation

$$Ax^2 + Bxy + Cy^2 + Dx + Ey + F = 0$$

will be transformed into an equation with no xy term.

To determine the required rotation equations, we must use $\cot 2\theta$ to determine $\sin \theta$ and $\cos \theta$.

Example 6 Transform the equation

$$17x^2 - 48xy + 31y^2 + 49 = 0$$

into an equation with no xy term. Graph the resulting equation.

Solution Rotate the coordinate system through an acute angle θ, where $\cot 2\theta = \frac{A-C}{B}$. To do so, first determine $\cot 2\theta$.

$$\cot 2\theta = \frac{A - C}{B}$$

$$= \frac{17 - 31}{-48} \qquad \text{Substitute 17 for } A, \text{ 31 for } C, \text{ and } -48 \text{ for } B.$$

$$= \frac{7}{24}$$

To determine the values of $\sin \theta$ and $\cos \theta$ required by the rotation equations, first determine the value of $\cos 2\theta$ and then use the identities

$$\sin \theta = \sqrt{\frac{1 - \cos 2\theta}{2}} \qquad \text{and} \qquad \cos \theta = \sqrt{\frac{1 + \cos 2\theta}{2}}$$

To evaluate $\cos 2\theta$, sketch a right triangle such as the one in Figure 9–31 and use the fact that $\cot 2\theta = \frac{7}{24}$ and the Pythagorean theorem to determine the sides of the triangle. You can then see that $\cos 2\theta = \frac{7}{25}$. Thus,

$$\sin \theta = \sqrt{\frac{1 - \cos 2\theta}{2}} \qquad \text{and} \qquad \cos \theta = \sqrt{\frac{1 + \cos 2\theta}{2}}$$

$$= \sqrt{\frac{1 - \dfrac{7}{25}}{2}} \qquad\qquad\qquad = \sqrt{\frac{1 + \dfrac{7}{25}}{2}}$$

$$= \sqrt{\frac{9}{25}} \qquad\qquad\qquad\qquad = \sqrt{\frac{16}{25}}$$

$$= \frac{3}{5} \qquad\qquad\qquad\qquad\quad = \frac{4}{5}$$

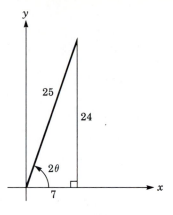

Figure 9–31

The required rotation is given by the equations

$$x = x' \cos \theta - y' \sin \theta$$
$$y = x' \sin \theta + y' \cos \theta$$

where $\frac{3}{5} = \sin \theta$ and $\frac{4}{5} = \cos \theta$:

$$x = x'\left(\frac{4}{5}\right) - y'\left(\frac{3}{5}\right)$$

$$y = x'\left(\frac{3}{5}\right) + y'\left(\frac{4}{5}\right)$$

To transform the given equation $17x^2 - 48xy + 31y^2 + 49 = 0$ into an equation that has no xy term, substitute $\frac{4}{5}x' - \frac{3}{5}y'$ for x and $\frac{3}{5}x' + \frac{4}{5}y'$ for y in the given equation:

$$17x^2 - 48xy + 31y^2 + 49 = 0$$

$$17\left(\frac{4}{5}x' - \frac{3}{5}y'\right)^2 - 48\left(\frac{4}{5}x' - \frac{3}{5}y'\right)\left(\frac{3}{5}x' + \frac{4}{5}y'\right) + 31\left(\frac{3}{5}x' + \frac{4}{5}y'\right)^2 + 49 = 0$$

Perform the indicated operations to obtain

$$17\left[\frac{16}{25}x'^2 - 2\left(\frac{12}{25}\right)x'y' + \frac{9}{25}y'^2\right] - 48\left(\frac{12}{25}x'^2 + \frac{16}{25}x'y' - \frac{9}{25}x'y' - \frac{12}{25}y'^2\right)$$

$$+ 31\left[\frac{9}{25}x'^2 + 2\left(\frac{12}{25}\right)x'y' + \frac{16}{25}y'^2\right] + 49 = 0$$

Then remove parentheses and combine terms to obtain

$$-\frac{25}{25}x'^2 + 0x'y' + \frac{1225}{25}y'^2 + 49 = 0$$

or

$$\frac{x'^2}{49} - y'^2 = 1$$

The graph of this transformed equation is the hyperbola that appears in Figure 9–32.

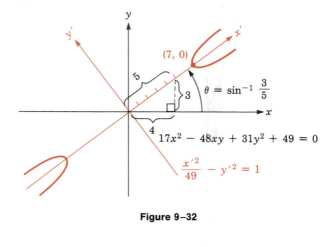

Figure 9–32

Exercise 9.6

In Exercises 1–4 the origin of the $x'y'$-system is at the point $(1, -3)$ of the xy-system. Find the xy-coordinates of the point whose $x'y'$-coordinates are given.

1. $P(2, 4)$ **2.** $Q(-1, 3)$ **3.** $R(0, 0)$ **4.** $S(1, -3)$

In Exercises 5–8 the origin of the $x'y'$-system is at the point $(2, -5)$ of the xy-system. Express each equation in terms of the variables x' and y'. Draw a sketch that shows the relation of the graphs of the equations to both coordinate systems.

5. $y = 3x - 11$ **6.** $y = -3x + 1$ **7.** $y = x^2 - 4x - 1$ **8.** $x = y^2 + 10y + 27$

In Exercises 9–12 the origin of the $x'y'$-system is at the point $(3, -2)$ of the xy-system. Express each equation in terms of the variables x' and y'. Draw a sketch that shows the relation of the graphs of the equations to both coordinate systems.

9. $x^2 + y^2 - 6x + 4y + 9 = 0$ **10.** $y^2 + 4y + x + 1 = 0$
11. $x^2 - 6x + 2y + 13 = 0$ **12.** $2y^2 + 8y + x + 5 = 0$

In Exercises 13–18 determine the point to which the origin of the $x'y'$-system should be translated to eliminate the first-degree terms of the given equation. Draw a sketch that shows the relation of the graphs of the equations to both coordinate systems.

13. $x^2 + y^2 + 4x - 10y - 6 = 0$ **14.** $x^2 + y^2 - 8x - 2y + 1 = 0$ **15.** $2x^2 + y^2 + 4x + 2y - 1 = 0$
16. $4x^2 + 9y^2 + 24x - 18y + 9 = 0$ **17.** $x^2 - y^2 - 6x - 4y + 4 = 0$ **18.** $4x^2 - 9y^2 + 8x - 36y = 68$

In Exercises 19–24 the xy-coordinate system is rotated counterclockwise through an angle θ to form the $x'y'$-coordinate system. Find the xy-coordinates of the point whose $x'y'$-coordinates are given.

19. $P(2\sqrt{3}, 0);\ \theta = 30°$ **20.** $Q(\sqrt{3} - 1, \sqrt{3} + 1);\ \theta = 30°$ **21.** $R(2, 6);\ \theta = 45°$
22. $S(\sqrt{2} - 1, \sqrt{2} + 1);\ \theta = 45°$ **23.** $T(7, 4);\ \theta = 60°$ **24.** $U(0, -8);\ \theta = 60°$

In Exercises 25–28 determine the angle θ through which the axes must be rotated to remove the xy term. **Do not transform the equation.**

25. $3x^2 - \sqrt{3}xy + 2y^2 - 3x + y - 10 = 0$ **26.** $5x^2 - 8xy + 5y^2 + y + 3 = 0$
27. $5\sqrt{3}x^2 + 8xy - 3\sqrt{3}y^2 - 9 = 0$ **28.** $x^2 - 6xy - 5y^2 - 27 = 0$

In Exercises 29–38 perform a rotation of axes to remove the xy term. Graph the equation.

29. $xy = 2$ **30.** $xy = 18$

31. $x^2 + 3xy + y^2 = 2$ **32.** $5x^2 + 26xy + 5y^2 - 72 = 0$

33. $x^2 + 2xy + y^2 - 2\sqrt{2}x + 2\sqrt{2}y = 0$ **34.** $x^2 - 2\sqrt{3}xy - y^2 - 2 = 0$

35. $2x^2 - 4xy + 5y^2 - 6 = 0$ **36.** $7x^2 + 2\sqrt{3}xy + 5y^2 = 8$

37. $6x^2 + 24xy - y^2 - 30 = 0$ **38.** $5x^2 - 4xy + 8y^2 - 36 = 0$

39.   Show that the radius of a circle is not affected by a translation of axes.

40.   Show that no translation of axes will remove the xy term of the equation $xy = 1$.

In Exercises 41–42 suppose that $Ax^2 + Bxy + Cy^2 + Dx + Ey + F = 0$ is changed into $A'x'^2 + B'x'y' + C'y'^2 + D'x' + E'y' + F' = 0$ by substituting $x' + h$ for x and $y' + k$ for k and then simplifying.

41.   Show that $A + C = A' + C'$.   **42.**   Show that $B^2 - 4AC = B'^2 - 4A'C'$.

In Exercises 43–44 suppose that $Ax^2 + Bxy + Cy^2 + Dx + Ey + F = 0$ is changed into $A'x'^2 + B'x'y' + C'y'^2 + D'x' + E'y' + F' = 0$ by rotating the axes through an angle θ.

43.   Show that $A + C = A' + C'$.   **44.**   Show that $B^2 - 4AC = B'^2 - 4A'C'$.

9.7   POLAR COORDINATES

Some equations such as $(x^2 + y^2)^{3/2} = x$ are difficult to graph using the x- and y-coordinates of the Cartesian rectangular coordinate system. However, these equations can often be written in a form using the variables r (a radius) and θ (an angle). These coordinates enable us to graph such equations on an alternative coordinate system called the **polar coordinate system**.

The polar coordinate system is based on a ray, called the **polar axis**, and its source, called the **pole**. In Figure 9–33, ray OA is the polar axis and point O is the pole. Any point $P(r, \theta)$ in the plane can be located if the length of a radius

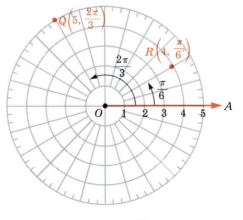

Figure 9–33

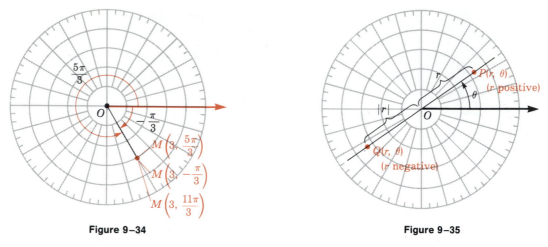

Figure 9–34   Figure 9–35

and an angle in standard position are known. For example, the polar coordinates $(4, \frac{\pi}{6})$ determine the position of point R in Figure 9–33, and the coordinates $(5, \frac{2\pi}{3})$ determine the point Q.

In Figure 9–34, point M is determined by the coordinates $(3, \frac{5\pi}{3})$. Note that point M is also determined by many other sets of polar coordinates, such as $(3, -\frac{\pi}{3})$ or $(3, \frac{11\pi}{3})$. Any set of polar coordinates locates a single point. However, any point has infinitely many sets of polar coordinates.

To plot point P with coordinates (r, θ) when r is positive, we draw angle θ in standard position and count r units along the terminal side of θ. See Figure 9–35. To plot point Q with coordinates (r, θ) when r is negative, we draw angle θ in standard position and count $|r|$ units along the *extension* through the origin of the terminal side of θ. For example, to graph the point $P(-2, \frac{\pi}{4})$, we first draw an angle of $\frac{\pi}{4}$ in standard position, as in Figure 9–36. We then draw the extension of ray OC in the opposite direction to obtain ray OB and count 2 units

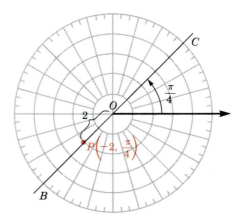

Figure 9–36

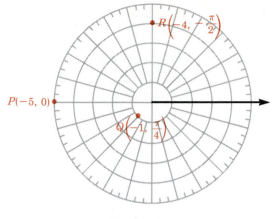

Figure 9–37

along ray OB to find point P. The graphs of the three points $P(-5, 0)$, $Q(-1, \frac{\pi}{4})$, and $R(-4, -\frac{\pi}{2})$ are shown in Figure 9–37.

There is a relationship between the rectangular coordinates (x, y) and the polar coordinates (r, θ) of a point. Suppose point P in Figure 9–38 has rectangular coordinates (x, y) and polar coordinates (r, θ). Draw PA perpendicular to the x-axis to form right triangle OAP with $OA = x$, $AP = y$, and $OP = r$. Because angle θ is in standard position, the hypotenuse OP is the terminal side of angle θ. Thus,

$$\cos \theta = \frac{x}{r} \quad \text{and} \quad \sin \theta = \frac{y}{r}$$

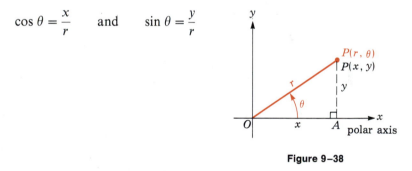

Figure 9–38

From these, we obtain the two relations

$$x = r \cos \theta \quad \text{and} \quad y = r \sin \theta$$

Hence, to change from polar to rectangular coordinates, we use the following equations.

Formulas to Convert from Polar to Rectangular Coordinates.

$$x = r \cos \theta$$
$$y = r \sin \theta$$

Example 1   If the polar coordinates of point P are $\left(10, \dfrac{5\pi}{6}\right)$, find the rectangular coordinates.

Solution $x = r \cos \theta = 10 \cos\left(\dfrac{5\pi}{6}\right) = 10\left(\dfrac{-\sqrt{3}}{2}\right) = -5\sqrt{3}$

$y = r \sin \theta = 10 \sin\left(\dfrac{5\pi}{6}\right) = 10\left(\dfrac{1}{2}\right) = 5$

The rectangular coordinates of point P are $(-5\sqrt{3}, 5)$.   ∎

To find the polar coordinates from the rectangular coordinates of point P, again refer to Figure 9–38. From the right triangle OAP, it follows that

$$r^2 = x^2 + y^2 \qquad \text{and} \qquad \tan \theta = \frac{y}{x}$$

To find possible polar coordinates for point P, let $r = \sqrt{x^2 + y^2}$. Then, find an angle θ (equal to $\tan^{-1}\frac{y}{x}$) whose terminal side passes through the point (x, y). That is, if x is negative, for example, and y is positive, choose θ to be a second-quadrant angle. If x and y are both negative, choose θ to be a third-quadrant angle.

Formulas to Convert from Rectangular to Polar Coordinates.

$$r = \sqrt{x^2 + y^2}$$

$$\theta = \tan^{-1}\frac{y}{x}$$

where the terminal side of θ passes through the point (x, y). If $x = 0$, choose θ to be $\frac{\pi}{2}$ (if $y > 0$) or $\frac{3\pi}{2}$ (if $y < 0$).

Example 2   If the rectangular coordinates of point P are $(\sqrt{3}, 1)$, find a pair of polar coordinates for P.

Solution   Refer to Figure 9–39. First find the r-coordinate of point P, as follows.

$r = \sqrt{x^2 + y^2}$

$= \sqrt{(\sqrt{3})^2 + 1^2}$

$= \sqrt{3 + 1}$

$= 2$

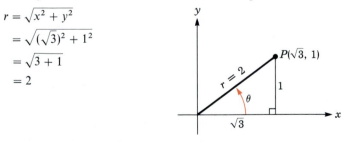

Figure 9–39

Secondly, determine the θ-coordinate of point P. Because x and y are both positive, θ is a first-quadrant angle.

$$\theta = \tan^{-1} \frac{y}{x}$$

$$= \tan^{-1} \frac{1}{\sqrt{3}}$$

$$= \tan^{-1} \frac{\sqrt{3}}{3}$$

$$= \frac{\pi}{6}$$

One possible choice of polar coordinates for point P is $(2, \frac{\pi}{6})$. ∎

Equations in the variables r and θ are called **polar equations**. The **graph** of a polar equation is the set of all points (r, θ) on the polar plane that satisfy the equation. The next several examples discuss the graphing of polar equations.

Example 3 Graph the polar equation $r = 1 - \sin \theta$.

Solution Compute several pairs (r, θ) that satisfy the polar equation, and join them in order of increasing values of θ. The easiest values of r to compute are those associated with the quadrantal-angle values of θ. When $\theta = 0$ or π, then $r = 1$. When $\theta = \frac{\pi}{2}$, then $r = 0$. When $\theta = \frac{3\pi}{2}$, then $r = 2$. These four points—$(1, 0)$, $(0, \frac{\pi}{2})$, $(1, \pi)$, and $(2, \frac{3\pi}{2})$—are the intercepts of the curve with the polar axis, or its extension, and with a line drawn perpendicular to the polar axis at the pole. Other pairs (r, θ) appear in the table in Figure 9–40.

$r = 1 - \sin \theta$

r	θ
1	0
1	π
0	$\frac{\pi}{2}$
2	$\frac{3\pi}{2}$
$\frac{1}{2}$	$\frac{\pi}{6}$
$\frac{1}{2}$	$\frac{5\pi}{6}$
$\frac{3}{2}$	$\frac{7\pi}{6}$
$\frac{3}{2}$	$\frac{11\pi}{6}$

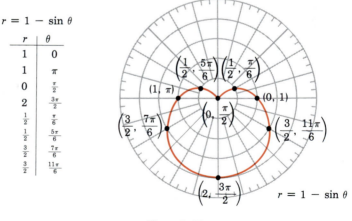

Figure 9–40

The points must be joined in order of increasing θ. As θ increases from 0 to π, the value of $1 - \sin \theta$, and hence the value of r, decreases from 1 to 0, and then increases back to 1. This accounts for the two "bumps" in the curve. As θ increases from π to $\frac{3\pi}{2}$, and finally to 2π, r increases from 1 to 2 and decreases to 1 again. This forms the large loop in the third and fourth quadrants. The curve of Figure 9–40 is called a **cardioid**.

◼

Example 4   Graph the curve $r\theta = \pi$.

Solution   Because the product of variables r and θ is a constant, r and θ are inversely proportional— as θ increases, then r decreases. Write the equation as $r = \frac{\pi}{\theta}$, and calculate the intercepts of the curve with the polar axis and a line perpendicular to the axis through the pole.

$$\text{At } \theta = \frac{\pi}{2}, \quad r = \frac{\pi}{\frac{\pi}{2}} = 2; \quad \text{at } \theta = \pi, \quad r = 1; \quad \text{and at} \quad \theta = \frac{3\pi}{2}, \quad r = \frac{2}{3}$$

These points and a portion of the curve are plotted in Figure 9–41a; the curve is a spiral. The difficulty is in determining the shape of the curve as θ approaches 0 (and r approaches infinity). If θ is close to zero, and r consequently is very large, the distance PQ of Figure 9–41b is very close to the length of arc PQ', centered at the pole. Because this arc length is $r\theta$, or π, the distance PQ is approximately π. The horizontal line that is π units above the polar axis is an asymptote of this curve. The graph of $r\theta = \pi$ is called a **hyperbolic spiral** and its graph appears in Figure 9–41c. The "tail" heading off to the right approaches the horizontal line that is parallel to the polar axis and π units above it. Since r cannot be zero, the hyperbolic spiral does not pass through the pole.

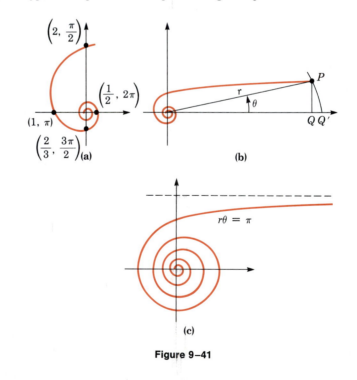

Figure 9–41

◼

Example 5   Change the rectangular equation $(x^2 + y^2)^{3/2} = x$ to an equation having variables of r and θ. Then graph the curve using polar coordinates.

Solution Because $x^2 + y^2 = r^2$ and $x = r \cos \theta$, you have

$$(x^2 + y^2)^{3/2} = x$$
$$r^3 = r \cos \theta$$

If $r = 0$, the graph is the pole for all θ. If $r \neq 0$, you can divide both sides by r and obtain

$$r^2 = \cos \theta$$
$$r = \pm\sqrt{\cos \theta}$$

Make a table of values, and plot the points as in Figure 9–42. Because r^2 is positive, $\cos \theta$ must be positive. Therefore, θ is in quadrant I or IV.

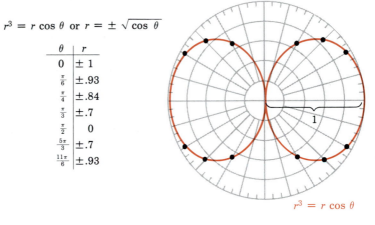

$r^3 = r \cos \theta$ or $r = \pm \sqrt{\cos \theta}$

θ	r
0	± 1
$\frac{\pi}{6}$	$\pm.93$
$\frac{\pi}{4}$	$\pm.84$
$\frac{\pi}{3}$	$\pm.7$
$\frac{\pi}{2}$	0
$\frac{5\pi}{3}$	$\pm.7$
$\frac{11\pi}{6}$	$\pm.93$

$r^3 = r \cos \theta$

Figure 9–42

Example 6 Change the polar equation $r(3 \cos \theta + 2 \sin \theta) = 7$ to an equation with rectangular coordinates.

Solution Use the distributive law to remove parentheses.

$$r(3 \cos \theta + 2 \sin \theta) = 7$$
$$3r \cos \theta + 2r \sin \theta = 7$$

Because $r \cos \theta = x$ and $r \sin \theta = y$, you can write this equation as

$$3x + 2y = 7$$

This is the equation of a line that can easily be graphed.

Exercise 9.7

In Exercises 1–8 plot each point with the given polar coordinates.

1. $\left(3, \dfrac{\pi}{6}\right)$

2. $\left(-3, \dfrac{\pi}{4}\right)$

3. $\left(5, -\dfrac{\pi}{3}\right)$

4. $\left(-5, -\dfrac{3\pi}{4}\right)$

5. $(7, \pi)$

6. $(-4, \pi)$

7. $\left(4, -\dfrac{5\pi}{4}\right)$

8. $\left(-3, -\dfrac{5\pi}{3}\right)$

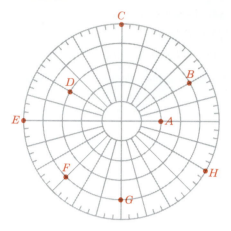

Illustration 1

In Exercises 9–16 refer to Illustration 1, and give two possible sets of polar coordinates for each point.

9. *A*
10. *B*
11. *C*
12. *D*
13. *F*
14. *E*
15. *G*
16. *H*

In Exercises 17–36 the polar coordinates of point P are given. Find the rectangular coordinates of point P.

17. $\left(2, \dfrac{\pi}{6}\right)$
18. $\left(5, \dfrac{3\pi}{4}\right)$
19. $\left(7, \dfrac{5\pi}{3}\right)$
20. $\left(20, \dfrac{5\pi}{4}\right)$

21. $\left(-3, \dfrac{\pi}{3}\right)$
22. $\left(-7, \dfrac{7\pi}{3}\right)$
23. $\left(2, \dfrac{\pi}{2}\right)$
24. $\left(4, \dfrac{3\pi}{2}\right)$

25. $\left(-2, \dfrac{13\pi}{6}\right)$
26. $(-5, 3\pi)$
27. $\left(5, \dfrac{13\pi}{4}\right)$
28. $\left(3, \dfrac{17\pi}{6}\right)$

29. $\left(2, -\dfrac{\pi}{6}\right)$
30. $\left(6, -\dfrac{5\pi}{4}\right)$
31. $\left(-10, -\dfrac{\pi}{2}\right)$
32. $\left(-15, -\dfrac{\pi}{4}\right)$

33. $(0, 17\pi)$
34. $(0, 0)$
35. $(6, 24\pi)$
36. $(35, 11.5\pi)$

In Exercises 37–52 the rectangular coordinates of point P are given. Find a set of polar coordinates for point P.

37. $(1, 1)$
38. $(1, \sqrt{3})$
39. $(2\sqrt{3}, -2)$
40. $(-2, -2\sqrt{3})$

41. $(-\sqrt{3}, -1)$
42. $(-1, \sqrt{3})$
43. $(-\sqrt{3}, 1)$
44. $(0, 3)$

45. $(0, 0)$
46. $(7, 0)$
47. $(-5, 0)$
48. $(7, 7)$

49. $(3, -3)$
50. (π, π)
51. $(7, 7\sqrt{3})$
52. $(-\sqrt{2}, -\sqrt{6})$

In Exercises 53–66 graph the given equation.

53. $r = -\theta, \theta \geq 0$
54. $r \sin \theta = 3$
55. $r \cos \theta = -3$

56. $r \cos \theta + r \sin \theta = 1$
57. $r = \cos 2\theta$
58. $r = \sin 2\theta$

59. $r = \sin 3\theta$
60. $r = 3 \cos 3\theta$
61. $r = 2(1 + \sin \theta)$

62. $r = \sqrt{2 \cos \theta}$
63. $r = 2 + \cos \theta$
64. $r = \frac{1}{2} + \cos \theta$

65. $r^2\theta = \pi$ (This curve has a horizontal asymptote.)

66. $r = \tan\theta$ (This curve has a vertical asymptote.)

In Exercises 67–78 each equation contains rectangular coordinates. Change each equation to an equation containing polar coordinates.

67. $x = 3$

68. $y = -7$

69. $3x + 2y = 3$

70. $2x - y = 7$

71. $x^2 + y^2 = 9x$

72. $yx = 12$

73. $(x^2 + y^2)^3 = 4x^2y^2$

74. $x^2 + y^2 = 9$

75. $x^2 = 2x - x^2$

76. $(x^2 + y^2)^2 = x^2 - y^2$

77. $x^2 = 2y + 1$

78. $y^2 = 2x + 1$

In Exercises 79–90 each equation contains polar coordinates. Change each equation to an equation containing rectangular coordinates.

79. $r = 3$

80. $r\sin\theta = 4$

81. $\cos\theta = \dfrac{5}{r}$

82. $3r\cos\theta + 2r\sin\theta = 2$

83. $r = \dfrac{1}{1 + \sin\theta}$

84. $r = \dfrac{1}{1 - \cos\theta}$

85. $r^2 = \sin 2\theta$

86. $r^2 = \cos 2\theta$

87. $\theta = \pi$

88. $\theta = 90°$

89. $r(2 - \cos\theta) = 2$

90. $r = 3\csc\theta + 2\sec\theta$

9.8 POLAR EQUATIONS OF THE CONICS

In this section we will use polar coordinates to write equations for the parabola, ellipse, and hyperbola. The discussion is based on the fact that each of these conics is completely determined by a point, a directrix, and a positive real number called the **eccentricity** of the conic.

There are four standard forms for the polar equation of a conic. We derive one of these by letting point F be the pole and the directrix d be a line perpendicular to an extension of the polar axis, as in Figure 9–43. Furthermore, we let p ($p > 0$) be the distance from point F to the directrix d, and we let e be a positive constant representing the eccentricity. It can be shown that, if $P(r, \theta)$ lies on the plane in any location for which the ratio of $d(FP)$ to $d(QP)$ is equal

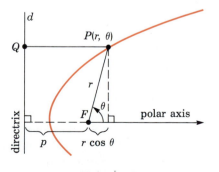

Figure 9–43

to the constant e, then P lies on a conic with focus at F. Thus, for any point P on the conic,

$$e = \frac{d(FP)}{d(PQ)}$$

To derive the polar equation for the conic, we note that $d(FP) = r$ and that $d(QP) = p + r \cos \theta$. Then we have

$$e = \frac{r}{p + r \cos \theta}$$

or, by solving for r,

1. $r = \dfrac{ep}{1 - e \cos \theta}$

Equation 1 is a standard form of the equation of a conic with eccentricity of e, with focus at the pole, and with directrix perpendicular to an extension of the polar axis.

If $e = 1$, we can show that

$$r = \frac{ep}{1 - e \cos \theta}$$

is the equation of a parabola. To do so, we substitute 1 for e, convert the equation into rectangular form, and reduce it to the equation of a parabola:

$$r = \frac{1p}{1 - 1 \cos \theta}$$

$$r(1 - \cos \theta) = p$$

$$r - r \cos \theta = p$$

$$\sqrt{x^2 + y^2} - x = p$$

$$\sqrt{x^2 + y^2} = x + p$$

$$x^2 + y^2 = (x + p)^2 \qquad \textcolor{red}{\text{Square both sides.}}$$

$$x^2 + y^2 = x^2 + 2px + p^2$$

$$y^2 = 2px + p^2$$

$$y^2 = 2p\left(x + \frac{p}{2}\right) \qquad \textcolor{red}{\text{Factor out } 2p \text{ from the right side.}}$$

This result is the equation of a parabola with vertex (in the rectangular system) at the point $(-\frac{p}{2}, 0)$. Thus, if $e = 1$, Equation 1 is the equation of a parabola.

It can also be shown that, if $0 < e < 1$, Equation 1 is the equation of an ellipse, and, if $e > 1$, Equation 1 is the equation of a hyperbola.

There are three other standard forms for the equation of a conic. They are determined by rotating the directrix and graph in Figure 9–43 through angles of 90°, 180°, and 270°. The four possibilities for the standard equations of a conic written in polar coordinates are summarized in Table 9–2.

Table 9–2

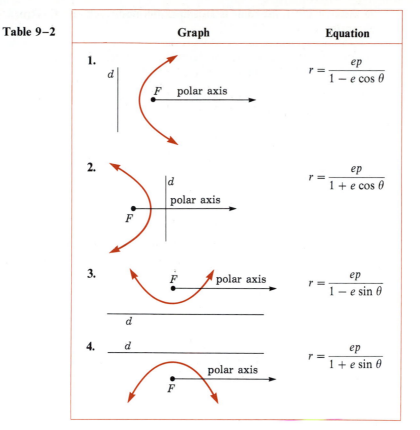

	Graph	Equation
1.	d, F polar axis	$r = \dfrac{ep}{1 - e \cos \theta}$
2.	d, polar axis, F	$r = \dfrac{ep}{1 + e \cos \theta}$
3.	$\dot{F}$ polar axis, d	$r = \dfrac{ep}{1 - e \sin \theta}$
4.	d, polar axis, F	$r = \dfrac{ep}{1 + e \sin \theta}$

Example 1 Graph the conic determined by $r = \dfrac{3}{4 - 2 \cos \theta}$.

Solution By dividing numerator and denominator of the fraction by 4, you can write the equation in the form

$$r = \frac{\dfrac{3}{4}}{1 - \dfrac{1}{2} \cos \theta}$$

Compare this to Equation 1 and determine that $e = \frac{1}{2}$ and that $ep = \frac{3}{4}$. Then calculate p as follows:

$$ep = \frac{3}{4}$$

$$\frac{1}{2} p = \frac{3}{4} \qquad \text{Substitute } \tfrac{1}{2} \text{ for } e.$$

$$p = \frac{3}{2}$$

Because $e = \frac{1}{2} < 1$, the conic is an ellipse. Because $p = \frac{3}{2}$, the directrix is $\frac{3}{2}$ units to the left of the focus.

To graph the ellipse, first find its vertices by determining r for $\theta = 0$ and for $\theta = \pi$. Then find its intercepts with the vertical axis by letting $\theta = \frac{\pi}{2}$ and $\theta = \frac{3\pi}{2}$. Plot these points and sketch the graph, as in Figure 9–44.

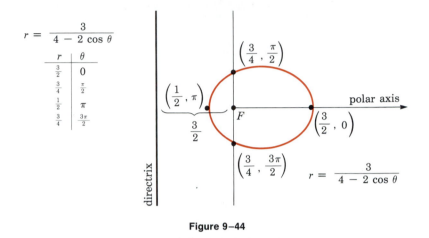

$$r = \frac{3}{4 - 2\cos\theta}$$

r	θ
$\frac{3}{2}$	0
$\frac{3}{4}$	$\frac{\pi}{2}$
$\frac{1}{2}$	π
$\frac{3}{4}$	$\frac{3\pi}{2}$

Figure 9–44

Example 2   Graph the conic determined by $r = \dfrac{5}{3 + 3\cos\theta}$.

Solution   Divide the numerator and the denominator of the fraction by 3, and write the equation in the form

$$r = \frac{\dfrac{5}{3}}{1 + 1\cos\theta}$$

Compare this result with Equation 2 in Table 9–2 and determine that $e = 1$ and that the graph of the equation is a parabola opening to the left. Because $p = \frac{5}{3}$, the directrix is $\frac{5}{3}$ units to the right of the pole.

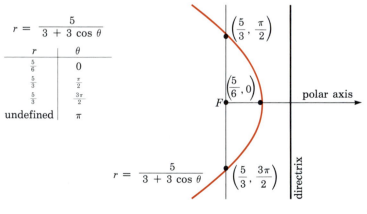

$$r = \frac{5}{3 + 3\cos\theta}$$

r	θ
$\frac{5}{6}$	0
$\frac{5}{3}$	$\frac{\pi}{2}$
$\frac{5}{3}$	$\frac{3\pi}{2}$
undefined	π

Figure 9–45

To graph the parabola, determine the coordinates of the vertex by letting $\theta = 0$: the vertex is the point $V(\frac{5}{6}, 0)$. Determine the intercepts with the vertical axis by letting $\theta = \frac{\pi}{2}$ and $\theta = \frac{3\pi}{2}$. From these three points it is easy to sketch the graph, as in Figure 9–45.

Note that the given function is not defined at $\theta = \pi$, because the denominator would be 0. Thus, no point on the parabola lies in the direction $\theta = \pi$. This is another indication that the parabola opens to the left. ∎

Example 3   Classify and sketch the conic $r = \dfrac{\sqrt{3}}{\sqrt{3} - 2 \sin \theta}$.

Solution   Divide numerator and denominator of the fraction by $\sqrt{3}$ and write the equation in the form

$$r = \frac{1}{1 - \dfrac{2}{\sqrt{3}} \sin \theta}$$

Compare this result with Equation 3 in Table 9–2 and determine that $e = \frac{2}{\sqrt{3}}$ and that $ep = 1$. From this, determine that $p = \frac{\sqrt{3}}{2}$. Because $e > 1$, the conic is a hyperbola. Its directrix is parallel to the polar axis and $\frac{\sqrt{3}}{2}$ units below it. The hyperbola opens upward and downward.

To graph the hyperbola, first find its vertices by determining r for $\theta = \frac{\pi}{2}$ and for $\theta = \frac{3\pi}{2}$. Then find the intercepts with the polar axis by letting $\theta = 0$ and $\theta = \pi$. Plot these points and sketch the graph, as in Figure 9–46.

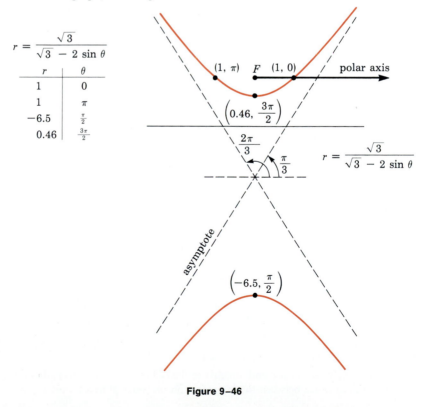

$$r = \frac{\sqrt{3}}{\sqrt{3} - 2 \sin \theta}$$

r	θ
1	0
1	π
-6.5	$\frac{\pi}{2}$
0.46	$\frac{3\pi}{2}$

Figure 9–46

Note that the given function is not defined if its denominator is 0. Determine the values excluded from the domain of the function by setting the denominator equal to 0 and solving for θ:

$$1 - \frac{2}{\sqrt{3}} \sin \theta = 0$$

$$\frac{2}{\sqrt{3}} \sin \theta = 1$$

$$\sin \theta = \frac{\sqrt{3}}{2}$$

$$\theta = \frac{\pi}{3} \quad \text{or} \quad \frac{2\pi}{3}$$

No point on the hyperbola can be found in the directions $\theta = \frac{\pi}{3}$ or $\theta = \frac{2\pi}{3}$; these directions are parallel to the asymptotes of the hyperbola.

Note also that for $0 \le \theta < 2\pi$, the *lower* branch of the hyperbola is traced as θ increases from $\frac{\pi}{3}$ to $\frac{2\pi}{3}$. Other values of θ determine the upper branch.   ∎

Example 4   Determine the polar equation of the conic with eccentricity 2 and having a directrix with equation $r \sin \theta = 3$.

Solution   Because the eccentricity is given as 2, $e = 2$. Because $2 > 1$, the conic is a hyperbola. Convert the equation of the directrix to rectangular form to see that it is the line given by the equation $y = 3$. Thus, the directrix is a line parallel to, and 3 units above, the polar axis. Because p is the distance from the pole to the directrix, $p = 3$. Because the directrix is *above* the polar axis, substitute the values for e and p into Equation 4 of Table 9–2 and simplify. The result is the polar equation of the given conic:

$$r = \frac{ep}{1 + e \sin \theta}$$

$$r = \frac{2 \cdot 3}{1 + 2 \sin \theta}$$

$$r = \frac{6}{1 + 2 \sin \theta}$$ ∎

Exercise 9.8

In Exercises 1–6 classify and sketch the conic whose polar equation is given.

1. $r = \dfrac{3}{1 + \cos \theta}$

2. $r = \dfrac{5}{1 - \sin \theta}$

3. $r = \dfrac{4}{2 - \cos \theta}$

4. $r = \dfrac{3}{3 + 2 \sin \theta}$

5. $r = \dfrac{4}{2 - 3 \sin \theta}$

6. $r = \dfrac{2}{1 + \sqrt{2} \cos \theta}$

In Exercises 7–10 write the polar equation of the conic described.

7. Eccentricity $\frac{2}{3}$, equation of directrix $r \cos \theta = 5$

8. Eccentricity 3, equation of directrix $r \sin \theta = -2$

9. Eccentricity $\frac{2}{3}$, focal width $\frac{7}{3}$, and directrix to the right of the pole, perpendicular to the polar axis (*Hint:* The focal width is the distance between the intercepts on the vertical axis.)

10. Eccentricity $\frac{4}{3}$, focal width 4, and directrix to the left of the pole, perpendicular to the polar axis

11. Assume that $e < 1$. Express the length of the major and minor axes of the ellipse in terms of e and p.

12. Assume that $e > 1$. Express the length of the transverse axes of the hyperbola in terms of e and p.

13. In Equation 1 of Table 9–2, assume that $e = 1$. Find polar coordinates of the vertex of the parabola.

14. In Equation 1 of Table 9–2, assume that $e > 1$. Find polar coordinates of the center of the hyperbola.

15. Assuming that $0 < e < 1$, determine $2a$, the length of the major axis, and c, the distance between a focus and the center of the ellipse. Then prove that $e = \frac{c}{a}$.

16. Assuming that $e > 1$, determine $2a$, the length of the transverse axis, and c, the distance between a focus and the center of the hyperbola. Then prove that $e = \frac{c}{a}$.

9.9 PARAMETRIC EQUATIONS

Graphs of equations in two variables drawn on the Cartesian plane are called **plane curves**. If such a curve passes the vertical line test, its equation defines y to be a function of x. However, a plane curve such as an ellipse or hyperbola does not pass the vertical line test, and its equation does not define y to be a function of x. Thus, it is too restrictive to define a plane curve as the graph of an equation of the form $y = f(x)$. A more general approach defines x and y to be functions of yet another variable, say t. Such curves are defined by a system of **parametric equations**.

> **Definition.** If f and g are functions of the real variable t, then the set of ordered pairs (x, y) determined by the equations
>
> $$\begin{cases} x = f(t) \\ y = g(t) \end{cases}$$
>
> defines a plane curve. The curve is said to be **defined parametrically**; the equations are called **parametric equations**; and the variable t is called the **parameter**.

Example 1 Graph the curve defined parametrically by the equations

$$\begin{cases} x = t^2 - 1 \\ y = t + 3 \end{cases}$$

where the domain of the parameter t is $\mathscr{R}$.

Solution As t varies over the domain $\mathscr{R}$, the point (x, y) traces the curve. When $t = 1$, for example, the values of x and y are

$$\begin{aligned} x &= t^2 - 1 & y &= t + 3 \\ x &= 1^2 - 1 & y &= 1 + 3 \\ &= 0 & &= 4 \end{aligned}$$

Thus, $(0, 4)$ is one point on the curve. Others appear in the table and graph shown in Figure 9–47.

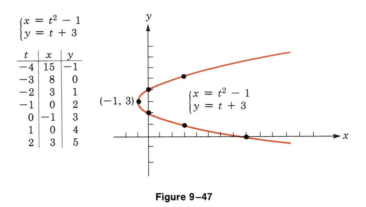

Figure 9–47

The two parametric equations of this curve can be combined into a single equation in the variables x and y by a method called **eliminating the parameter**. To do so, solve the second equation for t, and then substitute the result for t in the first equation:

$$\begin{cases} y = t + 3 \implies t = (y - 3) \\ x = t^2 - 1 \end{cases}$$
$$x = (y - 3)^2 - 1$$

or

$$x + 1 = (y - 3)^2$$

The curve represented by the parametric equations is a parabola with vertex at the point $(-1, 3)$ and opening to the right.   ■

The next two examples illustrate that a curve defined parametrically might be only a portion of the graph of the rectangular equation obtained by eliminating the parameter.

Example 2   Graph the curve determined by the parametric equations

$$\begin{cases} x = t^2 \\ y = -2t^2 + 1 \end{cases}$$

for $t \in \mathcal{R}$.

Solution   Eliminate the parameter t by substitution:

$$\begin{cases} x = t^2 \\ y = -2t^2 + 1 \end{cases}$$
$$y = -2x + 1$$

The graph of this equation is the line with slope -2 and y-intercept 1. The graph of the parametric equations, however, is not the entire line. Since t^2 is nonnegative and

$x = t^2$, it follows that $x \geq 0$. Thus the graph of the given parametric equations is only the color portion of the line $y = -2x + 1$ shown in Figure 9–48.

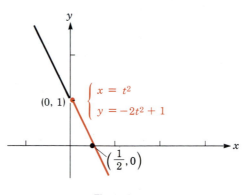

$$\begin{cases} x = t^2 \\ y = -2t^2 + 1 \end{cases}$$

(0, 1)

$\left(\frac{1}{2}, 0\right)$

Figure 9–48

Example 3 Graph the curve given by the parametric equations

$$\begin{cases} x = \dfrac{1}{2}\sin^2 t \\ y = \cos^2 t \end{cases}$$

for $t \in \mathscr{R}$.

Solution Eliminate the parameter t:

$$x = \frac{1}{2}\sin^2 t \qquad\qquad y = \cos^2 t$$

1. $2x = \sin^2 t$ **2.** $y = \cos^2 t$ Clear the first equation of fractions.

$2x + y = \sin^2 t + \cos^2 t$ Add Equations 1 and 2.

$2x + y = 1$ Use the identity $\sin^2 t + \cos^2 t = 1$.

or

$$y = -2x + 1$$

This is again the equation of a line with slope -2 and y-intercept 1. The graph of the given parametric equations is only a *portion* of this line, however, because, for any real number t,

$0 \leq \sin^2 t \leq 1$ and $0 \leq \cos^2 t \leq 1$

$0 \leq \dfrac{1}{2}\sin^2 t \leq \dfrac{1}{2}(1)$ $0 \leq y \leq 1$

$0 \leq x \leq \dfrac{1}{2}$

Thus, the graph of the given parametric equations is only that portion of the line $y = -2x + 1$ determined by numbers x between 0 and $\frac{1}{2}$, inclusive. For these numbers x, the values of y lie between 0 and 1, inclusive. The graph is that portion of the line indicated in color in Figure 9–49.

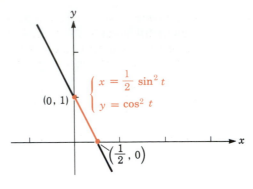

Figure 9–49

Example 4   Identify and graph the curve defined by the parametric equations

$$\begin{cases} x = a \sin t \\ y = b \cos t \end{cases}$$

Solution   Use the identity $\sin^2 t + \cos^2 t = 1$ to eliminate the parameter t:

$$x = a \sin t \qquad\qquad y = b \cos t$$

$$\frac{x}{a} = \sin t \qquad\qquad \frac{y}{b} = \cos t$$

1. $\dfrac{x^2}{a^2} = \sin^2 t$ **2.** $\dfrac{y^2}{b^2} = \cos^2 t$ Square both sides.

$$\frac{x^2}{a^2} + \frac{y^2}{b^2} = \sin^2 t + \cos^2 t$$ Add Equations 1 and 2.

$$\frac{x^2}{a^2} + \frac{y^2}{b^2} = 1$$ Use the identity $\sin^2 t + \cos^2 t = 1$.

Thus, the given parametric equations determine an ellipse. Its graph appears in Figure 9–50.

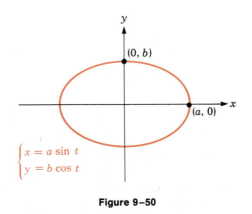

Figure 9–50

Example 5   If a circle with radius a rolls on the x-axis without slipping, a point P on the circle traces a curve called a **cycloid**. Let C be the center of the circle, and assume that P is initially at the origin. Using as the parameter the angle t (in radians) through which the radius CP turns as the circle rolls, find parametric equations for the curve traced by P. See Figure 9–51.

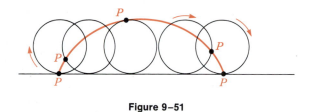

Figure 9–51

Solution   Let Q be the point at which the circle touches the x-axis. See Figure 9–52. Draw a vertical line through P meeting the x-axis at A and a horizontal line through P meeting $d(CQ)$ at B. Because the circle has rolled without slipping, $d(OQ)$ is equal to the arc length PQ. Because t is measured in radians,

$$d(OQ) = at$$

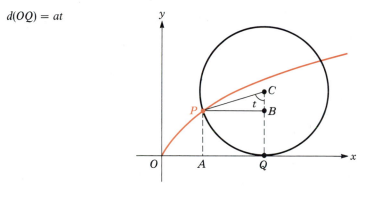

Figure 9–52

From the geometry of the figure,

$$x = d(OA) \qquad \text{and} \qquad y = d(AP)$$
$$ = d(OQ) - d(QA) \qquad\qquad = d(QB)$$
$$ = at - a \sin t \qquad\qquad\quad = d(QC) - d(CB)$$
$$ = a(t - \sin t) \qquad\qquad\quad\; = a - a \cos t$$
$$ \qquad\qquad\qquad\qquad\qquad\qquad = a(1 - \cos t)$$

Thus, the parametric equations for the cycloid are

$$\begin{cases} x = a(t - \sin t) \\ y = a(1 - \cos t) \end{cases}$$

The cycloid is also known as the *curve of quickest descent* because, of all possible curves joining P and Q in Figure 9–53, an object will slide (under the

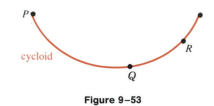

Figure 9–53

influence of gravity) from P to Q most quickly if the curve is a cycloid. The cycloid is also the *curve of equal descent* because, if R is any point on the cycloid, objects starting simultaneously from points P and R will slide to Q in equal times. These properties of the cycloid have been used in the design of clocks: a clock with a pendulum bob that traces a cycloid keeps excellent time.

▬ Exercise 9.9 ▬

In Exercises 1–8 sketch the curve determined by the given parametric equations.

1. $x = 2t, \ y = t - 5$ **2.** $x = 5t, \ y = 2 - t$

3. $x = 2(t - 3), \ y = 3(t - 2)$ **4.** $x = 3(t - 3), \ y = \frac{1}{2}(t - 2)$

5. $x = t^2, \ y = t - 2$ **6.** $x = 3t + 1, \ y = t^2$

7. $x = \sin t, \ y = \cos t$ **8.** $x = \sec t, \ y = \tan t$

In Exercises 9–16 eliminate the parameter and identify the curve defined by the given parametric equations.

9. $x = 3t + 2, \ y = 2t$ **10.** $x = t^2 - 1, \ y = t + 1$

11. $x = 3t, \ y = 5$ **12.** $x = t^3, \ y = t^3 + 3$

13. $x = 3 \cos t, \ y = 5 \sin t$ **14.** $x = 2 \tan t, \ y = 6 \cot t$

15. $x = \cos t, \ y = \sec t$ **16.** $x = \sin t, \ y = \csc t$

In Exercises 17–28 the graph of the parametric equations is only a portion of the graph of the rectangular equation obtained by eliminating the parameter. Determine the graph of the given parametric equations.

17. $x = t^2, \ y = 2t^2 + 1$ **18.** $x = 3t^2, \ y = 5t^2$

19. $x = 2t^2 - 1, \ y = 1 - t^2$ **20.** $x = 2(t^2 + 1), \ y = 1 - t^2$

21. $x = \sqrt{t}, \ y = t$ **22.** $x = \sqrt{t - 1}, \ y = \sqrt{t}$

23. $x = |t|, \ y = |t|$ **24.** $x = |t| - 2, \ y = |t| - 1$

25. $x = 2 \sin t, \ y = 3 \sin t$ **26.** $x = \sin t, \ y = \cos^2 t$

27. $x = 3 \sin^2 t, \ y = 4 \cos^2 t$ **28.** $x = \sec^2 t, \ y = \tan^2 t$

In Exercises 29–30 calculate pairs (x, y) determined by the given parametric equations, and sketch the curve. Do not attempt to eliminate the parameter.

29. $x = \sin t, \ y = \sin 2t$ Lissajous curve   **30.** $x = \cos^3 t, \ y = \sin^3 t$ Hypocycloid

31. Find parametric equations for a circle with radius a and center at the origin. Let P be a point on the circle, and let the parameter t be the angle that OP makes with the x-axis. See Illustration 1.

32. Find parametric equations of the parabola $x^2 = 4py$. Let the parameter t be the slope of the line OP. See Illustration 2.

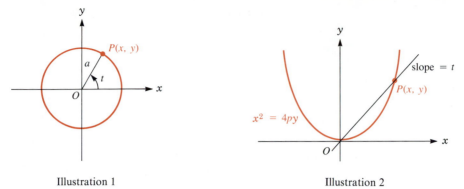

Illustration 1   Illustration 2

33. Circles of radius a and b are centered at the origin. See Illustration 3. Line OBA makes an angle t with the x-axis. Line PB is horizontal, and line PA is vertical. Using t as the parameter, determine parametric equations for the curve traced by P.

34. The line segment AB in Illustration 4 is $2k$ units long. Point A remains on the positive x-axis, and B lies on the positive y-axis. Determine parametric equations for the midpoint of segment AB. Use as the parameter the angle OM makes with the x-axis.

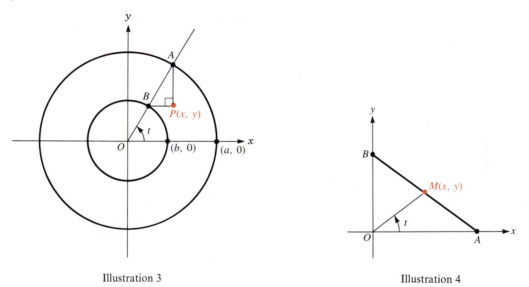

Illustration 3   Illustration 4

35. A projectile fired at an angle θ with the horizontal and with an initial velocity v_0 has position $P(x, y)$ at any time t given by

$$x = v_0 t \cos \theta, \quad y = v_0 t \sin \theta - 16t^2$$

Eliminate t, and identify the curve.

36. In Exercise 35, determine the range of the projectile (the distance it travels before it strikes the ground) and its maximum height.

37. Find parametric equations for the line through $P_1(x_1, y_1)$ and $P_2(x_2, y_2)$. See Illustration 5.

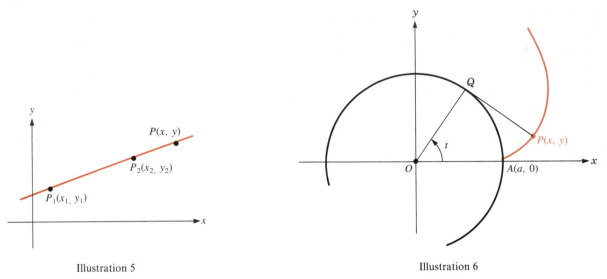

Illustration 5   Illustration 6

38. A string is wound around a circle of radius a centered at the origin. The string is held taut, and, as it is unwound, point P traces a curve known as an **involute**. See Illustration 6. Let t be the angle (in radians) that OQ makes with the x-axis. Show that parametric equations for the curve are

$$x = a(\cos t + t \sin t), \quad y = a(\sin t - t \cos t)$$

(*Hint: OQ and QP are perpendicular, and $QP = $ arc QA.*)

REVIEW EXERCISES

1. Write the equation of the circle with center at the origin and curve passing through point $(5, 5)$.

2. Write the equation of the circle with center at the origin and curve passing through point $(6, 8)$.

3. Write the equation of a circle with endpoints of its diameter at $(-2, 4)$ and $(12, 16)$.

4. Write the equation of a circle with endpoints of its diameter at $(-3, -6)$ and $(7, 10)$.

5. Write in standard form the equation of the circle $x^2 + y^2 - 6x + 4y = 3$ and graph the circle.

6. Write in standard form the equation of the circle $x^2 + 4x + y^2 - 10y = -13$ and graph the circle.

7. Write the equation of the parabola with vertex at the origin and curve passing through $(-8, 4)$ and $(-8, -4)$.

8. Write the equation of the parabola with vertex at the origin and curve passing through $(-8, 4)$ and $(8, 4)$.

9. Write the equation of the parabola $y = ax^2 + bx + c$ in standard form to show that the x-coordinate of the vertex of the parabola is $-\frac{b}{2a}$.

10. Find the equation of the parabola with vertex at $(-2, 3)$, curve passing through point $(-4, -8)$, and opening downward.

11. Graph the parabola $x^2 - 4y - 2x + 9 = 0$.

12. Graph the parabola $y^2 - 6y = 4x - 13$.

13. Graph $y^2 - 4x - 2y + 13 = 0$.

14. Write the equation of the ellipse with center at the origin, major axis that is horizontal and 12 units long, and minor axis 8 units long.

15. Write the equation of the ellipse with center at point $(-2, 3)$ and curve passing through points $(-2, 0)$ and $(2, 3)$.

16. Graph the ellipse $4x^2 + y^2 - 16x + 2y = -13$.

17. Graph the curve $x^2 + 9y^2 - 6x - 18y + 9 = 0$.

18. Write the equation of the hyperbola with vertices at points $(-3, 3)$ and $(3, 3)$ and a focus at point $(5, 3)$.

19. Graph the hyperbola $9x^2 - 4y^2 - 16y - 18x = 43$.

20. Graph the hyperbola $-2xy = 9$.

21. Solve the following system of equations by graphing:

$$\begin{cases} 3x^2 + y^2 = 52 \\ x^2 - y^2 = 12 \end{cases}$$

22. Solve the system in Review Exercise 21 algebraically.

23. Solve the following system of equations by graphing:

$$\begin{cases} x^2 + y^2 = 16 \\ -\sqrt{3}y + 4\sqrt{3} = 3x \end{cases}$$

24. Solve the system in Review Exercise 23 algebraically.

25. Solve the following system of equations by graphing:

$$\begin{cases} \dfrac{x^2}{16} + \dfrac{y^2}{12} = 1 \\ \dfrac{x^2}{1} - \dfrac{y^2}{3} = 1 \end{cases}$$

26. Solve the system in Review Exercise 25 algebraically.

In Review Exercises 27–30 the origin of the $x'y'$-system is at the point $(2, -3)$ of the xy-system. Express each equation in terms of the variables x' and y'. Draw a sketch that shows the relation of the graphs to both coordinate systems.

27. $x^2 + y^2 = 25$ **28.** $x^2 - y^2 = 4$ **29.** $4x^2 + 9y^2 = 36$ **30.** $x^2 - 4x - 3y = 5$

In Review Exercises 31–32 determine the point at which the origin of the $x'y'$-system should be located to eliminate the first-degree terms of the given equation. Draw a sketch that shows the relation of the graphs of the equations to both coordinate systems.

31. $x^2 + y^2 - 6x - 4y + 12 = 0$ **32.** $4x^2 - y^2 - 8x + 4y = 4$

In Review Exercises 33–34 determine the rotation of axes needed to remove the xy-term of the given equation. Draw a sketch that shows the relation of the graphs of the equations to both coordinate systems.

33. $13x^2 + 6\sqrt{3}xy + 7y^2 - 16 = 0$ **34.** $5x^2 + 2\sqrt{3}xy + 7y^2 - 8 = 0$

In Review Exercises 35–38 change the polar coordinates to rectangular coordinates.

35. $\left(5, \dfrac{\pi}{3}\right)$ **36.** $\left(-2, \dfrac{13\pi}{6}\right)$ **37.** $\left(-1, \dfrac{7\pi}{6}\right)$ **38.** $\left(10, -\dfrac{5\pi}{4}\right)$

In Review Exercises 39–42 change the rectangular coordinates to polar coordinates.

39. $(-\sqrt{2}, \sqrt{2})$ **40.** $(-\sqrt{3}, 1)$ **41.** $(1, 0)$ **42.** $(1, -\sqrt{3})$

In Review Exercises 43–46 change the rectangular equation to a polar equation.

43. $4xy = 4$ **44.** $x + 2y = 2$ **45.** $x^2 = 3y$ **46.** $(x^2 + y^2)^2 = 4xy$

In Review Exercises 47–50 change the polar equation to a rectangular equation.

47. $r^2 = 9 \cos 2\theta$ **48.** $r = 5 \sin \theta$ **49.** $r = \dfrac{1}{4 + \sin \theta}$ **50.** $r = \dfrac{2}{1 - \cos \theta}$

In Review Exercises 51–54 graph each polar equation.

51. $r = 4(1 + \cos \theta)$ **52.** $r = 8 - 4 \cos \theta$ **53.** $r = \dfrac{6}{1 + \sin \theta}$ **54.** $r = \dfrac{2}{1 - 2 \sin \theta}$

In Review Exercises 55–60 sketch the curve determined by the given parametric equations, and eliminate the parameter to determine an equation for the curve in x and y alone.

55. $x = 3t + 1$, $y = t - 1$ **56.** $x = 3t$, $y = t + 3$ **57.** $x = \sin t$, $y = 2 \cos t$

58. $x = 2 \sin t$, $y = 2 \csc t$ **59.** $x = 2 \sin t$, $y = \sin t$ **60.** $x = \cos t$, $y = 2 + \cos t$

10

NATURAL NUMBER FUNCTIONS

The brilliant German mathematician Carl Friedrich Gauss (1777–1855) was once a student of a very strict teacher. To keep the class busy one day, the teacher asked each student to add together the first 100 natural numbers. Almost immediately, Gauss knew the answer.

In the sum $1 + 2 + 3 + \cdots + 98 + 99 + 100$, Gauss recognized that the first number (1) added to the last number (100) was 101, that the second number (2) added to the second from last number (99) was 101, and that the third number (3) added to the third from last number (98) was 101 also. Gauss realized that this pattern continued, and, because there were 50 pairs of numbers, there were 50 sums of 101. Gauss multiplied 101 by 50 to obtain the correct answer of 5050.

This story illustrates a class of problems involving long strings of numbers called *sequences*. We begin the discussion of sequences by considering a method of proof, called **mathematical induction**, that can be used to prove many sequence formulas.

10.1 MATHEMATICAL INDUCTION

Recall the children's game in which several dominoes are placed on end fairly close together. If the first domino is knocked over, it falls against the second domino, knocking it over. The second domino, in turn, knocks over the third, which knocks over the fourth, and so on, until all of the dominoes fall. Two things must happen to guarantee that all of the dominoes fall: (1) The first domino must be knocked over, and (2) every domino that falls must topple the next one. When both of these conditions are met, we can be certain that all of the dominoes will fall.

This example illustrates the idea that underlies the principle of mathematical induction.

> **The Axiom of Mathematical Induction.** If a statement involving the natural number n has the two properties that
>
> **1.** the statement is true for $n = 1$, and
> **2.** the statement is true for $n = k + 1$ whenever it is true for $n = k$,
>
> then the statement is true for all natural numbers.

The axiom of mathematical induction provides a method for proving many theorems. Note that any such proof by induction involves two parts: We must first show that the formula is true for the natural number 1, and then show that, *if* the formula is true for any natural number k, then it also is true for the natural number $k + 1$. A proof by induction is complete only if both of the required properties are established.

Let us return to Gauss's problem of finding the sum of the first 100 natural numbers. There is a formula for finding the sum of the first n natural numbers:

$$1 + 2 + 3 + \cdots + n = \frac{n(n + 1)}{2}$$

To show that this formula is true for particular values of n, such as $n = 1$, 2, 3, and 4, for example, we proceed as follows:

Substituting **1** for n, we have

$$\mathbf{1} = \frac{\mathbf{1}(\mathbf{1} + 1)}{2} \qquad \text{or} \qquad 1 = 1$$

Substituting **2** for n, we have

$$1 + \mathbf{2} = \frac{\mathbf{2}(\mathbf{2} + 1)}{2} \qquad \text{or} \qquad 3 = 3$$

Substituting **3** for n, we have

$$1 + 2 + \mathbf{3} = \frac{\mathbf{3}(\mathbf{3} + 1)}{2} \qquad \text{or} \qquad 6 = 6$$

Substituting **4** for n, we have

$$1 + 2 + 3 + \mathbf{4} = \frac{\mathbf{4}(\mathbf{4} + 1)}{2} \qquad \text{or} \qquad 10 = 10$$

To prove that the formula is true for all natural numbers n, we use the axiom of mathematical induction as in Example 1.

Example 1   Use mathematical induction to prove that the formula

$$1 + 2 + 3 + \cdots + n = \frac{n(n + 1)}{2}$$

is true for every natural number n.

Solution   The proof has two parts.

Part 1.   As you have seen, substituting the number **1** for n gives

$$1 = \frac{\mathbf{n}(\mathbf{n} + 1)}{2}$$

$$1 = \frac{(\mathbf{1})(\mathbf{1} + 1)}{2}$$

$$1 = 1$$

Thus, the formula is true for $n = 1$, and Part 1 of the proof is complete.

Part 2. Assume that the given formula is true when n is replaced by *some* natural number k. By this assumption, called the **induction hypothesis**, you accept that

1. $\quad 1 + 2 + 3 + \cdots + k = \dfrac{k(k+1)}{2}$

is a true statement. The plan is to show that the given formula is true for the next natural number, $k + 1$. Do this by verifying the equation

2. $\quad 1 + 2 + 3 + \cdots + k + (k+1) = \dfrac{(k+1)[(k+1)+1]}{2}$

obtained from the given formula by replacing n with $k + 1$.

Compare the left sides of Equations 1 and 2 and note that the left side of Equation 2 contains an extra term of $k + 1$. Hence, add $k + 1$ to both sides of Equation 1 (which was assumed to be true) to obtain the equation

$$1 + 2 + 3 + \cdots + k + (k+1) = \frac{k(k+1)}{2} + (k+1)$$

Because both terms on the right side of this equation have a common factor of $k + 1$, the right side factors and the equation can be rewritten as follows:

$$1 + 2 + 3 + \cdots + k + (k+1) = (k+1)\left(\frac{k}{2} + 1\right)$$

$$= \frac{(k+1)(k+2)}{2}$$

$$= \frac{(k+1)[(k+1)+1]}{2}$$

This final result is Equation 2. Because the truth of Equation 1 implies the truth of Equation 2, Part 2 of the proof is complete. Parts 1 and 2 together establish that the formula

$$1 + 2 + 3 + \cdots + n = \frac{n(n+1)}{2}$$

is true for any natural number n. ■

Here is a brief overview of Example 1.

1. Did the first domino fall? That is, is the formula

$$1 + 2 + 3 + \cdots + n = \frac{n(n+1)}{2}$$

true for $n = 1$? Yes, Part 1 verified this.

2. Will toppling any domino knock over the next domino? If the given formula is true for the value $n = k$, is it also true for the value $n = k + 1$? Yes, Part 2 of the proof verified this.

Because both of the induction requirements were verified, the formula is true for all natural numbers n.

Example 2   Use mathematical induction to prove the formula

$$1 + 5 + 9 + \cdots + (4n - 3) = n(2n - 1)$$

for all natural numbers n.

Solution   The proof has two parts.

Part 1.   First verify the formula for the value $n = 1$. Substituting the value $n = 1$ into the term $4n - 3$ on the left side of the formula gives the single term 1. After substituting the same value into the right side, the equation becomes

$$1 = 1[2(1) - 1]$$
$$1 = 1$$

Thus, the formula is true for $n = 1$, and Part 1 of the proof is complete.

Part 2.   The induction hypothesis is the assumption that the formula is true for $n = k$. Hence, you assume that

$$1 + 5 + 9 + \cdots + (4k - 3) = k(2k - 1)$$

is a true statement. Because the truth of this assumption must guarantee the truth of the formula for $k + 1$ terms, add the $(k + 1)$th term to both sides of the induction hypothesis formula. In this example, the terms on the left side increase by 4, so the $(k + 1)$th term is $(4k - 3) + 4$, or $4k + 1$. Adding $4k + 1$ to both sides of the induction hypothesis formula gives

$$1 + 5 + 9 + \cdots + (4k - 3) + (4k + 1) = k(2k - 1) + (4k + 1)$$

Simplify the right side and rewrite the equation as follows:

$$1 + 5 + 9 + \cdots + (4k - 3) + [4(k + 1) - 3] = 2k^2 + 3k + 1$$
$$= (k + 1)(2k + 1)$$
$$= (k + 1)[2(k + 1) - 1]$$

Because the above equation has the same form as the given formula, except that $k + 1$ appears in place of n, the truth of the formula for $n = k$ implies the truth of the formula for $n = k + 1$. Part 2 of the proof is complete.

Because both of the induction requirements have been verified, the given formula is proved for all natural numbers.   ■

Example 3   Prove that $\dfrac{1}{2} + \dfrac{1}{4} + \dfrac{1}{8} + \cdots + \dfrac{1}{2^n} < 1$.

Solution   The proof is by induction.

Part 1.   Verify the formula for $n = 1$. Substituting 1 for n on the left side of the inequality gives $\frac{1}{2} < 1$. Thus, the formula is true for $n = 1$, and Part 1 of the proof is complete.

Part 2.   The induction hypothesis is the assumption that the inequality is true for $n = k$. Thus, assume that

$$\frac{1}{2} + \frac{1}{4} + \frac{1}{8} + \cdots + \frac{1}{2^k} < 1$$

Multiply both sides of the above inequality by $\frac{1}{2}$ to obtain the inequality

$$\frac{1}{2}\left(\frac{1}{2} + \frac{1}{4} + \frac{1}{8} + \cdots + \frac{1}{2^k}\right) < 1\left(\frac{1}{2}\right)$$

or

$$\frac{1}{4} + \frac{1}{8} + \frac{1}{16} + \cdots + \frac{1}{2^{k+1}} < \frac{1}{2}$$

Now add $\frac{1}{2}$ to both sides of this inequality to obtain

$$\frac{1}{2} + \frac{1}{4} + \frac{1}{8} + \frac{1}{16} + \cdots + \frac{1}{2^{k+1}} < \frac{1}{2} + \frac{1}{2}$$

or

$$\frac{1}{2} + \frac{1}{4} + \frac{1}{8} + \frac{1}{16} + \cdots + \frac{1}{2^{k+1}} < 1$$

The resulting inequality is the same as the original except that $k + 1$ appears in place of n. The truth of the inequality for $n = k$ implies the truth of the inequality for $n = k + 1$. Part 2 of the proof is complete.

Because both of the induction requirements have been verified, this inequality is true for all natural numbers. ∎

There are statements that are not true when $n = 1$, but that are true for all natural numbers equal to or greater than some given natural number, say q. In these cases, verify the given statements for $n = q$ in Part 1 of the induction proof. After establishing Part 2 of the induction proof, the given statement is proved for all natural numbers that are greater than or equal to q.

▬▬ Exercise 10.1 ▬▬

In Exercises 1–4 verify each given formula for $n = 1, 2, 3,$ and 4.

1. $5 + 10 + 15 + \cdots + 5n = \dfrac{5n(n + 1)}{2}$

2. $1^2 + 2^2 + 3^2 + \cdots + n^2 = \dfrac{n(n + 1)(2n + 1)}{6}$

3. $7 + 10 + 13 + \cdots + (3n + 4) = \dfrac{n(3n + 11)}{2}$

4. $1(3) + 2(4) + 3(5) + \cdots + n(n + 2) = \dfrac{n}{6}(n + 1)(2n + 7)$

In Exercises 5–20 prove each of the following formulas by mathematical induction, if possible.

5. $2 + 4 + 6 + \cdots + 2n = n(n + 1)$

6. $1 + 3 + 5 + \cdots + (2n - 1) = n^2$

7. $3 + 7 + 11 + \cdots + (4n - 1) = n(2n + 1)$

8. $4 + 8 + 12 + \cdots + 4n = 2n(n + 1)$

9. $10 + 6 + 2 + \cdots + (14 - 4n) = 12n - 2n^2$

10. $8 + 6 + 4 + \cdots + (10 - 2n) = 9n - n^2$

11. $2 + 5 + 8 + \cdots + (3n - 1) = \dfrac{n(3n + 1)}{2}$

12. $3 + 6 + 9 + \cdots + 3n = \dfrac{3n(n + 1)}{2}$

13. $1^2 + 2^2 + 3^2 + \cdots + n^2 = \dfrac{n(n + 1)(2n + 1)}{6}$

14. $1 + 2 + 3 + \cdots + (n - 1) + n + (n - 1) + \cdots + 3 + 2 + 1 = n^2$

15. $\frac{1}{3} + 2 + \frac{11}{3} + \cdots + \left(\frac{5}{3}n - \frac{4}{3}\right) = n\left(\frac{5}{6}n - \frac{1}{2}\right)$

16. $\frac{1}{1 \cdot 2} + \frac{1}{2 \cdot 3} + \frac{1}{3 \cdot 4} + \cdots + \frac{1}{n(n+1)} = \frac{n}{n+1}$

17. $\frac{1}{2} + \frac{1}{4} + \frac{1}{8} + \cdots + \left(\frac{1}{2}\right)^n = 1 - \left(\frac{1}{2}\right)^n$

18. $\frac{1}{3} + \frac{2}{9} + \frac{4}{27} + \cdots + \frac{1}{3}\left(\frac{2}{3}\right)^{n-1} = 1 - \left(\frac{2}{3}\right)^n$

19. $2^0 + 2^1 + 2^2 + 2^3 + \cdots + 2^{n-1} = 2^n - 1$

20. $1^3 + 2^3 + 3^3 + \cdots + n^3 = \left[\frac{n(n+1)}{2}\right]^2$

21. Prove that $x - y$ is a factor of $x^n - y^n$. (*Hint:* Consider subtracting and adding xy^k to the binomial $x^{k+1} - y^{k+1}$.)

22. Prove that $n < 2^n$.

23. There are $180°$ in the sum of the angles of any triangle. Prove by induction that $(n - 2)180°$ gives the sum of the angles of any simple polygon when n is the number of sides of that polygon. (*Hint:* If a polygon has $k + 1$ sides, it has $k - 2$ sides plus three more sides.)

24. Consider the equation $1 + 3 + 5 + \cdots + 2n - 1 = 3n - 2$.
   a. Is the equation true for $n = 1$?
   b. Is the equation true for $n = 2$?
   c. Is the equation true for all natural numbers n?

25. If $1 + 2 + 3 + \cdots + n = \frac{n}{2}(n + 1) + 1$ were true for $n = k$, show that it would be true for $n = k + 1$. Is it true for $n = 1$?

26. Prove that $n + 1 = 1 + n$ for each natural number n.

27. If n is any natural number, prove that $7^n - 1$ is divisible by 6.

28. Prove that $1 + 2n < 3^n$ for $n > 1$.

29. Prove that, if r is a real number where $r \neq 1$, then $1 + r + r^2 + \cdots + r^n = \frac{1 - r^{n+1}}{1 - r}$.

30. The expression a^m where m is a natural number was defined in Section 1.2. An alternative definition of a^m, useful in proofs by induction, is (Part 1) $a^1 = a$ and (Part 2) $a^{m+1} = a^m \cdot a$. Use mathematical induction on n to prove the familiar law of exponents, $a^m a^n = a^{m+n}$.

In Exercises 31–34 use mathematical induction to prove each formula for all n, where n is a natural number.

31. $\sin(x + n\pi) = (-1)^n \sin x$

32. $\cos(x + n\pi) = (-1)^n \cos x$

33. $\tan(x + n\pi) = \tan x$

34. $(r \operatorname{cis} \theta)^n = r^n \operatorname{cis} n\theta$

10.2   SEQUENCES, SERIES, AND SUMMATION NOTATION

We now formally define a **sequence**.

> **Definition.**   A **sequence** is a function whose domain is the set of natural numbers.

Since a sequence is a function whose domain is the set of natural numbers, we can write its values as a list of numbers. For example, if n is a natural number, the function defined by $f(n) = 2n - 1$ generates the list

$$1, 3, 5, \ldots, 2n - 1, \ldots$$

It is common to call such a list, as well as the function, a sequence. The number 1 is the first term of this sequence, the number 3 is the second term, and the expression $2n - 1$ represents the **general**, or **nth**, **term** of the sequence. Likewise, if n is a natural number, then the function defined by $f(n) = 3n^2 + 1$ generates the list

$$4, 13, 28, \ldots, 3n^2 + 1, \ldots$$

The number 4 is the first term, 13 is the second term, 28 is the third term, and $3n^2 + 1$ is the general term.

Because the domain of any sequence is the infinite set of natural numbers, the sequence itself is an unending list of numbers. Note that a constant function such as $g(n) = 1$ is a sequence also; it generates the list $1, 1, 1, \ldots$.

Many times, sequences do not lend themselves to functional notation because it is difficult or even impossible to write the general term—the expression that shows how the terms are constructed. In such cases, if there is a pattern that is assumed to be continued, it is acceptable simply to list several terms of the sequence. Some examples of sequences follow:

$$1^2, 2^2, 3^2, \ldots, n^2, \ldots$$

$$3, 9, 19, 33, \ldots, 2n^2 + 1, \ldots$$

$$1, 3, 6, 10, 15, 21, \ldots, \frac{n(n + 1)}{2}, \ldots$$

$$1, 1, 2, 3, 5, 8, 13, 21, \ldots \quad \text{(Fibonacci sequence)}$$

$$2, 3, 5, 7, 11, 13, 17, 19, 23, \ldots \quad \text{(prime numbers)}$$

The fourth example listed is called the **Fibonacci sequence**, after the twelfth-century mathematician Leonardo of Pisa—known to his friends as Fibonacci. After the two 1's in the Fibonacci sequence, each term is the sum of the two terms that immediately precede it. The Fibonacci sequence occurs in the study of botany, for example, in the growth patterns of certain plants.

Sometimes a sequence is defined by giving its first term, a_1, and a rule showing how to obtain the $(n + 1)$th term, a_{n+1}, from the nth term, a_n. Such a definition is called a **recursive definition**. For example, the information

$$a_1 = 5 \quad \text{and} \quad a_{n+1} = 3a_n - 2$$

defines a sequence recursively. To find the first five terms of this sequence, we proceed as follows:

$$a_1 = 5$$
$$a_2 = 3(a_1) - 2 = 3(5) - 2 = 13$$
$$a_3 = 3(a_2) - 2 = 3(13) - 2 = 37$$
$$a_4 = 3(a_3) - 2 = 3(37) - 2 = 109$$
$$a_5 = 3(a_4) - 2 = 3(109) - 2 = 325$$

To add the terms of a sequence, we replace each comma between the terms with a + sign, forming what is called a **series**. Because each sequence is infinite, the number of terms in the series associated with it is infinite also. Two examples of infinite series are

$$1^2 + 2^2 + 3^2 + \cdots + n^2 + \cdots$$

and

$$1 + 2 + 3 + 5 + 8 + 13 + 21 + \cdots$$

If the signs between successive terms of an infinite series alternate, the series is often called an **alternating infinite series**. Two examples of alternating infinite series are

$$-3 + 6 - 9 + 12 - \cdots + (-1)^n 3n + \cdots$$

and

$$2 - 4 + 8 - 16 + \cdots + (-1)^{n+1} 2^n + \cdots$$

There is a shorthand method of indicating the sum of the first n terms, or the **nth partial sum** of a series. This method, called **summation notation**, involves the symbol $\sum$, which is capital sigma in the Greek alphabet. The expression

$$\sum_{n=1}^{3} (2n^2 + 1)$$

designates the sum of the three terms obtained if we successively substitute the natural numbers 1, 2, and 3 for n in the expression $2n^2 + 1$. Hence,

$$\sum_{n=1}^{3} (2n^2 + 1) = [2(1)^2 + 1] + [2(2)^2 + 1] + [2(3)^2 + 1]$$
$$= 3 + 9 + 19$$
$$= 31$$

Example 1   Evaluate $\displaystyle\sum_{n=1}^{4} (n^2 - 1)$.

Solution   In this example, n is said to run from 1 to 4. Hence, substitute 1, 2, 3, and 4 for n in the expression $n^2 - 1$, and find the sum of the resulting values:

$$\sum_{n=1}^{4} (n^2 - 1) = (1^2 - 1) + (2^2 - 1) + (3^2 - 1) + (4^2 - 1)$$
$$= 0 + 3 + 8 + 15$$
$$= 26$$ ■

Example 2   Evaluate $\displaystyle\sum_{n=3}^{5} (3n + 2)$.

Solution   In this example, n runs from 3 to 5. Hence, substitute 3, 4, and 5 for n in the expression $3n + 2$, and find the sum of the resulting values:

$$\sum_{n=3}^{5} (3n + 2) = [3(3) + 2] + [3(4) + 2] + [3(5) + 2]$$
$$= 11 + 14 + 17$$
$$= 42$$ ■

The following theorems give three properties of summations.

> **Theorem.**   If c is a constant, then $\displaystyle\sum_{k=1}^{n} c = nc$.

Proof   Because c is a constant, each term is c for each value of k as k runs from 1 to n.

$$\sum_{k=1}^{n} c = \overbrace{c + c + c + c + \cdots + c}^{n \text{ number of } c\text{'s}} = nc$$

☐

In words, this theorem states that the summation of a constant as k runs from 1 to n is n times that constant.

Example 3   Evaluate $\displaystyle\sum_{n=1}^{5} 13$.

Solution
$$\sum_{n=1}^{5} 13 = 13 + 13 + 13 + 13 + 13$$
$$= 5(13)$$
$$= 65$$

■

> **Theorem.**   If c is a constant, then $\displaystyle\sum_{k=1}^{n} cf(k) = c\sum_{k=1}^{n} f(k)$.

Proof
$$\sum_{k=1}^{n} cf(k) = cf(1) + cf(2) + cf(3) + \cdots + cf(n)$$
$$= c[f(1) + f(2) + f(3) + \cdots + f(n)] \qquad \text{Factor out } c.$$
$$= c\sum_{k=1}^{n} f(k)$$

☐

In words, this theorem states that a constant factor may be brought outside a summation sign.

Example 4   Show that $\displaystyle\sum_{k=1}^{3} 5k^2 = 5\sum_{k=1}^{3} k^2$.

Solution

$$\sum_{k=1}^{3} 5k^2 = 5(1)^2 + 5(2)^2 + 5(3)^2 \qquad\qquad 5\sum_{k=1}^{3} k^2 = 5[(1)^2 + (2)^2 + (3)^2]$$
$$= 5 + 20 + 45 \qquad\qquad\qquad\qquad\qquad = 5[1 + 4 + 9]$$
$$= 70 \qquad\qquad\qquad\qquad\qquad\qquad\quad = 5(14)$$
$$\qquad\qquad\qquad\qquad\qquad\qquad\qquad\qquad = 70$$

■

Theorem. $\displaystyle\sum_{k=1}^{n} [f(k) + g(k)] = \sum_{k=1}^{n} f(k) + \sum_{k=1}^{n} g(k)$

Proof

$$\sum_{k=1}^{n} [f(k) + g(k)] = [f(1) + g(1)] + [f(2) + g(2)]$$
$$+ [f(3) + g(3)] + \cdots + [f(n) + g(n)]$$
$$= [f(1) + f(2) + f(3) + \cdots + f(n)]$$
$$+ [g(1) + g(2) + g(3) + \cdots + g(n)]$$
$$= \sum_{k=1}^{n} f(k) + \sum_{k=1}^{n} g(k) \qquad \square$$

In words, this theorem states that the summation of a sum is equal to the sum of the summations.

Example 5   Show that $\displaystyle\sum_{k=1}^{3} (k + k^2) = \sum_{k=1}^{3} k + \sum_{k=1}^{3} k^2$.

Solution

$$\sum_{k=1}^{3} (k + k^2) = (1 + 1^2) + (2 + 2^2) + (3 + 3^2)$$
$$= 2 + 6 + 12$$
$$= 20$$

$$\sum_{k=1}^{3} k + \sum_{k=1}^{3} k^2 = (1 + 2 + 3) + (1^2 + 2^2 + 3^2)$$
$$= 6 + 14$$
$$= 20 \qquad \blacksquare$$

In Section 10.1, we proved a formula for finding the sum of the first n natural numbers. Written with summation notation, that formula is

---

Theorem. $\displaystyle\sum_{k=1}^{n} k = \frac{n(n + 1)}{2}$

In Exercise 10.1, you proved a formula for finding the sum of the squares of the first n natural numbers. Using summation notation, the formula is

---

Theorem. $\displaystyle\sum_{k=1}^{n} k^2 = \frac{n(n + 1)(2n + 1)}{6}$

These two theorems are used in the next example.

Example 6   Evaluate $\sum_{k=1}^{5} (2k-1)^2$ directly. Then expand the binomial, apply the previous theorems, and evaluate the expression again.

Solution   *Part 1.* $\sum_{k=1}^{5} (2k-1)^2 = 1 + 9 + 25 + 49 + 81 = 165$

Part 2. $\sum_{k=1}^{5} (2k-1)^2 = \sum_{k=1}^{5} (4k^2 - 4k + 1)$

$$= \sum_{k=1}^{5} 4k^2 + \sum_{k=1}^{5} (-4k) + \sum_{k=1}^{5} 1$$

$$= 4\sum_{k=1}^{5} k^2 - 4\sum_{k=1}^{5} k + \sum_{k=1}^{5} 1$$

Use the previous theorems with $n = 5$ to get

$$= 4\left[\frac{5(5+1)(2 \cdot 5 + 1)}{6}\right] - 4\left[\frac{5(5+1)}{2}\right] + 5 \cdot 1$$

$$= 4(55) - 4(15) + 5$$

$$= 165$$

Note that the sum is 165, regardless of the method used.   ∎

━━ Exercise 10.2 ━━━━━━━━

1. Write the first eight terms of the sequence defined by the function $f(n) = 5n(n-1)$.

2. Write the first six terms of the sequence defined by the function $f(n) = n\left(\dfrac{n-1}{2}\right)\left(\dfrac{n-2}{3}\right)$.

In Exercises 3–8 write the fifth term in each of the given sequences.

3. $1, 6, 11, 16, \ldots$

4. $1, 8, 27, 64, \ldots$

5. $a, a+d, a+2d, a+3d, \ldots$

6. $a, ar, ar^2, ar^3, \ldots$

7. $1, 3, 6, 10, \ldots$

8. $20, 17, 13, 8, \ldots$

In Exercises 9–16 find the sum of the first five terms of the sequence with the given general term.

9. n

10. $2k$

11. 3

12. $4k^0$

13. $2\left(\dfrac{1}{3}\right)^n$

14. $(-1)^n$

15. $3n-2$

16. $2k+1$

In Exercises 17–24 a sequence is defined recursively. Find the first four terms of each sequence.

17. $a_1 = 3$ and $a_{n+1} = 2a_n + 1$

18. $a_1 = -5$ and $a_{n+1} = -a_n - 3$

19. $a_1 = -4$ and $a_{n+1} = \dfrac{a_n}{2}$

20. $a_1 = 0$ and $a_{n+1} = 2a_n^2$

21. $a_1 = k$ and $a_{n+1} = a_n^2$

22. $a_1 = 3$ and $a_{n+1} = ka_n$

23. $a_1 = 8$ and $a_{n+1} = \dfrac{2a_n}{k}$

24. $a_1 = m$ and $a_{n+1} = \dfrac{(a_n)^2}{m}$

In Exercises 25–28 tell whether each series is an alternating infinite series.

25. $-1 + 2 - 3 + \cdots + (-1)^n n + \cdots$

26. $a + \dfrac{a}{b} + \dfrac{a}{b^2} + \cdots + a\left(\dfrac{1}{b}\right)^{n-1} + \cdots; b = 4$

27. $a + a^2 + a^3 + \cdots + a^n + \cdots; a = 3$

28. $a + a^2 + a^3 + \cdots + a^n + \cdots; a = -2$

In Exercises 29–40 evaluate each sum.

29. $\displaystyle\sum_{k=1}^{5} 2k$ **30.** $\displaystyle\sum_{k=3}^{6} 3k$ **31.** $\displaystyle\sum_{k=3}^{4} (-2k^2)$ **32.** $\displaystyle\sum_{k=1}^{100} 5$

33. $\displaystyle\sum_{k=1}^{5} (3k - 1)$ **34.** $\displaystyle\sum_{n=2}^{5} (n^2 + 3n)$ **35.** $\displaystyle\sum_{n=1}^{1000} \frac{1}{2}$ **36.** $\displaystyle\sum_{x=4}^{5} \frac{2}{x}$

37. $\displaystyle\sum_{x=3}^{4} \frac{1}{x}$ **38.** $\displaystyle\sum_{x=2}^{6} (3x^2 + 2x) - 3\sum_{x=2}^{6} x^2$

39. $\displaystyle\sum_{x=1}^{4} (4x + 1)^2 - \sum_{x=1}^{4} (4x - 1)^2$ **40.** $\displaystyle\sum_{x=0}^{10} (2x - 1)^2 + 4\sum_{x=0}^{10} x(1 - x)$

In Exercises 41–43 use mathematical induction to prove each formula.

41. $\displaystyle\sum_{k=1}^{n} (4k - 3) = n(2n - 1)$ **42.** $\displaystyle\sum_{k=1}^{n} (5k - 3) = \frac{n(5n - 1)}{2}$ **43.** $\displaystyle\sum_{k=1}^{n} (6k + 4) = n(3n + 7)$

44. Construct an example to disprove the proposition that the summation of a product is the product of the summations. In other words, prove that

$$\sum_{k=1}^{n} f(k)g(k) \quad \text{is not always equal to} \quad \sum_{k=1}^{n} f(k) \sum_{k=1}^{n} g(k)$$

10.3 ARITHMETIC AND GEOMETRIC SEQUENCES

Two important sequences are the arithmetic and geometric sequences. These sequences are often called **progressions**.

> **Definition.** An **arithmetic sequence** is a sequence of the form
>
> $$a, a + d, a + 2d, a + 3d, \ldots, a + (n - 1)d, \ldots$$
>
> where a is the first term, $a + (n - 1)d$ is the nth term, and d is the common difference.

In this definition, note that the second term of the sequence has an addend of d, the third term has an addend of $2d$, the fourth term has an addend of $3d$, and so on. This is why the nth term has an addend of $(n - 1)d$.

Example 1   For an arithmetic sequence that has a first term of 7 and a common difference of 5, write the first six terms and the 21st term of the sequence.

Solution   Because the first term, a, is 7 and the common difference, d, is 5, the first six terms are

$$7, \quad 7 + 5, \quad 7 + 2(5), \quad 7 + 3(5), \quad 7 + 4(5), \quad 7 + 5(5)$$

or

$$7, \quad 12, \quad 17, \quad 22, \quad 27, \quad 32$$

The nth term is $a + (n - 1)d$, and, because you are looking for the 21st term, $n = 21$.

$$n\text{th term} = a + (n - 1)d$$
$$21\text{st term} = 7 + (21 - 1)5$$
$$= 7 + (20)5$$
$$= 107$$

The 21st term is 107.   ■

Example 2   For an arithmetic sequence with the first three terms 2, 6, and 10, find the 98th term.

Solution   In this example, $a = 2$, $n = 98$, and $d = 6 - 2 = 10 - 6 = 4$. The nth term is given by the formula $a + (n - 1)d$. Therefore, the 98th term is

$$n\text{th term} = a + (n - 1)d$$
$$98\text{th term} = 2 + (98 - 1)4$$
$$= 2 + (97)4$$
$$= 390$$

   ■

Numbers inserted between a first and last term to form a segment of an arithmetic sequence are called **arithmetic means**. In this type of problem, the last term, l, is considered the nth term:

$$l = a + (n - 1)d$$

Example 3   Insert three arithmetic means between the numbers -3 and 12.

Solution   Begin by finding the common difference d. In this example the first term is -3 and the last term is 12. Because you are inserting three terms, the total number of terms is five. Thus, $a = -3$, $l = 12$, and $n = 5$. The formula for the nth (or last) term is

$$l = a + (n - 1)d$$

Substituting 12 for l, -3 for a, and 5 for n in the formula and solving for d gives

$$12 = -3 + (5 - 1)d$$
$$15 = 4d$$
$$\frac{15}{4} = d$$

Once the common difference has been found, the arithmetic means are the second, third, and fourth terms of the arithmetic sequence with a first term of -3 and a fifth term of 12:

$$a + d = -3 + \frac{15}{4} = \frac{3}{4}$$
$$a + 2d = -3 + \frac{30}{4} = 4\frac{1}{2}$$
$$a + 3d = -3 + \frac{45}{4} = 8\frac{1}{4}$$

The three arithmetic means are $\frac{3}{4}$, $4\frac{1}{2}$, and $8\frac{1}{4}$.   ■

The formula stated in the following theorem gives the sum of the first n terms of an arithmetic series.

> **Theorem.** The formula
> $$S_n = \frac{n(a + l)}{2}$$
> gives the sum of the first n terms of an arithmetic series. In this formula, a is the first term, l is the last (or nth) term, and n is the number of terms.

Proof   We write the first n terms of an arithmetic series letting S_n represent their sum, rewrite the same sum in reverse order, and add the equations together term by term:

$$
\begin{aligned}
S_n &= \quad a \quad + \quad (a+d) \quad + \cdots + [a+(n-2)d] + [a+(n-1)d] \\
S_n &= [a+(n-1)d] + [a+(n-2)d] + \cdots + \quad (a+d) \quad + \quad a \\
\hline
2S_n &= [2a+(n-1)d] + [2a+(n-1)d] + \cdots + [2a+(n-1)d] + [2a+(n-1)d]
\end{aligned}
$$

Because there are n equal terms on the right side of the previous equation,

$$2S_n = n[2a + (n-1)d]$$

or

$$2S_n = n\{a + [a + (n-1)d]\}$$

Because $a + (n-1)d = l$, we make that substitution in the right side of the above equation and divide both sides by 2 to obtain

$$S_n = \frac{n(a + l)}{2}$$

The theorem is proved. (Exercise 45 will ask you to prove this theorem again using mathematical induction.)   □

Example 4   Find the sum of the first 30 terms of the arithmetic series $5, 8, 11, \ldots$.

Solution   In this example, $a = 5$, $n = 30$, $d = 3$, and $l = 5 + 29(3) = 92$. Substituting these values into the formula $S_n = \dfrac{n(a + l)}{2}$ and simplifying gives

$$S_{30} = \frac{30(5 + 92)}{2} = 15(97) = 1455$$

The sum of the first 30 terms is 1455.   ∎

Another important sequence is the **geometric sequence**.

> **Definition.**   A **geometric sequence** is a sequence of the form
>
> $$a, ar, ar^2, ar^3, \ldots, ar^{n-1}, \ldots$$
>
> where a is the first term, ar^{n-1} is the nth term, and r is the common ratio.

In this definition, note that the second term of the sequence has a factor of r^1, the third term has a factor of r^2, the fourth term has a factor of r^3, and so on. This explains why the nth term has a factor of r^{n-1}.

Example 5   For a geometric sequence with a first term of 3 and a common ratio of 2, write the first six terms and the 15th term.

Solution   Write the first six terms of the geometric sequence.

$$3, \quad 3(2), \quad 3(2)^2, \quad 3(2)^3, \quad 3(2)^4, \quad 3(2)^5$$

or

$$3, \quad 6, \quad 12, \quad 24, \quad 48, \quad 96$$

To obtain the 15th term, substitute **15** for n, **3** for a, and **2** for r in the formula for the nth term:

$$n\text{th term} = ar^{n-1}$$
$$15\text{th term} = 3(2)^{15-1}$$
$$= 3(2)^{14}$$
$$= 3(16{,}384)$$
$$= 49{,}152 \qquad \blacksquare$$

Example 6   For a geometric sequence with the first three terms 9, 3, and 1, find the eighth term.

Solution   In this example, $a = 9$, $r = \frac{1}{3}$, $n = 8$, and the nth term is ar^{n-1}. To obtain the eighth term, substitute these values into the expression for the nth term:

$$n\text{th term} = ar^{n-1}$$
$$8\text{th term} = 9\left(\frac{1}{3}\right)^{8-1}$$
$$= 9\left(\frac{1}{3}\right)^{7}$$
$$= \frac{1}{243} \qquad \blacksquare$$

As with an arithmetic sequence, numbers may be inserted between a first and last term to form a segment of a geometric sequence. The numbers inserted are called **geometric means**. In this type of problem, the last term, l, is considered to be the nth term: $l = ar^{n-1}$.

Example 7   Insert two geometric means between the numbers 4 and 256.

Solution   Begin by finding the common ratio. The first term, a, is **4**. Because 256 is to be the fourth term, $n = $ **4** and $l = $ **256**. Substituting these values into the formula for the nth term of a geometric progression and solving for r gives

$$ar^{n-1} = l$$
$$4r^{4-1} = 256$$
$$r^3 = 64$$
$$r = 4$$

The common ratio is 4. The two geometric means are the second and third terms of the geometric sequence:

$$ar = 4 \cdot 4 = 16$$
$$ar^2 = 4 \cdot 4^2 = 4 \cdot 16 = 64$$

The first four terms of the geometric sequence are 4, 16, 64, and 256; 16 and 64 are geometric means between 4 and 256.   ■

There is a formula that gives the sum of the first n terms of a geometric series.

Theorem.   The formula

$$S_n = \frac{a - ar^n}{1 - r} \quad (r \neq 1)$$

gives the sum of the first n terms of a geometric series. In this formula, S_n is the sum, a is the first term, r is the common ratio, and n is the number of terms.

Proof   We write out the sum of the first n terms of the geometric series:

1. $S_n = a + ar + ar^2 + \cdots + ar^{n-3} + ar^{n-2} + ar^{n-1}$

Multiplying both sides of this equation by r gives

2. $S_n r = \quad\;\; ar + ar^2 + \quad \cdots \quad + ar^{n-2} + ar^{n-1} + ar^n$

We now subtract Equation 2 from Equation 1 and solve for S_n:

$$S_n - S_n r = a - ar^n$$
$$S_n(1 - r) = a - ar^n$$
$$S_n = \frac{a - ar^n}{1 - r}$$

The theorem is proved. (Exercise 46 will ask you to prove this theorem using mathematical induction.)   □

Example 8   Find the sum of the first six terms of the geometric series 8, 4, 2, ...

Solution   In this example, $a = 8$, $n = 6$, and $r = \frac{1}{2}$. Substituting these values into the formula for the sum of the first n terms of a geometric series gives

$$S_n = \frac{a - ar^n}{1 - r}$$

$$S_6 = \frac{8 - 8\left(\frac{1}{2}\right)^6}{1 - \frac{1}{2}}$$

$$= 2\left(\frac{63}{8}\right)$$

$$= \frac{63}{4}$$

The sum of the first six terms is $\frac{63}{4}$.   ∎

Under certain conditions, it is possible to find the sum of all the terms in an infinite series. To define this sum, we consider the series

$$a_1 + a_2 + a_3 + \cdots$$

The first partial sum, S_1, of the series is a_1. Hence,

$$S_1 = a_1$$

The second partial sum, S_2, of this series is $a_1 + a_2$. Hence,

$$S_2 = a_1 + a_2$$

In general, the nth partial sum, S_n, of this series is

$$S_n = a_1 + a_2 + a_3 + \cdots + a_n$$

If the nth partial sum, S_n, of an infinite series approaches some number S as n becomes large without bound, we say that S is the **limit** of S_n as n approaches infinity, and we write

$$\lim_{n \to \infty} S_n = S$$

If this limit exists, we say that the series is **convergent**, and write

$$S = \sum_{n=1}^{\infty} a_n$$

If this limit does not exist, we say that the series is **divergent**. To develop a formula for the sum of the infinite geometric series

$$a + ar + ar^2 + \cdots$$

where a and r are constants, we consider

$$\lim_{n \to \infty} S_n = \lim_{n \to \infty} \frac{a - ar^n}{1 - r}$$

If $|r| \geq 1$, the limit does not exist, and the series is divergent. If $|r| < 1$, then the term ar^n in the formula approaches 0 as n becomes large without bound. Thus, S_n approaches $\dfrac{a}{1-r}$ as n approaches infinity:

$$\lim_{n \to \infty} S_n = \frac{a}{1-r}$$

This argument leads to the following theorem.

Theorem. If $|r| < 1$, then the sum of an infinite geometric series is given by the formula

$$S = \frac{a}{1-r}$$

where a is the first term and r is the common ratio.

Every repeating decimal fraction can be written as an infinite geometric series.

Example 9 Change 0.444 . . . to a common fraction.

Solution Write the decimal as an infinite geometric series and find its sum:

$$S = \frac{4}{10} + \frac{4}{100} + \frac{4}{1000} + \frac{4}{10,000} + \cdots$$

$$S = \frac{4}{10} + \frac{4}{10}\left(\frac{1}{10}\right) + \frac{4}{10}\left(\frac{1}{10}\right)^2 + \frac{4}{10}\left(\frac{1}{10}\right)^3 + \cdots$$

Because the common ratio is $\frac{1}{10}$ and $\left|\frac{1}{10}\right| < 1$, use the formula for the sum of an infinite geometric series:

$$S = \frac{a}{1-r} = \frac{\dfrac{4}{10}}{1 - \dfrac{1}{10}} = \frac{\dfrac{4}{10}}{\dfrac{9}{10}} = \frac{4}{9}$$

Long division will verify that $\frac{4}{9}$ equals 0.444 . . . .   ■

▬▬ Exercise 10.3 ▬▬▬▬▬▬▬▬▬▬▬▬▬▬▬▬▬▬▬▬▬▬▬▬▬▬

In Exercises 1–6 write the first six terms of the arithmetic sequence with the given properties.

1. $a = 1$ and $d = 2$
2. $a = -12$ and $d = -5$
3. $a = 5$ and the third term is 2
4. $a = 4$ and the fifth term is 12
5. The seventh term is 24, and the common difference is $\frac{5}{2}$.
6. The 20th term is -49, and the common difference is -3.

In Exercises 7–10 find the sum of the first n terms of each arithmetic series.

7. $5 + 7 + 9 + \cdots$ (to 15 terms)

8. $-3 + (-4) + (-5) + \cdots$ (to 10 terms)

9. $\displaystyle\sum_{n=1}^{20} \left(\frac{3}{2}n + 12 \right)$

10. $\displaystyle\sum_{n=1}^{10} \left(\frac{2}{3}n + \frac{1}{3} \right)$

11. In an arithmetic sequence, the 25th term is 10 and the common difference is $\frac{1}{2}$. Find the sum of the first 30 terms.

12. In an arithmetic sequence, the 15th term is 86 and the first term is 2. Find the sum of the first 100 terms.

13. If the fifth term of an arithmetic sequence is 14 and the second term is 5, find the 15th term.

14. Can an arithmetic sequence have a first term of 4, a 25th term of 126, and a common difference of $4\frac{1}{4}$? If not, explain why.

15. Insert three arithmetic means between 10 and 20.

16. Insert five arithmetic means between 5 and 15.

17. Insert four arithmetic means between -7 and $\frac{2}{3}$.

18. Insert three arithmetic means between -11 and -2.

In Exercises 19–26 write the first four terms of each geometric sequence with the given properties.

19. $a = 10$ and $r = 2$

20. $a = -3$ and $r = 2$

21. $a = -2$ and $r = 3$

22. $a = 64$ and $r = \frac{1}{2}$

23. $a = 3$ and $r = \sqrt{2}$

24. $a = 2$ and $r = \sqrt{3}$

25. $a = 2$ and the fourth term is 54

26. The third term is 4, and $r = \frac{1}{2}$.

In Exercises 27–32 find the sum of the indicated terms of each geometric series.

27. $4 + 8 + 16 + \cdots$ (to 5 terms)

28. $9 + 27 + 81 + \cdots$ (to 6 terms)

29. $2 - 6 + 18 - \cdots$ (to 10 terms)

30. $\dfrac{1}{8} + \dfrac{1}{4} + \dfrac{1}{2} + \cdots$ (to 12 terms)

31. $\displaystyle\sum_{n=1}^{6} 3 \left(\frac{3}{2} \right)^{n-1}$

32. $\displaystyle\sum_{n=1}^{6} 12 \left(-\frac{1}{2} \right)^{n-1}$

In Exercises 33–36 find the sum of each infinite geometric series.

33. $6 + 4 + \dfrac{8}{3} + \cdots$

34. $8 + 4 + 2 + 1 + \cdots$

35. $\displaystyle\sum_{n=1}^{\infty} 12 \left(-\frac{1}{2} \right)^{n-1}$

36. $\displaystyle\sum_{n=1}^{\infty} 1 \left(\frac{1}{3} \right)^{n-1}$

37. Insert three geometric means between 10 and 20.

38. Insert five geometric means between -5 and 5, if possible.

39. Insert four geometric means between 2 and 2048.

40. Insert three geometric means between 162 and 2.

In Exercises 41–44 change each decimal to a common fraction.

41. $0.555\ldots$ **42.** $0.666\ldots$ **43.** $0.252525\ldots$ **44.** $0.373737\ldots$

45. Use mathematical induction to prove the formula for finding the sum of the first n terms of an arithmetic series.

46. Use mathematical induction to prove the formula for finding the sum of the first n terms of a geometric series.

47. If Justin earns 1¢ on the first day of May, 2¢ on the second day, 4¢ on the third day, and the pay continues to double each day throughout the month, what will his total earnings be for the month?

48. A single arithmetic mean between two numbers is called *the* arithmetic mean of the two numbers. Similarly, a single geometric mean between two numbers is called *the* geometric mean of the two numbers. Find the arithmetic mean and the geometric mean between the numbers 4 and 64. Which is larger, the arithmetic or the geometric mean?

49. Use the definitions in Exercise 48 to compute the arithmetic mean and the geometric mean between $\frac{1}{2}$ and $\frac{7}{8}$. Which is larger, the arithmetic or the geometric mean?

50. If a and b are positive numbers and $a \neq b$, prove that their arithmetic mean is greater than their geometric mean. (*Hint:* See Exercises 48 and 49.)

51. Find the indicated sum: $\displaystyle\sum_{n=1}^{100} \frac{1}{n(n+1)}$. (*Hint:* Use partial fractions first.)

52. Find the indicated sum: $\displaystyle\sum_{k=1}^{100} \ln\left(\frac{k}{k+1}\right)$.

10.4 APPLICATIONS OF SEQUENCES

The following examples illustrate some applications of arithmetic and geometric sequences.

Example 1   A town with a population of 3500 people has a predicted growth rate of 6% over the preceding year for the next 20 years. How many people are expected to live in the town 20 years from now?

Solution   Let p_0 be the initial population of the town. After 1 year, there will be a different population, p_1. The initial population (p_0) plus the growth (the product of p_0 and the rate of growth, r) will equal the new population after 1 year (p_1):

$$p_1 = p_0 + p_0 r = p_0(1 + r)$$

The population of the town at the end of 2 years will be p_2, and

$$p_2 = p_1 + p_1 r$$
$$p_2 = p_1(1 + r)$$
$$p_2 = p_0(1 + r)(1 + r)$$
$$p_2 = p_0(1 + r)^2$$

The population at the end of the third year will be $p_3 = p_0(1 + r)^3$. Writing the terms in a sequence yields

$$p_0, \quad p_0(1+r), \quad p_0(1+r)^2, \quad p_0(1+r)^3, \quad p_0(1+r)^4, \quad \dots$$

This is a geometric sequence with p_0 as the first term and $1 + r$ as the common ratio. Recall that the nth term is given by the formula $l = ar^{n-1}$. In this example $p_0 = 3500$, $1 + r = 1.06$, and, because the population after 20 years will be the value of the 21st term of the geometric sequence, $n = 21$. The population after 20 years is $p = 3500(1.06)^{20}$. Use a calculator to find that $p \approx 11{,}225$. ∎

Example 2   A woman deposits $2500 in a bank at 9% annual interest compounded daily. If the investment is left untouched for 6 years, how much money will be in the account?

Solution   This problem is similar to that of Example 1. Let the initial amount in the account be A_0. At the end of the first day, the amount in the account is

$$A_1 = A_0 + A_0\left(\frac{r}{365}\right) = A_0\left(1 + \frac{r}{365}\right)$$

The amount in the bank after the second day is

$$A_2 = A_1 + A_1\left(\frac{r}{365}\right) = A_1\left(1 + \frac{r}{365}\right) = A_0\left(1 + \frac{r}{365}\right)^2$$

Just as in Example 1, the amounts in the account each day form a geometric sequence:

$$A_0, \quad A_0\left(1 + \frac{r}{365}\right), \quad A_0\left(1 + \frac{r}{365}\right)^2, \quad A_0\left(1 + \frac{r}{365}\right)^3, \quad \ldots$$

where A_0 is the initial deposit and r is the annual rate of interest.

Because the interest is compounded daily for 6 years, the amount in the bank at the end of 6 years will be the 2191th term $(6 \cdot 365 + 1)$ of the sequence. The amount in the account at the end of 6 years is

$$A_{2191} = 2500\left(1 + \frac{0.09}{365}\right)^{2190}$$

Use a calculator to find that $A_{2191} \approx \$4289.73$.   ■

Example 3   The equation $S = 16t^2$ represents the distance in feet, S, that a dropped object will fall in t seconds. After 1 second, the object has fallen 16 feet. After 2 seconds, the object has fallen 64 feet. After 3 seconds, the object has fallen 144 feet. In other words, the object fell 16 feet during the first second, 48 feet during the next second, and 80 feet during the third second. Thus, the sequence 16, 48, 80, ... represents the distance an object will fall during the first second, second second, third second, and so forth. Find the distance the object falls during the 12th second.

Solution   The sequence 16, 48, 80, ... is an arithmetic sequence with $a = 16$ and $d = 32$. To find the 12th term, substitute these values into the formula $l = a + (n - 1)d$ and simplify:

$$l = a + (n - 1)d$$
$$l = 16 + 11(32)$$
$$l = 16 + 352$$
$$l = 368$$

During the 12th second, the object falls 368 feet.   ■

Example 4   A pump can remove 20% of the gas in a container with each stroke. What percent of the gas will remain in the container after six strokes?

Solution   Let V represent the volume of the container. Because each stroke of the pump removes 20% of the gas, 80% of the gas remains after each stroke, and you have the sequence

$$V, \quad 0.80V, \quad 0.80(0.80V), \quad 0.80[0.80(0.80V)], \quad \ldots$$

This can be written as the geometric sequence

$$V, \quad 0.8V, \quad (0.8)^2V, \quad (0.8)^3V, \quad (0.8)^4V, \quad \ldots$$

You wish to know the amount of gas remaining after six strokes. This amount is the seventh term, l, of the sequence:

$$l = ar^{n-1}$$
$$l = V(0.8)^6$$

Use a calculator to find that approximately 26% of the gas remains after six strokes of the pump.   ■

■ Exercise 10.4 ■

Decide whether each of the following exercises involves an arithmetic or geometric sequence and then solve each problem. **You may use a calculator.**

1. The number of students studying college algebra this year at State College is 623. If a trend has been established that the following year's enrollment is always 10% higher than the preceding year, how many professors will be needed in 8 years to teach college algebra if one professor can handle 60 students?

2. If Amelia borrows $5500 interest free from her mother to buy a new car and agrees to pay her mother back at the rate of $105 per month, how much does she still owe after four years?

3. A Super Ball can always rebound to 95% of the height from which it was dropped. How high will the ball rise after the 13th bounce if it was dropped from a height of 10 meters?

4. If Philip invests $1000 in a 1-year certificate of deposit at $6\frac{3}{4}$% annual interest compounded daily, how much interest will be earned that year?

5. If a single cell divides into two cells every 30 minutes, how many cells will there be at the end of ten hours?

6. If a lawn tractor, which cost c dollars when new, depreciates 20% of its previous year's value each year, how much is the lawn tractor worth after 5 years?

7. Maria can invest $1000 at $7\frac{1}{2}$% compounded annually or at $7\frac{1}{4}$% compounded daily. If she invests the money for a year, which is the best investment?

8. Find how many feet a brick will travel during the 10th second of its fall.

9. If the population of the world were to double every 30 years, approximately how many people would be on earth in the year 3000? (Consider the population in 1980 to be 4 billion, and use 1980 as the base year.)

10. If Linda deposits $1300 in a bank at 7% interest compounded annually, how much will be in the bank 17 years later? (Assume that there are no other transactions on the account.)

11. If a house purchased for $50,000 in 1978 appreciates in value by 6% each year, how much will the house be worth in the year 2000?

12. Calculate the value of $1000 left on deposit for 10 years at an annual rate of 7% compounded annually.

13. Calculate the value of $1000 left on deposit for 10 years at an annual rate of 7% compounded quarterly.

14. Calculate the value of $1000 left on deposit for 10 years at an annual rate of 7% compounded monthly.

15. Calculate the value of $1000 left on deposit for 10 years at an annual rate of 7% compounded daily.

16. Calculate the value of $1000 left on deposit for 10 years at an annual rate of 7% compounded hourly.

17. When John was 20 years old, he opened an individual retirement account by investing $2000 that will earn 11% interest compounded quarterly. How much will his investment be worth when John is 65 years old?

18. One lone bacterium divides to form two bacteria every 5 minutes. If two bacteria multiply enough to fill a petri dish completely in 2 hours, how long will it take one bacterium to fill the dish?

19. A legend tells of an ancient king who was grateful to the inventor of the game of chess and offered to grant him any request. The man was shrewd, and he said, "My request is modest, Your Majesty. Simply place one grain of wheat on the first square on the chessboard, two grains on the second, four on the third, and so on,

with each square holding double that of the square before. Do this until each of the 64 squares is covered." The king, thinking he'd gotten off lightly, readily agreed. How many grains did the king need to fill the chessboard?

20. Estimate the size of the wheat pile in Exercise 19. (*Hint:* There are about one-half million grains of wheat in a bushel.)

21. Does $0.999999 = 1$? Explain.   **22.** Does $0.999\ldots = 1$? Explain.

10.5 THE BINOMIAL THEOREM

In this section we discuss a method to raise binomials to positive integral powers. To this end, we consider the following binomial expansions:

$$(a + b)^0 = 1$$
$$(a + b)^1 = a + b$$
$$(a + b)^2 = a^2 + 2ab + b^2$$
$$(a + b)^3 = a^3 + 3a^2b + 3ab^2 + b^3$$
$$(a + b)^4 = a^4 + 4a^3b + 6a^2b^2 + 4ab^3 + b^4$$
$$(a + b)^5 = a^5 + 5a^4b + 10a^3b^2 + 10a^2b^3 + 5ab^4 + b^5$$
$$(a + b)^6 = a^6 + 6a^5b + 15a^4b^2 + 20a^3b^3 + 15a^2b^4 + 6ab^5 + b^6$$

Four patterns are apparent in the above expansions.

1. Each expansion has one more term than the power of the binomial.
2. The degree of each term in each expansion equals the exponent of the binomial.
3. The first term in each expansion is *a* raised to the power of the binomial, and the last term is *b* raised to the power of the binomial.
4. The exponents of *a* decrease by one in each successive term, and the exponents of *b*, beginning with b^0 in the first term, increase by one in each successive term.

Blaise Pascal (1623–1662) Pascal made several contributions to the field of probability.

To make another pattern apparent, we write the coefficients of each of the binomial expansions in a triangular array:

$(a + b)^0$ 1
$(a + b)^1$ 1 1
$(a + b)^2$ 1 2 1
$(a + b)^3$ 1 3 3 1
$(a + b)^4$ 1 4 6 4 1
$(a + b)^5$ 1 5 10 10 5 1
$(a + b)^6$ 1 6 15 20 15 6 1

In this triangular array, each entry other than the 1's is the sum of the closest pair of numbers in the line immediately above it. For example, the 6 in the bottom row is the sum of the 1 and the 5 above it, and the 15 is the sum of the 10 and 5 above it.

The triangular array, called **Pascal's triangle** after the French mathematician Blaise Pascal (1623–1662), continues with the same pattern forever. The next two lines are shown below.

$(a + b)^7$ 1   7   21   35   35   21   7   1
$(a + b)^8$ 1   8   28   56   70   56   28   8   1

Example 1   Expand $(x + y)^6$.

Solution   The first term in the expansion is x^6, and the exponents of x decrease by one in each successive term. The y will appear in the second term, and the exponents of y will increase in each successive term, concluding when the term y^6 is reached. The variables in the expansion are

x^6 x^5y x^4y^2 x^3y^3 x^2y^4 xy^5 y^6

Using Pascal's triangle, you can find the coefficients of these variables. Because the binomial is raised to the sixth power, choose the row in Pascal's triangle whose second entry is 6. The coefficients of the variables are the numbers in that row:

1   6   15   20   15   6   1

Putting these two pieces of information together, the expansion is

$$(x + y)^6 = x^6 + 6x^5y + 15x^4y^2 + 20x^3y^3 + 15x^2y^4 + 6xy^5 + y^6$$ ■

Example 2   Expand $(x - y)^6$.

Solution   To expand $(x - y)^6$, rewrite the binomial in the form

$[x + (-y)]^6$

The expansion is

$$[x + (-y)]^6 = x^6 + 6x^5(-y) + 15x^4(-y)^2 + 20x^3(-y)^3 + 15x^2(-y)^4$$
$$+ 6x(-y)^5 + (-y)^6$$
$$= x^6 - 6x^5y + 15x^4y^2 - 20x^3y^3 + 15x^2y^4 - 6xy^5 + y^6$$

In general, in the binomial expansion of $(x - y)^n$, the sign of the first term, x^n, is $+$, the sign of the second term is $-$, and the signs continue to alternate.   ■

Another method for expanding a binomial, called the **binomial theorem**, uses **factorial notation**.

Definition.   The symbol $n!$ (read either as "n factorial" or as "factorial n") is defined as

$$n! = n(n - 1)(n - 2)(n - 3) \cdots (3)(2)(1)$$

where n is a positive integer.

Thus, $n!$ is the product of the first n positive integers.

Example 3   Evaluate **a.** 3!, **b.** 6!, and **c.** 10!.

Solution   **a.** $3! = 3 \cdot 2 \cdot 1 = 6$

b. $6! = 6 \cdot 5 \cdot 4 \cdot 3 \cdot 2 \cdot 1 = 720$

c. $10! = 10 \cdot 9 \cdot 8 \cdot 7 \cdot 6 \cdot 5 \cdot 4 \cdot 3 \cdot 2 \cdot 1 = 3,628,800$ ∎

To extend the definition of $n!$ to include all nonnegative integers, we make the following definition.

Definition. $0! = 1$

A result of the previous two definitions is this fact:

$$n! = n(n - 1)! \qquad \text{for } n = 1, 2, 3, \ldots$$

Example 4   Show that $6! = 6 \cdot 5!$.

Solution $6! = 6 \cdot 5 \cdot 4 \cdot 3 \cdot 2 \cdot 1$

$= 6(\mathbf{5 \cdot 4 \cdot 3 \cdot 2 \cdot 1})$

$= 6 \cdot \mathbf{5!}$ ∎

We now state the binomial theorem.

The Binomial Theorem.   If n is any positive integer, then

$$(a + b)^n = a^n + \frac{n!}{1!(n - 1)!} a^{n-1}b + \frac{n!}{2!(n - 2)!} a^{n-2}b^2$$

$$+ \frac{n!}{3!(n - 3)!} a^{n-3}b^3 + \cdots + \frac{n!}{r!(n - r)!} a^{n-r}b^r + \cdots + b^n$$

The binomial theorem can be proved for positive integral exponents by using mathematical induction.

Proof   As in all induction proofs, there are two parts.

Part 1.   Substituting the number **1** for n on both sides of the equation, we have

$$(a + b)^{\mathbf{1}} = \boldsymbol{a}^{\mathbf{1}} + \frac{\mathbf{1!}}{1!(\mathbf{1} - 1)!} a^{\mathbf{1} - 1}b^{\mathbf{1}}$$

$a + b = a + a^0 b$

$a + b = a + b$

and the theorem is true when $n = 1$. Part 1 is complete.

Part 2. We write expressions for two general terms in the statement of the induction hypothesis. We assume that the theorem is true for $n = k$:

$$(a + b)^k = a^k + \frac{k!}{1!(k-1)!} a^{k-1}b + \frac{k!}{2!(k-2)!} a^{k-2}b^2 + \cdots$$

$$+ \frac{k!}{(r-1)!(k-r+1)!} a^{k-r+1}b^{r-1}$$

$$+ \frac{k!}{r!(k-r)!} a^{k-r}b^r + \cdots + b^k$$

We multiply both sides of the equation above by $a + b$ and hope to obtain a similar equation in which the quantity $k + 1$ replaces all occurrences of the variable n in the binomial theorem:

$$(a+b)^k(a+b) = (a+b)\left[a^k + \frac{k!}{1!(k-1)!} a^{k-1}b + \frac{k!}{2!(k-2)!} a^{k-2}b^2 + \cdots \right.$$

$$\left. + \frac{k!}{(r-1)!(k-r+1)!} a^{k-r+1}b^{r-1} + \frac{k!}{r!(k-r)!} a^{k-r}b^r + \cdots + b^k \right]$$

We distribute the multiplication first by a and then by b:

$$(a+b)^{k+1} = \left[a^{k+1} + \frac{k!}{1!(k-1)!} a^k b + \frac{k!}{2!(k-2)!} a^{k-1}b^2 + \cdots \right.$$

$$\left. + \frac{k!}{(r-1)!(k-r+1)!} a^{k-r+2}b^{r-1} + \frac{k!}{r!(k-r)!} a^{k-r+1}b^r + \cdots + ab^k \right]$$

$$+ \left[a^k b + \frac{k!}{1!(k-1)!} a^{k-1}b^2 + \frac{k!}{2!(k-2)!} a^{k-2}b^3 + \cdots \right.$$

$$\left. + \frac{k!}{(r-1)!(k-r+1)!} a^{k-r+1}b^r + \frac{k!}{r!(k-r)!} a^{k-r}b^{r+1} + \cdots + b^{k+1} \right]$$

Combining like terms, we have

$$(a+b)^{k+1} = a^{k+1} + \left[\frac{k!}{1!(k-1)!} + 1 \right] a^k b$$

$$+ \left[\frac{k!}{2!(k-2)!} + \frac{k!}{1!(k-1)!} \right] a^{k-1}b^2 + \cdots$$

$$+ \left[\frac{k!}{r!(k-r)!} + \frac{k!}{(r-1)!(k-r+1)!} \right] a^{k-r+1}b^r + \cdots + b^{k+1}$$

These results may be written as

$$(a+b)^{k+1} = a^{k+1} + \frac{(k+1)!}{1!(k+1-1)!} a^{(k+1)-1}b + \frac{(k+1)!}{2!(k+1-2)!} a^{(k+1)-2}b^2$$

$$+ \cdots + \frac{(k+1)!}{r!(k+1-r)!} a^{(k+1)-r}b^r + \cdots + b^{k+1}$$

This formula has precisely the same form as the binomial theorem with the quantity $k + 1$ replacing all occurrences of n. Therefore, the truth of the theorem for $n = k$ implies the truth of the theorem for $n = k + 1$. Because both parts of the axiom of mathematical induction are verified, the theorem is proved. $\qquad\qquad\qquad\qquad\qquad$ □

In the binomial theorem, the exponents of the variables in each term on the right side follow familiar patterns: The sum of the exponents of a and b in each term is n, the exponents of a decrease, and the exponents of b increase. Only the method of finding the coefficients is different. Except for the first and last terms, $n!$ is the numerator of each coefficient. If the exponent of b is r in a particular term, the two factors $r!$ and $(n - r)!$ form the denominator of the fractional coefficient.

Example 5   Use the binomial theorem to expand $(a + b)^5$.

Solution   Substituting directly into the binomial theorem gives

$$(a + b)^5 = a^5 + \frac{5!}{1!(5 - 1)!} a^4b + \frac{5!}{2!(5 - 2)!} a^3b^2 + \frac{5!}{3!(5 - 3)!} a^2b^3$$

$$+ \frac{5!}{4!(5 - 4)!} ab^4 + b^5$$

$$= a^5 + \frac{5 \cdot 4!}{1 \cdot 4!} a^4b + \frac{5 \cdot 4 \cdot 3!}{2 \cdot 1 \cdot 3!} a^3b^2 + \frac{5 \cdot 4 \cdot 3!}{3! \cdot 2 \cdot 1} a^2b^3 + \frac{5 \cdot 4!}{4! \cdot 1} ab^4 + b^5$$

$$= a^5 + 5a^4b + 10a^3b^2 + 10a^2b^3 + 5ab^4 + b^5$$

Note that the coefficients in this example are the same numbers that appear in the sixth row of Pascal's triangle (the row whose second entry is 5). $\qquad\qquad\qquad$ ■

Example 6   Find the expansion of $(2x - 3y)^4$.

Solution   Note that $(2x - 3y)^4 = (2x + [-3y])^4$. To find the expansion of $(2x + [-3y])^4$, let $a = 2x$ and $b = -3y$. Then find the expansion of $(a + b)^4$. Substituting **4** for n in the binomial theorem gives

$$(a + b)^4 = a^4 + \frac{4!}{1!(4 - 1)!} a^3b + \frac{4!}{2!(4 - 2)!} a^2b^2 + \frac{4!}{3!(4 - 3)!} ab^3 + b^4$$

$$= a^4 + \frac{4 \cdot 3!}{3!} a^3b + \frac{4 \cdot 3 \cdot 2!}{2 \cdot 1 \cdot 2!} a^2b^2 + \frac{4 \cdot 3!}{3!} ab^3 + b^4$$

$$= a^4 + 4a^3b + 6a^2b^2 + 4ab^3 + b^4$$

In this expansion, substitute **2x** for a and **$-3y$** for b, and simplify to obtain

$$(2x - 3y)^4 = (2x)^4 + 4(2x)^3(-3y) + 6(2x)^2(-3y)^2 + 4(2x)(-3y)^3 + (-3y)^4$$

$$= 16x^4 - 96x^3y + 216x^2y^2 - 216xy^3 + 81y^4 \qquad\qquad ■$$

Suppose we wish to find the fifth term of the expansion of $(a + b)^{11}$. It is possible to raise the binomial $a + b$ to the 11th power and then look at the fifth term, but that would be tedious. However, this task is easy if we use the binomial theorem.

Example 7   Find the fifth term of the expansion of $(a + b)^{11}$.

Solution   The exponent of b in the fifth term of this expansion is 4, because the exponent for b is always 1 less than the number of the term. Because the exponent of b added to the exponent of a must equal 11, the exponent of a must be 7. The variables of the fifth term appear as a^7b^4.

   Because of the binomial theorem, the number in the numerator of the coefficient is $n!$, which in this case is 11!. The factors in the denominator are 4! and $(11 - 4)!$. Thus, the complete fifth term of the expansion of $(a + b)^{11}$ is

$$\frac{11!}{4!(11 - 4)!}\, a^7b^4 = \frac{11!}{4!7!}\, a^7b^4 = \frac{11 \cdot 10 \cdot 9 \cdot 8 \cdot 7!}{4 \cdot 3 \cdot 2 \cdot 1 \cdot 7!}\, a^7b^4 = 330a^7b^4$$

■

Example 8   Find the sixth term of the expansion of $(a + b)^9$.

Solution   The exponent of b is 5, and the exponent of a is $9 - 5$ or 4 in the sixth term. The factors in the denominator of the coefficient are 5! and $(9 - 5)!$, and 9! is the numerator. Thus, the sixth term of the expansion of $(a + b)^9$ is

$$\frac{9!}{5!(9 - 5)!}\, a^4b^5 = \frac{9 \cdot 8 \cdot 7 \cdot 6 \cdot 5!}{5!4!}\, a^4b^5 = \frac{9 \cdot 8 \cdot 7 \cdot 6}{4 \cdot 3 \cdot 2 \cdot 1}\, a^4b^5 = 126a^4b^5$$

■

Example 9   Find the third term of the expansion of $(3x - 2y)^6$.

Solution   Let $a = 3x$ and $b = -2y$ and use the binomial theorem to find the third term in the expansion of $(a + b)^6$:

$$\frac{6!}{2!(6 - 2)!}\, a^4b^2 = \frac{6 \cdot 5 \cdot 4!}{2 \cdot 1 \cdot 4!}\, a^4b^2 = 15a^4b^2$$

Replacing a with $3x$ and b with $-2y$ in the term $15a^4b^2$ gives the third term of the expansion of $(3x - 2y)^6$.

$$15a^4b^2 = 15(3x)^4(-2y)^2$$
$$= 15(3)^4(-2)^2x^4y^2$$
$$= 4860x^4y^2$$

■

▬ Exercise 10.5 ▬

In Exercises 1–10 evaluate each expression.

1. $4!$

2. $-5!$

3. $3! \cdot 5!$

4. $0! \cdot 7!$

5. $6! + 6!$

6. $5! - 2!$

7. $\dfrac{9!}{12!}$

8. $\dfrac{8!}{5!}$

9. $\dfrac{18!}{6!(18 - 6)!}$

10. $\dfrac{15!}{9!(15 - 9)!}$

In Exercises 11–22 use the binomial theorem to expand each binomial.

11. $(a + b)^4$

12. $(a + b)^3$

13. $(a - b)^5$

14. $(x - y)^6$

15. $(2x - y)^3$

16. $(x + 2y)^5$

17. $(2x + y)^4$

18. $(2x - y)^4$

19. $(4x - 3y)^4$ **20.** $(5x + 2y)^5$ **21.** $(6x - 3y)^2$ **22.** $\left(\dfrac{x}{2} + \dfrac{y}{3}\right)^4$

In Exercises 23–38 find the required term in the expansion of the given expression.

23. $(a + b)^4$; third term   **24.** $(a - b)^4$; second term

25. $(a + b)^7$; fifth term   **26.** $(a + b)^5$; fourth term

27. $(a - b)^5$; sixth term   **28.** $(a + b)^{12}$; twelfth term

29. $(x - y)^8$; seventh term   **30.** $(2x - y)^4$; second term

31. $(\sqrt{2}x + y)^5$; third term   **32.** $(\sqrt{2}x - 3y)^5$; third term

33. $(x + 2y)^9$; eighth term   **34.** $(3x - 5y)^6$; fourth term

35. $(a + b)^r$; fourth term   **36.** $(a - b)^r$; fifth term

37. $(a + b)^n$; rth term   **38.** $(a + b)^n$; $(r + 1)$th term

39. Find the sum of the numbers in each row of the first ten rows of Pascal's triangle. What is the pattern?

40. Show that the sum of the coefficients in the binomial expansion of $(x + y)^n$ is 2^n. (*Hint:* Let $x = y = 1$.)

41. Find the constant term in the expansion of $\left(a - \dfrac{1}{a}\right)^{10}$.

42. Find the coefficient of x^5 in the expansion of $\left(x + \dfrac{1}{x}\right)^9$.

43. Find the coefficient of x^8 in the expansion of $\left(\sqrt{x} + \dfrac{1}{2x}\right)^{25}$.

44. Find the constant term in the expansion of $\left(\dfrac{1}{a} + a\right)^8$.

REVIEW EXERCISES

1. Verify the following formula for $n = 1$, $n = 2$, $n = 3$, and $n = 4$, and then prove the formula by mathematical induction:

$$1^3 + 2^3 + 3^3 + \cdots + n^3 = \frac{n^2(n + 1)^2}{4}$$

2. Evaluate $\displaystyle\sum_{k=1}^{4} 3k^2$.   **3.** Evaluate $\displaystyle\sum_{k=5}^{8} (k^3 + 3k^2)$.   **4.** Evaluate $\displaystyle\sum_{k=1}^{30} \left(\frac{3}{2}k - 12\right) - \frac{3}{2}\sum_{k=1}^{30} k$.

In Review Exercises 5–8 find the required term of each arithmetic sequence.

5. $5, 9, 13, \ldots$; 29th term   **6.** $8, 15, 22, \ldots$; 40th term

7. $6, -1, -8, \ldots$; 15th term   **8.** $\dfrac{1}{2}, -\dfrac{3}{2}, -\dfrac{7}{2}, \ldots$; 35th term

In Review Exercises 9–12 find the required term of each geometric sequence.

9. $81, 27, 9, \ldots$; 11th term   **10.** $2, 6, 18, \ldots$; 9th term

11. $9, \dfrac{9}{2}, \dfrac{9}{4}, \ldots$; 15th term   **12.** $8, -\dfrac{8}{5}, \dfrac{8}{25}, \ldots$; 7th term

In Review Exercises 13–16 find the sum of the first 40 terms in each series.

13. $5 + 9 + 13 + \cdots$ **14.** $8 + 15 + 22 + \cdots$ **15.** $6 - 1 - 8 - \cdots$ **16.** $\dfrac{1}{2} - \dfrac{3}{2} - \dfrac{7}{2} - \cdots$

In Review Exercises 17–20 find the sum of the first eight terms in each series.

17. $81 + 27 + 9 + \cdots$ **18.** $2 + 6 + 18 + \cdots$ **19.** $9 + \dfrac{9}{2} + \dfrac{9}{4} + \cdots$ **20.** $8 - \dfrac{8}{5} + \dfrac{8}{25} - \cdots$

In Review Exercises 21–24 find the sum of each infinite series, if possible.

21. $\dfrac{1}{3} + \dfrac{1}{6} + \dfrac{1}{12} + \cdots$ **22.** $\dfrac{1}{5} - \dfrac{2}{15} + \dfrac{4}{45} - \cdots$

23. $1 + \dfrac{3}{2} + \dfrac{9}{4} + \cdots$ **24.** $0.5 + 0.25 + 0.125 + \cdots$

In Review Exercises 25–28 use the formula for the sum of the terms of an infinite geometric series to change each decimal into a common fraction.

25. $0.333\ldots$ **26.** $0.999\ldots$ **27.** $0.171717\ldots$ **28.** $0.454545\ldots$

29. Insert three arithmetic means between 2 and 8.

30. Insert five arithmetic means between 10 and 100.

31. Insert three geometric means between 2 and 8.

32. Insert four geometric means between -2 and 64.

33. Find the sum of the first 8 terms of the series $\frac{1}{3} + 1 + 3 + \cdots$.

34. Find the seventh term of the series $2\sqrt{2} + 4 + 4\sqrt{2} + \cdots$.

35. Find the single geometric mean between 4 and 64.

36. If Leonard invests $3000 in a 6-year certificate of deposit at the annual rate of 7.75% compounded daily, how much money will be in the account when it matures?

37. The enrollment at Hometown College is growing at the rate of 5% over each previous year's enrollment. If the enrollment is currently 4000 students, what will the enrollment be 10 years from now? What was it 5 years ago?

38. A house trailer that originally cost $10,000 depreciates in value at the rate of 10% per year. How much will the trailer be worth after 10 years?

In Review Exercises 39–42 use the binomial theorem to find the expansion of each expression.

39. $(x - y)^3$ **40.** $(u + 2v)^3$ **41.** $(4a - 5b)^5$ **42.** $(\sqrt{7}r + \sqrt{3}s)^4$

In Review Exercises 43–46 find the required term of each expansion.

43. $(a + b)^8$; fourth term   **44.** $(2x - y)^5$; third term

45. $(x - y)^9$; seventh term   **46.** $(4x + 7)^6$; fourth term

Exercise 1.1 (page 8)

1. positive, prime, odd, and rational **3.** even and rational **5.** positive and rational
7. negative and rational **9.** positive, composite, odd, and rational **11.** 17 **13.** 5
15. $\pi - 2$ **17.** -8 **19.** $-\frac{1}{8}$ **21.** -4 **23.** closure property of addition
25. closure property of multiplication **27.** associative property of addition **29.** distributive property
31. additive inverse property **33.** ![number line from -3 to 5] **35.** ![number line -4 to 1] **37.** ![number line -2 to 4]

39. ![number line 0 to 2] **41.** 3 **43.** 8

Exercise 1.2 (page 18)

1. 1 **3.** $-x^3$ **5.** x^5 **7.** $\dfrac{1}{a^{10}}$ **9.** $-243x^{10}y^5$ **11.** $-\dfrac{9}{x^4y^4}$ **13.** $\dfrac{x^6}{y^9}$ **15.** $\dfrac{9x^4}{z^8}$

17. $\dfrac{9x^{10}}{25}$ **19.** $9x^8y^8$ **21.** 43 **23.** 0 **25.** 2 **27.** 7 **29.** -3 **31.** 2 **33.** -7

35. -7 **37.** -17 **39.** 3 **41.** 4 **43.** 1,000 **45.** $\frac{1}{5}$ **47.** $\frac{1}{8}$ **49.** $-\frac{1}{9}$ **51.** $|xy|$

53. $2|x|y^2$ **55.** $|x|y^2$ **57.** $|x|y^2$ **59.** $5|x|\sqrt{x}$ **61.** $2xy\sqrt[3]{2x}$ **63.** $-2|y|\sqrt[4]{2x^3y}$ **65.** xy

67. $\dfrac{3|y|}{|x|}$ **69.** $12\sqrt{3}$ **71.** $17\sqrt{2x}$ **73.** $2y^2\sqrt{3y}$ **75.** $6z\sqrt[4]{3z}$ **77.** x^6y^3 **79.** $\dfrac{1}{x^6y^8}$

81. $\dfrac{s^2}{r}$ **83.** $\dfrac{y^4}{4z^2}$ **85.** $\dfrac{256y^8}{625x^{12}}$ **87.** $\sqrt{5}$ **89.** $\dfrac{2\sqrt{x}}{x}$ **91.** $\sqrt[3]{16}$ **93.** $\sqrt[3]{5a^2}$ **95.** $\dfrac{\sqrt[5]{12x^3y^3}}{2y}$

97. $\sqrt[6]{32}$ **99.** $\sqrt[4]{3}$ **101.** all nonnegative values of x **103.** 2.3×10^4 **105.** 2.47×10^4
107. 0.0000257 **109.** 300 **111.** 670 **113.** 62.7 **115.** 32.0

Exercise 1.3 (page 26)

1. polynomial, second degree, trinomial **3.** not a polynomial
5. polynomial, undefined degree, monomial **7.** $4x^3 + 14$ **9.** $-x^2 + 14$ **11.** $-28x + 96$
13. $-4x^2 + x$ **15.** $8x^3y^7$ **17.** $-4r^3s - 4rs^3$ **19.** $x^2 - 25$ **21.** $3x^2 + 4x - 4$
23. $2x^2 - 5x - 12$ **25.** $10x^2 + 13x - 3$ **27.** $6 - 16x + 8x^2$ **29.** $-6x^2 + 7x + 3$
31. $9x^3 - x^2 - 27x + 3$ **33.** $5x^3 + 2x^2 - 5x - 2$ **35.** $9x^2 + 54x + 81$ **37.** $27x^3 - 27x^2 + 9x - 1$
39. $x^3y^3 + 2xy^2 - x^2y - 2$ **41.** $6x^3 + 14x^2 - 5x - 3$ **43.** $x^4 + 2x^3 + x^2 - 1$ **45.** $xy + x^{3/2}y^{1/2}$

47. $a - b$ **49.** $\sqrt{3} + 1$ **51.** $2(\sqrt{5} + \sqrt{2})$ **53.** $\dfrac{x(x + \sqrt{3})}{x^2 - 3}$ **55.** $\dfrac{y^2 + 2\sqrt{2}y + 2}{y^2 - 2}$

57. $\dfrac{1}{\sqrt{x + 3} + \sqrt{x}}$ **59.** $2x + 4 + \dfrac{3}{x}$ **61.** $x - 7$ **63.** $x - 3$ **65.** $x^3 - 3 + \dfrac{3}{x^2 - 2}$

67. $3(x-2)$ **69.** $4x^2(2+x)$ **71.** $7xy(y+2x)$ **73.** $3abc(a+2b+3c)$ **75.** $(x+y)(b-a)$
77. $(4a+b)(1-3a)$ **79.** $(x+1)(3x^2-1)$ **81.** $(2x+3)(2x-3)$ **83.** $(x+z+5)(x+z-5)$
85. $(x^2+y^2)(x+y)(x-y)$ **87.** $3(x+2)(x-2)$ **89.** $(x-5)(x+3)$ **91.** $(6a+5)(4a-3)$
93. prime **95.** $x(6x+7)(x-5)$ **97.** $(6x-5)(x-7)$ **99.** $(2z-3)(4z^2+6z+9)$
101. $(2a+y)(2a-y)(4a^2-2ay+y^2)(4a^2+2ay+y^2)$ **103.** $(x-1)(y+x+2)$
105. $(a+b-5)(a+b+2)$ **107.** $(x+2)(x-1)(x^2-2x+4)(x^2+x+1)$ **109.** $(x^2+x+2)(x^2-x+2)$
111. $(2a^2-a+1)(2a^2+a+1)$ **113.** $2(\frac{3}{2}x+1)$ **115.** $x^{1/2}(x^{1/2}+1)$ **117.** $ab(b^{1/2}-a^{1/2})$
119. $x^{-1/2}y(x+y)$

Exercise 1.4 (page 31)

1. $\dfrac{x+4}{x-4}$ **3.** $\dfrac{3x-4}{2x-1}$ **5.** $\dfrac{x(x-1)}{x+1}$ **7.** $\frac{1}{2}$ **9.** $\dfrac{z-4}{(z+2)(z-2)}$ **11.** $\dfrac{x-1}{x}$ **13.** $\dfrac{x(x+1)(x+1)}{x+2}$

15. 1 **17.** $\dfrac{(x+12)(2x-1)}{(2x+1)(2x-3)}$ **19.** $\dfrac{x(x+3)}{x+1}$ **21.** $\dfrac{x+5}{x+3}$ **23.** $\dfrac{-3x^2+7x+4}{(x-1)(x+1)}$

25. $\dfrac{2a-4}{(a+4)(a-4)}$ **27.** $\dfrac{-2}{(3a+4)(a+1)}$ **29.** $\dfrac{1}{x+2}$ **31.** $\dfrac{2x-5}{2x(x-2)}$ **33.** $\dfrac{2x^2+19x+1}{(x-4)(x+4)}$ **35.** 0

37. $\dfrac{-x^4+3x^3-18x^2-58x+72}{(x+5)(x-5)(x-4)(x+4)}$ **39.** $\dfrac{b}{2c}$ **41.** $81a$ **43.** -1 **45.** $\dfrac{y+x}{x^2y^2}$ **47.** $\dfrac{y+x}{y-x}$

49. $\dfrac{a^2(3x-4ab)}{ax+b}$ **51.** $\dfrac{x-2}{x+2}$ **53.** $\dfrac{3x^2y^2}{xy-1}$ **55.** $\dfrac{3x^2}{x^2+1}$ **57.** $\dfrac{x^2-3x-4}{x^2+5x-3}$ **59.** $\dfrac{x}{x+1}$

61. $\dfrac{5x+1}{x-1}$ **63.** $\dfrac{8}{2+x}$

Exercise 1.5 (page 40)

1. 5 **3.** 8 **5.** no solution **7.** 3 **9.** 6 **11.** an identity; all numbers y are solutions

13. 2 **15.** 1 **17.** 0 **19.** no solution **21.** 3 **23.** no solution **25.** $w=\dfrac{p-2l}{2}$

27. $\mu=x-\sigma z$ **29.** $r=\dfrac{3V+\pi h^3}{3\pi h^2}$ **31.** $h=\dfrac{6V}{B+B'+4M}$ **33.** $r=\dfrac{a-S}{l-S}$ or $r=\dfrac{S-a}{S-l}$

35. $x=\dfrac{r-y}{1+ry}$ **37.** $0,1$ **39.** $3,-3$ **41.** $7,2$ **43.** $2,\frac{3}{5}$ **45.** $\frac{3}{5},-\frac{5}{3}$ **47.** $\frac{1}{2},\frac{3}{2}$

49. $0,3,5$ **51.** $0,1,-\frac{2}{3}$ **53.** $0,1,-1$ **55.** $1,-1,3,-3$ **57.** $2,-2,3,-3$ **59.** $3,4,-1,-2$
61. 7 **63.** \$10,000 **65.** 327 **67.** 8 of each kind **69.** $\frac{1}{15}$ liter **71.** 4 liters **73.** 65%
75. 39 and 65 mph **77.** $\frac{4}{3}$ hours **79.** 4 feet by 8 feet **81.** 9 centimeters
83. 20 mph going, 10 mph returning **85.** 7 hours **87.** 4 hours **89.** $\frac{5}{3}$ hours
91. Matilda at 8%; Maude at 7% **93.** 10 **95.** 25 seconds

Exercise 1.6 (page 54)

1. i **3.** -1 **5.** $(x+4i)(x-4i)$ **7.** $5-6i$ **9.** $-2-10i$ **11.** $4+10i$ **13.** $6-17i$
15. $52+56i$ **17.** $-6+17i$ **19.** $-5+12i$ **21.** $\frac{2}{5}-\frac{1}{5}i$ **23.** $\frac{1}{25}+\frac{7}{25}i$ **27.** $3,-3$
29. $5\sqrt{2},-5\sqrt{2}$ **31.** $3,-1$ **33.** $2,-4$ **35.** $5,3$ **37.** $2,-3$ **39.** $0,25$ **41.** $\frac{2}{3},-2$
43. $\dfrac{-5+\sqrt{5}}{2},\dfrac{-5-\sqrt{5}}{2}$ **45.** $2\sqrt{3},-2\sqrt{3}$ **47.** $3,-\frac{5}{2}$ **49.** $\dfrac{-3+\sqrt{7}}{2},\dfrac{-3-\sqrt{7}}{2}$

51. $\dfrac{-1+\sqrt{61}}{10}, \dfrac{-1-\sqrt{61}}{10}$ **53.** $-1+i, -1-i$ **55.** $-2+i, -2-i$ **57.** $2, -1+i\sqrt{3}, -1-i\sqrt{3}$

59. $3, -4$ **61.** $\frac{3}{2}, -\frac{1}{4}$ **63.** $\frac{5}{6}, -\frac{2}{5}$ **65.** $\dfrac{-a \pm \sqrt{-3a^2 + 4ay}}{2a}$ **67.** $\dfrac{-x \pm \sqrt{x^2 + 4ax}}{2}$ **69.** 2

71. 5 **73.** 2 **75.** $3, 5$ **77.** $1, 144$ **79.** $\frac{1}{64}$ **81.** $1, -1$ **83.** $-1, -2, 3, 4$ **85.** $1, -1$
87. $-\frac{1}{2}, 5$ **89.** $\frac{1}{3}, -\frac{1}{3}, \frac{1}{2}, -\frac{1}{2}$ **91.** rational and equal   **93.** not real numbers

95. rational and unequal   **97.** $2, 10$ **99.** yes   **101.** $r_1 + r_2 = -\dfrac{b}{a}; r_1 r_2 = \dfrac{c}{a}$

Exercise 1.7   (page 64)

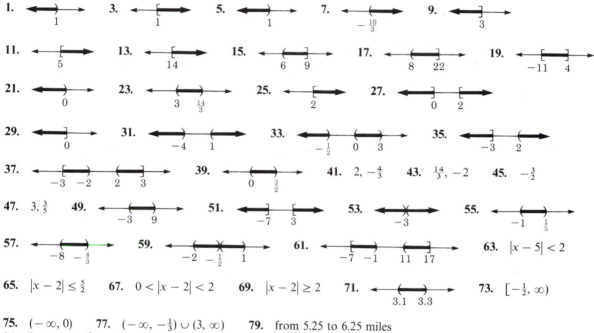

1. (graph, 1)   **3.** (graph, 1)   **5.** (graph, 1)   **7.** (graph, $-\frac{10}{3}$)   **9.** (graph, 3)

11. (graph, 5)   **13.** (graph, 14)   **15.** (graph, $6\ \ 9$)   **17.** (graph, $8\ \ 22$)   **19.** (graph, $-11\ \ 4$)

21. (graph, 0)   **23.** (graph, $3\ \ \frac{14}{3}$)   **25.** (graph, 2)   **27.** (graph, $0\ \ 2$)

29. (graph, 0)   **31.** (graph, $-4\ \ 1$)   **33.** (graph, $-\frac{1}{2}\ \ 0\ \ 3$)   **35.** (graph, $-3\ \ 2$)

37. (graph, $-3\ -2\ \ 2\ \ 3$)   **39.** (graph, $0\ \ \frac{3}{2}$)   **41.** $2, -\frac{4}{3}$ **43.** $\frac{14}{3}, -2$ **45.** $-\frac{3}{2}$

47. $3, \frac{3}{5}$ **49.** (graph, $-3\ \ 9$)   **51.** (graph, $-7\ \ 3$)   **53.** (graph, -3)   **55.** (graph, $-1\ \ \frac{1}{5}$)

57. (graph, $-8\ -\frac{4}{3}$)   **59.** (graph, $-2\ -\frac{1}{2}\ \ 1$)   **61.** (graph, $-7\ -1\ \ 11\ \ 17$)   **63.** $|x - 5| < 2$

65. $|x - 2| \le \frac{5}{2}$ **67.** $0 < |x - 2| < 2$ **69.** $|x - 2| \ge 2$ **71.** (graph, $3.1\ \ 3.3$)   **73.** $[-\frac{1}{2}, \infty)$

75. $(-\infty, 0)$ **77.** $(-\infty, -\frac{1}{3}) \cup (3, \infty)$ **79.** from 5.25 to 6.25 miles
81. between $16\frac{2}{3}$ and 20 centimeters   **83.** $40 + 2w < P \le 60 + 2w$
85. $\$4.00 \le c \le \$5.50$, from $54\frac{6}{11}$ to 75 hours

REVIEW EXERCISES   (page 66)

1. negative, composite, odd, and rational   **2.** positive and irrational   **3.** 6 **4.** 1
5. associative property of addition   **6.** distributive property   **7.** (graph, $-3\ \ \frac{1}{3}$)   **8.** (graph, $0\ \ 4$)

9. 9 **10.** 5 **11.** $x^4 y^6$ **12.** $\dfrac{y^6}{27x^{12}}$ **13.** $\dfrac{y^2}{9}$ **14.** $-\dfrac{y^6}{27x^6}$ **15.** 1 **16.** 2 **17.** -6

18. 2 **19.** $\frac{1}{6}$ **20.** -64 **21.** $|x|y^2$ **22.** $x^2 y$ **23.** $7\sqrt{2}$ **24.** $-2xy\sqrt[3]{2x}$ **25.** $x^6 y$

26. $\dfrac{y^2}{x^7}$ **27.** $\dfrac{9x^2}{16y^4}$ **28.** $\dfrac{64a^6}{b^3}$ **29.** $\dfrac{4\sqrt{6}}{3}$ **30.** $\dfrac{y\sqrt[4]{2x^3}}{2x}$ **31.** $\sqrt[6]{243}$ **32.** $\sqrt[6]{2}$

33. 3.45×10^{-4} **34.** $621{,}000$ **35.** $6x^3 + 8x^2 - 12x + 8$ **36.** $-6x^2 + 6x + 9$

37. $12x^2 + x - 6$ **38.** $2x^5 + 4x^4 + 6x^3 + 11x^2 - 3$ **39.** $\dfrac{2(\sqrt{x} + \sqrt{2})}{x - 2}$ **40.** $\dfrac{9x - h^2}{h(3\sqrt{x} + h)}$

41. $\dfrac{xy}{2} + y^2 + \dfrac{3y^3}{2x}$ **42.** $x - 3 + \dfrac{8x + 5}{x^2 + 3x}$ **43.** $3x(x + 1)(x - 1)$ **44.** $5(x^2 + y^2)(x + y)(x - y)$

45. $(3x + 8)(2x - 3)$ **46.** $(3a + x)(a - 1)$ **47.** $(2x + 5)(4x^2 - 10x + 25)$ **48.** $2(3x + 2)(x - 4)$

49. $(x + 3 + 2t)(x + 3 - 2t)$ **50.** prime   **51.** $(11z - 2)(11z - 2)$ **52.** $16(2y - 5)(4y^2 + 10y + 25)$

53. $(y - 2z)(2x - w)$ **54.** $(x^2 + x + 1)(x^2 - x + 1)$ **55.** $(x - 2)(x + 3)$ **56.** $\dfrac{x + 1}{5}$

57. $\dfrac{3x^2 - 10x + 10}{(x - 4)(x + 5)}$ **58.** $\dfrac{2x^2 + 20x + 2}{(x - 2)(x + 3)}$ **59.** $\dfrac{-5x - 6}{(x + 1)(x + 2)}$ **60.** $\dfrac{3x}{x + 1}$ **61.** $\dfrac{20}{3x}$

62. $\dfrac{y + x}{x - y}$ **63.** 4   **64.** no solution   **65.** 7   **66.** $\frac{1}{3}$ **67.** $f_1 = \dfrac{ff_2}{f_2 - f}$

68. $\dfrac{-y \pm \sqrt{y^2 + 8y}}{2}$ **69.** $0, \frac{8}{5}$ **70.** $0, \frac{2}{3}, \frac{4}{9}$ **71.** $6 + i$ **72.** $21 + 9i$ **73.** $\frac{3}{2} - \frac{3}{2}i$

74. $1 + 2i$ **75.** $-3, -5$ **76.** $-1 + 2i, -1 - 2i$ **77.** $2, -3$ **78.** $4, -2$ **79.** $1, 1, -1, -1$

80. $1, -1, 6, -6$ **81.** 5   **82.** $4, -4$ **83.** 0   **84.** no solution   **85.** $\frac{1}{3}$ **86.** 2, 10

87. 1.5 liters   **88.** $3\frac{27}{31}$ hours   **89.** $5\frac{1}{7}$ hours   **90.** $3\frac{1}{3}$ ounces   **91.** \$4500 at 11%, \$5500 at 14%

92. either 95 by 110 yards or 55 by 190 yards   **93.**   **94.**   **95.**

96.   **97.**   **98.**   **99.**

100.   **101.**   **102.**   **103.**

104.

Exercise 2.1   (page 76)

1. I   **3.** III   **5.** I   **7.** positive x-axis   **9.** positive y-axis   **11.** negative x-axis

13.   **15.**   **17.**

19.   **21.**   **23.**

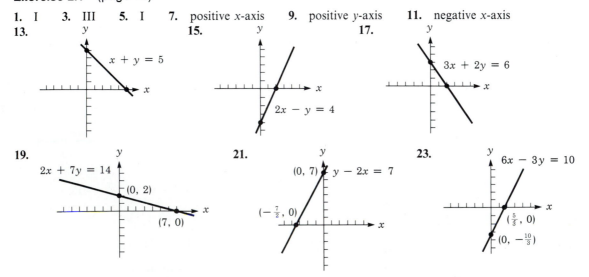

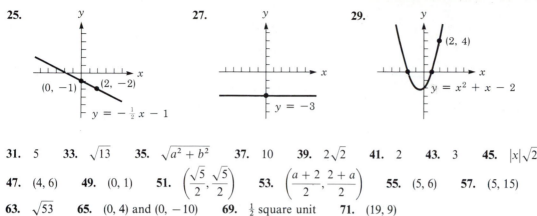

25. $y = -\frac{1}{2}x - 1$ $(0, -1)$, $(2, -2)$

27. $y = -3$

29. $(2, 4)$, $y = x^2 + x - 2$

31. 5   **33.** $\sqrt{13}$ **35.** $\sqrt{a^2 + b^2}$ **37.** 10   **39.** $2\sqrt{2}$ **41.** 2   **43.** 3   **45.** $|x|\sqrt{2}$

47. $(4, 6)$ **49.** $(0, 1)$ **51.** $\left(\dfrac{\sqrt{5}}{2}, \dfrac{\sqrt{5}}{2}\right)$ **53.** $\left(\dfrac{a+2}{2}, \dfrac{2+a}{2}\right)$ **55.** $(5, 6)$ **57.** $(5, 15)$

63. $\sqrt{53}$ **65.** $(0, 4)$ and $(0, -10)$ **69.** $\frac{1}{2}$ square unit   **71.** $(19, 9)$

Exercise 2.2   (page 88)

1. 5   **3.** -2 **5.** undefined   **7.** -1 **9.** $\frac{1}{2}$ **11.** $\frac{2}{3}$ **13.** 0   **15.** perpendicular
17. parallel   **19.** perpendicular   **21.** parallel   **23.** perpendicular   **25.** They do not lie on a straight
line.   **27.** They lie on a straight line.   **29.** None are perpendicular.   **31.** PQ and PR are perpendicular.

33. $y = 2x$ **35.** $y = 2x + \frac{7}{2}$ **37.** $3x - y = 2$ **39.** $ax - y = -\dfrac{1}{a}$ **41.** $m = \frac{42}{5}; b = \frac{21}{5}$

43. no defined slope; no y-intercept   **45.** $y = \frac{3}{2}x + \frac{5}{2}$ **47.** $y = \frac{13}{8}x - \frac{1}{4}$ **49.** $2x + 3y = 1$
51. $8x + 13y = 55$ **53.** $x = 1$ **55.** $y = 5$ **57.** $m = -1; b = 2$ **59.** $m = 3; b = 11$
61. $m = \frac{1}{3}; b = \frac{7}{3}$

Exercise 2.3   (page 98)

1. function   **3.** not a function   **5.** function   **7.** function   **9.** Domain and range are the set of
real numbers. $f(0) = 5$; $f(3) = 14$; $f(-4) = -7$ **11.** Domain is the set of real numbers; range is the set of
nonnegative real numbers. $f(0) = 0$; $f(3) = 9$; $f(-4) = 16$ **13.** Domain is the set of real numbers except -1;
range is the set of real numbers except 0. $f(0) = 3$; $f(3) = \frac{3}{4}$; $f(-4) = -1$ **15.** Domain is the set of real
numbers; range is the set of nonnegative real numbers. $f(0) = 3$; $f(3) = 0$; $f(-4) = 7$ **17.** Domain is the set of
real numbers x such that $x \geq 4$ or $x \leq -4$. Range is the set of nonnegative real numbers. $f(0)$ and $f(3)$ are not
real numbers; $f(-4) = 0$ **19.** Domain is the set of real numbers except 3; range is the set of real numbers
except 1. $f(0) = 0$; $f(3)$ is undefined; $f(-4) = \frac{4}{7}$ **21.** a function   **23.** not a function   **25.** not a function

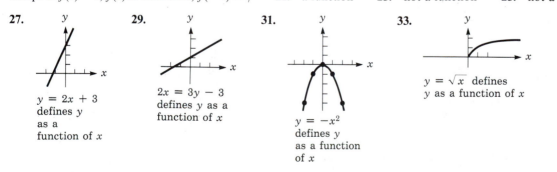

27. $y = 2x + 3$ defines y as a function of x

29. $2x = 3y - 3$ defines y as a function of x

31. $y = -x^2$ defines y as a function of x

33. $y = \sqrt{x}$ defines y as a function of x

35.

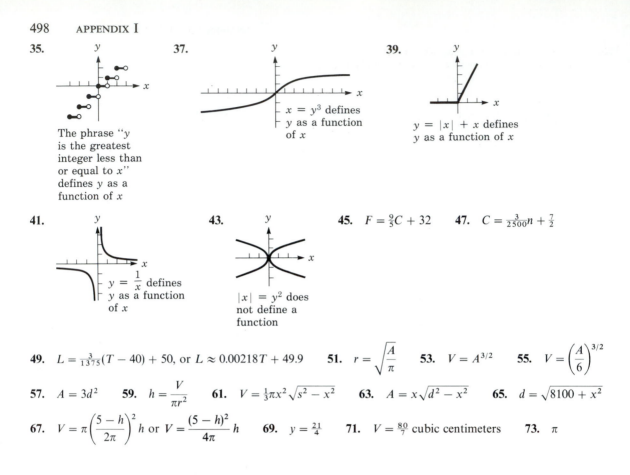

The phrase "y is the greatest integer less than or equal to x" defines y as a function of x

37.

$x = y^3$ defines y as a function of x

39.

$y = |x| + x$ defines y as a function of x

41.

$y = \frac{1}{x}$ defines y as a function of x

43.

$|x| = y^2$ does not define a function

45. $F = \frac{9}{5}C + 32$ **47.** $C = \frac{3}{2500}n + \frac{7}{2}$

49. $L = \frac{3}{1375}(T - 40) + 50$, or $L \approx 0.00218T + 49.9$ **51.** $r = \sqrt{\dfrac{A}{\pi}}$ **53.** $V = A^{3/2}$ **55.** $V = \left(\dfrac{A}{6}\right)^{3/2}$

57. $A = 3d^2$ **59.** $h = \dfrac{V}{\pi r^2}$ **61.** $V = \frac{1}{3}\pi x^2 \sqrt{s^2 - x^2}$ **63.** $A = x\sqrt{d^2 - x^2}$ **65.** $d = \sqrt{8100 + x^2}$

67. $V = \pi\left(\dfrac{5 - h}{2\pi}\right)^2 h$ or $V = \dfrac{(5 - h)^2}{4\pi} h$ **69.** $y = \frac{21}{4}$ **71.** $V = \frac{80}{7}$ cubic centimeters   **73.** π

Exercise 2.4   (page 115)

1.

$f(x) = x^2 - 4x + 1$

3.

$f(x) = -(x - 1)^2$

5.

$f(x) = x(3 - x)$

7.

$f(x) = (x + 2)^2 + 3$

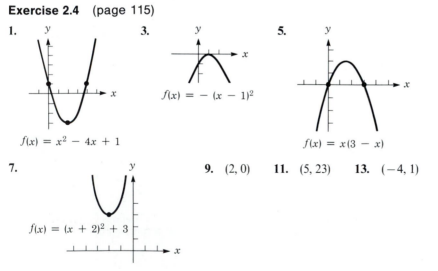

9. $(2, 0)$ **11.** $(5, 23)$ **13.** $(-4, 1)$

15. symmetric about the origin; odd function   **17.** symmetric about the x-axis; not a function
19. not symmetric; neither   **21.** symmetric about the y-axis; even function

23. symmetric about the *x*-axis; not a function
25. symmetric about the *y*-axis; no *x*-intercept; *y*-intercept is 1

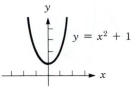

$y = x^2 + 1$

27. symmetric about the origin; *x*-intercept 0; *y*-intercept 0

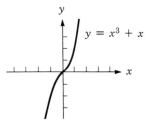

$y = x^3 + x$

29. symmetric about the *y*-axis; *x*-intercept 0; *y*-intercept 0

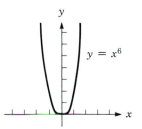

$y = x^6$

31. 2.5 seconds   **33.** 100 feet   **37.** 25 feet by 25 feet   **39.** Both numbers are 3.   **41.** $\frac{1}{2}$
43. 6 units by $4\frac{1}{2}$ units   **45.** $m(h) = 12 + 6h + h^2$ **47.** increasing when $x \le 0$ or $x \ge 1$; decreasing when $0 \le x \le 1$ **49.** Shift $y = x^3$ down 6 units.   **51.** Shift $y = x^3$ to left 2 units and down 1 unit.   **53.** Shift left 2 units and up 3 units.   **55.** Stretch vertically by a factor of -1—that is, reflect in *x*-axis; or stretch horizontally by a factor of -1—that is, reflect in *y*-axis.

Exercise 2.5 (page 124)

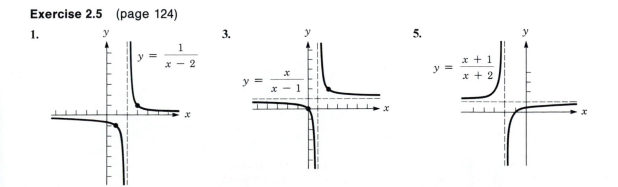

1. $y = \dfrac{1}{x - 2}$ **3.** $y = \dfrac{x}{x - 1}$ **5.** $y = \dfrac{x + 1}{x + 2}$

7.

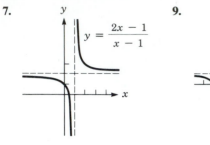

$$y = \frac{2x - 1}{x - 1}$$

9.

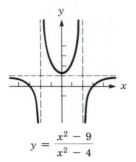

$$y = \frac{x^2 - 9}{x^2 - 4}$$

11.

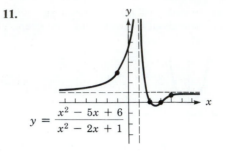

$$y = \frac{x^2 - 5x + 6}{x^2 - 2x + 1}$$

13.

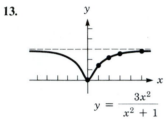

$$y = \frac{3x^2}{x^2 + 1}$$

15.

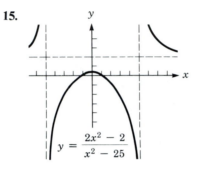

$$y = \frac{2x^2 - 2}{x^2 - 25}$$

17.

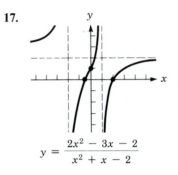

$$y = \frac{2x^2 - 3x - 2}{x^2 + x - 2}$$

19.

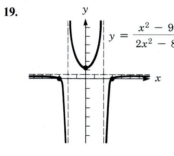

$$y = \frac{x^2 - 9}{2x^2 - 8}$$

21.

$$y = \frac{x^2 - 2x - 8}{x - 1}$$

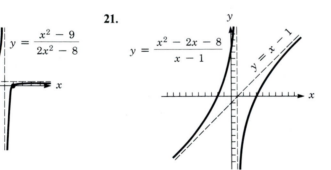

$y = x - 1$

23.

$$y = \frac{x^3 + x^2 + 6x}{x^2 - 1}$$

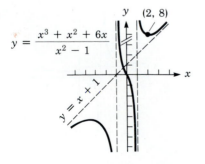

(2, 8)

$y = x + 1$

25.

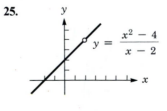

$$y = \frac{x^2 - 4}{x - 2}$$

27.

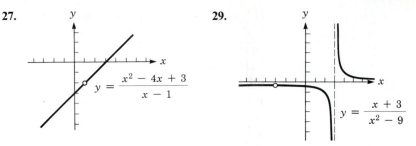

$$y = \frac{x^2 - 4x + 3}{x - 1}$$

29.

$$y = \frac{x + 3}{x^2 - 9}$$

Exercise 2.6   (page 133)

1. $(f + g)(x) = 5x - 1$; all real numbers   **3.** $(f \cdot g)(x) = 6x^2 - x - 2$; all real numbers

5. $(f - g)(x) = x + 1$; all real numbers   **7.** $(f/g)(x) = \dfrac{x^2 + x}{x^2 - 1} = \dfrac{x(x + 1)}{(x + 1)(x - 1)} = \dfrac{x}{x - 1}$; all real numbers

except 1 and -1 **9.** 7   **11.** no value   **13.** $f(x) = 3x^2; g(x) = 2x$ **15.** $f(x) = 3x^2; g(x) = x^2 - 1$

17. $f(x) = 3x^3; g(x) = -x$ **19.** $f(x) = x + 9; g(x) = x - 2$ **21.** 11   **23.** -17 **25.** 104

27. 145   **29.** $(f \circ g)(x) = 3x + 3$; all real numbers   **31.** $(g \circ g)(x) = x + 2$; all real numbers

33. $(g \circ f)(x) = 2x^2$; all real numbers   **35.** $(f \circ f)(x) = x^4$; all real numbers

37. $(g \circ f)(x) = x + 1$; nonnegative real numbers   **39.** $(g \circ g)(x) = x^4 + 2x^2 + 2$; all real numbers

41. $(f \circ g)(x) = \sqrt{x^2} = |x|$; all real numbers   **43.** $(f \circ f)(x) = \sqrt{\sqrt{x + 1} + 1}$; real numbers greater than or

equal to -1 **45.** $f(x) = x - 2; g(x) = 3x$ **47.** $f(x) = x - 2, g(x) = x^2$ **49.** $f(x) = x^2, g(x) = x - 2$

51. $f(x) = \sqrt{x}, g(x) = x + 2$ **53.** $f(x) = x + 2, g(x) = \sqrt{x}$ **55.** $f(x) = x, g(x) = x$ **59.** $(f \circ f)(x) = -\dfrac{1}{x}$

61. one-to-one   **63.** not one-to-one   **65.** one-to-one   **67.** not one-to-one   **73.** $f^{-1}(x) = \dfrac{x - 2}{3}$

75. $f^{-1}(x) = \dfrac{1}{\sqrt[3]{x}}$ **77.**

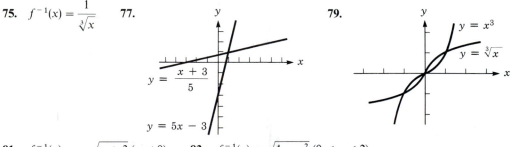

$$y = \frac{x + 3}{5}$$

$$y = 5x - 3$$

79.

$$y = x^3$$

$$y = \sqrt[3]{x}$$

81. $f^{-1}(x) = -\sqrt{x + 3}\ (y \le 0)$ **83.** $f^{-1}(x) = \sqrt{4 - x^2}\ (0 \le y \le 2)$

85. Domain is the set of all real numbers except 1; range is the set of all real numbers except 1.

87. Domain is the set of all real numbers except 0; range is the set of all real numbers except -2.

REVIEW EXERCISES   (page 135)

1.

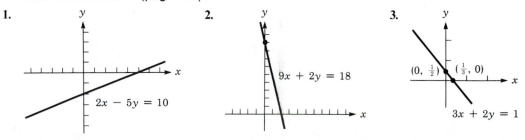

$$2x - 5y = 10$$

2.

$$9x + 2y = 18$$

3.

$(0, \tfrac{1}{2})$ $(\tfrac{1}{3}, 0)$

$$3x + 2y = 1$$

4.

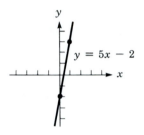

$3x - 7y = 5$

$(\frac{5}{3}, 0)$

$(0, -\frac{5}{7})$

5. $10; (0, 3)$ **6.** $13; (-6, \frac{15}{2})$ **7.** $16; (-\sqrt{3}, 1)$ **8.** $2a\sqrt{2}; (0, 0)$

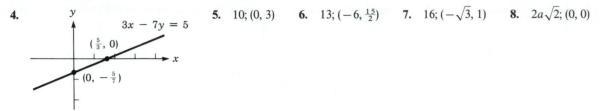

9. -6 **10.** 2 **11.** -1 **12.** 1 **13.** $7x + 5y = 0$ **14.** $4x + y = -7$ **15.** $2x + y = 9$
16. $2x - 3y = -9$ **17.** $y = 17$ **18.** $x = -5$ **19.** $7x + y = 54$ **20.** $x - 7y = 22$
21. $3x - 4y = 6$ **22.** $3x - y = 21$
23. A function; both domain and range are the set of real numbers.

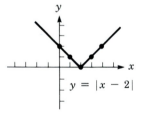

$y = 5x - 2$

24. A function; domain is the set of real numbers, range is the set of nonnegative real numbers.

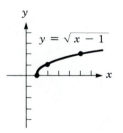

$y = |x - 2|$

25. A function; domain is the set of all x such that $x \geq 1$, range is the set of nonnegative real numbers.

$y = \sqrt{x - 1}$

26. not a function

27. A function; domain is the set of all real numbers, range is the set of all real numbers greater than or equal to $\frac{3}{4}$.

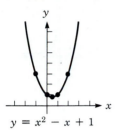

$y = x^2 - x + 1$

28. A function; domain is the set of all real numbers except 0, range is the set $\{-1, 1\}$.

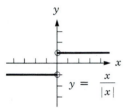

$y = \dfrac{x}{|x|}$

29.

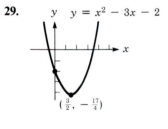

$y = x^2 - 3x - 2$

$\left(\frac{3}{2}, -\frac{17}{4}\right)$

30.

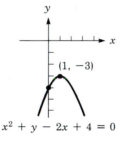

$(1, -3)$

$x^2 + y - 2x + 4 = 0$

31.

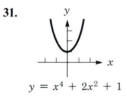

$y = x^4 + 2x^2 + 1$

32. **33.** **34.**

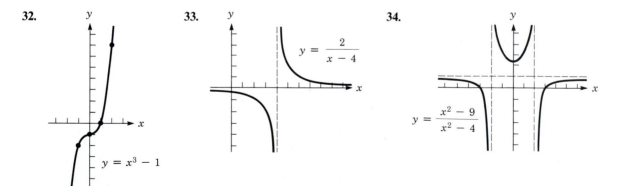

$y = x^3 - 1$

$y = \dfrac{2}{x - 4}$

$y = \dfrac{x^2 - 9}{x^2 - 4}$

35.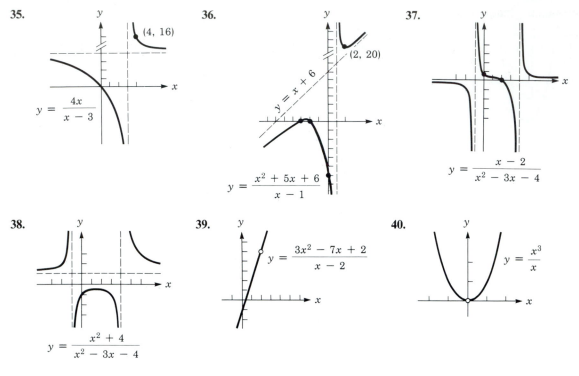

36.

37.

$y = \dfrac{4x}{x-3}$

$y = \dfrac{x^2 + 5x + 6}{x - 1}$

$y = \dfrac{x - 2}{x^2 - 3x - 4}$

38.

39.

40.

$y = \dfrac{x^2 + 4}{x^2 - 3x - 4}$

$y = \dfrac{3x^2 - 7x + 2}{x - 2}$

$y = \dfrac{x^3}{x}$

41. $f(2) = 6$ **42.** $g(-3) = 0$ **43.** $(f + g)(2) = 11$ **44.** $(f - g)(-1) = 1$ **45.** $(f \cdot g)(0) = 0$
46. $(f/g)(2) = \frac{6}{5}$ **47.** $(f \circ g)(-2) = 1$ **48.** $(g \circ f)(4) = 31$ **49.** $(f + g)(x) = \sqrt{x + 1} + x^2 - 1$
50. $(f \cdot g)(x) = \sqrt{x + 1}(x^2 - 1)$ **51.** $(f \circ g)(x) = \sqrt{x^2} = |x|$ **52.** $(g \circ f)(x) = x$
53. all x such that $x > -1$ and $x \neq 1$ **54.** all real numbers x **55.** one-to-one; $f^{-1}(x) = \frac{1}{7}x$
56. not one-to-one   **57.** one-to-one; $f^{-1}(x) = \dfrac{1 + x}{x}$ **58.** one-to-one; $f^{-1}(x) = \sqrt[3]{\dfrac{3}{x}}$
59. one-to-one; $f^{-1}(x) = \dfrac{2 + 3x}{x - 1}$ **60.** not one-to-one
61. The domain is the set of real numbers except $\frac{1}{2}$; the range is the set of real numbers except 1.
62. The domain is the set of real numbers except 0; the range is the set of real numbers except $\frac{1}{2}$.
63. $m(h) = 3$ **64.** $m(h) = 6 + h$ **65.** $\frac{9}{5}$ pounds   **66.** $\frac{1000}{3}$ cubic centimeters   **67.** about 2.8
68. about 117 ohms

Exercise 3.1   (page 142)

1. $P(2) = 35$ **3.** $P(0) = -1$ **5.** $P(-4) = 719$ **7.** $P(2) = 31$ **9.** true   **11.** true   **13.** false
15. true   **17.** true   **19.** $-1, 3, -5$ **21.** $2, 3, i, -i$ **23.** $x^3 - 3x^2 + 3x - 1$
25. $x^3 - 11x^2 + 38x - 40$ **27.** $x^4 - 3x^2 + 2$ **29.** $x^3 - \sqrt{2}x^2 + x - \sqrt{2}$
31. $1, -\dfrac{1}{2} + \dfrac{\sqrt{3}}{2}i, -\dfrac{1}{2} - \dfrac{\sqrt{3}}{2}i$ **33.** $-5, \dfrac{5}{2} + \dfrac{5\sqrt{3}}{2}i, \dfrac{5}{2} - \dfrac{5\sqrt{3}}{2}i$ **35.** $a_0 = 0$
37. The fundamental theorem guarantees at least one zero. That zero will be a root of the polynomial equation.
39. No. For example, $P(x) = 4$ is a polynomial of degree 0, and it has no zeros.
41. 3 and -1

Exercise 3.2   (page 146)

1. $P(2) = 47$ 3. $P(-5) = -569$ 5. $P(0) = 1$ 7. $P(-i) = -1 + 6i$ 9. $P(1) = 3$
11. $P(-2) = 30$ 13. $P(\frac{1}{2}) = \frac{15}{8}$ 15. $P(i) = 5$ 17. $P(1) = 0$ 19. $P(-3) = 384$ 21. $P(3) = 0$
23. $P(2) = 9$ 25. $P(0) = 8$ 27. $P(2) = 0$ 29. $P(i) = 16 + 2i$ 31. $P(-2i) = 40 - 40i$
33. $(x + 1)(3x^2 - 5x - 1) - 3$ 35. $(x - 2)(3x^2 + 4x + 2) + 0$ 37. $x(3x^2 - 2x - 6) - 4$

39. $7x^2 - 10x + 5 + \dfrac{-4}{x + 1}$ 41. $4x^3 + 9x^2 + 27x + 80 + \dfrac{245}{x - 3}$ 43. $3x^4 + 12x^3 + 48x^2 + 192x$

45. $3^5 = 243$ 47. $2^7 = 128$ 49.   51.

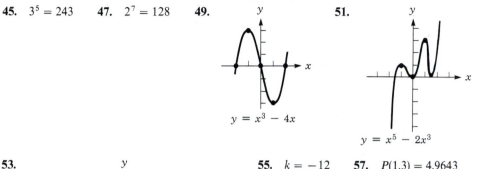

$y = x^3 - 4x$

$y = x^5 - 2x^3$

53.   55. $k = -12$ 57. $P(1.3) = 4.9643$

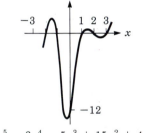

$y = x^5 - 3x^4 - 5x^3 + 15x^2 + 4x - 12$

Exercise 3.3   (page 154)

1. 10 3. 4 5. $x^3 - 2x^2 - 15x = 0$ 7. $x^3 - 5x^2 + 7x - 3 = 0$ 9. $x^4 + 8x^2 + 16 = 0$
11. $x^2 - 4x + 5 = 0$ 13. $x^4 + 5x^2 + 4 = 0$ 15. $xxx(x - 2)(x + 1)$ 17. $x(x^2 + 1)$
19. $(x + 2)(x + 3)(x^2 + 1)$ 21. 0 or 2 positive; 1 negative; 0 or 2 nonreal
23. 0 positive; 1 or 3 negative; 0 or 2 nonreal   25. 0 positive; 0 negative; 4 nonreal
27. 1 positive; 1 negative; 2 nonreal   29. 0 positive; 0 negative; 10 nonreal
31. 0 positive; 0 negative; 8 nonreal; 1 root of 0   33. 1 positive; 1 negative; 2 nonreal
35. $-1, 6$ 37. $-4, 6$ 39. $-4, 3$ 41. $-5, 2$
43. An odd-degree polynomial equation must have an odd number of roots. Since complex roots occur in conjugate pairs, one root must be left over, and it is real.

Exercise 3.4   (page 159)

1. $-1, 1, 2$ 3. $1, 2, 3, 4$ 5. $-1, -1, 1, 1, 2$ 7. $-5, 2, \sqrt{3}, -\sqrt{3}$ 9. $0, -2, -2, -2, 2, 2, 2$
11. $-3, -\frac{1}{3}, \frac{1}{2}, 2, 2$ 13. $-\frac{1}{2}, \frac{1}{2}, 3, 2i, -2i$ 15. $-1, \frac{2}{3}, 2, 3$ 17. $-1, \frac{1}{2}, 1, \frac{3}{2}$

19. $-\frac{3}{5}, \frac{2}{3}, \frac{3}{2}$ 21. $2, -\dfrac{1}{2} + \dfrac{\sqrt{3}}{2}i, -\dfrac{1}{2} - \dfrac{\sqrt{3}}{2}i$

23. $\sqrt{3}$ is a real root of $x^3 - 3 = 0$. The only possible rational solutions are ± 1 and ± 3. However, none satisfies the equation. Thus the solution, $\sqrt{3}$, must be irrational.

25. Because there is 1 variation in sign of $P(x)$, there is exactly 1 positive real root. Because $P(-x)$ has 1 variation in sign, there is exactly 1 negative root. Because 0 is not a root, there are exactly 2 real roots.

Exercise 3.5   (page 163)

1. If $a = 1$, then $P(a) = 3$. If $b = 4$, then $P(4) = 15$. Thus $P(a) < 5 < P(b)$. If $c = 2$, then $P(c) = 5$.
3. $P(1) = 8$; $P(2) = -1$ **5.** $P(2) = -72$; $P(3) = 154$ **7.** $P(0) = 10$, $P(1) = -60$ **9.** 1.7   **11.** 2.24
13. 0.4   **15.** $x \approx 0.50$

REVIEW EXERCISES   (page 163)

1. $P(2) = 16$ **2.** $P(-3) = -8$ **3.** $P(2) = 72$ **4.** $P(\frac{1}{2}) = -3$ **5.** false   **6.** true   **7.** false

8. true   **9.** true   **10.** false   **11.** $-4, 2 + 2\sqrt{3}i, 2 - 2\sqrt{3}i$ **12.** $7, -\dfrac{7}{2} + \dfrac{7\sqrt{3}}{2}i, -\dfrac{7}{2} - \dfrac{7\sqrt{3}}{2}i$

13. $2x^3 - 5x^2 - x + 6$ **14.** $x^4 + 2x^3 + 2x^2 + 2x + 1 = 0$
15. $3x^3 + 9x^2 + 29x + 90$ with a remainder of 277
16. $5x^4 - 14x^3 + 31x^2 - 64x + 129$ with a remainder of -259 **17.** 6   **18.** 1984
19. 2 or 0 positive; 2 or 0 negative; 0, 2, or 4 nonreal
20. 1 positive; 0, 2, or 4 negative; 4, 2, or 0 nonreal   **21.** 0 positive; 0 negative; 4 nonreal

22. 0 positive; 1 negative; 6 nonreal   **23.** $-5, -\frac{3}{2}, -2$ **24.** $\dfrac{1}{3}, -\dfrac{1}{2} + \dfrac{\sqrt{3}}{2}i, -\dfrac{1}{2} - \dfrac{\sqrt{3}}{2}i$

25. $P(0) = 18$; $P(-1) = -9$ **26.** $P(1) = -8$; $P(2) = 21$ **27.** $\sqrt{7} \approx 2.65$ **28.** $0.33; \frac{1}{3}$

Exercise 4.1   (page 174)

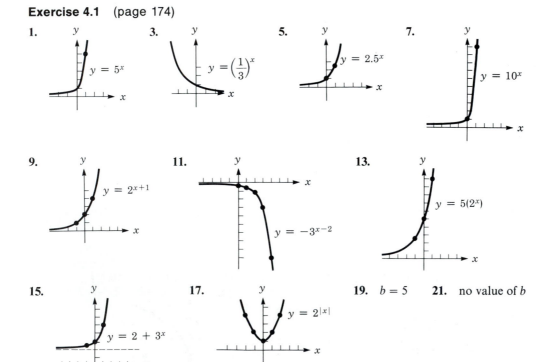

1. $y = 5^x$ **3.** $y = \left(\dfrac{1}{3}\right)^x$ **5.** $y = 2.5^x$ **7.** $y = 10^x$

9. $y = 2^{x+1}$ **11.** $y = -3^{x-2}$ **13.** $y = 5(2^x)$

15. $y = 2 + 3^x$ **17.** $y = 2^{|x|}$ **19.** $b = 5$ **21.** no value of b **23.** $b = \frac{1}{2}$

25.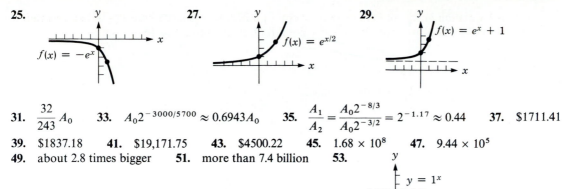

27.

29.

31. $\dfrac{32}{243} A_0$ **33.** $A_0 2^{-3000/5700} \approx 0.6943 A_0$ **35.** $\dfrac{A_1}{A_2} = \dfrac{A_0 2^{-8/3}}{A_0 2^{-3/2}} = 2^{-1.17} \approx 0.44$ **37.** $1711.41

39. $1837.18   **41.** $19,171.75   **43.** $4500.22   **45.** 1.68×10^8 **47.** 9.44×10^5
49. about 2.8 times bigger   **51.** more than 7.4 billion   **53.**

$y = 1^x$

The function is not one-to-one.

55. $y = \dfrac{1}{2}(e^x + e^{-x})$

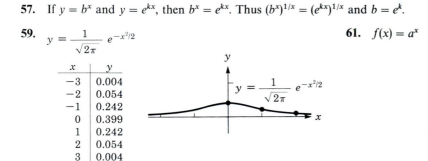

x	y
-2	3.76
-1.5	2.35
-1	1.54
-.5	1.13
0	1
.5	1.13
1	1.54
1.5	2.35
2	3.76

$y = \dfrac{1}{2}(e^x + e^{-x})$

57. If $y = b^x$ and $y = e^{kx}$, then $b^x = e^{kx}$. Thus $(b^x)^{1/x} = (e^{kx})^{1/x}$ and $b = e^k$.

59. $y = \dfrac{1}{\sqrt{2\pi}} e^{-x^2/2}$ **61.** $f(x) = a^x$

x	y
-3	0.004
-2	0.054
-1	0.242
0	0.399
1	0.242
2	0.054
3	0.004

$y = \dfrac{1}{\sqrt{2\pi}} e^{-x^2/2}$

63. Because $f(x + y) = f(x)f(y)$, you have $f(0 + y) = f(0)f(y)$ or $f(y) = f(0)f(y)$. After dividing both sides by $f(y)$, you have $1 = f(0)$.

Exercise 4.2   (page 183)

1. 3   **3.** -3 **5.** 2   **7.** 2   **9.** 7   **11.** 4   **13.** $-\frac{3}{2}$ **15.** $\frac{2}{3}$ **17.** 5   **19.** $\frac{3}{2}$
21. $\frac{1}{9}$ **23.** 8   **25.**   **27.**   **29.**

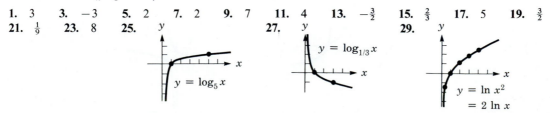

$y = \log_5 x$

$y = \log_{1/3} x$

$y = \ln x^2$
$= 2 \ln x$

31.

33.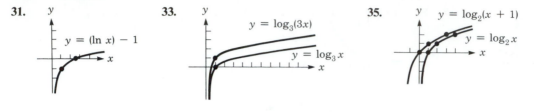

35.

37.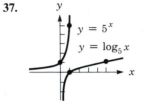

39. no value of b **41.** 3   **43.** no value of b **45.** true   **47.** false

49. true   **51.** false   **53.** true   **55.** true   **57.** false   **59.** true   **61.** false   **63.** false
65. true   **67.** 1.4472   **69.** 0.3521   **71.** 1.1972   **73.** 0.4771   **75.** 0.5563   **77.** 0.5119
79. 69.4079   **81.** 40.6853   **83.** 1573 joules   **85.** 3.2   **87.** from 1.26×10^{-3} to 5.01×10^{-4}
89. approximately 207,944 volts   **91.** 38

Exercise 4.3   (page 191)

1. $x = \dfrac{\log 5}{\log 4} \approx 1.16$ **3.** $x = \dfrac{\log 2}{\log 13} + 1 \approx 1.27$ **5.** $x = \dfrac{\log 2}{\log 3 - \log 2} \approx 1.71$ **7.** $x = 0$

9. $x = \pm\sqrt{\dfrac{1}{\log 7}} \approx \pm 1.09$ **11.** $x = 0$ or $x = \dfrac{\log 9}{\log 8} \approx 1.06$ **13.** $x = \dfrac{\log 2}{\log 3} \approx 0.6309$ **15.** $x = \dfrac{\log 1}{\log 2} = 0$

17. 0   **19.** 7   **21.** 4   **23.** 10, -10 **25.** 50   **27.** 20   **29.** 10   **31.** 3, 4   **33.** 1, 100
35. 9   **37.** 1.771   **39.** 2.322   **41.** about 4200 years old   **43.** about 5.146 years
45. about 42.7 days   **47.** about 3.15 days   **49.** in about 6.58 years   **51.** 0.71 volt   **53.** 4.4
55. 2500 micrometers   **57.** The intensity must be cubed.   **59.** 4.03 years   **61.** about 6.96%
65. about 27 ft   **67.** $x = \ln(y \pm \sqrt{y^2 - 1})$ **69.** $x = \ln\sqrt{\dfrac{y + 1}{y - 1}}$

REVIEW EXERCISES   (page 194)

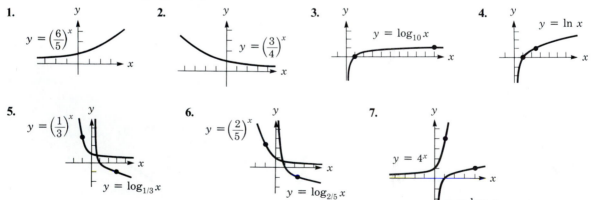

1.   **2.**   **3.**   **4.**

5.   **6.**   **7.**

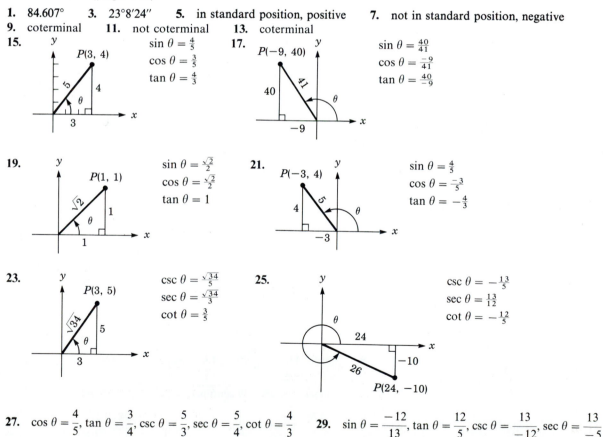

8. $y = 3^x$ $y = \log_3 x$

9. 8   **10.** $\frac{1}{9}$ **11.** 3   **12.** 2   **13.** 1   **14.** $\frac{1}{2}$ **15.** $\frac{1}{6}$ **16.** 2

17. -2 **18.** 0   **19.** 27   **20.** $\frac{1}{5}$ **21.** 32   **22.** 9   **23.** 27   **24.** -1 **25.** $\frac{1}{8}$ **26.** 2

27. 4   **28.** 2   **29.** 10   **30.** $\frac{1}{25}$ **31.** $x = 2.8665$ **32.** no value   **33.** $z = -2.0149$

34. $M = 209.5579$ **35.** $x = \dfrac{\log 7}{\log 3}$ **36.** $x = \dfrac{\log 1.2}{5.6 \log 3.4}$ **37.** $x = \dfrac{\log 3}{\log 3 - \log 2}$

38. $x = 4$ or $x = 25$ **39.** $x = 4$ **40.** $x = 2$ **41.** $x = \dfrac{\ln 9}{\ln 2}$ **42.** no value   **43.** $x = \dfrac{e}{e - 1}$

44. $x = 1$ **45.** approximately 9034 years old   **46.** 7.94×10^{-4}

47. $\text{pH} = \log_{10} \dfrac{1}{[\text{H}^+]} = \log_{10}[\text{H}^+]^{-1} = -\log_{10}[\text{H}^+]$ **48.** 34.19 years

Exercise 5.1   (page 203)

1. 84.607°   **3.** 23°8′24″   **5.** in standard position, positive   **7.** not in standard position, negative

9. coterminal   **11.** not coterminal   **13.** coterminal

15. $P(3, 4)$ $\sin \theta = \frac{4}{5}$ $\cos \theta = \frac{3}{5}$ $\tan \theta = \frac{4}{3}$

17. $P(-9, 40)$ $\sin \theta = \frac{40}{41}$ $\cos \theta = \frac{-9}{41}$ $\tan \theta = \frac{40}{-9}$

19. $P(1, 1)$ $\sin \theta = \frac{\sqrt{2}}{2}$ $\cos \theta = \frac{\sqrt{2}}{2}$ $\tan \theta = 1$

21. $P(-3, 4)$ $\sin \theta = \frac{4}{5}$ $\cos \theta = \frac{-3}{5}$ $\tan \theta = -\frac{4}{3}$

23. $P(3, 5)$ $\csc \theta = \frac{\sqrt{34}}{5}$ $\sec \theta = \frac{\sqrt{34}}{3}$ $\cot \theta = \frac{3}{5}$

25. $P(24, -10)$ $\csc \theta = -\frac{13}{5}$ $\sec \theta = \frac{13}{12}$ $\cot \theta = -\frac{12}{5}$

27. $\cos \theta = \dfrac{4}{5}$, $\tan \theta = \dfrac{3}{4}$, $\csc \theta = \dfrac{5}{3}$, $\sec \theta = \dfrac{5}{4}$, $\cot \theta = \dfrac{4}{3}$ **29.** $\sin \theta = \dfrac{-12}{13}$, $\tan \theta = \dfrac{12}{5}$, $\csc \theta = \dfrac{13}{-12}$, $\sec \theta = \dfrac{13}{-5}$

510   APPENDIX I

31. $\sin\theta = \dfrac{4}{5}$, $\cos\theta = \dfrac{-3}{5}$, $\tan\theta = \dfrac{-4}{3}$, $\cot\theta = \dfrac{-3}{4}$

33. $\tan^2\theta + 1 = \left(\dfrac{y}{x}\right)^2 + 1 = \dfrac{y^2}{x^2} + \dfrac{x^2}{x^2} = \dfrac{y^2 + x^2}{x^2} = \left(\dfrac{r}{x}\right)^2 = \sec^2\theta$

35. $\cot^2\theta + \sin^2\theta = \dfrac{x^2}{y^2} + \dfrac{y^2}{r^2} = \dfrac{r^2 - y^2}{y^2} + \dfrac{r^2 - x^2}{r^2} = \dfrac{r^2}{y^2} - 1 + 1 - \dfrac{x^2}{r^2} = \dfrac{r^2}{y^2} - \dfrac{x^2}{r^2} = \csc^2\theta - \cos^2\theta$

37.   **39.**   **41.** negative **43.** positive

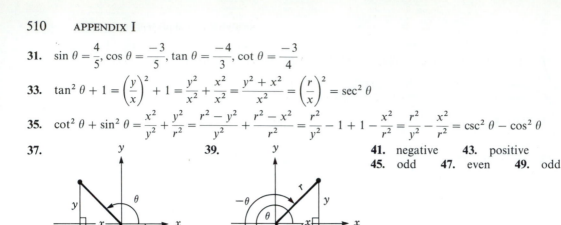

45. odd **47.** even **49.** odd

Exercise 5.2   (page 212)

1. $\sin 135° = \dfrac{\sqrt{2}}{2}$; $\cos 135° = -\dfrac{\sqrt{2}}{2}$; $\tan 135° = -1$ **3.** $\sin(-450°) = -\sin 450° = -\sin 90° = -1$;
$\cos(-450°) = \cos(450°) = \cos 90° = 0$; $\tan(-450°) = -\tan 450° = -\tan 90°$, but $\tan 90°$ is undefined

5. $\sin(-240°) = -\sin 240° = \dfrac{\sqrt{3}}{2}$; $\cos(-240°) = \cos 240° = -\dfrac{1}{2}$; $\tan(-240°) = -\tan 240° = -\sqrt{3}$

7. $\sin 540° = \sin 180° = 0$; $\cos 540° = \cos 180° = -1$; $\tan 540° = \tan 180° = 0$
9. $\csc 225° = -\sqrt{2}$; $\sec 225° = -\sqrt{2}$; $\cot 225° = 1$
11. $\csc(-1080°) = -\csc 1080° = -\csc 0°$, but $\csc 0°$ is undefined; $\sec(-1080°) = \sec 1080° = \sec 0° = 1$;
$\cot(-1080°) = -\cot 1080° = -\cot 0°$, but $\cot 0°$ is undefined

13. $\csc(-210°) = -\csc 210° = 2$; $\sec(-210°) = \sec 210° = -\dfrac{2\sqrt{3}}{3}$; $\cot(-210°) = -\cot 210° = -\sqrt{3}$

15. $\csc 585° = \csc 225° = -\sqrt{2}$; $\sec 585° = \sec 225° = -\sqrt{2}$; $\cot 585° = \cot 225° = 1$ **17.** 1   **19.** 0

21. 2   **23.** $-\dfrac{23}{4}$ **25.** $\dfrac{\sqrt{6}}{4} + 1$ **27.** $\dfrac{8\sqrt{3} - \sqrt{6}}{3}$ **29.** $\theta = 30°$ **31.** $\theta = 300°$

33. $\theta = 210°$ **35.** $\theta = 315°$ **37.** $\theta = 135°$ **39.** $\theta = 90°, 270°$ **41.** $\sin 17° = 0.2924$;
$\cos 17° = 0.9563$; $\tan 17° = 0.3057$ **43.** $\sin 73° = 0.9563$; $\cos 73° = 0.2924$; $\tan 73° = 3.271$
45. $\sin 89° = 0.9998$; $\cos 89° = 0.0175$; $\tan 89° = 57.29$ **47.** $\sin 23.1° = 0.3923$ **49.** $\cos 133.7° = -0.6909$
51. $\tan 223.5° = 0.9490$ **53.** $\csc 312.4° = -1.3542$ **55.** $\sec(-47.4°) = \sec 47.4° = 1.4774$
57. $\cot 640.6° = -0.1871$ **59.** $14.0°$ **61.** $110.0°$ **63.** $207.0°$ **65.** $280.0°$ **67.** $49.0°$
69. $220.0°$

Exercise 5.3   (page 222)

1. $\dfrac{12}{5}$ **3.** 1   **5.** $\dfrac{1}{12}\pi$ **7.** $\dfrac{2}{3}\pi$ **9.** $\dfrac{11}{6}\pi$ **11.** $-\dfrac{26}{9}\pi$ **13.** $135°$ **15.** $450°$

17. $-240°$ **19.** $\dfrac{1080°}{\pi} \approx 343.77°$ **21.** 12π meters   **23.** 39 centimeters   **25.** 2970 miles

27. 104.72 square units   **29.** 2π rad/hr   **31.** $\dfrac{176}{3}$ rad/sec   **33.** 1000π ft/min   **35.** $\dfrac{1}{2}$

37. $-\dfrac{\sqrt{3}}{2}$ **39.** 1   **41.** 2   **43.** 0.9093   **45.** -3.3805 **47.** -0.8163 **49.** -7.0862

51. $(0, -1)$ **53.** $(-1, 0)$ **55.** $\left(\dfrac{\sqrt{2}}{2}, -\dfrac{\sqrt{2}}{2}\right)$ **57.** $\left(-\dfrac{1}{2}, -\dfrac{\sqrt{3}}{2}\right)$

Exercise 5.4   (page 230)

1. $2; 2\pi$ **3.** $1; \dfrac{2\pi}{9}$ **5.** $1; 6\pi$ **7.** $1; 10\pi$ **9.** $3; 4\pi$ **11.** $\dfrac{1}{2}; 2$ **13.** $3; 1$ **15.** $\dfrac{1}{3}; \dfrac{2\pi^2}{3}$

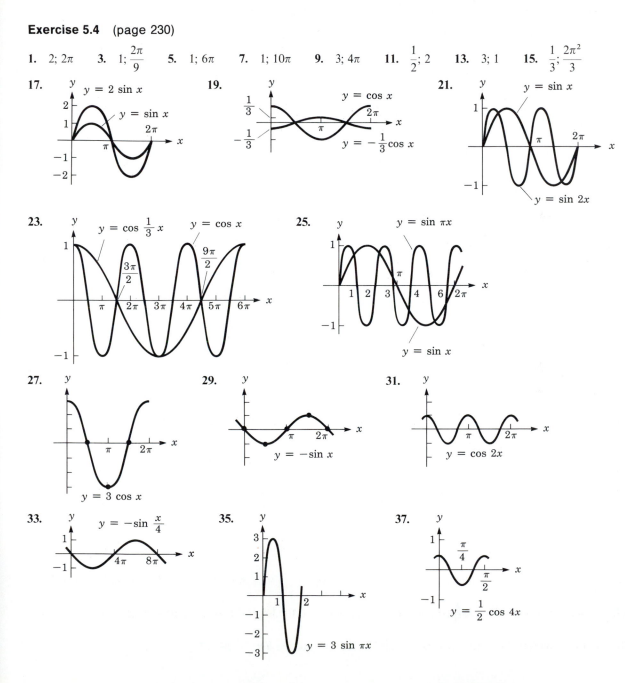

512   APPENDIX I

39.

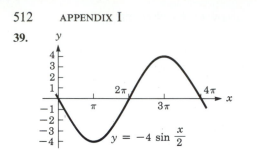

$y = -4 \sin \dfrac{x}{2}$

Exercise 5.5   (page 234)

1. π **3.** 2π **5.** $\dfrac{\pi}{3}$ **7.** 2   **9.** 6π **11.** $\dfrac{3}{2}$ **13.** 4   **15.** $2\pi^2$ **17.** $\dfrac{2}{3}\pi^2$

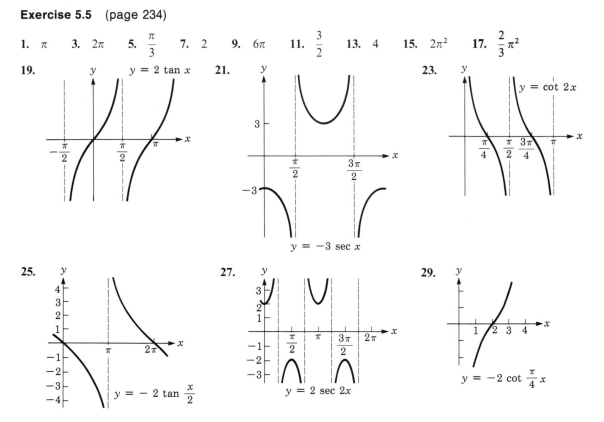

31. The value of y can be as large as you want.   **33.** $\ldots, -\pi, 0, \pi, 2\pi, 3\pi, \ldots$

Exercise 5.6   (page 241)

1. 2 units up; 2π **3.** 1 unit down; π **5.** 7 units up; $\dfrac{2\pi}{5}$ **7.** 3 units up; 2π **9.** 5 units down; π

11. 6 units up; 1   **13.** $2\pi; \dfrac{\pi}{3}$ to the right   **15.** $2\pi; \dfrac{\pi}{6}$ to the left   **17.** 1; no phase shift

19. $\pi; \pi$ to the right   **21.** $2\pi; \dfrac{\pi}{4}$ to the left   **23.** $\pi; \dfrac{\pi}{2}$ to the left   **25.** $2; \dfrac{1}{2}$ to the left

27. 6π; 18π to the right **29.** $\dfrac{\pi}{7}$; $\dfrac{3}{2}\pi$ to the right **31.**

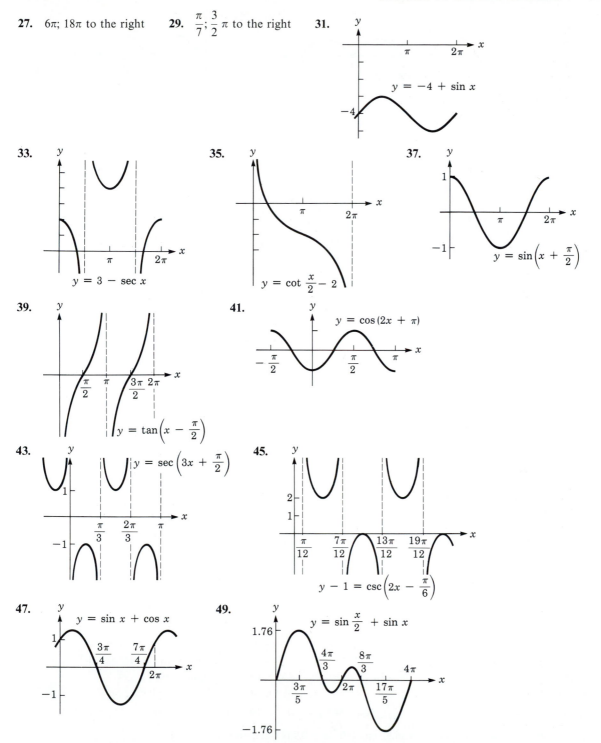

33.

$y = 3 - \sec x$

35.

$y = \cot \dfrac{x}{2} - 2$

37.

$y = \sin\left(x + \dfrac{\pi}{2}\right)$

39.

$y = \tan\left(x - \dfrac{\pi}{2}\right)$

41.

$y = \cos(2x + \pi)$

43.

$y = \sec\left(3x + \dfrac{\pi}{2}\right)$

45.

$y - 1 = \csc\left(2x - \dfrac{\pi}{6}\right)$

47.

$y = \sin x + \cos x$

49.

$y = \sin \dfrac{x}{2} + \sin x$

51.

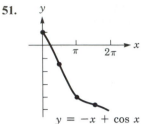

$y = -x + \cos x$

53.

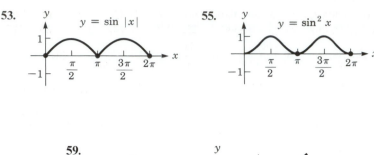

$y = \sin |x|$

55.

$y = \sin^2 x$

57.

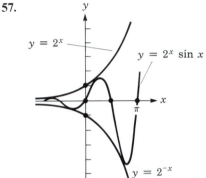

$y = 2^x$

$y = 2^x \sin x$

$y = 2^{-x}$

59.

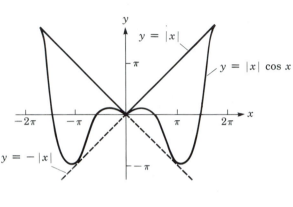

$y = |x|$

$y = |x| \cos x$

$y = -|x|$

61.

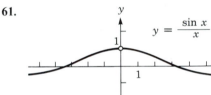

$y = \dfrac{\sin x}{x}$

REVIEW EXERCISES　(page 242)

1. yes　**2.** no　**3.** no　**4.** yes　**5.** yes　**6.** no　**7.** $\cos \theta = -\dfrac{\sqrt{51}}{10}$, $\tan \theta = \dfrac{7\sqrt{51}}{51}$,

$\csc \theta = -\dfrac{10}{7}$, $\sec \theta = -\dfrac{10\sqrt{51}}{51}$, $\cot \theta = \dfrac{\sqrt{51}}{7}$　**8.** $\sin \theta = -\dfrac{7\sqrt{130}}{130}$, $\cos \theta = -\dfrac{9\sqrt{130}}{130}$, $\csc \theta = -\dfrac{\sqrt{130}}{7}$,

$\sec \theta = -\dfrac{\sqrt{130}}{9}$, $\cot \theta = \dfrac{9}{7}$　**9.** $\sin \theta = \dfrac{\sqrt{51}}{10}$, $\tan \theta = -\dfrac{\sqrt{51}}{7}$, $\csc \theta = \dfrac{10\sqrt{51}}{51}$, $\sec \theta = -\dfrac{10}{7}$, $\cot \theta = -\dfrac{7\sqrt{51}}{51}$

10. $\sin \theta = -\dfrac{8\sqrt{145}}{145}$, $\cos \theta = \dfrac{9\sqrt{145}}{145}$, $\tan \theta = -\dfrac{8}{9}$, $\csc \theta = -\dfrac{\sqrt{145}}{8}$, $\sec \theta = \dfrac{\sqrt{145}}{9}$　**11.** $\dfrac{\sqrt{6}}{4}$　**12.** $\dfrac{1}{2}$

13. $\dfrac{9}{16}$　**14.** 0　**15.** $\sin 930° = -\dfrac{1}{2}$, $\cos 930° = -\dfrac{\sqrt{3}}{2}$, $\tan 930° = \dfrac{\sqrt{3}}{3}$　**16.** $\sin 1380° = -\dfrac{\sqrt{3}}{2}$,

$\cos 1380° = \dfrac{1}{2}$, $\tan 1380° = -\sqrt{3}$　**17.** $\sin(-300°) = \dfrac{\sqrt{3}}{2}$, $\cos(-300°) = \dfrac{1}{2}$, $\tan(-300°) = \sqrt{3}$

18. $\sin(-585°) = \dfrac{\sqrt{2}}{2}$, $\cos(-585°) = -\dfrac{\sqrt{2}}{2}$, $\tan(-585°) = -1$　**19.** $\sin 15° \approx 0.2588$, $\cos 15° \approx 0.9659$,

$\tan 15° \approx 0.2679$　**20.** $\sin 160° \approx 0.3420$, $\cos 160° \approx -0.9397$, $\tan 160° \approx -0.3640$

21. $\sin 265° \approx -0.9962$, $\cos 265° \approx -0.0872$, $\tan 265° \approx 11.430$ **22.** $\sin(-340°) \approx 0.3420$, $\cos(-340°) \approx 0.9397$, $\tan(-340°) \approx 0.3640$ **23.** $119°$ **24.** $211°$ **25.** $317°$ **26.** $57.7°$ **27.** $100°$

28. $287°$ **29.** $\dfrac{7}{12}\pi$ **30.** $\dfrac{65}{36}\pi$ **31.** $\dfrac{53}{30}\pi$ **32.** $-\dfrac{7}{12}\pi$ **33.** $570°$ **34.** $-150°$ **35.** $1260°$

36. $458.366°$ **37.** $\dfrac{1}{2}$ **38.** $\dfrac{\sqrt{3}}{2}$ **39.** $-\sqrt{3}$ **40.** 2 **41.** 2750 miles   **42.** $41.5°$ N

43. 19 square centimeters   **44.** $\dfrac{\pi}{43{,}200}\dfrac{\text{rad}}{\text{sec}}$ **45.** $\dfrac{1125}{\pi}$ rpm   **46.** $\dfrac{1000\pi}{3}\dfrac{\text{ft}}{\text{min}}$ **47.** $\left(-\dfrac{\sqrt{3}}{2}, -\dfrac{1}{2}\right)$

48. $\left(-\dfrac{\sqrt{2}}{2}, -\dfrac{\sqrt{2}}{2}\right)$ **49.** 0.7539 **50.** -2.1850 **51.** undefined   **52.** -0.3983 **53.** $4; \dfrac{2\pi}{3}$

54. $\dfrac{1}{8}; \dfrac{\pi}{2}$ **55.** $\dfrac{1}{3}; 6\pi$ **56.** $0.875; 8\pi$ **57.** 2 units up   **58.** $\dfrac{\pi}{6}$ to the left

59. 4 units up; $\dfrac{21}{2}$ to the left   **60.** 1 unit down; $\dfrac{5}{2}\pi$ to the right

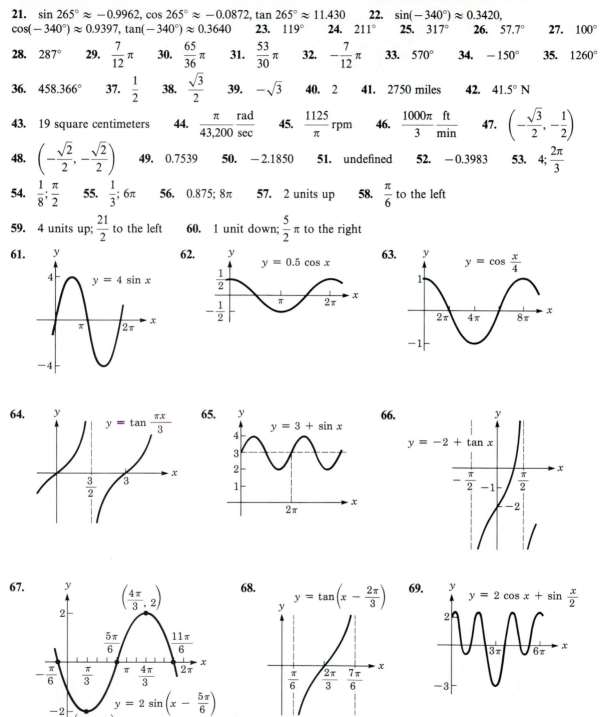

61. $y = 4 \sin x$

62. $y = 0.5 \cos x$

63. $y = \cos \dfrac{x}{4}$

64. $y = \tan \dfrac{\pi x}{3}$

65. $y = 3 + \sin x$

66. $y = -2 + \tan x$

67. $\left(\dfrac{4\pi}{3}, 2\right)$ $y = 2 \sin\left(x - \dfrac{5\pi}{6}\right)$ $\left(\dfrac{\pi}{3}, -2\right)$

68. $y = \tan\left(x - \dfrac{2\pi}{3}\right)$

69. $y = 2 \cos x + \sin \dfrac{x}{2}$

70.

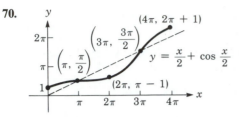

Exercise 6.1 (page 253)

1. angle $B = 53°$, $a = 12$, $b = 16$ **3.** 15 feet   **5.** 721 feet   **7.** 4.0°   **9.** 28.95 feet   **11.** 319 miles
13. S 62.9° W   **15.** 5.8 miles   **17.** 685 mph   **19.** 60.7 feet   **21.** 9180 feet; 1.74 miles   **23.** 48.3°

25. 1450 feet   **27.** 556 feet   **29.** 631 feet   **31.** 123 feet   **33.** $\dfrac{r}{\tan\frac{\theta}{2}}$ **35.** $b = 2k \cos \alpha$

37. $H = k(\tan \alpha + \tan \beta)$ **39.** $D(\tan \phi - \tan \theta)$ **41.** $a = d \tan \beta$ **43.** 3 mph; 7 mph   **45.** 63°
47. 352 mph   **49.** N 11.6° W   **51.** 384 pounds   **53.** 1160 pounds   **55.** 13.7°
57. 224 pounds; 48.1°   **59.** N 79.9° E

Exercise 6.2 (page 261)

1. 61.0 centimeters   **3.** 1410 kilometers   **5.** 65.9 centimeters   **7.** 54°   **9.** 90°   **11.** 37.85°
13. 210 pounds   **15.** 5° if 210 pounds is used for the resultant force   **17.** 1090 pounds
19. 36.7 nautical miles   **21.** 131 meters   **23.** 69.3°, 64.4°, 46.3°   **25.** 86°   **27.** 85.2°
29. 90.0°   **31.** 1 hour

Exercise 6.3 (page 269)

1. 67 kilometers   **3.** 256 meters   **5.** 305 meters   **7.** 218 inches   **9.** 3.97 meters   **11.** 180 yards
13. 420 feet   **15.** 7.1°   **17.** 2.5 nautical miles   **19.** 31.5°   **21.** 61.6°   **23.** 12 or 2.1 meters
25. not a triangle   **27.** 2900 feet or 780 feet   **29.** 957 feet

Exercise 6.4 (page 275)

1. $x = -4$ **3.** $y = \sqrt{3}$ **5.** $\dfrac{3}{2}$; 4; $\dfrac{1}{4}$ **7.** 5; $\dfrac{8}{3}$; $\dfrac{3}{8}$ **9.** 12 centimeters; 4π; $\dfrac{1}{4\pi}$

11. The period increases by a factor of $\sqrt{2}$.   **13.** approximately 238 cycles per second
15. The period is approximately 2 seconds per cycle, and the frequency is $\frac{1}{2}$ cycle per second.
17. approximately 0.76 meter   **19.** It is cut in half.   **21.** 214 meters

23.

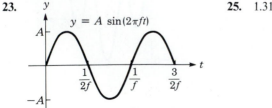

25. 1.31   **27.** about 1.66 inches

REVIEW EXERCISES   (page 277)

1. $17.3°$ **2.** 59.7 feet   **3.** 55 miles   **4.** 150 miles   **5.** 10 pounds   **6.** $11.5°$ **7.** 7.6
8. 32   **9.** 0.6   **10.** 11.1   **11.** $25°$ **12.** $37°, 143°$ **13.** $18.6°, 161.4°$ **14.** $51.31°, 128.69°$
15. 65.7   **16.** 70.8   **17.** $14°, 82°$ **18.** $42.4°$ **19.** 558 miles   **20.** about $104.0°$ **21.** 180 feet
22. 611 feet   **23.** $x = -1$; amplitude is 2; period is 6; frequency is $\frac{1}{6}$ **24.** It is divided by $\sqrt{2}$.
25. $\frac{1}{440}$ sec   **26.** approximately $6.4°$

Exercise 7.1   (page 283)

1. identity   **3.** not an identity   **5.** identity   **7.** not an identity   **9.** identity

Exercise 7.2   (page 289)

1. $\sin 195° = \sin(45° + 150°) = -\frac{\sqrt{6}+\sqrt{2}}{4}$ **3.** $\tan 195° = \tan(225° - 30°) = \frac{3-\sqrt{3}}{3+\sqrt{3}} = 2 - \sqrt{3}$ **5.** $\cos \frac{11\pi}{12} =$
$\cos(\frac{\pi}{6} + \frac{3\pi}{4}) = \frac{-\sqrt{6}-\sqrt{2}}{4}$ **7.** $\cos \frac{19\pi}{12} = \cos(\frac{11\pi}{6} - \frac{\pi}{4}) = \frac{\sqrt{6}-\sqrt{2}}{4}$ **9.** $\sin 255° = \sin(210° + 45°) = \frac{-\sqrt{2}+\sqrt{6}}{4}$
11. $\tan 105° = \tan(60° + 45°) = \frac{\sqrt{3}+1}{1-\sqrt{3}} = -2 - \sqrt{3}$ **13.** $\cos \frac{\pi}{12} = \cos(\frac{\pi}{3} - \frac{\pi}{4}) = \frac{\sqrt{2}+\sqrt{6}}{4}$
15. $\sin \frac{5\pi}{12} = \sin(\frac{2\pi}{3} - \frac{\pi}{4}) = \frac{\sqrt{6}+\sqrt{2}}{4}$ **17.** $\sin(60° + \theta) = \sin 60° \cos \theta + \cos 60° \sin \theta = \frac{\sqrt{3}}{2}\cos \theta + \frac{1}{2}\sin \theta$
19. $\tan(\pi + x) = \dfrac{\tan \pi + \tan x}{1 - \tan \pi \tan x} = \dfrac{\tan x}{1} = \tan x$ **21.** $\cos(\pi - x) = \cos \pi \cos x + \sin \pi \sin x = -\cos x$
23. $\sin(10° + 30°) = \sin 40°$ **25.** $\tan(75° + 40°) = \tan 115°$ **27.** $\cos(120° - 40°) = \cos 80°$
29. $\sin(x + 2x) = \sin 3x$ **31.** $\sin(\alpha + \beta) = -\frac{56}{65}$; $\cos(\alpha - \beta) = \frac{63}{65}$ **33.** $\tan(\alpha + \beta) = \frac{140}{171}$; $\tan(\alpha - \beta) = -\frac{220}{21}$
35. $\sin \alpha = \frac{416}{425}$; $\cos \alpha = \frac{87}{425}$

Exercise 7.3   (page 294)

1. $\sin 2\alpha$ **3.** $\sin 6\theta$ **5.** $\cos 2\beta$ **7.** $\cos \beta$ **9.** $2 \sin 2\theta$ **11.** $(\sin 4\theta)^2 = \sin^2 4\theta$ **13.** $\frac{1}{2}\cos 2\alpha$
15. $\cos 18\theta$ **17.** $\sin^2 10\theta$ **19.** $\tan 8C$ **21.** $\tan A$ **23.** $\cos 8x$ **25.** $-\cos 10x$ **27.** $\frac{\sqrt{3}}{2}$
29. undefined   **31.** -0.5 **33.** 0   **35.** $\frac{\sqrt{3}}{2}$ **37.** 0.5   **39.** $\sin 2\theta = \frac{120}{169}$; $\cos 2\theta = -\frac{119}{169}$;
$\tan 2\theta = -\frac{120}{119}$ **41.** $\sin 2\theta = \frac{120}{169}$; $\cos 2\theta = -\frac{119}{169}$; $\tan 2\theta = -\frac{120}{169}$ **43.** $\sin 2\theta = \frac{24}{25}$; $\cos 2\theta = \frac{7}{25}$; $\tan 2\theta = \frac{24}{7}$
45. $\sin 2\theta = -\frac{336}{625}$; $\cos 2\theta = -\frac{527}{625}$; $\tan 2\theta = \frac{336}{527}$ **47.** $\sin 2\theta = \frac{720}{1681}$; $\cos 2\theta = -\frac{1519}{1681}$; $\tan 2\theta = -\frac{720}{1519}$
49. $\sin 2\theta = -\frac{720}{1681}$; $\cos 2\theta = \frac{1519}{1681}$; $\tan 2\theta = -\frac{720}{1519}$

Exercise 7.4   (page 301)

1. $\dfrac{\sqrt{2 + \sqrt{3}}}{2}$ **3.** $-2 - \sqrt{3}$ **5.** $\dfrac{\sqrt{2 - \sqrt{2}}}{2}$ **7.** $\dfrac{\sqrt{2 + \sqrt{3}}}{2}$ **9.** $\sqrt{3} - 2$ **11.** 1
13. $\sin \frac{\theta}{2} = \frac{\sqrt{10}}{10}$; $\cos \frac{\theta}{2} = \frac{3\sqrt{10}}{10}$; $\tan \frac{\theta}{2} = \frac{1}{3}$ **15.** $\sin \frac{\theta}{2} = \frac{2\sqrt{5}}{5}$; $\cos \frac{\theta}{2} = -\frac{\sqrt{5}}{5}$; $\tan \frac{\theta}{2} = -2$ **17.** $\sin \frac{\theta}{2} = \frac{3\sqrt{34}}{34}$;
$\cos \frac{\theta}{2} = -\frac{5\sqrt{34}}{34}$; $\tan \frac{\theta}{2} = -\frac{3}{5}$ **19.** $\sin \frac{\theta}{2} = \frac{\sqrt{82}}{82}$; $\cos \frac{\theta}{2} = \frac{9\sqrt{82}}{82}$; $\tan \frac{\theta}{2} = \frac{1}{9}$
21. $\sin \frac{\theta}{2} = \frac{4\sqrt{17}}{17}$; $\cos \frac{\theta}{2} = \frac{\sqrt{17}}{17}$; $\tan \frac{\theta}{2} = 4$ **23.** $\sin \frac{\theta}{2} = \frac{\sqrt{6}}{6}$; $\cos \frac{\theta}{2} = \frac{-\sqrt{30}}{6}$; $\tan \frac{\theta}{2} = \frac{-\sqrt{5}}{5}$ **25.** $\cos 15°$
27. $\tan 100°$ **29.** $\tan 40°$ **31.** $\tan \pi = 0$ **33.** $\tan \frac{x}{4}$ **35.** $\tan 5A$

Exercise 7.5   (page 307)

1. $\frac{1}{4}$ **3.** $\frac{1}{4}$ **5.** $\frac{\sqrt{2}}{4}$ **7.** $\frac{1}{4}$ **9.** $\frac{\sqrt{3}-2}{4}$ **11.** $\frac{1}{2}(1 + \frac{\sqrt{3}}{2})$ **13.** $\frac{\sqrt{6}}{2}$ **15.** $-\frac{\sqrt{2}}{2}$ **17.** $\frac{-\sqrt{2}}{2}$
19. $\frac{\sqrt{6}}{2}$ **21.** 0   **23.** $\frac{\sqrt{2}}{2}$ **25.** $10 \sin(x + 53.1°)$ **27.** $10 \sin(x - 53.1°)$ **29.** $\sqrt{5} \sin(x + 26.6°)$

31. $\sqrt{2}\sin(x+45°)$ **33.** $\sqrt{26}\sin(x+101.3°)$ or $-\sqrt{26}\sin(x-78.7°)$ **35.** $2\sqrt{3}\sin(x-60°)$

37.

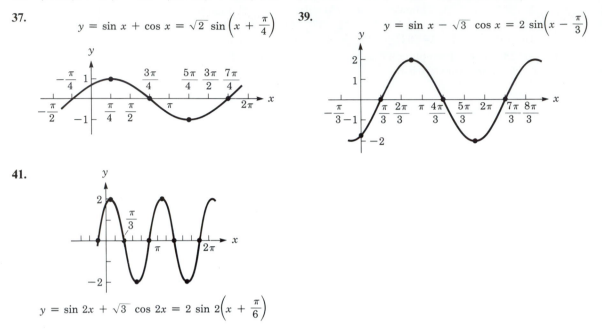

$$y = \sin x + \cos x = \sqrt{2}\,\sin\left(x+\frac{\pi}{4}\right)$$

39.

$$y = \sin x - \sqrt{3}\,\cos x = 2\sin\left(x-\frac{\pi}{3}\right)$$

41.

$$y = \sin 2x + \sqrt{3}\,\cos 2x = 2\sin 2\left(x+\frac{\pi}{6}\right)$$

Exercise 7.6   (page 312)

1. $\ldots,\ -300°,\ -240°,\ 60°,\ 120°,\ 420°,\ 480°,\ldots$ **3.** $\ldots,\ -180°,\ -90°,\ 0°,\ 90°,\ 180°,\ 270°,\ 360°,\ 450°,\ldots$
5. $120°$ **7.** $0°,\ 90°,\ 180°,\ 270°$ **9.** $90°,\ 270°$ **11.** $0°,\ 180°$ **13.** $45°,\ 135°,\ 225°,\ 315°$
15. $0°,\ 120°,\ 240°$ **17.** $210°,\ 330°$ **19.** $90°,\ 210°,\ 270°,\ 330°$ **21.** $45°,\ 225°$ **23.** $0°,\ 240°$
25. $0°,\ 120°,\ 240°$ **27.** $0°,\ 180°$ **29.** $30°,\ 150°,\ 210°,\ 330°$ **31.** $45°$ **33.** $0°,\ 30°,\ 150°,\ 180°$
35. $90°,\ 120°,\ 240°,\ 270°$ **37.** $0°,\ 30°,\ 90°,\ 150°,\ 180°,\ 210°,\ 270°,\ 330°$ **39.** $0°,\ 240°$
41. $60°,\ 120°,\ 240°,\ 300°$ **43.** $0°,\ 180°$ **45.** $45°,\ 135°,\ 315°$ **47.** $0°,\ 120°$ **49.** $30°$
51. $0°,\ 240°$ **53.** $\frac{\pi}{4},\frac{5\pi}{4}$ **55.** $0,\frac{\pi}{2},\pi,\frac{3\pi}{2}$ **57.** $0,\frac{\pi}{4},\frac{\pi}{2},\frac{3\pi}{4},\pi,\frac{5\pi}{4},\frac{3\pi}{2},\frac{7\pi}{4}$ **59.** $0,\frac{\pi}{3},\frac{\pi}{2},\frac{2\pi}{3},\pi,\frac{4\pi}{3},\frac{3\pi}{2},\frac{5\pi}{3}$
61. $\frac{\pi}{12},\frac{\pi}{6},\frac{\pi}{4},\frac{5\pi}{12},\frac{\pi}{2},\frac{7\pi}{12},\frac{3\pi}{4},\frac{5\pi}{6},\frac{11\pi}{12},\frac{13\pi}{12},\frac{7\pi}{6},\frac{5\pi}{4},\frac{17\pi}{12},\frac{3\pi}{2},\frac{19\pi}{12},\frac{7\pi}{4},\frac{11\pi}{6},\frac{23\pi}{12}$ **63.** $\frac{\pi}{4},\frac{3\pi}{4},\frac{5\pi}{4},\frac{7\pi}{4}$ **65.** $\frac{\pi}{6},\frac{\pi}{2},\frac{5\pi}{6},$
$\frac{7\pi}{6},\frac{3\pi}{2},\frac{11\pi}{6}$ **67.** $\frac{\pi}{2},\frac{3\pi}{2}$ **69.** $\frac{7\pi}{6},\frac{11\pi}{6}$ **71.** $\frac{\pi}{20},\frac{3\pi}{20},\frac{\pi}{4},\frac{7\pi}{20},\frac{9\pi}{20},\frac{11\pi}{20},\frac{13\pi}{20},\frac{3\pi}{4},\frac{17\pi}{20},\frac{19\pi}{20},\frac{21\pi}{20},\frac{23\pi}{20},\frac{5\pi}{4},\frac{27\pi}{20},\frac{29\pi}{20},$
$\frac{31\pi}{20},\frac{33\pi}{20},\frac{7\pi}{4},\frac{37\pi}{20},\frac{39\pi}{20},$ **73.** $0,\frac{\pi}{4},\pi,\frac{5\pi}{4}$ **75.** $\frac{\pi}{6},\frac{5\pi}{6},\frac{7\pi}{6},\frac{11\pi}{6}$

Exercise 7.7   (page 320)

1. $\frac{\pi}{6}$ **3.** $\frac{\pi}{2}$ **5.** $\frac{\pi}{4}$ **7.** no value   **9.** $\frac{3\pi}{4}$ **11.** $\frac{\pi}{3}$ **13.** $\sin\frac{\pi}{6}=\frac{1}{2}$; $\cos\frac{\pi}{6}=\frac{\sqrt{3}}{2}$; $\tan\frac{\pi}{6}=\frac{\sqrt{3}}{3}$
15. $\sin 0=0$; $\cos 0=1$; $\tan 0=0$ **17.** $\sin\frac{5\pi}{6}=\frac{1}{2}$; $\cos\frac{5\pi}{6}=-\frac{\sqrt{3}}{2}$; $\tan\frac{5\pi}{6}=-\frac{\sqrt{3}}{3}$ **19.** $\sin\frac{\pi}{2}=1$;
$\cos\frac{\pi}{2}=0$; $\tan\frac{\pi}{2}$ is undefined   **21.** $\sin\pi=0$; $\cos\pi=-1$; $\tan\pi=0$ **23.** $\sin\frac{\pi}{4}=\frac{\sqrt{2}}{2}$; $\cos\frac{\pi}{4}=\frac{\sqrt{2}}{2}$; $\tan\frac{\pi}{4}=1$
25. $\frac{1}{2}$ **27.** 1 **29.** 1 **31.** $\frac{1}{2}$ **33.** $-\sqrt{3}$ **35.** $\frac{\sqrt{2}}{2}$ **37.** $\frac{3}{5}$ **39.** $\frac{12}{13}$ **41.** $\frac{-4}{3}$ **43.** $\frac{12}{5}$
45. $\frac{12}{13}$ **47.** $\frac{40}{41}$ **49.** $\sin(\frac{\pi}{6}+\frac{\pi}{3})=1$ **51.** $\frac{24}{25}$ **53.** $\sin 2(\frac{\pi}{4})=1$ **55.** $\tan 2(\frac{\pi}{4})$ is undefined

57. $\sin\frac{1}{2}(\frac{\pi}{3})=\frac{1}{2}$ **59.** $\frac{1}{4}$ **61.** $\dfrac{x}{\sqrt{1+x^2}}$ **63.** $\dfrac{x}{\sqrt{1-x^2}}$ **65.** $\sqrt{1-x^2}$ **67.** $2x\sqrt{1-x^2}$

69. $\dfrac{2x}{1-x^2}$ **71.** $1-2x^2$ **73.** $-\frac{\pi}{6}$ **75.** $\frac{\pi}{4}$ **77.** $\frac{\pi}{6}$ **79.** π **81.** $-\frac{\pi}{2}$

Exercise 7.8   (page 326)

1. imaginaries

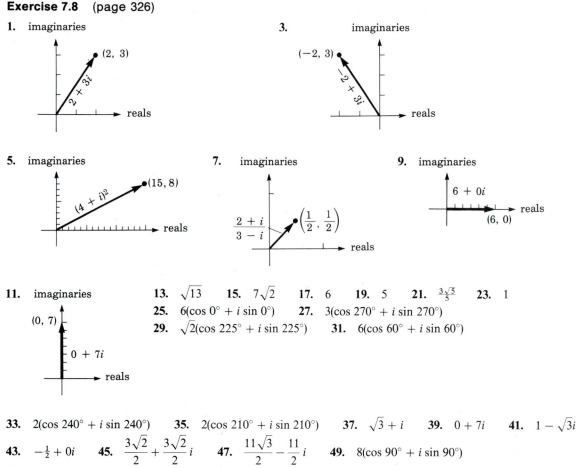

3. imaginaries

5. imaginaries

7. imaginaries

9. imaginaries

11. imaginaries

13. $\sqrt{13}$ **15.** $7\sqrt{2}$ **17.** 6   **19.** 5   **21.** $\frac{3\sqrt{5}}{5}$ **23.** 1
25. $6(\cos 0° + i \sin 0°)$ **27.** $3(\cos 270° + i \sin 270°)$
29. $\sqrt{2}(\cos 225° + i \sin 225°)$ **31.** $6(\cos 60° + i \sin 60°)$

33. $2(\cos 240° + i \sin 240°)$ **35.** $2(\cos 210° + i \sin 210°)$ **37.** $\sqrt{3} + i$ **39.** $0 + 7i$ **41.** $1 - \sqrt{3}i$
43. $-\frac{1}{2} + 0i$ **45.** $\frac{3\sqrt{2}}{2} + \frac{3\sqrt{2}}{2} i$ **47.** $\frac{11\sqrt{3}}{2} - \frac{11}{2} i$ **49.** $8(\cos 90° + i \sin 90°)$
51. $\cos 300° + i \sin 300°$ **53.** $6(\cos 2\pi + i \sin 2\pi)$ **55.** $6(\cos \frac{\pi}{2} + i \sin \frac{\pi}{2})$ **57.** $30 \text{ cis } 116°$
59. $36 \text{ cis } \frac{13\pi}{12}$ **61.** $6(\cos 30° + i \sin 30°)$ **63.** $\frac{3}{2}(\cos \frac{\pi}{2} + i \sin \frac{\pi}{2})$ **65.** $\frac{12}{5} \text{ cis } 130°$ **67.** $\frac{1}{2} \text{ cis } \frac{\pi}{2}$
69. $\text{cis } 40°$ **71.** $4 \text{ cis } \frac{5\pi}{6}$

Exercise 7.9   (page 331)

1. $27(\cos 90° + i \sin 90°)$ **3.** $\cos 180° + i \sin 180°$ **5.** $3125 \text{ cis } 10°$ **7.** $81(\cos \pi + i \sin \pi)$
9. $256(\cos 12 + i \sin 12)$ **11.** $\frac{1}{27} \text{ cis } \frac{3\pi}{2}$ **13.** $1 + \sqrt{3}i$ **15.** $\frac{1}{2} + \frac{\sqrt{3}}{2}i$ **17.** $1 + \sqrt{3}i$
19. $1 + \sqrt{3}i, -2, 1 - \sqrt{3}i$ **21.** $\frac{\sqrt{2}}{2} + \frac{\sqrt{2}}{2}i, -\frac{\sqrt{2}}{2} - \frac{\sqrt{2}}{2}i$

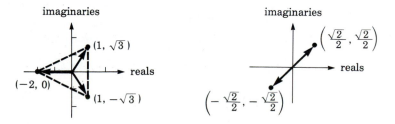

23.

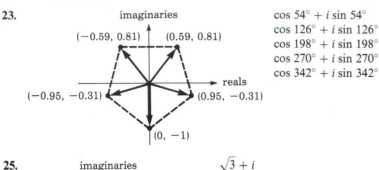

imaginaries

(−0.59, 0.81) (0.59, 0.81)

reals

(−0.95, −0.31) (0.95, −0.31)

(0, −1)

$\cos 54° + i \sin 54°$
$\cos 126° + i \sin 126°$
$\cos 198° + i \sin 198°$
$\cos 270° + i \sin 270°$
$\cos 342° + i \sin 342°$

25.

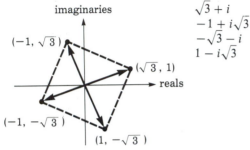

imaginaries

$(-1, \sqrt{3})$

$(\sqrt{3}, 1)$

reals

$(-1, -\sqrt{3})$

$(1, -\sqrt{3})$

$\sqrt{3} + i$
$-1 + i\sqrt{3}$
$-\sqrt{3} - i$
$1 - i\sqrt{3}$

Exercise 7.10 (page 339)

1. $\langle 7, -5 \rangle$ **3.** $\langle 6, -9 \rangle$ **5.** $\langle 9, -8 \rangle$ **7.** $\sqrt{13}$ **9.** $\sqrt{5}$ **11.** $\sqrt{13} + \sqrt{2}$ **13.** $8\mathbf{i} + 8\mathbf{j}$
15. $5\sqrt{3}\mathbf{i} + 5\mathbf{j}$ **17.** $18.6\mathbf{i} + 14.1\mathbf{j}$ **19.** 9 **21.** 7 **23.** 0 **25.** 45° **27.** 150° **29.** 36.9°
31. perpendicular **33.** not perpendicular **35.** perpendicular **37.** $\frac{63}{13}$ **39.** 0

REVIEW EXERCISES (page 340)

5. $\sin 71°$ **6.** $\tan(-31°) = -\tan 31°$ **7.** $\cos \frac{4\pi}{11}$ **8.** $\cos \frac{2\pi}{3} = -\frac{1}{2}$ **13.** $\sin 2\theta = -\frac{120}{169}$;
$\cos 2\theta = \frac{119}{169}$; $\tan 2\theta = -\frac{120}{119}$ **14.** $\sin 2\theta = \frac{120}{169}$; $\cos 2\theta = \frac{119}{169}$; $\tan 2\theta = \frac{120}{119}$ **15.** $\sin 2\theta = -\frac{840}{841}$;
$\cos 2\theta = \frac{41}{841}$; $\tan 2\theta = -\frac{840}{41}$ **16.** $\sin 2\theta = \frac{840}{841}$; $\cos 2\theta = \frac{41}{841}$; $\tan 2\theta = \frac{840}{41}$ **17.** $\sin 2\theta = \frac{24}{25}$; $\cos 2\theta = \frac{-7}{25}$;
$\tan 2\theta = \frac{-24}{7}$ **18.** $\sin 2\theta = -\frac{24}{25}$; $\cos 2\theta = -\frac{7}{25}$; $\tan 2\theta = \frac{24}{7}$ **19.** $\sin \frac{\theta}{2} = \frac{\sqrt{26}}{26}$; $\cos \frac{\theta}{2} = \frac{5\sqrt{26}}{26}$; $\tan \frac{\theta}{2} = \frac{1}{5}$
20. $\sin \frac{\theta}{2} = \frac{\sqrt{26}}{26}$; $\cos \frac{\theta}{2} = -\frac{5\sqrt{26}}{26}$; $\tan \frac{\theta}{2} = -\frac{1}{5}$ **21.** $\sin \frac{\theta}{2} = \frac{\sqrt{26}}{26}$; $\cos \frac{\theta}{2} = -\frac{5\sqrt{26}}{26}$; $\tan \frac{\theta}{2} = -\frac{1}{5}$
22. $\sin \frac{\theta}{2} = \frac{2\sqrt{5}}{5}$; $\cos \frac{\theta}{2} = \frac{\sqrt{5}}{5}$; $\tan \frac{\theta}{2} = 2$ **23.** $\sin \frac{\theta}{2} = \frac{2\sqrt{5}}{5}$; $\cos \frac{\theta}{2} = \frac{\sqrt{5}}{5}$; $\tan \frac{\theta}{2} = 2$ **24.** $\sin \frac{\theta}{2} = \frac{2\sqrt{5}}{5}$;
$\cos \frac{\theta}{2} = -\frac{\sqrt{5}}{5}$; $\tan \frac{\theta}{2} = -2$ **25.** $-\frac{2+\sqrt{3}}{4}$ **26.** $\frac{1}{2}(-\sqrt{3} + 1)$ **27.** $\frac{1-\sqrt{3}}{4}$ **28.** $\frac{1}{2}(-\frac{1}{2} - \frac{\sqrt{3}}{2})$
29. $2 \sin 6° \cos 1°$ **30.** $2 \cos 226° \sin 86°$ **31.** $-2 \sin \frac{2\pi}{5} \sin \frac{\pi}{5}$ **32.** $2 \cos \frac{5\pi}{14} \cos \frac{\pi}{14}$ **33.** $-\frac{\sqrt{2}}{2}$
34. $\frac{\sqrt{6}}{2}$ **35.** $-\frac{\sqrt{2}}{2}$ **36.** $\frac{\sqrt{6}}{2}$ **37.** $y = \sqrt{5} \sin(x + 63.4°)$ **38.** $y = -\sqrt{2} \sin(x + 315°)$ **39.** $\frac{\pi}{4}, \frac{5\pi}{4}$
40. $0, \frac{\pi}{4}, \frac{5\pi}{4}$ **41.** $\frac{\pi}{4}, \frac{3\pi}{4}, \frac{5\pi}{4}, \frac{7\pi}{4}$ **42.** no solutions **43.** $\frac{3}{5}$ **44.** $\frac{2\pi}{3}$ **45.** $-\frac{\pi}{3}$ **46.** no value
47. $\frac{\pi}{2}$ **48.** no value **49.** $-\frac{\pi}{6}$ **50.** $\frac{\pi}{6}$ **51.** $\frac{\pi}{2}$ **52.** π **53.** 0 **54.** 0 **55.** 1
56. $\frac{\sqrt{3}}{2}$ **57.** $\frac{\sqrt{3}}{2}$ **58.** 0 **59.** $\sqrt{1 - u^2}$ **60.** $\frac{2u}{1-u^2}$ **61.** imaginaries

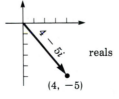

reals

(4, −5)

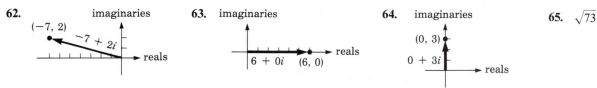

62. imaginaries
$(-7, 2)$
$-7 + 2i$
reals

63. imaginaries
$6 + 0i$ $(6, 0)$
reals

64. imaginaries
$(0, 3)$
$0 + 3i$
reals

65. $\sqrt{73}$

66. $10\sqrt{2}$ **67.** $\frac{3\sqrt{10}}{10}$ **68.** 1   **69.** $2\sqrt{2}(\cos 135° + i \sin 135°)$ **70.** $5\sqrt{2}(\cos 315° + i \sin 315°)$
71. $6(\cos 60° + i \sin 60°)$ **72.** $4(\cos 0° + i \sin 0°)$ **73.** $\frac{3}{2} + \frac{3\sqrt{3}}{2}i$ **74.** $\sqrt{3} - i$ **75.** $-\frac{3}{2} - \frac{3\sqrt{3}}{2}i$
76. $-\frac{7\sqrt{3}}{2} + \frac{7}{2}i$ **77.** cis 110°   **78.** 6 cis 570°   **79.** $6(\cos \frac{\pi}{4} + i \sin \frac{\pi}{4})$ **80.** $21(\cos \pi + i \sin \pi)$
81. 2 cis 50°   **82.** $\frac{2}{3}(\cos 10° + i \sin 10°)$ **83.** $\cos 20° + i \sin 20°$ **84.** $\sqrt[4]{14}(\cos \frac{45°}{4} + i \sin \frac{45°}{4})$
85. $5, \frac{-5+5\sqrt{3}i}{2}, \frac{-5-5\sqrt{3}i}{2}$ **86.** $3, 3i, -3, -3i$ **87.** $\langle 0, 29 \rangle$ **88.** $2\sqrt{58}$ **89.** $\langle 25, 10 \rangle$
90. $\sqrt{377}$ **91.** 90°   **92.** 0°   **93.** 75°   **94.** 120°

Exercise 8.1   (page 350)

1.

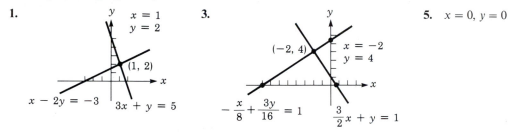

y
$x = 1$
$y = 2$
$(1, 2)$
x
$x - 2y = -3$ $3x + y = 5$

3.

y
$(-2, 4)$
$x = -2$
$y = 4$
x
$-\frac{x}{8} + \frac{3y}{16} = 1$ $\frac{3}{2}x + y = 1$

5. $x = 0, y = 0$

7. $x = 3, y = -2$ **9.** $x = 2, y = 3$ **11.** no solution   **13.** $x = 3, y = 1$ **15.** $x = 3, y = 2$
17. $x = -3, y = 0$ **19.** $x = 1, y = -\frac{1}{2}$ **21.** all (x, y) such that $x = 2y$ **23.** $x = 2, y = 2$
25. $x = 1, y = 2, z = 0$ **27.** $x = 0, y = -\frac{1}{3}, z = -\frac{1}{3}$ **29.** $x = 1, y = 2, z = -1$
31. $x = 1, y = 0, z = 5$ **33.** no solution   **35.** $x = 2 - y, y =$ any number, $z = 2 - y$
37. $x = 1, y = 1, z = 2$ **39.** $x = 2 - y, y =$ any number, $z = 1$ **41.** $x = 3, y = 10$
43. $(\frac{\pi}{3}, \frac{1}{2}), (\frac{5\pi}{3}, \frac{1}{2})$ **45.** 50 units from A, the rest from B and C in any combination
47. 15 hours cooking hamburgers, 10 hours pumping gas, and 5 hours as a janitor
49. 14 nickels, 12 dimes, 6 quarters

Exercise 8.2   (page 359)

1. $x = 2, y = -1$ **3.** $x = -2, y = 0$ **5.** $x = 3, y = 1$ **7.** no solution   **9.** $x = 1, y = 0, z = 2$
11. $x = 2, y = -2, z = 1$ **13.** $x = 1, y = 1, z = 2$ **15.** $x = -1, y = 3, z = 1$ **17.** $x = 1, y = -3$
19. $x = \frac{8}{7} + \frac{1}{7}z, y = \frac{10}{7} - \frac{4}{7}z, z =$ any number   **21.** $w = y, x = 1, y =$ any number, $z = 1$ **23.** no solution
25. $x = 1, y = 2, z = 1, t = 1$ **27.** $x = 1, y = 2, z = 0, t = 1$ **29.** $x = \pm 2, y = \pm 1, z = \pm 3$
31. $x = 16, y = 1, z = 0$

Exercise 8.3   (page 367)

1. $x = 2, y = 5$ **3.** no values   **5.** $x = 1, y = 2$ **7.** $x = 2, y = 2$ **9.** $\begin{bmatrix} -1 & 2 & 1 \\ -6 & 0 & 0 \end{bmatrix}$

11. $\begin{bmatrix} 5 & -4 & 3 \\ -7 & -4 & -2 \\ 0 & 4 & -8 \end{bmatrix}$ **13.** not possible   **15.** $\begin{bmatrix} 5 & 7 & 9 \end{bmatrix}$ **17.** $\begin{bmatrix} 3 & 2 & -7 \\ 9 & 19 & 15 \\ 0 & -4 & -1 \end{bmatrix}$

19. $\begin{bmatrix} 2 & -2 \\ 3 & 10 \end{bmatrix}$ **21.** $\begin{bmatrix} -22 & -22 \\ -105 & 126 \end{bmatrix}$ **23.** $\begin{bmatrix} 4 & 2 & 10 \\ 5 & -2 & 4 \\ 2 & -2 & 1 \end{bmatrix}$ **25.** $[32]$ **27.** not possible

29. $[30 \quad 36 \quad 42]$ **31.** $\begin{bmatrix} 2 & 0 & 3 \\ 3 & 0 & 5 \end{bmatrix}$ **33.** $\begin{bmatrix} 12 & -4 & 9 \\ 16 & -2 & 2 \\ 10 & -2 & 1 \end{bmatrix}$ **35.** $\begin{bmatrix} 7 \\ 6 \end{bmatrix}$

37. $\begin{bmatrix} 4 & 5 \\ -7 & -1 \end{bmatrix}$ **39.** not possible   **41.** $\begin{bmatrix} 24 & 16 \\ 39 & 26 \end{bmatrix}$ **43.** $\begin{bmatrix} 64 & 64 \\ 64 & 64 \end{bmatrix}$

45. One example is $\begin{bmatrix} 1 & 0 \\ 0 & 0 \end{bmatrix}$.

Exercise 8.4   (page 373)

1. $\begin{bmatrix} 3 & 4 \\ 2 & 3 \end{bmatrix}$ **3.** $\begin{bmatrix} 5 & -7 \\ -2 & 3 \end{bmatrix}$ **5.** $\begin{bmatrix} -40 & 16 & 9 \\ 13 & -5 & -3 \\ 5 & -2 & -1 \end{bmatrix}$ **7.** $\begin{bmatrix} 4 & 1 & -3 \\ -5 & -1 & 4 \\ -1 & -1 & 1 \end{bmatrix}$

9. no inverse   **11.** $\begin{bmatrix} 1 & -2 & 1 \\ 0 & 1 & -2 \\ 0 & 0 & 1 \end{bmatrix}$ **13.** no inverse   **15.** $\begin{bmatrix} 1 & -2 & 1 & 0 \\ 0 & 1 & -2 & 1 \\ 0 & 0 & 1 & -2 \\ 0 & 0 & 0 & 1 \end{bmatrix}$

17. $x = 23, y = 17$ **19.** $x = 70, y = -30$ **21.** $x = -10, y = 4, z = 1$ **23.** $x = 7, y = -9, z = -1$
25. $x = 4, y = -6, z = 3$
29. $A^2 = \begin{bmatrix} 1 & 0 \\ 0 & 1 \end{bmatrix}, A^3 = \begin{bmatrix} 0 & 1 \\ 1 & 0 \end{bmatrix}, A^4 = \begin{bmatrix} 1 & 0 \\ 0 & 1 \end{bmatrix}. A^n = \begin{bmatrix} 0 & 1 \\ 1 & 0 \end{bmatrix}$ if n is odd, $A^n = \begin{bmatrix} 1 & 0 \\ 0 & 1 \end{bmatrix}$ if n is even.

33. $A^2 = \begin{bmatrix} 1 & 0 \\ 2 & 1 \end{bmatrix}, A^3 = \begin{bmatrix} 1 & 0 \\ 3 & 1 \end{bmatrix}, A^n = \begin{bmatrix} 1 & 0 \\ n & 1 \end{bmatrix}$

35. $x = \pm 4$ **39.** $E = \begin{bmatrix} 1 & 0 & 0 \\ 0 & 1 & 0 \\ 3 & 0 & 1 \end{bmatrix}, E^{-1} = \begin{bmatrix} 1 & 0 & 0 \\ 0 & 1 & 0 \\ -3 & 0 & 1 \end{bmatrix}$

Exercise 8.5   (page 383)

1. 8 **3.** 1 **5.** -54 **7.** -7 **9.** 86 **11.** -2 **13.** 12 **15.** 1 **17.** $x = 1, y = 2$
19. $x = 3, y = 0$ **21.** $x = 1, y = 0, z = 1$ **23.** $x = 1, y = -1, z = 2$ **25.** $x = 6, y = 6, z = 12$
27. $p = 1, q = 1, r = 1, s = 1$

31. For example, $\begin{vmatrix} 1 & 0 \\ 0 & 1 \end{vmatrix} + \begin{vmatrix} 1 & 0 \\ 0 & 1 \end{vmatrix} = 1 + 1 = 2$, however $\begin{vmatrix} \begin{bmatrix} 1 & 0 \\ 0 & 1 \end{bmatrix} + \begin{bmatrix} 1 & 0 \\ 0 & 1 \end{bmatrix} \end{vmatrix} = \begin{vmatrix} 2 & 0 \\ 0 & 2 \end{vmatrix} = 4.$

33. $\begin{vmatrix} a & b & c \\ 0 & d & e \\ 0 & 0 & f \end{vmatrix} = f \begin{vmatrix} a & b \\ 0 & d \end{vmatrix} = adf$ **39.** $x = 8$ **41.** $x = -1$

43. domain is the set of $n \times n$ matrices, range is the set of real numbers
45. Yes, because $|AB| = |A||B|$, and $|AB| = 0$ implies that $|A||B| = 0$, and therefore $|A| = 0$ or $|B| = 0$.

Exercise 8.6   (page 389)

1. $\dfrac{1}{x + 1} + \dfrac{2}{x - 1}$ **3.** $\dfrac{1}{x^2 + 2} - \dfrac{3}{x + 1}$ **5.** $\dfrac{1}{x} + \dfrac{2}{x^2} - \dfrac{3}{x - 1}$ **7.** $\dfrac{1}{x^2} + \dfrac{1}{x^2 + 1}$ **9.** $\dfrac{2}{x} + \dfrac{3x + 2}{x^2 + 1}$

11. $\dfrac{1}{x} + \dfrac{1}{x^2} + \dfrac{2}{x^2 + x + 1}$ **13.** $\dfrac{1}{x^2 + x + 5} + \dfrac{x + 1}{x^2 + 1}$ **15.** $\dfrac{-1}{x^2 + 1} - \dfrac{x}{(x^2 + 1)^2} + \dfrac{1}{x}$

17. $\dfrac{1}{x^2 + 1} + \dfrac{x + 2}{x^2 + x + 2}$ **19.** $\dfrac{1}{x} + \dfrac{x}{x^2 + 2x + 5} + \dfrac{x + 2}{(x^2 + 2x + 5)^2}$ **21.** $2 + \dfrac{1}{x} + \dfrac{2}{x^2} + \dfrac{3}{x + 1}$

23. $x^2 + x + 1 + \dfrac{1}{x} + \dfrac{1}{x^2}$

Exercise 8.7   (page 395)

1. $y = -2x + 3$

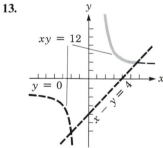

3.

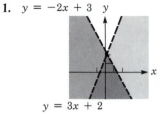

5.

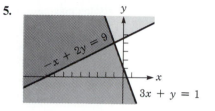

7.

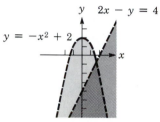

9.

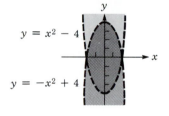

11.

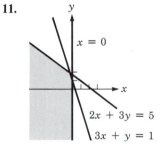

13.

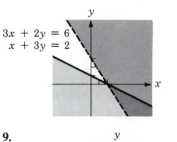

15.

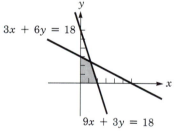

17.

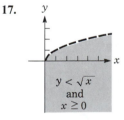

19.

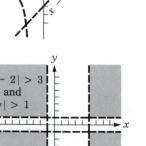

21.

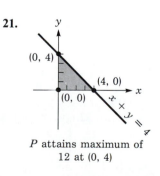

P attains maximum of
12 at $(0, 4)$

23.

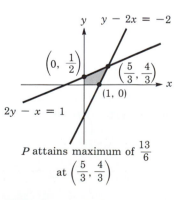

P attains maximum of $\dfrac{13}{6}$
at $\left(\dfrac{5}{3}, \dfrac{4}{3}\right)$

25.

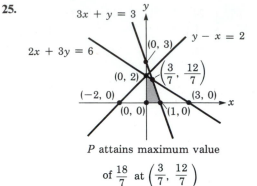

$3x + y = 3$
$y - x = 2$
$2x + 3y = 6$
$(0, 3)$
$\left(\dfrac{3}{7}, \dfrac{12}{7}\right)$
$(0, 2)$
$(-2, 0)$
$(3, 0)$
$(0, 0)$
$(1, 0)$

P attains maximum value

of $\dfrac{18}{7}$ at $\left(\dfrac{3}{7}, \dfrac{12}{7}\right)$

27. 2 tables, no chairs

29. 10 square meters of strawberries, 30 square meters of pumpkins
31. 10 ounces of X and 25 ounces of Y per day

REVIEW EXERCISES (page 396)

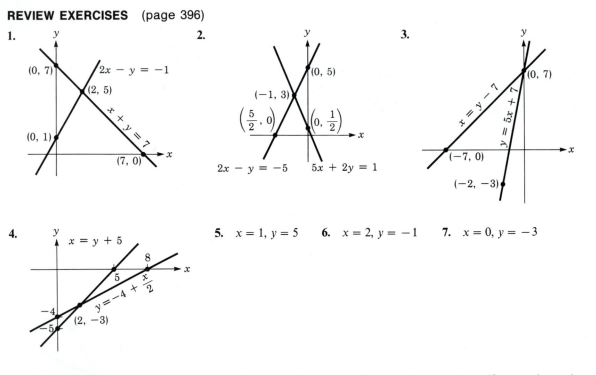

1.

$(0, 7)$
$2x - y = -1$
$(2, 5)$
$x + y = 7$
$(0, 1)$
$(7, 0)$

2.

$(0, 5)$
$(-1, 3)$
$\left(\dfrac{5}{2}, 0\right)$
$\left(0, \dfrac{1}{2}\right)$
$2x - y = -5$ $5x + 2y = 1$

3.

$(0, 7)$
$x = y - 7$
$y = 5x + 7$
$(-7, 0)$
$(-2, -3)$

4.

$x = y + 5$
8
5
$y = -4 + \dfrac{x}{2}$
-4
-5
$(2, -3)$

5. $x = 1, y = 5$ **6.** $x = 2, y = -1$ **7.** $x = 0, y = -3$

8. $x = 1, y = 1$ **9.** $x = -3, y = 2$ **10.** $x = -2, y = 5$ **11.** $x = 2, y = -1$ **12.** $x = 4, y = 1$
13. $x = 1, y = 0, z = 1$ **14.** $x = 1, y = 1, z = -1$ **15.** $x = 0, y = 1, z = 2$ **16.** $x = 1, y = -2, z = 3$
17. $x = 1, y = 1$ **18.** $x = 3, y = 1, z = -2$ **19.** $x = 1 - \frac{4}{7}z, y = 2 - \frac{1}{7}z, z = $ any number
20. no solution   **21.** $w = 1, x = 2, y = 0, z = -1$ **22.** $w = 2 - z, x = 1, y = -2 + z, z = $ any number
23. $\begin{bmatrix} 1 & 3 & 4 \\ 4 & 0 & 2 \end{bmatrix}$ **24.** $\begin{bmatrix} 2 & 5 & 4 \\ -2 & -6 & 6 \\ -4 & 5 & -3 \end{bmatrix}$ **25.** $\begin{bmatrix} -7 & -1 \\ -7 & -4 \end{bmatrix}$ **26.** $\begin{bmatrix} -17 & 19 \\ 10 & -12 \end{bmatrix}$ **27.** $[5]$

28. $\begin{bmatrix} -3 & 14 & -12 \\ 1 & 2 & -4 \end{bmatrix}$ **29.** $\begin{bmatrix} 2 & -1 & 1 & 3 \\ 4 & -2 & 2 & 6 \\ 2 & -1 & 1 & 3 \\ 10 & -5 & 5 & 15 \end{bmatrix}$ **30.** $\begin{bmatrix} 24 \\ -8 \end{bmatrix}$ **31.** $[-18]$ **32.** $\begin{bmatrix} 0 \\ -6 \end{bmatrix}$

33. $\begin{bmatrix} \frac{5}{14} & -\frac{3}{14} \\ \frac{3}{14} & \frac{1}{14} \end{bmatrix}$ **34.** $\begin{bmatrix} 9 & -7 \\ -5 & 4 \end{bmatrix}$ **35.** $\begin{bmatrix} 1 & -3 & 32 \\ 0 & 1 & -9 \\ 0 & 0 & 1 \end{bmatrix}$ **36.** $\begin{bmatrix} 1 & 0 & 0 \\ -\frac{3}{2} & \frac{1}{2} & \frac{1}{2} \\ 1 & -\frac{1}{2} & 0 \end{bmatrix}$

37. $\begin{bmatrix} 9 & 16 & -56 \\ -3 & -5 & 18 \\ -1 & -2 & 7 \end{bmatrix}$ **38.** $\begin{bmatrix} 1 & -1 & 0 \\ 2 & -1 & 0 \\ 1 & -2 & -1 \end{bmatrix}$ **39.** $x = 1, y = 2, z = -1$

40. $w = 1, x = 1, y = 0, z = -1$ **41.** -7 **42.** 3 **43.** -6 **44.** -25 **45.** $x = 1, y = -2$

46. $x = 1, y = 0, z = -2$ **47.** $x = 1, y = -1, z = 3$ **48.** $w = 1, x = 0, y = -1, z = 2$

49. $\dfrac{1}{x} + \dfrac{3x + 4}{x^2 + 1}$ **50.** $\dfrac{3}{x} + \dfrac{2}{x^2} + \dfrac{x - 1}{x^2 + 1}$ **51.** $\dfrac{1}{x} + \dfrac{-1}{x^2 + x + 5}$ **52.** $\dfrac{1}{x + 1} - \dfrac{2}{(x + 1)^2} + \dfrac{2}{(x + 1)^3}$

53.   **54.**   **55.**

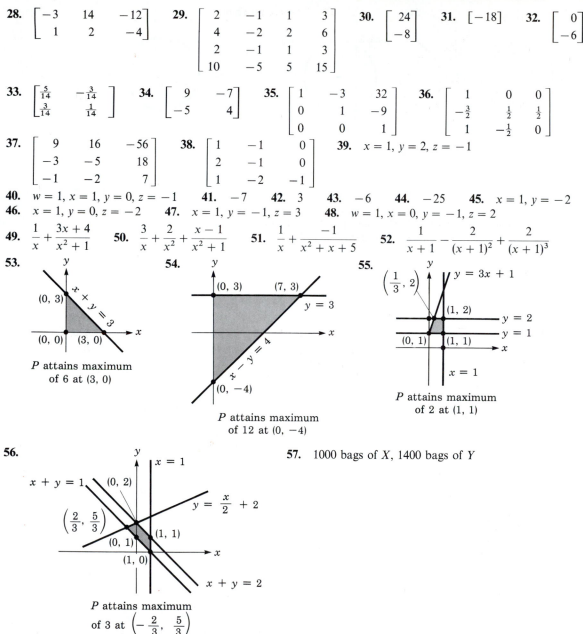

53. P attains maximum of 6 at $(3, 0)$

54. P attains maximum of 12 at $(0, -4)$

55. P attains maximum of 2 at $(1, 1)$

56.

57. 1000 bags of X, 1400 bags of Y

P attains maximum of 3 at $\left(-\dfrac{2}{3}, \dfrac{5}{3} \right)$

Exercise 9.1   (page 403)

1. $x^2 + y^2 = 1$ **3.** $(x - 6)^2 + (y - 8)^2 = 16$ **5.** $(x + 5)^2 + (y - 3)^2 = 25$ **7.** $(x - 3)^2 + (y + 4)^2 = 2$
9. $(x - 3)^2 + (y - 3)^2 = 25$ **11.** $(x + 5)^2 + (y - 1)^2 = 65$ **13.** $(x + 3)^2 + (y - 4)^2 = 25$
15. $(x + 2)^2 + (y + 6)^2 = 40$ **17.** $x^2 + (y + 3)^2 = 157$ **19.** $(x - 5)^2 + (y - 8)^2 = 338$
21. $(x + 4)^2 + (y + 2)^2 = 98$ **23.** $(x - 1)^2 + (y + 2)^2 = 36$ **25.** $x^2 + (y + 12)^2 = 10$ **27.** no

29.

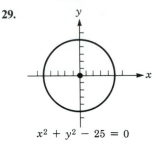

$$x^2 + y^2 - 25 = 0$$

31.

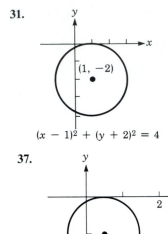

$$(x - 1)^2 + (y + 2)^2 = 4$$

33.

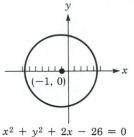

$$x^2 + y^2 + 2x - 26 = 0$$

35.

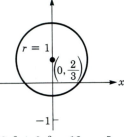

$$9x^2 + 9y^2 - 12y = 5$$

37.

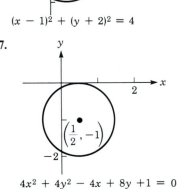

$$4x^2 + 4y^2 - 4x + 8y + 1 = 0$$

39. $x^2 + (y - 3)^2 = 25$

41. $A = 5\pi$ square units

43. $A = 2a\sqrt{r^2 - a^2}$ square units

Exercise 9.2 (page 409)

1. $x^2 = 12y$ **3.** $y^2 = 12x$ **5.** $(x - 3)^2 = -12(y - 5)$ **7.** $(x - 3)^2 = -28(y - 5)$

9. $(x - 2)^2 = -2(y - 2)$ or $(y - 2)^2 = -2(x - 2)$ **11.** $(x + 4)^2 = -\frac{16}{3}(y - 6)$ or $(y - 6)^2 = \frac{9}{4}(x + 4)$

13. $(y - 8)^2 = -4(x - 6)$ **15.** $(x - 3)^2 = \frac{1}{2}(y - 1)$ **17.**

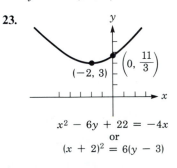

$$y = x^2 + 4x + 5$$
or
$$y - 1 = (x + 2)^2$$

19.

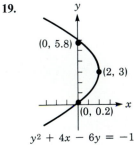

$$y^2 + 4x - 6y = -1$$
or
$$(y - 3)^2 = -4(x - 2)$$

21.

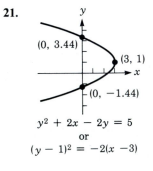

$$y^2 + 2x - 2y = 5$$
or
$$(y - 1)^2 = -2(x - 3)$$

23.

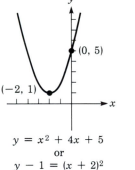

$$x^2 - 6y + 22 = -4x$$
or
$$(x + 2)^2 = 6(y - 3)$$

25.

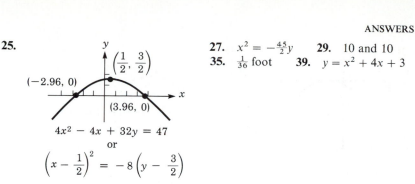

27. $x^2 = -\frac{45}{2}y$ **29.** 10 and 10   **31.** 8 cabins   **33.** $\frac{80}{7}$ meters
35. $\frac{1}{36}$ foot   **39.** $y = x^2 + 4x + 3$

Exercise 9.3 (page 417)

1. $\dfrac{x^2}{25} + \dfrac{y^2}{16} = 1$ **3.** $\dfrac{9x^2}{16} + \dfrac{9y^2}{25} = 1$ **5.** $\dfrac{x^2}{7} + \dfrac{y^2}{16} = 1$ **7.** $\dfrac{(x-3)^2}{4} + \dfrac{(y-4)^2}{9} = 1$

9. $\dfrac{(x-3)^2}{9} + \dfrac{(y-4)^2}{4} = 1$ **11.** $\dfrac{(x-3)^2}{41} + \dfrac{(y-4)^2}{16} = 1$ **13.** $\dfrac{x^2}{36} + \dfrac{(y-4)^2}{20} = 1$ **15.** $\dfrac{x^2}{100} + \dfrac{y^2}{64} = 1$

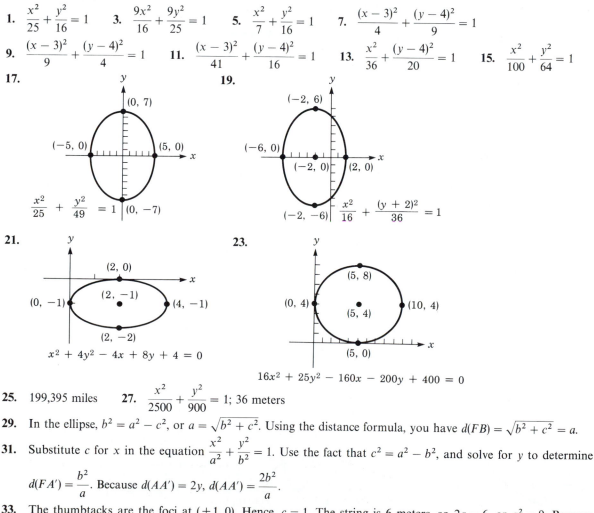

25. 199,395 miles   **27.** $\dfrac{x^2}{2500} + \dfrac{y^2}{900} = 1$; 36 meters

29. In the ellipse, $b^2 = a^2 - c^2$, or $a = \sqrt{b^2 + c^2}$. Using the distance formula, you have $d(FB) = \sqrt{b^2 + c^2} = a$.

31. Substitute c for x in the equation $\dfrac{x^2}{a^2} + \dfrac{y^2}{b^2} = 1$. Use the fact that $c^2 = a^2 - b^2$, and solve for y to determine

$d(FA') = \dfrac{b^2}{a}$. Because $d(AA') = 2y$, $d(AA') = \dfrac{2b^2}{a}$.

33. The thumbtacks are the foci at $(\pm 1, 0)$. Hence, $c = 1$. The string is 6 meters, so $2a = 6$, or $a^2 = 9$. Because $b^2 = a^2 - c^2$, you have $b^2 = 8$. The equation is $\dfrac{x^2}{9} + \dfrac{y^2}{8} = 1$.

Exercise 9.4   (page 424)

1. $\dfrac{x^2}{25} - \dfrac{y^2}{24} = 1$ 3. $\dfrac{(x-2)^2}{4} - \dfrac{(y-4)^2}{9} = 1$ 5. $\dfrac{(y-3)^2}{9} - \dfrac{(x-5)^2}{9} = 1$ 7. $\dfrac{y^2}{9} - \dfrac{x^2}{16} = 1$

9. $\dfrac{(x-1)^2}{4} - \dfrac{(y+3)^2}{16} = 1$ or $\dfrac{(y+3)^2}{4} - \dfrac{(x-1)^2}{16} = 1$ 11. $\dfrac{x^2}{10} - \dfrac{3y^2}{20} = 1$ 13. 24 square units

15. 12 square units   17. $\dfrac{(x+2)^2}{4} - \dfrac{4(y+4)^2}{81} = 1$ 19. $\dfrac{x^2}{36} - \dfrac{16y^2}{25} = 1$

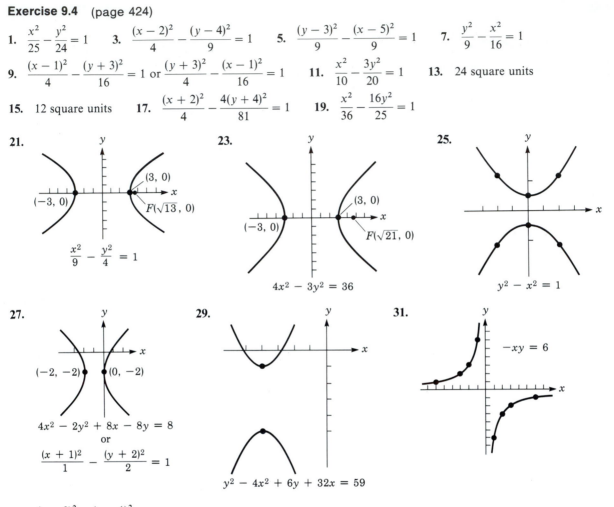

21.

$\dfrac{x^2}{9} - \dfrac{y^2}{4} = 1$

(−3, 0)   (3, 0) $F(\sqrt{13}, 0)$

23.

$4x^2 - 3y^2 = 36$

(−3, 0)   (3, 0) $F(\sqrt{21}, 0)$

25.

$y^2 - x^2 = 1$

27.

$4x^2 - 2y^2 + 8x - 8y = 8$
or
$\dfrac{(x+1)^2}{1} - \dfrac{(y+2)^2}{2} = 1$

(−2, −2)   (0, −2)

29.

$y^2 - 4x^2 + 6y + 32x = 59$

31.

$-xy = 6$

33. $\dfrac{(x-3)^2}{9} - \dfrac{(y-1)^2}{16} = 1$ 35. $4x^2 - 5y^2 - 60y = 0$

37. From the geometry of Figure 9-17, $d(PF) + d(F'F) > d(F'P)$. This is equivalent to $d(F'P) - d(PF) < d(F'F)$, or $2a < 2c$, which implies that $c > a$.

Exercise 9.5   (page 428)

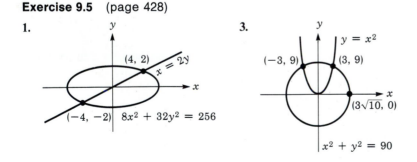

1.

(4, 2) $x = 2y$

(−4, −2) $8x^2 + 32y^2 = 256$

3.

$y = x^2$

(−3, 9)   (3, 9)

$(3\sqrt{10}, 0)$

$x^2 + y^2 = 90$

5.

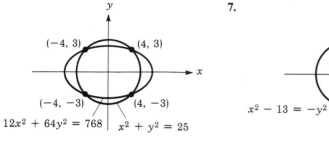

$12x^2 + 64y^2 = 768$ $x^2 + y^2 = 25$

7.

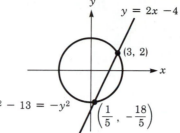

$x^2 - 13 = -y^2$

9.

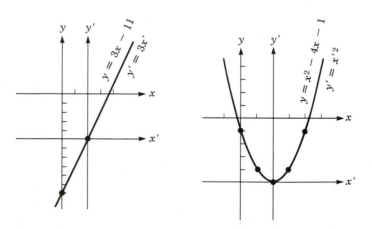

$x^2 - 6x - y = -5$

$x^2 - 6x + y = -5$

11. $(3, 0), (0, 5)$ **13.** $(1, 1)$ **15.** $(1, 2), (2, 1)$ **17.** $(-2, 3), (2, 3)$

19. $(\sqrt{5}, 5), (-\sqrt{5}, 5)$ **21.** $(3, 2), (3, -2), (-3, 2), (-3, -2)$ **23.** $(2, 4), (2, -4), (-2, 4), (-2, -4)$

25. $(-\sqrt{15}, 5), (\sqrt{15}, 5), (-2, -6), (2, -6)$ **27.** $(0, -4), (-3, 5), (3, 5)$

29. $(-2, 3), (2, 3), (-2, -3), (2, -3)$ **31.** $(3, 3)$ **33.** $(6, 2), (-6, -2), (\sqrt{42}, 0), (-\sqrt{42}, 0)$

35. $(\frac{1}{2}, \frac{1}{3}), (\frac{1}{3}, \frac{1}{2})$ **37.** 7 by 9 centimeters   **39.** 14 and -5 **41.** either \$750 at 9% or \$900 at 7.5%

Exercise 9.6   (page 438)

1. $P(3, 1)$ **3.** $R(1, -3)$ **5.** $y' = 3x'$ **7.** $y' = x'^2$

9. $x'^2 + y'^2 = 4$

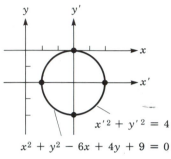

$x'^2 + y'^2 = 4$

$x^2 + y^2 - 6x + 4y + 9 = 0$

11. $x'^2 = -2y'$

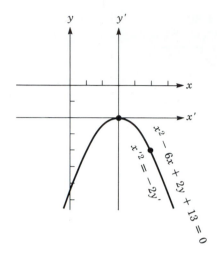

$x'^2 = -2y'$

$x^2 - 6x + 2y + 13 = 0$

13.

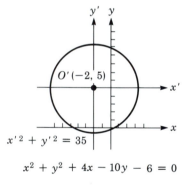

$O'(-2, 5)$

$x'^2 + y'^2 = 35$

$x^2 + y^2 + 4x - 10y - 6 = 0$

15.

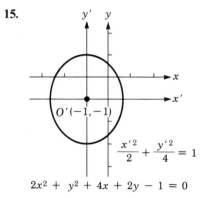

$O'(-1, -1)$

$\dfrac{x'^2}{2} + \dfrac{y'^2}{4} = 1$

$2x^2 + y^2 + 4x + 2y - 1 = 0$

17.

$x^2 - y^2 - 6x - 4y + 4 = 0$

$x'^2 - y'^2 = 1$

19. $(3, \sqrt{3})$ **21.** $(-2\sqrt{2}, 4\sqrt{2})$ **23.** $\left(\dfrac{7 - 4\sqrt{3}}{2}, \dfrac{7\sqrt{3} + 4}{2}\right)$

25. 60°   **27.** 15°   **29.**

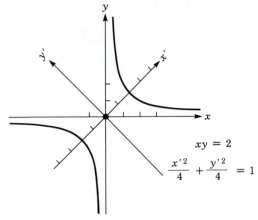

$$xy = 2$$
$$\frac{x'^2}{4} + \frac{y'^2}{4} = 1$$

31.

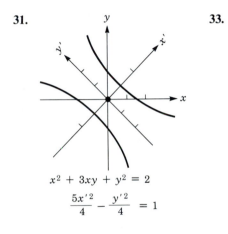

$$x^2 + 3xy + y^2 = 2$$
$$\frac{5x'^2}{4} - \frac{y'^2}{4} = 1$$

33.

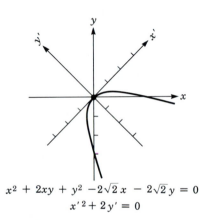

$$x^2 + 2xy + y^2 - 2\sqrt{2}\,x - 2\sqrt{2}\,y = 0$$
$$x'^2 + 2y' = 0$$

35.

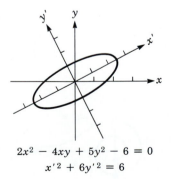

$$2x^2 - 4xy + 5y^2 - 6 = 0$$
$$x'^2 + 6y'^2 = 6$$

37.

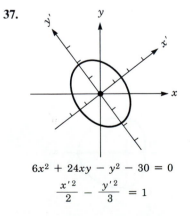

$$6x^2 + 24xy - y^2 - 30 = 0$$
$$\frac{x'^2}{2} - \frac{y'^2}{3} = 1$$

Exercise 9.7   (page 445)

1, 3, 5, 7.

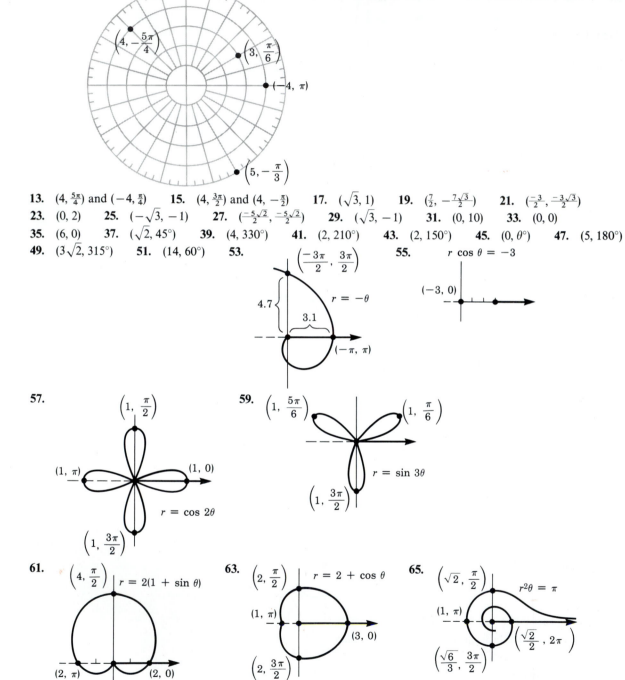

9. $(2, 0)$ and $(2, 2\pi)$ **11.** $(5, \frac{\pi}{2})$ and $(5, -\frac{3\pi}{2})$

13. $(4, \frac{5\pi}{4})$ and $(-4, \frac{\pi}{4})$ **15.** $(4, \frac{3\pi}{2})$ and $(4, -\frac{\pi}{2})$ **17.** $(\sqrt{3}, 1)$ **19.** $(\frac{7}{2}, -\frac{7\sqrt{3}}{2})$ **21.** $(\frac{-3}{2}, \frac{-3\sqrt{3}}{2})$

23. $(0, 2)$ **25.** $(-\sqrt{3}, -1)$ **27.** $(\frac{-5\sqrt{2}}{2}, \frac{-5\sqrt{2}}{2})$ **29.** $(\sqrt{3}, -1)$ **31.** $(0, 10)$ **33.** $(0, 0)$

35. $(6, 0)$ **37.** $(\sqrt{2}, 45°)$ **39.** $(4, 330°)$ **41.** $(2, 210°)$ **43.** $(2, 150°)$ **45.** $(0, \theta°)$ **47.** $(5, 180°)$

49. $(3\sqrt{2}, 315°)$ **51.** $(14, 60°)$ **53.**   **55.**

67. $r \cos \theta = 3$ **69.** $r(3 \cos \theta + 2 \sin \theta) = 3$ **71.** $r = 9 \cos \theta$ **73.** $r^2 = 4 \cos^2 \theta \sin^2 \theta$
75. $\theta = \frac{\pi}{2}$ or $r = \sec \theta$ **77.** $r^2 \cos^2 \theta - 2r \sin \theta = 1$ **79.** $x^2 + y^2 = 9$ **81.** $x = 5$
83. $\sqrt{x^2 + y^2} + y = 1$ **85.** $(x^2 + y^2)^2 = 2xy$ **87.** $y = 0$ **89.** $2\sqrt{x^2 + y^2} - x = 2$

Exercise 9.8   (page 452)

1. a parabola

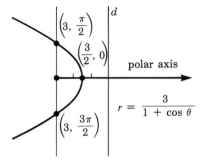

$$r = \frac{3}{1 + \cos \theta}$$

3. an ellipse

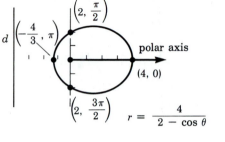

$$r = \frac{4}{2 - \cos \theta}$$

5. a hyperbola

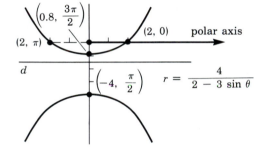

$$r = \frac{4}{2 - 3 \sin \theta}$$

7. $r = \dfrac{10}{3 + 2 \cos \theta}$ **9.** $r = \dfrac{14}{12 + 8 \cos \theta} = \dfrac{7}{6 + 4 \cos \theta}$

11. major axis $= \dfrac{ep}{1 + e} + \dfrac{ep}{1 - e}$
$$= \dfrac{2ep}{1 - e^2}$$
minor axis $= 2ep$

13. $\left(\dfrac{p}{2}, \pi\right)$ **15.** $2a = \dfrac{2ep}{1 - e^2}; c = \dfrac{e^2 p}{1 - e^2}; \dfrac{c}{a} = e$

Exercise 9.9   (page 458)

1.

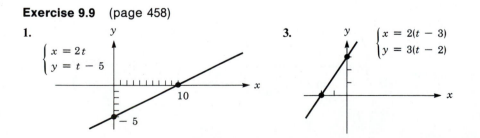

$\begin{cases} x = 2t \\ y = t - 5 \end{cases}$

3.

$\begin{cases} x = 2(t - 3) \\ y = 3(t - 2) \end{cases}$

5.

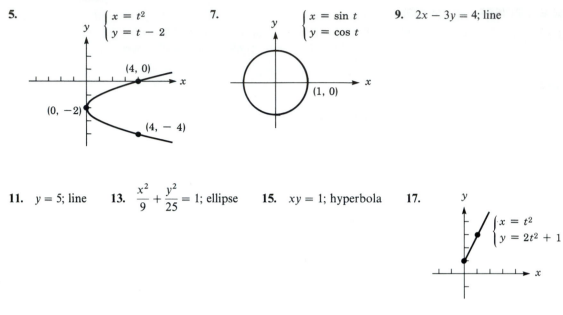

$$\begin{cases} x = t^2 \\ y = t - 2 \end{cases}$$

(4, 0)

(0, −2)

(4, − 4)

7.

$$\begin{cases} x = \sin t \\ y = \cos t \end{cases}$$

(1, 0)

9. $2x - 3y = 4$; line

11. $y = 5$; line   **13.** $\dfrac{x^2}{9} + \dfrac{y^2}{25} = 1$; ellipse   **15.** $xy = 1$; hyperbola   **17.**

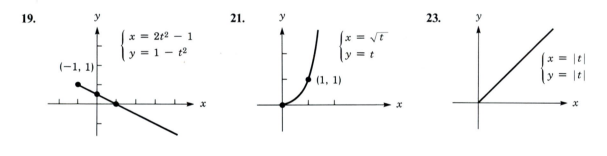

$$\begin{cases} x = t^2 \\ y = 2t^2 + 1 \end{cases}$$

19.

$$\begin{cases} x = 2t^2 - 1 \\ y = 1 - t^2 \end{cases}$$

(−1, 1)

21.

$$\begin{cases} x = \sqrt{t} \\ y = t \end{cases}$$

(1, 1)

23.

$$\begin{cases} x = |t| \\ y = |t| \end{cases}$$

25.

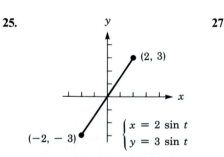

(2, 3)

$$\begin{cases} x = 2 \sin t \\ y = 3 \sin t \end{cases}$$

(−2, − 3)

27.

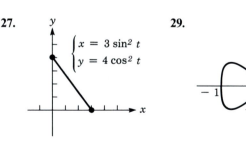

$$\begin{cases} x = 3 \sin^2 t \\ y = 4 \cos^2 t \end{cases}$$

29.

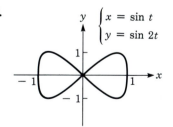

$$\begin{cases} x = \sin t \\ y = \sin 2t \end{cases}$$

1

− 1   1

− 1

31. $\begin{cases} x = a \cos t \\ y = a \sin t \end{cases}$ **33.** $\begin{cases} x = a \cos t \\ y = b \sin t \end{cases}$

35. $y = x \tan \theta - \dfrac{16x^2}{v_0^2 \cos^2 \theta}$; parabola

37. 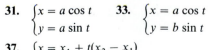 $\begin{cases} x = x_1 + t(x_2 - x_1) \\ y = y_1 + t(y_2 - y_1) \end{cases}$

REVIEW EXERCISES   (page 460)

1. $x^2 + y^2 = 50$ **2.** $x^2 + y^2 = 100$ **3.** $(x - 5)^2 + (y - 10)^2 = 85$ **4.** $(x - 2)^2 + (y - 2)^2 = 89$

5.

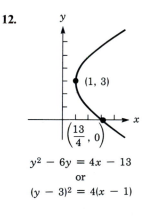

$(x - 3)^2 + (y + 2)^2 = 16$

6.

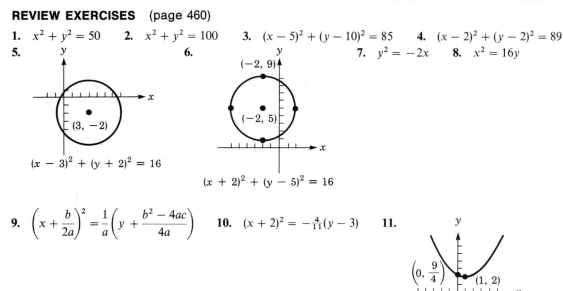

$(x + 2)^2 + (y - 5)^2 = 16$

7. $y^2 = -2x$ **8.** $x^2 = 16y$

9. $\left(x + \dfrac{b}{2a}\right)^2 = \dfrac{1}{a}\left(y + \dfrac{b^2 - 4ac}{4a}\right)$ **10.** $(x + 2)^2 = -\frac{4}{11}(y - 3)$ **11.**

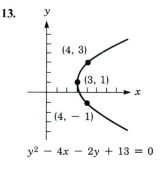

$x^2 - 4y - 2x + 9 = 0$

12.

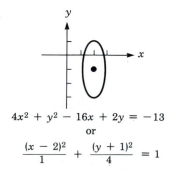

$y^2 - 6y = 4x - 13$
or
$(y - 3)^2 = 4(x - 1)$

13.

$y^2 - 4x - 2y + 13 = 0$

14. $\dfrac{x^2}{36} + \dfrac{y^2}{16} = 1$

15. $\dfrac{(x + 2)^2}{16} + \dfrac{(y - 3)^2}{9} = 1$ **16.**

$4x^2 + y^2 - 16x + 2y = -13$
or
$\dfrac{(x - 2)^2}{1} + \dfrac{(y + 1)^2}{4} = 1$

17.

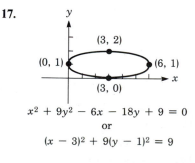

$x^2 + 9y^2 - 6x - 18y + 9 = 0$
or
$(x - 3)^2 + 9(y - 1)^2 = 9$

18. $\dfrac{x^2}{9} - \dfrac{(y-3)^2}{16} = 1$

19.

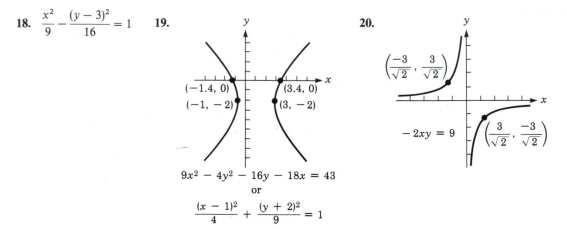

$9x^2 - 4y^2 - 16y - 18x = 43$

or

$\dfrac{(x-1)^2}{4} + \dfrac{(y+2)^2}{9} = 1$

20.

$\left(\dfrac{-3}{\sqrt{2}}, \dfrac{3}{\sqrt{2}}\right)$

$-2xy = 9$

$\left(\dfrac{3}{\sqrt{2}}, \dfrac{-3}{\sqrt{2}}\right)$

21.

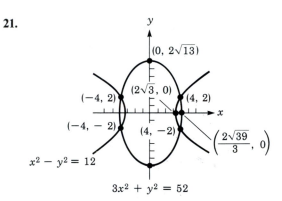

$(0, 2\sqrt{13})$

$(2\sqrt{3}, 0)$

$(-4, 2)$ $(4, 2)$

$(-4, -2)$ $(4, -2)$

$\left(\dfrac{2\sqrt{39}}{3}, 0\right)$

$x^2 - y^2 = 12$

$3x^2 + y^2 = 52$

22. $(4, 2), (4, -2), (-4, 2), (-4, -2)$

23.

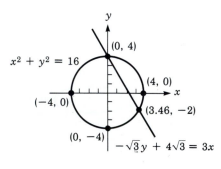

$x^2 + y^2 = 16$

$(0, 4)$

$(4, 0)$

$(-4, 0)$

$(3.46, -2)$

$(0, -4)$

$-\sqrt{3}\,y + 4\sqrt{3} = 3x$

24. $(0, 4), (2\sqrt{3}, -2)$

25.

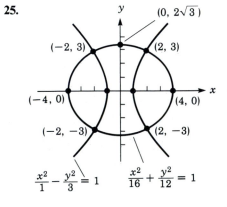

$(0, 2\sqrt{3})$

$(-2, 3)$ $(2, 3)$

$(-4, 0)$ $(4, 0)$

$(-2, -3)$ $(2, -3)$

$\dfrac{x^2}{1} - \dfrac{y^2}{3} = 1$ $\dfrac{x^2}{16} + \dfrac{y^2}{12} = 1$

26. $(2, 3), (2, -3), (-2, 3), (-2, -3)$

27.

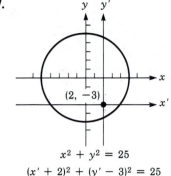

$$x^2 + y^2 = 25$$
$$(x' + 2)^2 + (y' - 3)^2 = 25$$

28.

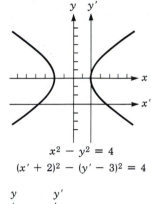

$$x^2 - y^2 = 4$$
$$(x' + 2)^2 - (y' - 3)^2 = 4$$

29.

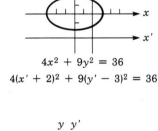

$$4x^2 + 9y^2 = 36$$
$$4(x' + 2)^2 + 9(y' - 3)^2 = 36$$

30.

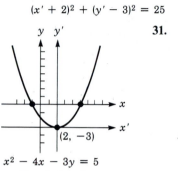

$$x^2 - 4x - 3y = 5$$
$$x'^2 - 3y' = 0$$

31.

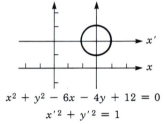

$$x^2 + y^2 - 6x - 4y + 12 = 0$$
$$x'^2 + y'^2 = 1$$

32.

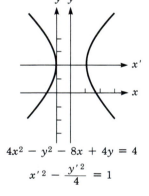

$$4x^2 - y^2 - 8x + 4y = 4$$
$$x'^2 - \frac{y'^2}{4} = 1$$

33.

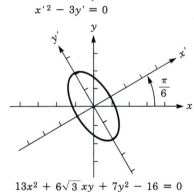

$$13x^2 + 6\sqrt{3}\,xy + 7y^2 - 16 = 0$$
$$x'^2 + \frac{y'^2}{4} = 1$$

34.

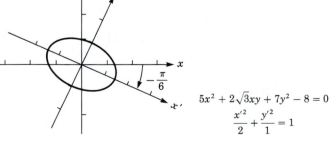

35. $\left(\dfrac{5}{2}, \dfrac{5\sqrt{3}}{2}\right)$ **36.** $(-\sqrt{3}, -1)$ **37.** $\left(\dfrac{\sqrt{3}}{2}, \dfrac{1}{2}\right)$

$$5x^2 + 2\sqrt{3}xy + 7y^2 - 8 = 0$$
$$\frac{x'^2}{2} + \frac{y'^2}{1} = 1$$

38. $(-5\sqrt{2}, 5\sqrt{2})$ **39.** $\left(2, \dfrac{3\pi}{4}\right)$ **40.** $\left(2, \dfrac{5\pi}{6}\right)$ **41.** $(1, 0)$ **42.** $\left(2, \dfrac{5\pi}{3}\right)$ **43.** $r^2 \cos\theta \sin\theta = 1$

44. $r(\cos\theta + 2\sin\theta) = 2$ **45.** $r\cos^2\theta = 3\sin\theta$ **46.** $r^2 = 4\cos\theta\sin\theta$ **47.** $(x^2 + y^2)^2 = 9x^2 - 9y^2$

48. $x^2 + y^2 = 5y$ **49.** $4\sqrt{x^2 + y^2} + y = 1$ **50.** $\sqrt{x^2 + y^2} - x = 2$

51.

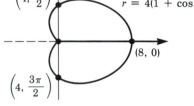

$\left(4, \dfrac{\pi}{2}\right)$ $r = 4(1 + \cos\theta)$

$(8, 0)$

$\left(4, \dfrac{3\pi}{2}\right)$

52.

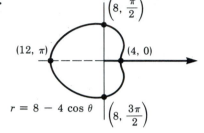

$\left(8, \dfrac{\pi}{2}\right)$

$(12, \pi)$ $(4, 0)$

$r = 8 - 4\cos\theta$ $\left(8, \dfrac{3\pi}{2}\right)$

53.

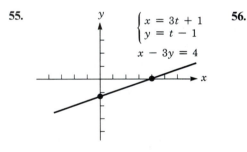

$\left(3, \dfrac{\pi}{2}\right)$

$(6, \pi)$ $(6, 0)$

$r = \dfrac{6}{1 + \sin\theta}$

54.

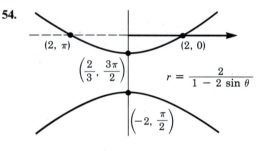

$(2, \pi)$ $(2, 0)$

$\left(\dfrac{2}{3}, \dfrac{3\pi}{2}\right)$ $r = \dfrac{2}{1 - 2\sin\theta}$

$\left(-2, \dfrac{\pi}{2}\right)$

55.

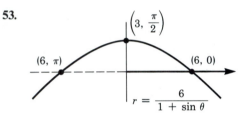

$\begin{cases} x = 3t + 1 \\ y = t - 1 \end{cases}$

$x - 3y = 4$

56.

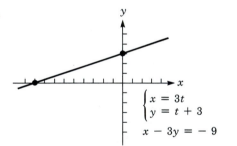

$\begin{cases} x = 3t \\ y = t + 3 \end{cases}$

$x - 3y = -9$

57.

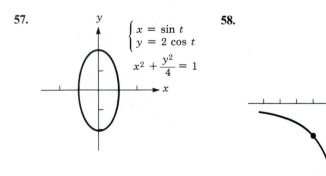

$\begin{cases} x = \sin t \\ y = 2\cos t \end{cases}$

$x^2 + \dfrac{y^2}{4} = 1$

58.

$\begin{cases} x = 2\sin t \\ y = 2\csc t \end{cases}$

$xy = 4$

59.

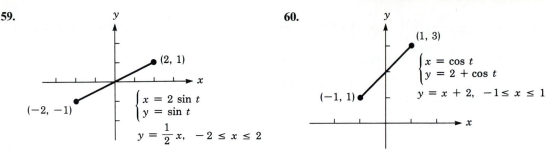

$$\begin{cases} x = 2\sin t \\ y = \sin t \end{cases}$$

$$y = \frac{1}{2}x, \quad -2 \le x \le 2$$

60.

$$\begin{cases} x = \cos t \\ y = 2 + \cos t \end{cases}$$

$$y = x + 2, \quad -1 \le x \le 1$$

Exercise 10.1   (page 467)

25. no

Exercise 10.2   (page 473)

1. 0, 10, 30, 60, 100, 150, 210, 280   **3.** 21   **5.** $a + 4d$ **7.** 15   **9.** 15   **11.** 15   **13.** $\frac{242}{243}$

15. 35   **17.** 3, 7, 15, 31   **19.** $-4, -2, -1, -\frac{1}{2}$ **21.** k, k^2, k^4, k^8 **23.** $8, \frac{16}{k}, \frac{32}{k^2}, \frac{64}{k^3},$

25. an alternating infinite series   **27.** not an alternating infinite series   **29.** 30   **31.** -50
33. 40   **35.** 500   **37.** $\frac{7}{12}$ **39.** 160

Exercise 10.3   (page 480)

1. 1, 3, 5, 7, 9, 11   **3.** $5, \frac{7}{2}, 2, \frac{1}{2}, -1, -\frac{5}{2}$ **5.** $9, \frac{23}{2}, 14, \frac{33}{2}, 19, \frac{43}{2}$ **7.** 285   **9.** 555   **11.** $157\frac{1}{2}$
13. 44   **15.** $\frac{25}{2}, 15, \frac{35}{2}$ **17.** $-\frac{82}{15}, -\frac{59}{15}, -\frac{36}{15}, -\frac{13}{15}$ **19.** 10, 20, 40, 80   **21.** $-2, -6, -18, -54$
23. $3, 3\sqrt{2}, 6, 6\sqrt{2}$ **25.** 2, 6, 18, 54   **27.** 124   **29.** $-29,524$ **31.** $\frac{1995}{32}$ **33.** 18   **35.** 8
37. $10\sqrt[4]{2}, 10\sqrt{2}, 10\sqrt[4]{8}$ **39.** 8, 32, 128, 512   **41.** $\frac{5}{9}$ **43.** $\frac{25}{99}$ **47.** \$21,474,836.47
49. Arithmetic mean is $\frac{11}{16}$; geometric mean is $\sqrt{7}/4$; the arithmetic mean is larger.   **51.** $1 - \frac{1}{101} = \frac{100}{101}$

Exercise 10.4   (page 484)

1. 23   **3.** 5.13 meters   **5.** 1,048,576   **7.** She will earn \$0.19 more on the $7\frac{1}{4}\%$ investment.
9. approximately 6.87×10^{19} **11.** \$180,176.87   **13.** \$2001.60   **15.** \$2013.62   **17.** \$264,094.58

19. 1.8447×10^{19} grains   **21.** No, because $0.999999 = \dfrac{999,999}{1,000,000}$, which does not equal 1.

Exercise 10.5   (page 490)

1. 24   **3.** 720   **5.** 1440   **7.** $\frac{1}{1320}$ **9.** 18,564   **11.** $a^4 + 4a^3b + 6a^2b^2 + 4ab^3 + b^4$
13. $a^5 - 5a^4b + 10a^3b^2 - 10a^2b^3 + 5ab^4 - b^5$ **15.** $8x^3 - 12x^2y + 6xy^2 - y^3$
17. $16x^4 + 32x^3y + 24x^2y^2 + 8xy^3 + y^4$ **19.** $256x^4 - 768x^3y + 864x^2y^2 - 432xy^3 + 81y^4$
21. $36x^2 - 36xy + 9y^2$ **23.** $6a^2b^2$ **25.** $35a^3b^4$ **27.** $-b^5$ **29.** $28x^2y^6$ **31.** $20\sqrt{2}x^3y^2$

33. $4608x^2y^7$ **35.** $\dfrac{r!}{3!(r-3)!}a^{r-3}b^3$ **37.** $\dfrac{n!}{(r-1)!(n-r+1)!}a^{n-r+1}b^{r-1}$

39. 1, 2, 4, 8, 16, 32, 64, 128, 256, 512. They are sequential powers of 2.

41. The sixth term is $\dfrac{10!}{5!(10-5)!}\,a^5\left(-\dfrac{1}{a}\right)^5$, or -252.   **43.** $(x^{1/2})^{25-r}\left(\dfrac{1}{x}\right)^r$ must be x^8. Hence, $r=3$.

Coefficient is $\dfrac{25!}{3!(25-3)!}\cdot\dfrac{1}{2^3}=\dfrac{575}{2}$.

REVIEW EXERCISES   (page 491)

2. 90   **3.** 1718   **4.** -360 **5.** 117   **6.** 281   **7.** -92 **8.** -67.5 **9.** $\dfrac{1}{3^6}$ **10.** 13,122

11. $\dfrac{9}{2^{14}}$ **12.** $\dfrac{8}{15,625}$ **13.** 3320   **14.** 5780   **15.** -5220 **16.** -1540 **17.** $\dfrac{3280}{27}$

18. 6560   **19.** $\dfrac{2295}{128}$ **20.** $\dfrac{520,832}{78,125}$ **21.** $\frac{2}{3}$ **22.** $\frac{3}{25}$ **23.** no sum   **24.** 1   **25.** $\frac{1}{3}$ **26.** 1

27. $\frac{17}{99}$ **28.** $\frac{5}{11}$ **29.** $\frac{7}{2}, 5, \frac{13}{2}$ **30.** 25, 40, 55, 70, 85   **31.** $2\sqrt{2}, 4, 4\sqrt{2}$ **32.** $4, -8, 16, -32$

33. $\dfrac{3280}{3}$ **34.** $16\sqrt{2}$ **35.** 16   **36.** \$4775.81   **37.** 6516, 3134   **38.** \$3487

39. $x^3-3x^2y+3xy^2-y^3$ **40.** $u^3+6u^2v+12uv^2+8v^3$ **41.** $1024a^5-6400a^4b+16,000a^3b^2-20,000a^2b^3+12,500ab^4-3125b^5$ **42.** $49r^4+28\sqrt{21}r^3s+126r^2s^2+12\sqrt{21}rs^3+9s^4$ **43.** $56a^5b^3$
44. $80x^3y^2$ **45.** $84x^3y^6$ **46.** $439,040x^3$

APPENDIX II

TABLES

Table A Values of the Trigonometric Functions

Radians	Degrees	Sin	Cos	Tan	Cot	Degrees	Radians
.0000	0.0°	.0000	1.0000	.0000	—	90.0°	1.5708
.0017	0.1°	.0017	1.0000	.0017	573.0	89.9°	1.5691
.0035	0.2°	.0035	1.0000	.0035	286.5	89.8°	1.5673
.0052	0.3°	.0052	1.0000	.0052	191.0	89.7°	1.5656
.0070	0.4°	.0070	1.0000	.0070	143.2	89.6°	1.5638
.0087	0.5°	.0087	1.0000	.0087	114.6	89.5°	1.5621
.0105	0.6°	.0105	.9999	.0105	95.49	89.4°	1.5603
.0122	0.7°	.0122	.9999	.0122	81.85	89.3°	1.5586
.0140	0.8°	.0140	.9999	.0140	71.62	89.2°	1.5568
.0157	0.9°	.0157	.9999	.0157	63.66	89.1°	1.5551
.0175	1.0°	.0175	.9998	.0175	57.29	89.0°	1.5533
.0192	1.1°	.0192	.9998	.0192	52.08	88.9°	1.5516
.0209	1.2°	.0209	.9998	.0209	47.74	88.8°	1.5499
.0227	1.3°	.0227	.9997	.0227	44.07	88.7°	1.5481
.0244	1.4°	.0244	.9997	.0244	40.92	88.6°	1.5464
.0262	1.5°	.0262	.9997	.0262	38.19	88.5°	1.5446
.0279	1.6°	.0279	.9996	.0279	35.80	88.4°	1.5429
.0297	1.7°	.0297	.9996	.0297	33.69	88.3°	1.5411
.0314	1.8°	.0314	.9995	.0314	31.82	88.2°	1.5394
.0332	1.9°	.0332	.9995	.0332	30.14	88.1°	1.5376
.0349	2.0°	.0349	.9994	.0349	28.64	88.0°	1.5359
.0367	2.1°	.0366	.9993	.0367	27.27	87.9°	1.5341
.0384	2.2°	.0384	.9993	.0384	26.03	87.8°	1.5324
.0401	2.3°	.0401	.9992	.0402	24.90	87.7°	1.5307
.0419	2.4°	.0419	.9991	.0419	23.86	87.6°	1.5289
.0436	2.5°	.0436	.9990	.0437	22.90	87.5°	1.5272
.0454	2.6°	.0454	.9990	.0454	22.02	87.4°	1.5254
.0471	2.7°	.0471	.9989	.0472	21.20	87.3°	1.5237
.0489	2.8°	.0488	.9988	.0489	20.45	87.2°	1.5219
.0506	2.9°	.0506	.9987	.0507	19.74	87.1°	1.5202
.0524	3.0°	.0523	.9986	.0524	19.08	87.0°	1.5184
.0541	3.1°	.0541	.9985	.0542	18.46	86.9°	1.5167
.0559	3.2°	.0558	.9984	.0559	17.89	86.8°	1.5149
.0576	3.3°	.0576	.9983	.0577	17.34	86.7°	1.5132
.0593	3.4°	.0593	.9982	.0594	16.83	86.6°	1.5115
.0611	3.5°	.0610	.9981	.0612	16.35	86.5°	1.5097
.0628	3.6°	.0628	.9980	.0629	15.89	86.4°	1.5080
.0646	3.7°	.0645	.9979	.0647	15.46	86.3°	1.5062
.0663	3.8°	.0663	.9978	.0664	15.06	86.2°	1.5045
.0681	3.9°	.0680	.9977	.0682	14.67	86.1°	1.5027
.0698	4.0°	.0698	.9976	.0699	14.30	86.0°	1.5010
.0716	4.1°	.0715	.9974	.0717	13.95	85.9°	1.4992
.0733	4.2°	.0732	.9973	.0734	13.62	85.8°	1.4975
.0750	4.3°	.0750	.9972	.0752	13.30	85.7°	1.4957
.0768	4.4°	.0767	.9971	.0769	13.00	85.6°	1.4940
.0785	4.5°	.0785	.9969	.0787	12.71	85.5°	1.4923
.0803	4.6°	.0802	.9968	.0805	12.43	85.4°	1.4905
.0820	4.7°	.0819	.9966	.0822	12.16	85.3°	1.4888
.0838	4.8°	.0837	.9965	.0840	11.91	85.2°	1.4870
.0855	4.9°	.0854	.9963	.0857	11.66	85.1°	1.4853
		Cos	Sin	Cot	Tan	Degrees	Radians

Table A (Continued)

Radians	Degrees	Sin	Cos	Tan	Cot	Degrees	Radians
.0873	5.0°	.0872	.9962	.0875	11.43	85.0°	1.4835
.0890	5.1°	.0889	.9960	.0892	11.20	84.9°	1.4818
.0908	5.2°	.0906	.9959	.0910	10.99	84.8°	1.4800
.0925	5.3°	.0924	.9957	.0928	10.78	84.7°	1.4783
.0942	5.4°	.0941	.9956	.0945	10.58	84.6°	1.4765
.0960	5.5°	.0958	.9954	.0963	10.39	84.5°	1.4748
.0977	5.6°	.0976	.9952	.0981	10.20	84.4°	1.4731
.0995	5.7°	.0993	.9951	.0998	10.02	84.3°	1.4713
.1012	5.8°	.1011	.9949	.1016	9.845	84.2°	1.4696
.1030	5.9°	.1028	.9947	.1033	9.677	84.1°	1.4678
.1047	6.0°	.1045	.9945	.1051	9.514	84.0°	1.4661
.1065	6.1°	.1063	.9943	.1069	9.357	83.9°	1.4643
.1082	6.2°	.1080	.9942	.1086	9.205	83.8°	1.4626
.1100	6.3°	.1097	.9940	.1104	9.058	83.7°	1.4608
.1117	6.4°	.1115	.9938	.1122	8.915	83.6°	1.4591
.1134	6.5°	.1132	.9936	.1139	8.777	83.5°	1.4573
.1152	6.6°	.1149	.9934	.1157	8.643	83.4°	1.4556
.1169	6.7°	.1167	.9932	.1175	8.513	83.3°	1.4539
.1187	6.8°	.1184	.9930	.1192	8.386	83.2°	1.4521
.1204	6.9°	.1201	.9928	.1210	8.264	83.1°	1.4504
.1222	7.0°	.1219	.9925	.1228	8.144	83.0°	1.4486
.1239	7.1°	.1236	.9923	.1246	8.028	82.9°	1.4469
.1257	7.2°	.1253	.9921	.1263	7.916	82.8°	1.4451
.1274	7.3°	.1271	.9919	.1281	7.806	82.7°	1.4434
.1292	7.4°	.1288	.9917	.1299	7.700	82.6°	1.4416
.1309	7.5°	.1305	.9914	.1317	7.596	82.5°	1.4399
.1326	7.6°	.1323	.9912	.1334	7.495	82.4°	1.4382
.1344	7.7°	.1340	.9910	.1352	7.396	82.3°	1.4364
.1361	7.8°	.1357	.9907	.1370	7.300	82.2°	1.4347
.1379	7.9°	.1374	.9905	.1388	7.207	82.1°	1.4329
.1396	8.0°	.1392	.9903	.1405	7.115	82.0°	1.4312
.1414	8.1°	.1409	.9900	.1423	7.026	81.9°	1.4294
.1431	8.2°	.1426	.9898	.1441	6.940	81.8°	1.4277
.1449	8.3°	.1444	.9895	.1459	6.855	81.7°	1.4259
.1466	8.4°	.1461	.9893	.1477	6.772	81.6°	1.4242
.1484	8.5°	.1478	.9890	.1495	6.691	81.5°	1.4224
.1501	8.6°	.1495	.9888	.1512	6.612	81.4°	1.4207
.1518	8.7°	.1513	.9885	.1530	6.535	81.3°	1.4190
.1536	8.8°	.1530	.9882	.1548	6.460	81.2°	1.4172
.1553	8.9°	.1547	.9880	.1566	6.386	81.1°	1.4155
.1571	9.0°	.1564	.9877	.1584	6.314	81.0°	1.4137
.1588	9.1°	.1582	.9874	.1602	6.243	80.9°	1.4120
.1606	9.2°	.1599	.9871	.1620	6.174	80.8°	1.4102
.1623	9.3°	.1616	.9869	.1638	6.107	80.7°	1.4085
.1641	9.4°	.1633	.9866	.1655	6.041	80.6°	1.4067
.1658	9.5°	.1650	.9863	.1673	5.976	80.5°	1.4050
.1676	9.6°	.1668	.9860	.1691	5.912	80.4°	1.4032
.1693	9.7°	.1685	.9857	.1709	5.850	80.3°	1.4015
.1710	9.8°	.1702	.9854	.1727	5.789	80.2°	1.3998
.1728	9.9°	.1719	.9851	.1745	5.730	80.1°	1.3980
		Cos	Sin	Cot	Tan	Degrees	Radians

Table A (Continued)

Radians	Degrees	Sin	Cos	Tan	Cot	Degrees	Radians
.1745	10.0°	.1736	.9848	.1763	5.671	80.0°	1.3963
.1763	10.1°	.1754	.9845	.1781	5.614	79.9°	1.3945
.1780	10.2°	.1771	.9842	.1799	5.558	79.8°	1.3928
.1798	10.3°	.1788	.9839	.1817	5.503	79.7°	1.3910
.1815	10.4°	.1805	.9836	.1835	5.449	79.6°	1.3893
.1833	10.5°	.1822	.9833	.1853	5.396	79.5°	1.3875
.1850	10.6°	.1840	.9829	.1871	5.343	79.4°	1.3858
.1868	10.7°	.1857	.9826	.1890	5.292	79.3°	1.3840
.1885	10.8°	.1874	.9823	.1908	5.242	79.2°	1.3823
.1902	10.9°	.1891	.9820	.1926	5.193	79.1°	1.3806
.1920	11.0°	.1908	.9816	.1944	5.145	79.0°	1.3788
.1937	11.1°	.1925	.9813	.1962	5.097	78.9°	1.3771
.1955	11.2°	.1942	.9810	.1980	5.050	78.8°	1.3753
.1972	11.3°	.1959	.9806	.1998	5.005	78.7°	1.3736
.1990	11.4°	.1977	.9803	.2016	4.959	78.6°	1.3718
.2007	11.5°	.1994	.9799	.2035	4.915	78.5°	1.3701
.2025	11.6°	.2011	.9796	.2053	4.872	78.4°	1.3683
.2042	11.7°	.2028	.9792	.2071	4.829	78.3°	1.3666
.2059	11.8°	.2045	.9789	.2089	4.787	78.2°	1.3648
.2077	11.9°	.2062	.9785	.2107	4.745	78.1°	1.3631
.2094	12.0°	.2079	.9781	.2126	4.705	78.0°	1.3614
.2112	12.1°	.2096	.9778	.2144	4.665	77.9°	1.3596
.2129	12.2°	.2113	.9774	.2162	4.625	77.8°	1.3579
.2147	12.3°	.2130	.9770	.2180	4.586	77.7°	1.3561
.2164	12.4°	.2147	.9767	.2199	4.548	77.6°	1.3544
.2182	12.5°	.2164	.9763	.2217	4.511	77.5°	1.3526
.2199	12.6°	.2181	.9759	.2235	4.474	77.4°	1.3509
.2217	12.7°	.2198	.9755	.2254	4.437	77.3°	1.3491
.2234	12.8°	.2215	.9751	.2272	4.402	77.2°	1.3474
.2251	12.9°	.2233	.9748	.2290	4.366	77.1°	1.3456
.2269	13.0°	.2250	.9744	.2309	4.331	77.0°	1.3439
.2286	13.1°	.2267	.9740	.2327	4.297	76.9°	1.3422
.2304	13.2°	.2284	.9736	.2345	4.264	76.8°	1.3404
.2321	13.3°	.2300	.9732	.2364	4.230	76.7°	1.3387
.2339	13.4°	.2317	.9728	.2382	4.198	76.6°	1.3369
.2356	13.5°	.2334	.9724	.2401	4.165	76.5°	1.3352
.2374	13.6°	.2351	.9720	.2419	4.134	76.4°	1.3334
.2391	13.7°	.2368	.9715	.2438	4.102	76.3°	1.3317
.2409	13.8°	.2385	.9711	.2456	4.071	76.2°	1.3299
.2426	13.9°	.2402	.9707	.2475	4.041	76.1°	1.3282
.2443	14.0°	.2419	.9703	.2493	4.011	76.0°	1.3265
.2461	14.1°	.2436	.9699	.2512	3.981	75.9°	1.3247
.2478	14.2°	.2453	.9694	.2530	3.952	75.8°	1.3230
.2496	14.3°	.2470	.9690	.2549	3.923	75.7°	1.3212
.2513	14.4°	.2487	.9686	.2568	3.895	75.6°	1.3195
.2531	14.5°	.2504	.9681	.2586	3.867	75.5°	1.3177
.2548	14.6°	.2521	.9677	.2605	3.839	75.4°	1.3160
.2566	14.7°	.2538	.9673	.2623	3.812	75.3°	1.3142
.2583	14.8°	.2554	.9668	.2642	3.785	75.2°	1.3125
.2601	14.9°	.2571	.9664	.2661	3.758	75.1°	1.3107
		Cos	Sin	Cot	Tan	Degrees	Radians

Table A (Continued)

Radians	Degrees	Sin	Cos	Tan	Cot	Degrees	Radians
.2618	15.0°	.2588	.9659	.2679	3.732	75.0°	1.3090
.2635	15.1°	.2605	.9655	.2698	3.706	74.9°	1.3073
.2653	15.2°	.2622	.9650	.2717	3.681	74.8°	1.3055
.2670	15.3°	.2639	.9646	.2736	3.655	74.7°	1.3038
.2688	15.4°	.2656	.9641	.2754	3.630	74.6°	1.3020
.2705	15.5°	.2672	.9636	.2773	3.606	74.5°	1.3003
.2723	15.6°	.2689	.9632	.2792	3.582	74.4°	1.2985
.2740	15.7°	.2706	.9627	.2811	3.558	74.3°	1.2968
.2758	15.8°	.2723	.9622	.2830	3.534	74.2°	1.2950
.2775	15.9°	.2740	.9617	.2849	3.511	74.1°	1.2933
.2793	16.0°	.2756	.9613	.2867	3.487	74.0°	1.2915
.2810	16.1°	.2773	.9608	.2886	3.465	73.9°	1.2898
.2827	16.2°	.2790	.9603	.2905	3.442	73.8°	1.2881
.2845	16.3°	.2807	.9598	.2924	3.420	73.7°	1.2863
.2862	16.4°	.2823	.9593	.2943	3.398	73.6°	1.2846
.2880	16.5°	.2840	.9588	.2962	3.376	73.5°	1.2828
.2897	16.6°	.2857	.9583	.2981	3.354	73.4°	1.2811
.2915	16.7°	.2874	.9578	.3000	3.333	73.3°	1.2793
.2932	16.8°	.2890	.9573	.3019	3.312	73.2°	1.2776
.2950	16.9°	.2907	.9568	.3038	3.291	73.1°	1.2758
.2967	17.0°	.2924	.9563	.3057	3.271	73.0°	1.2741
.2985	17.1°	.2940	.9558	.3076	3.251	72.9°	1.2723
.3002	17.2°	.2957	.9553	.3096	3.230	72.8°	1.2706
.3019	17.3°	.2974	.9548	.3115	3.211	72.7°	1.2689
.3037	17.4°	.2990	.9542	.3134	3.191	72.6°	1.2671
.3054	17.5°	.3007	.9537	.3153	3.172	72.5°	1.2654
.3072	17.6°	.3024	.9532	.3172	3.152	72.4°	1.2636
.3089	17.7°	.3040	.9527	.3191	3.133	72.3°	1.2619
.3107	17.8°	.3057	.9521	.3211	3.115	72.2°	1.2601
.3124	17.9°	.3074	.9516	.3230	3.096	72.1°	1.2584
.3142	18.0°	.3090	.9511	.3249	3.078	72.0°	1.2566
.3159	18.1°	.3107	.9505	.3269	3.060	71.9°	1.2549
.3176	18.2°	.3123	.9500	.3288	3.042	71.8°	1.2531
.3194	18.3°	.3140	.9494	.3307	3.024	71.7°	1.2514
.3211	18.4°	.3156	.9489	.3327	3.006	71.6°	1.2497
.3229	18.5°	.3173	.9483	.3346	2.989	71.5°	1.2479
.3246	18.6°	.3190	.9478	.3365	2.971	71.4°	1.2462
.3264	18.7°	.3206	.9472	.3385	2.954	71.3°	1.2444
.3281	18.8°	.3223	.9466	.3404	2.937	71.2°	1.2427
.3299	18.9°	.3239	.9461	.3424	2.921	71.1°	1.2409
.3316	19.0°	.3256	.9455	.3443	2.904	71.0°	1.2392
.3334	19.1°	.3272	.9449	.3463	2.888	70.9°	1.2374
.3351	19.2°	.3289	.9444	.3482	2.872	70.8°	1.2357
.3368	19.3°	.3305	.9438	.3502	2.856	70.7°	1.2339
.3386	19.4°	.3322	.9432	.3522	2.840	70.6°	1.2322
.3403	19.5°	.3338	.9426	.3541	2.824	70.5°	1.2305
.3421	19.6°	.3355	.9421	.3561	2.808	70.4°	1.2287
.3438	19.7°	.3371	.9415	.3581	2.793	70.3°	1.2270
.3456	19.8°	.3387	.9409	.3600	2.778	70.2°	1.2252
.3473	19.9°	.3404	.9403	.3620	2.762	70.1°	1.2235
		Cos	Sin	Cot	Tan	Degrees	Radians

Table A (Continued)

Radians	Degrees	Cot	Tan	Cos	Sin	Degrees	Radians
1.1345	65.0°	2.145	.4663	.9063	.4226	25.0°	.4363
1.1327	64.9°	2.135	.4684	.9056	.4242	25.1°	.4381
1.1310	64.8°	2.125	.4706	.9048	.4258	25.2°	.4398
1.1292	64.7°	2.116	.4727	.9041	.4274	25.3°	.4416
1.1275	64.6°	2.106	.4748	.9033	.4289	25.4°	.4433
1.1257	64.5°	2.097	.4770	.9026	.4305	25.5°	.4451
1.1240	64.4°	2.087	.4791	.9018	.4321	25.6°	.4468
1.1222	64.3°	2.078	.4813	.9011	.4337	25.7°	.4485
1.1205	64.2°	2.069	.4834	.9003	.4352	25.8°	.4503
1.1188	64.1°	2.059	.4856	.8996	.4368	25.9°	.4520
1.1170	64.0°	2.050	.4877	.8988	.4384	26.0°	.4538
1.1153	63.9°	2.041	.4899	.8980	.4399	26.1°	.4555
1.1135	63.8°	2.032	.4921	.8973	.4415	26.2°	.4573
1.1118	63.7°	2.023	.4942	.8965	.4431	26.3°	.4590
1.1100	63.6°	2.014	.4964	.8957	.4446	26.4°	.4608
1.1083	63.5°	2.006	.4986	.8949	.4462	26.5°	.4625
1.1065	63.4°	1.997	.5008	.8942	.4478	26.6°	.4643
1.1048	63.3°	1.988	.5029	.8934	.4493	26.7°	.4660
1.1030	63.2°	1.980	.5051	.8926	.4509	26.8°	.4677
1.1013	63.1°	1.971	.5073	.8918	.4524	26.9°	.4695
1.0996	63.0°	1.963	.5095	.8910	.4540	27.0°	.4712
1.0978	62.9°	1.954	.5117	.8902	.4555	27.1°	.4730
1.0961	62.8°	1.946	.5139	.8894	.4571	27.2°	.4747
1.0943	62.7°	1.937	.5161	.8886	.4586	27.3°	.4765
1.0926	62.6°	1.929	.5184	.8878	.4602	27.4°	.4782
1.0908	62.5°	1.921	.5206	.8870	.4617	27.5°	.4800
1.0891	62.4°	1.913	.5228	.8862	.4633	27.6°	.4817
1.0873	62.3°	1.905	.5250	.8854	.4648	27.7°	.4835
1.0856	62.2°	1.897	.5272	.8846	.4664	27.8°	.4852
1.0838	62.1°	1.889	.5295	.8838	.4679	27.9°	.4869
1.0821	62.0°	1.881	.5317	.8829	.4695	28.0°	.4887
1.0804	61.9°	1.873	.5340	.8821	.4710	28.1°	.4904
1.0786	61.8°	1.865	.5362	.8813	.4726	28.2°	.4922
1.0769	61.7°	1.857	.5384	.8805	.4741	28.3°	.4939
1.0751	61.6°	1.849	.5407	.8796	.4756	28.4°	.4957
1.0734	61.5°	1.842	.5430	.8788	.4772	28.5°	.4974
1.0716	61.4°	1.834	.5452	.8780	.4787	28.6°	.4992
1.0699	61.3°	1.827	.5475	.8771	.4802	28.7°	.5009
1.0681	61.2°	1.819	.5498	.8763	.4818	28.8°	.5027
1.0664	61.1°	1.811	.5520	.8755	.4833	28.9°	.5044
1.0647	61.0°	1.804	.5543	.8746	.4848	29.0°	.5061
1.0629	60.9°	1.797	.5566	.8738	.4863	29.1°	.5079
1.0612	60.8°	1.789	.5589	.8729	.4879	29.2°	.5096
1.0594	60.7°	1.782	.5612	.8721	.4894	29.3°	.5114
1.0577	60.6°	1.775	.5635	.8712	.4909	29.4°	.5131
1.0559	60.5°	1.767	.5658	.8704	.4924	29.5°	.5149
1.0542	60.4°	1.760	.5681	.8695	.4939	29.6°	.5166
1.0524	60.3°	1.753	.5704	.8686	.4955	29.7°	.5184
1.0507	60.2°	1.746	.5727	.8678	.4970	29.8°	.5201
1.0489	60.1°	1.739	.5750	.8669	.4985	29.9°	.5219
Radians	**Degrees**	**Tan**	**Cot**	**Sin**	**Cos**	**Degrees**	**Radians**

Table A (Continued)

Radians	Degrees	Sin	Cos	Tan	Cot	Degrees	Radians
.3491	20.0°	.3420	.9397	.3640	2.747	70.0°	1.2217
.3508	20.1°	.3437	.9391	.3659	2.733	69.9°	1.2200
.3526	20.2°	.3453	.9385	.3679	2.718	69.8°	1.2182
.3543	20.3°	.3469	.9379	.3699	2.703	69.7°	1.2165
.3560	20.4°	.3486	.9373	.3719	2.689	69.6°	1.2147
.3578	20.5°	.3502	.9367	.3739	2.675	69.5°	1.2130
.3595	20.6°	.3518	.9361	.3759	2.660	69.4°	1.2113
.3613	20.7°	.3535	.9354	.3779	2.646	69.3°	1.2095
.3630	20.8°	.3551	.9348	.3799	2.633	69.2°	1.2078
.3648	20.9°	.3567	.9342	.3819	2.619	69.1°	1.2060
.3665	21.0°	.3584	.9336	.3839	2.605	69.0°	1.2043
.3683	21.1°	.3600	.9330	.3859	2.592	68.9°	1.2025
.3700	21.2°	.3616	.9323	.3879	2.578	68.8°	1.2008
.3718	21.3°	.3633	.9317	.3899	2.565	68.7°	1.1990
.3735	21.4°	.3649	.9311	.3919	2.552	68.6°	1.1973
.3752	21.5°	.3665	.9304	.3939	2.539	68.5°	1.1956
.3770	21.6°	.3681	.9298	.3959	2.526	68.4°	1.1938
.3787	21.7°	.3697	.9291	.3979	2.513	68.3°	1.1921
.3805	21.8°	.3714	.9285	.4000	2.500	68.2°	1.1903
.3822	21.9°	.3730	.9278	.4020	2.488	68.1°	1.1886
.3840	22.0°	.3746	.9272	.4040	2.475	68.0°	1.1868
.3857	22.1°	.3762	.9265	.4061	2.463	67.9°	1.1851
.3875	22.2°	.3778	.9259	.4081	2.450	67.8°	1.1833
.3892	22.3°	.3795	.9252	.4101	2.438	67.7°	1.1816
.3910	22.4°	.3811	.9245	.4122	2.426	67.6°	1.1798
.3927	22.5°	.3827	.9239	.4142	2.414	67.5°	1.1781
.3944	22.6°	.3843	.9232	.4163	2.402	67.4°	1.1764
.3962	22.7°	.3859	.9225	.4183	2.391	67.3°	1.1746
.3979	22.8°	.3875	.9219	.4204	2.379	67.2°	1.1729
.3997	22.9°	.3891	.9212	.4224	2.367	67.1°	1.1711
.4014	23.0°	.3907	.9205	.4245	2.356	67.0°	1.1694
.4032	23.1°	.3923	.9198	.4265	2.344	66.9°	1.1676
.4049	23.2°	.3939	.9191	.4286	2.333	66.8°	1.1659
.4067	23.3°	.3955	.9184	.4307	2.322	66.7°	1.1641
.4084	23.4°	.3971	.9178	.4327	2.311	66.6°	1.1624
.4102	23.5°	.3987	.9171	.4348	2.300	66.5°	1.1606
.4119	23.6°	.4003	.9164	.4369	2.289	66.4°	1.1589
.4136	23.7°	.4019	.9157	.4390	2.278	66.3°	1.1572
.4154	23.8°	.4035	.9150	.4411	2.267	66.2°	1.1554
.4171	23.9°	.4051	.9143	.4431	2.257	66.1°	1.1537
.4189	24.0°	.4067	.9135	.4452	2.246	66.0°	1.1519
.4206	24.1°	.4083	.9128	.4473	2.236	65.9°	1.1502
.4224	24.2°	.4099	.9121	.4494	2.225	65.8°	1.1484
.4241	24.3°	.4115	.9114	.4515	2.215	65.7°	1.1467
.4259	24.4°	.4131	.9107	.4536	2.204	65.6°	1.1449
.4276	24.5°	.4147	.9100	.4557	2.194	65.5°	1.1432
.4294	24.6°	.4163	.9092	.4578	2.184	65.4°	1.1414
.4311	24.7°	.4179	.9085	.4599	2.174	65.3°	1.1397
.4328	24.8°	.4195	.9078	.4621	2.164	65.2°	1.1380
.4346	24.9°	.4210	.9070	.4642	2.154	65.1°	1.1362
Radians	**Degrees**	**Cos**	**Sin**	**Cot**	**Tan**	**Degrees**	**Radians**

Table A (Continued)

Radians	Degrees	Sin	Cos	Tan	Cot	Degrees	Radians
.5236	30.0°	.5000	.8660	.5774	1.732	60.0°	1.0472
.5253	30.1°	.5015	.8652	.5797	1.725	59.9°	1.0455
.5271	30.2°	.5030	.8643	.5820	1.718	59.8°	1.0437
.5288	30.3°	.5045	.8634	.5844	1.711	59.7°	1.0420
.5306	30.4°	.5060	.8625	.5867	1.704	59.6°	1.0402
.5323	30.5°	.5075	.8616	.5890	1.698	59.5°	1.0385
.5341	30.6°	.5090	.8607	.5914	1.691	59.4°	1.0367
.5358	30.7°	.5105	.8599	.5938	1.684	59.3°	1.0350
.5376	30.8°	.5120	.8590	.5961	1.678	59.2°	1.0332
.5393	30.9°	.5135	.8581	.5985	1.671	59.1°	1.0315
.5411	31.0°	.5150	.8572	.6009	1.664	59.0°	1.0297
.5428	31.1°	.5165	.8563	.6032	1.658	58.9°	1.0280
.5445	31.2°	.5180	.8554	.6056	1.651	58.8°	1.0263
.5463	31.3°	.5195	.8545	.6080	1.645	58.7°	1.0245
.5480	31.4°	.5210	.8536	.6104	1.638	58.6°	1.0228
.5498	31.5°	.5225	.8526	.6128	1.632	58.5°	1.0210
.5515	31.6°	.5240	.8517	.6152	1.625	58.4°	1.0193
.5533	31.7°	.5255	.8508	.6176	1.619	58.3°	1.0175
.5550	31.8°	.5270	.8499	.6200	1.613	58.2°	1.0158
.5568	31.9°	.5284	.8490	.6224	1.607	58.1°	1.0140
.5585	32.0°	.5299	.8480	.6249	1.600	58.0°	1.0123
.5603	32.1°	.5314	.8471	.6273	1.594	57.9°	1.0105
.5620	32.2°	.5329	.8462	.6297	1.588	57.8°	1.0088
.5637	32.3°	.5344	.8453	.6322	1.582	57.7°	1.0071
.5655	32.4°	.5358	.8443	.6346	1.576	57.6°	1.0053
.5672	32.5°	.5373	.8434	.6371	1.570	57.5°	1.0036
.5690	32.6°	.5388	.8425	.6395	1.564	57.4°	1.0018
.5707	32.7°	.5402	.8415	.6420	1.558	57.3°	1.0001
.5725	32.8°	.5417	.8406	.6445	1.552	57.2°	.9983
.5742	32.9°	.5432	.8396	.6469	1.546	57.1°	.9966
.5760	33.0°	.5446	.8387	.6494	1.540	57.0°	.9948
.5777	33.1°	.5461	.8377	.6519	1.534	56.9°	.9931
.5794	33.2°	.5476	.8368	.6544	1.528	56.8°	.9913
.5812	33.3°	.5490	.8358	.6569	1.522	56.7°	.9896
.5829	33.4°	.5505	.8348	.6594	1.517	56.6°	.9879
.5847	33.5°	.5519	.8339	.6619	1.511	56.5°	.9861
.5864	33.6°	.5534	.8329	.6644	1.505	56.4°	.9844
.5882	33.7°	.5548	.8320	.6669	1.499	56.3°	.9826
.5899	33.8°	.5563	.8310	.6694	1.494	56.2°	.9809
.5917	33.9°	.5577	.8300	.6720	1.488	56.1°	.9791
.5934	34.0°	.5592	.8290	.6745	1.483	56.0°	.9774
.5952	34.1°	.5606	.8281	.6771	1.477	55.9°	.9756
.5969	34.2°	.5621	.8271	.6796	1.471	55.8°	.9739
.5986	34.3°	.5635	.8261	.6822	1.466	55.7°	.9721
.6004	34.4°	.5650	.8251	.6847	1.460	55.6°	.9704
.6021	34.5°	.5664	.8241	.6873	1.455	55.5°	.9687
.6039	34.6°	.5678	.8231	.6899	1.450	55.4°	.9669
.6056	34.7°	.5693	.8221	.6924	1.444	55.3°	.9652
.6074	34.8°	.5707	.8211	.6950	1.439	55.2°	.9634
.6091	34.9°	.5721	.8202	.6976	1.433	55.1°	.9617
		Cos	Sin	Cot	Tan	Degrees	Radians

Table A (Continued)

Radians	Degrees	Sin	Cos	Tan	Cot	Degrees	Radians
.6109	35.0°	.5736	.8192	.7002	1.428	55.0°	.9599
.6126	35.1°	.5750	.8181	.7028	1.423	54.9°	.9582
.6144	35.2°	.5764	.8171	.7054	1.418	54.8°	.9564
.6161	35.3°	.5779	.8161	.7080	1.412	54.7°	.9547
.6178	35.4°	.5793	.8151	.7107	1.407	54.6°	.9530
.6196	35.5°	.5807	.8141	.7133	1.402	54.5°	.9512
.6213	35.6°	.5821	.8131	.7159	1.397	54.4°	.9495
.6231	35.7°	.5835	.8121	.7186	1.392	54.3°	.9477
.6248	35.8°	.5850	.8111	.7212	1.387	54.2°	.9460
.6266	35.9°	.5864	.8100	.7239	1.381	54.1°	.9442
.6283	36.0°	.5878	.8090	.7265	1.376	54.0°	.9425
.6301	36.1°	.5892	.8080	.7292	1.371	53.9°	.9407
.6318	36.2°	.5906	.8070	.7319	1.366	53.8°	.9390
.6336	36.3°	.5920	.8059	.7346	1.361	53.7°	.9372
.6353	36.4°	.5934	.8049	.7373	1.356	53.6°	.9355
.6370	36.5°	.5948	.8039	.7400	1.351	53.5°	.9338
.6388	36.6°	.5962	.8028	.7427	1.347	53.4°	.9320
.6405	36.7°	.5976	.8018	.7454	1.342	53.3°	.9303
.6423	36.8°	.5990	.8007	.7481	1.337	53.2°	.9285
.6440	36.9°	.6004	.7997	.7508	1.332	53.1°	.9268
.6458	37.0°	.6018	.7986	.7536	1.327	53.0°	.9250
.6475	37.1°	.6032	.7976	.7563	1.322	52.9°	.9233
.6493	37.2°	.6046	.7965	.7590	1.317	52.8°	.9215
.6510	37.3°	.6060	.7955	.7618	1.313	52.7°	.9198
.6528	37.4°	.6074	.7944	.7646	1.308	52.6°	.9180
.6545	37.5°	.6088	.7934	.7673	1.303	52.5°	.9163
.6562	37.6°	.6101	.7923	.7701	1.299	52.4°	.9146
.6580	37.7°	.6115	.7912	.7729	1.294	52.3°	.9128
.6597	37.8°	.6129	.7902	.7757	1.289	52.2°	.9111
.6615	37.9°	.6143	.7891	.7785	1.285	52.1°	.9093
.6632	38.0°	.6157	.7880	.7813	1.280	52.0°	.9076
.6650	38.1°	.6170	.7869	.7841	1.275	51.9°	.9058
.6667	38.2°	.6184	.7859	.7869	1.271	51.8°	.9041
.6685	38.3°	.6198	.7848	.7898	1.266	51.7°	.9023
.6702	38.4°	.6211	.7837	.7926	1.262	51.6°	.9006
.6720	38.5°	.6225	.7826	.7954	1.257	51.5°	.8988
.6737	38.6°	.6239	.7815	.7983	1.253	51.4°	.8971
.6754	38.7°	.6252	.7804	.8012	1.248	51.3°	.8954
.6772	38.8°	.6266	.7793	.8040	1.244	51.2°	.8936
.6789	38.9°	.6280	.7782	.8069	1.239	51.1°	.8919
.6807	39.0°	.6293	.7771	.8098	1.235	51.0°	.8901
.6824	39.1°	.6307	.7760	.8127	1.230	50.9°	.8884
.6842	39.2°	.6320	.7749	.8156	1.226	50.8°	.8866
.6859	39.3°	.6334	.7738	.8185	1.222	50.7°	.8849
.6877	39.4°	.6347	.7727	.8214	1.217	50.6°	.8831
.6894	39.5°	.6361	.7716	.8243	1.213	50.5°	.8814
.6912	39.6°	.6374	.7705	.8273	1.209	50.4°	.8796
.6929	39.7°	.6388	.7694	.8302	1.205	50.3°	.8779
.6946	39.8°	.6401	.7683	.8332	1.200	50.2°	.8762
.6964	39.9°	.6414	.7672	.8361	1.196	50.1°	.8744
		Cos	Sin	Cot	Tan	Degrees	Radians

Table B Base-10 Logarithms

N	0	1	2	3	4	5	6	7	8	9
1.0	.0000	.0043	.0086	.0128	.0170	.0212	.0253	.0294	.0334	.0374
1.1	.0414	.0453	.0492	.0531	.0569	.0607	.0645	.0682	.0719	.0755
1.2	.0792	.0828	.0864	.0899	.0934	.0969	.1004	.1038	.1072	.1106
1.3	.1139	.1173	.1206	.1239	.1271	.1303	.1335	.1367	.1399	.1430
1.4	.1461	.1492	.1523	.1553	.1584	.1614	.1644	.1673	.1703	.1732
1.5	.1761	.1790	.1818	.1847	.1875	.1903	.1931	.1959	.1987	.2014
1.6	.2041	.2068	.2095	.2122	.2148	.2175	.2201	.2227	.2253	.2279
1.7	.2304	.2330	.2355	.2380	.2405	.2430	.2455	.2480	.2504	.2529
1.8	.2553	.2577	.2601	.2625	.2648	.2672	.2695	.2718	.2742	.2765
1.9	.2788	.2810	.2833	.2856	.2878	.2900	.2923	.2945	.2967	.2989
2.0	.3010	.3032	.3054	.3075	.3096	.3118	.3139	.3160	.3181	.3201
2.1	.3222	.3243	.3263	.3284	.3304	.3324	.3345	.3365	.3385	.3404
2.2	.3424	.3444	.3464	.3483	.3502	.3522	.3541	.3560	.3579	.3598
2.3	.3617	.3636	.3655	.3674	.3692	.3711	.3729	.3747	.3766	.3784
2.4	.3802	.3820	.3838	.3856	.3874	.3892	.3909	.3927	.3945	.3962
2.5	.3979	.3997	.4014	.4031	.4048	.4065	.4082	.4099	.4116	.4133
2.6	.4150	.4166	.4183	.4200	.4216	.4232	.4249	.4265	.4281	.4298
2.7	.4314	.4330	.4346	.4362	.4378	.4393	.4409	.4425	.4440	.4456
2.8	.4472	.4487	.4502	.4518	.4533	.4548	.4564	.4579	.4594	.4609
2.9	.4624	.4639	.4654	.4669	.4683	.4698	.4713	.4728	.4742	.4757
3.0	.4771	.4786	.4800	.4814	.4829	.4843	.4857	.4871	.4886	.4900
3.1	.4914	.4928	.4942	.4955	.4969	.4983	.4997	.5011	.5024	.5038
3.2	.5051	.5065	.5079	.5092	.5105	.5119	.5132	.5145	.5159	.5172
3.3	.5185	.5198	.5211	.5224	.5237	.5250	.5263	.5276	.5289	.5302
3.4	.5315	.5328	.5340	.5353	.5366	.5378	.5391	.5403	.5416	.5428
3.5	.5441	.5453	.5465	.5478	.5490	.5502	.5514	.5527	.5539	.5551
3.6	.5563	.5575	.5587	.5599	.5611	.5623	.5635	.5647	.5658	.5670
3.7	.5682	.5694	.5705	.5717	.5729	.5740	.5752	.5763	.5775	.5786
3.8	.5798	.5809	.5821	.5832	.5843	.5855	.5866	.5877	.5888	.5899
3.9	.5911	.5922	.5933	.5944	.5955	.5966	.5977	.5988	.5999	.6010
4.0	.6021	.6031	.6042	.6053	.6064	.6075	.6085	.6096	.6107	.6117
4.1	.6128	.6138	.6149	.6160	.6170	.6180	.6191	.6201	.6212	.6222
4.2	.6232	.6243	.6253	.6263	.6274	.6284	.6294	.6304	.6314	.6325
4.3	.6335	.6345	.6355	.6365	.6375	.6385	.6395	.6405	.6415	.6425
4.4	.6435	.6444	.6454	.6464	.6474	.6484	.6493	.6503	.6513	.6522
4.5	.6532	.6542	.6551	.6561	.6571	.6580	.6590	.6599	.6609	.6618
4.6	.6628	.6637	.6646	.6656	.6665	.6675	.6684	.6693	.6702	.6712
4.7	.6721	.6730	.6739	.6749	.6758	.6767	.6776	.6785	.6794	.6803
4.8	.6812	.6821	.6830	.6839	.6848	.6857	.6866	.6875	.6884	.6893
4.9	.6902	.6911	.6920	.6928	.6937	.6946	.6955	.6964	.6972	.6981
5.0	.6990	.6998	.7007	.7016	.7024	.7033	.7042	.7050	.7059	.7067
5.1	.7076	.7084	.7093	.7101	.7110	.7118	.7126	.7135	.7143	.7152
5.2	.7160	.7168	.7177	.7185	.7193	.7202	.7210	.7218	.7226	.7235
5.3	.7243	.7251	.7259	.7267	.7275	.7284	.7292	.7300	.7308	.7316
5.4	.7324	.7332	.7340	.7348	.7356	.7364	.7372	.7380	.7388	.7396

Table A (Continued)

Radians	Degrees	Sin	Cos	Tan	Cot	Degrees	Radians
.6981	40.0°	.6428	.7660	.8391	1.192	50.0°	.8727
.6999	40.1°	.6441	.7649	.8421	1.188	49.9°	.8709
.7016	40.2°	.6455	.7638	.8451	1.183	49.8°	.8692
.7034	40.3°	.6468	.7627	.8481	1.179	49.7°	.8674
.7051	40.4°	.6481	.7615	.8511	1.175	49.6°	.8657
.7069	40.5°	.6494	.7604	.8541	1.171	49.5°	.8639
.7086	40.6°	.6508	.7593	.8571	1.167	49.4°	.8622
.7103	40.7°	.6521	.7581	.8601	1.163	49.3°	.8604
.7121	40.8°	.6534	.7570	.8632	1.159	49.2°	.8587
.7138	40.9°	.6547	.7559	.8662	1.154	49.1°	.8570
.7156	41.0°	.6561	.7547	.8693	1.150	49.0°	.8552
.7173	41.1°	.6574	.7536	.8724	1.146	48.9°	.8535
.7191	41.2°	.6587	.7524	.8754	1.142	48.8°	.8517
.7208	41.3°	.6600	.7513	.8785	1.138	48.7°	.8500
.7226	41.4°	.6613	.7501	.8816	1.134	48.6°	.8482
.7243	41.5°	.6626	.7490	.8847	1.130	48.5°	.8465
.7261	41.6°	.6639	.7478	.8878	1.126	48.4°	.8447
.7278	41.7°	.6652	.7466	.8910	1.122	48.3°	.8430
.7295	41.8°	.6665	.7455	.8941	1.118	48.2°	.8412
.7313	41.9°	.6678	.7443	.8972	1.115	48.1°	.8395
.7330	42.0°	.6691	.7431	.9004	1.111	48.0°	.8378
.7348	42.1°	.6704	.7420	.9036	1.107	47.9°	.8360
.7365	42.2°	.6717	.7408	.9067	1.103	47.8°	.8343
.7383	42.3°	.6730	.7396	.9099	1.099	47.7°	.8325
.7400	42.4°	.6743	.7385	.9131	1.095	47.6°	.8308
.7418	42.5°	.6756	.7373	.9163	1.091	47.5°	.8290
.7435	42.6°	.6769	.7361	.9195	1.087	47.4°	.8273
.7453	42.7°	.6782	.7349	.9228	1.084	47.3°	.8255
.7470	42.8°	.6794	.7337	.9260	1.080	47.2°	.8238
.7487	42.9°	.6807	.7325	.9293	1.076	47.1°	.8221
.7505	43.0°	.6820	.7314	.9325	1.072	47.0°	.8203
.7522	43.1°	.6833	.7302	.9358	1.069	46.9°	.8186
.7540	43.2°	.6845	.7290	.9391	1.065	46.8°	.8168
.7557	43.3°	.6858	.7278	.9424	1.061	46.7°	.8151
.7575	43.4°	.6871	.7266	.9457	1.057	46.6°	.8133
.7592	43.5°	.6884	.7254	.9490	1.054	46.5°	.8116
.7610	43.6°	.6896	.7242	.9523	1.050	46.4°	.8098
.7627	43.7°	.6909	.7230	.9556	1.046	46.3°	.8081
.7645	43.8°	.6921	.7218	.9590	1.043	46.2°	.8063
.7662	43.9°	.6934	.7206	.9623	1.039	46.1°	.8046
.7679	44.0°	.6947	.7193	.9657	1.036	46.0°	.8029
.7697	44.1°	.6959	.7181	.9691	1.032	45.9°	.8011
.7714	44.2°	.6972	.7169	.9725	1.028	45.8°	.7994
.7732	44.3°	.6984	.7157	.9759	1.025	45.7°	.7976
.7749	44.4°	.6997	.7145	.9793	1.021	45.6°	.7959
.7767	44.5°	.7009	.7133	.9827	1.018	45.5°	.7941
.7784	44.6°	.7022	.7120	.9861	1.014	45.4°	.7924
.7802	44.7°	.7034	.7108	.9896	1.011	45.3°	.7906
.7819	44.8°	.7046	.7096	.9930	1.007	45.2°	.7889
.7837	44.9°	.7059	.7083	.9965	1.003	45.1°	.7871
.7854	45.0°	.7071	.7071	1.0000	1.000	45.0°	.7854
		Cos	Sin	Cot	Tan	Degrees	Radians

Table C Base-e Logarithms

N	0	1	2	3	4	5	6	7	8	9
1.0	.0000	.0100	.0198	.0296	.0392	.0488	.0583	.0677	.0770	.0862
1.1	.0953	.1044	.1133	.1222	.1310	.1398	.1484	.1570	.1655	.1740
1.2	.1823	.1906	.1989	.2070	.2151	.2231	.2311	.2390	.2469	.2546
1.3	.2624	.2700	.2776	.2852	.2927	.3001	.3075	.3148	.3221	.3293
1.4	.3365	.3436	.3507	.3577	.3646	.3716	.3784	.3853	.3920	.3988
1.5	.4055	.4121	.4187	.4253	.4318	.4383	.4447	.4511	.4574	.4637
1.6	.4700	.4762	.4824	.4886	.4947	.5008	.5068	.5128	.5188	.5247
1.7	.5306	.5365	.5423	.5481	.5539	.5596	.5653	.5710	.5766	.5822
1.8	.5878	.5933	.5988	.6043	.6098	.6152	.6206	.6259	.6313	.6366
1.9	.6419	.6471	.6523	.6575	.6627	.6678	.6729	.6780	.6831	.6881
2.0	.6931	.6981	.7031	.7080	.7129	.7178	.7227	.7275	.7324	.7372
2.1	.7419	.7467	.7514	.7561	.7608	.7655	.7701	.7747	.7793	.7839
2.2	.7885	.7930	.7975	.8020	.8065	.8109	.8154	.8198	.8242	.8286
2.3	.8329	.8372	.8416	.8459	.8502	.8544	.8587	.8629	.8671	.8713
2.4	.8755	.8796	.8838	.8879	.8920	.8961	.9002	.9042	.9083	.9123
2.5	.9163	.9203	.9243	.9282	.9322	.9361	.9400	.9439	.9478	.9517
2.6	.9555	.9594	.9632	.9670	.9708	.9746	.9783	.9821	.9858	.9895
2.7	.9933	.9969	1.0006	.0043	.0080	.0116	.0152	.0188	.0225	.0260
2.8	1.0296	.0332	.0367	.0403	.0438	.0473	.0508	.0543	.0578	.0613
2.9	.0647	.0682	.0716	.0750	.0784	.0818	.0852	.0886	.0919	.0953
3.0	1.0986	.1019	.1053	.1086	.1119	.1151	.1184	.1217	.1249	.1282
3.1	.1314	.1346	.1378	.1410	.1442	.1474	.1506	.1537	.1569	.1600
3.2	.1632	.1663	.1694	.1725	.1756	.1787	.1817	.1848	.1878	.1909
3.3	.1939	.1969	.2000	.2030	.2060	.2090	.2119	.2149	.2179	.2208
3.4	.2238	.2267	.2296	.2326	.2355	.2384	.2413	.2442	.2470	.2499
3.5	1.2528	.2556	.2585	.2613	.2641	.2669	.2698	.2726	.2754	.2782
3.6	.2809	.2837	.2865	.2892	.2920	.2947	.2975	.3002	.3029	.3056
3.7	.3083	.3110	.3137	.3164	.3191	.3218	.3244	.3271	.3297	.3324
3.8	.3350	.3376	.3403	.3429	.3455	.3481	.3507	.3533	.3558	.3584
3.9	.3610	.3635	.3661	.3686	.3712	.3737	.3762	.3788	.3813	.3838
4.0	1.3863	.3888	.3913	.3938	.3962	.3987	.4012	.4036	.4061	.4085
4.1	.4110	.4134	.4159	.4183	.4207	.4231	.4255	.4279	.4303	.4327
4.2	.4351	.4375	.4398	.4422	.4446	.4469	.4493	.4516	.4540	.4563
4.3	.4586	.4609	.4633	.4656	.4679	.4702	.4725	.4748	.4770	.4793
4.4	.4816	.4839	.4861	.4884	.4907	.4929	.4951	.4974	.4996	.5019
4.5	1.5041	.5063	.5085	.5107	.5129	.5151	.5173	.5195	.5217	.5239
4.6	.5261	.5282	.5304	.5326	.5347	.5369	.5390	.5412	.5433	.5454
4.7	.5476	.5497	.5518	.5539	.5560	.5581	.5602	.5623	.5644	.5665
4.8	.5686	.5707	.5728	.5748	.5769	.5790	.5810	.5831	.5851	.5872
4.9	.5892	.5913	.5933	.5953	.5974	.5994	.6014	.6034	.6054	.6074
5.0	1.6094	.6114	.6134	.6154	.6174	.6194	.6214	.6233	.6253	.6273
5.1	.6292	.6312	.6332	.6351	.6371	.6390	.6409	.6429	.6448	.6467
5.2	.6487	.6506	.6525	.6544	.6563	.6582	.6601	.6620	.6639	.6658
5.3	.6677	.6696	.6715	.6734	.6752	.6771	.6790	.6808	.6827	.6845
5.4	.6864	.6882	.6901	.6919	.6938	.6956	.6974	.6993	.7011	.7029

Use the properties of logarithms and ln 10 ≈ 2.3026 to find logarithms of numbers less than 1 or greater than 10.

Table B (Continued)

N	0	1	2	3	4	5	6	7	8	9
5.5	.7404	.7412	.7419	.7427	.7435	.7443	.7451	.7459	.7466	.7474
5.6	.7482	.7490	.7497	.7505	.7513	.7520	.7528	.7536	.7543	.7551
5.7	.7559	.7566	.7574	.7582	.7589	.7597	.7604	.7612	.7619	.7627
5.8	.7634	.7642	.7649	.7657	.7664	.7672	.7679	.7686	.7694	.7701
5.9	.7709	.7716	.7723	.7731	.7738	.7745	.7752	.7760	.7767	.7774
6.0	.7782	.7789	.7796	.7803	.7810	.7818	.7825	.7832	.7839	.7846
6.1	.7853	.7860	.7868	.7875	.7882	.7889	.7896	.7903	.7910	.7917
6.2	.7924	.7931	.7938	.7945	.7952	.7959	.7966	.7973	.7980	.7987
6.3	.7993	.8000	.8007	.8014	.8021	.8028	.8035	.8041	.8048	.8055
6.4	.8062	.8069	.8075	.8082	.8089	.8096	.8102	.8109	.8116	.8122
6.5	.8129	.8136	.8142	.8149	.8156	.8162	.8169	.8176	.8182	.8189
6.6	.8195	.8202	.8209	.8215	.8222	.8228	.8235	.8241	.8248	.8254
6.7	.8261	.8267	.8274	.8280	.8287	.8293	.8299	.8306	.8312	.8319
6.8	.8325	.8331	.8338	.8344	.8351	.8357	.8363	.8370	.8376	.8382
6.9	.8388	.8395	.8401	.8407	.8414	.8420	.8426	.8432	.8439	.8445
7.0	.8451	.8457	.8463	.8470	.8476	.8482	.8488	.8494	.8500	.8506
7.1	.8513	.8519	.8525	.8531	.8537	.8543	.8549	.8555	.8561	.8567
7.2	.8573	.8579	.8585	.8591	.8597	.8603	.8609	.8615	.8621	.8627
7.3	.8633	.8639	.8645	.8651	.8657	.8663	.8669	.8675	.8681	.8686
7.4	.8692	.8698	.8704	.8710	.8716	.8722	.8727	.8733	.8739	.8745
7.5	.8751	.8756	.8762	.8768	.8774	.8779	.8785	.8791	.8797	.8802
7.6	.8808	.8814	.8820	.8825	.8831	.8837	.8842	.8848	.8854	.8859
7.7	.8865	.8871	.8876	.8882	.8887	.8893	.8899	.8904	.8910	.8915
7.8	.8921	.8927	.8932	.8938	.8943	.8949	.8954	.8960	.8965	.8971
7.9	.8976	.8982	.8987	.8993	.8998	.9004	.9009	.9015	.9020	.9025
8.0	.9031	.9036	.9042	.9047	.9053	.9058	.9063	.9069	.9074	.9079
8.1	.9085	.9090	.9096	.9101	.9106	.9112	.9117	.9122	.9128	.9133
8.2	.9138	.9143	.9149	.9154	.9159	.9165	.9170	.9175	.9180	.9186
8.3	.9191	.9196	.9201	.9206	.9212	.9217	.9222	.9227	.9232	.9238
8.4	.9243	.9248	.9253	.9258	.9263	.9269	.9274	.9279	.9284	.9289
8.5	.9294	.9299	.9304	.9309	.9315	.9320	.9325	.9330	.9335	.9340
8.6	.9345	.9350	.9355	.9360	.9365	.9370	.9375	.9380	.9385	.9390
8.7	.9395	.9400	.9405	.9410	.9415	.9420	.9425	.9430	.9435	.9440
8.8	.9445	.9450	.9455	.9460	.9465	.9469	.9474	.9479	.9484	.9489
8.9	.9494	.9499	.9504	.9509	.9513	.9518	.9523	.9528	.9533	.9538
9.0	.9542	.9547	.9552	.9557	.9562	.9566	.9571	.9576	.9581	.9586
9.1	.9590	.9595	.9600	.9605	.9609	.9614	.9619	.9624	.9628	.9633
9.2	.9638	.9643	.9647	.9652	.9657	.9661	.9666	.9671	.9675	.9680
9.3	.9685	.9689	.9694	.9699	.9703	.9708	.9713	.9717	.9722	.9727
9.4	.9731	.9736	.9741	.9745	.9750	.9754	.9759	.9763	.9768	.9773
9.5	.9777	.9782	.9786	.9791	.9795	.9800	.9805	.9809	.9814	.9818
9.6	.9823	.9827	.9832	.9836	.9841	.9845	.9850	.9854	.9859	.9863
9.7	.9868	.9872	.9877	.9881	.9886	.9890	.9894	.9899	.9903	.9908
9.8	.9912	.9917	.9921	.9926	.9930	.9934	.9939	.9943	.9948	.9952
9.9	.9956	.9961	.9965	.9969	.9974	.9978	.9983	.9987	.9991	.9996

Table C (Continued)

N	0	1	2	3	4	5	6	7	8	9
5.5	.7047	.7066	.7084	.7102	.7120	.7138	.7156	.7174	.7192	.7210
5.6	.7228	.7246	.7263	.7281	.7299	.7317	.7334	.7352	.7370	.7387
5.7	.7405	.7422	.7440	.7457	.7475	.7492	.7509	.7527	.7544	.7561
5.8	.7579	.7596	.7613	.7630	.7647	.7664	.7681	.7699	.7716	.7733
5.9	.7750	.7766	.7783	.7800	.7817	.7834	.7851	.7867	.7884	.7901
6.0	.7918	.7934	.7951	.7967	.7984	.8001	.8017	.8034	.8050	.8066
6.1	.8083	.8099	.8116	.8132	.8148	.8165	.8181	.8197	.8213	.8229
6.2	.8245	.8262	.8278	.8294	.8310	.8326	.8342	.8358	.8374	.8390
6.3	.8405	.8421	.8437	.8453	.8469	.8485	.8500	.8516	.8532	.8547
6.4	.8563	.8579	.8594	.8610	.8625	.8641	.8656	.8672	.8687	.8703
6.5	.8718	.8733	.8749	.8764	.8779	.8795	.8810	.8825	.8840	.8856
6.6	.8871	.8886	.8901	.8916	.8931	.8946	.8961	.8976	.8991	.9006
6.7	.9021	.9036	.9051	.9066	.9081	.9095	.9110	.9125	.9140	.9155
6.8	.9169	.9184	.9199	.9213	.9228	.9242	.9257	.9272	.9286	.9301
6.9	.9315	.9330	.9344	.9359	.9373	.9387	.9402	.9416	.9430	.9445
7.0	.9459	.9473	.9488	.9502	.9516	.9530	.9544	.9559	.9573	.9587
7.1	.9601	.9615	.9629	.9643	.9657	.9671	.9685	.9699	.9713	.9727
7.2	.9741	.9755	.9769	.9782	.9796	.9810	.9824	.9838	.9851	.9865
7.3	.9879	.9892	.9906	.9920	.9933	.9947	.9961	.9974	.9988	2.0001
7.4	2.0015	.0028	.0042	.0055	.0069	.0082	.0096	.0109	.0122	.0136
7.5	2.0149	.0162	.0176	.0189	.0202	.0215	.0229	.0242	.0255	.0268
7.6	.0281	.0295	.0308	.0321	.0334	.0347	.0360	.0373	.0386	.0399
7.7	.0412	.0425	.0438	.0451	.0464	.0477	.0490	.0503	.0516	.0528
7.8	.0541	.0554	.0567	.0580	.0592	.0605	.0618	.0631	.0643	.0656
7.9	.0669	.0681	.0694	.0707	.0719	.0732	.0744	.0757	.0769	.0782
8.0	2.0794	.0807	.0819	.0832	.0844	.0857	.0869	.0882	.0894	.0906
8.1	.0919	.0931	.0943	.0956	.0968	.0980	.0992	.1005	.1017	.1029
8.2	.1041	.1054	.1066	.1078	.1090	.1102	.1114	.1126	.1138	.1150
8.3	.1163	.1175	.1187	.1199	.1211	.1223	.1235	.1247	.1258	.1270
8.4	.1282	.1294	.1306	.1318	.1330	.1342	.1353	.1365	.1377	.1389
8.5	2.1401	.1412	.1424	.1436	.1448	.1459	.1471	.1483	.1494	.1506
8.6	.1518	.1529	.1541	.1552	.1564	.1576	.1587	.1599	.1610	.1622
8.7	.1633	.1645	.1656	.1668	.1679	.1691	.1702	.1713	.1725	.1736
8.8	.1748	.1759	.1770	.1782	.1793	.1804	.1815	.1827	.1838	.1849
8.9	.1861	.1872	.1883	.1894	.1905	.1917	.1928	.1939	.1950	.1961
9.0	2.1972	.1983	.1994	.2006	.2017	.2028	.2039	.2050	.2061	.2072
9.1	.2083	.2094	.2105	.2116	.2127	.2138	.2148	.2159	.2170	.2181
9.2	.2192	.2203	.2214	.2225	.2235	.2246	.2257	.2268	.2279	.2289
9.3	.2300	.2311	.2322	.2332	.2343	.2354	.2364	.2375	.2386	.2396
9.4	.2407	.2418	.2428	.2439	.2450	.2460	.2471	.2481	.2492	.2502
9.5	2.2513	.2523	.2534	.2544	.2555	.2565	.2576	.2586	.2597	.2607
9.6	.2618	.2628	.2638	.2649	.2659	.2670	.2680	.2690	.2701	.2711
9.7	.2721	.2732	.2742	.2752	.2762	.2773	.2783	.2793	.2803	.2814
9.8	.2824	.2834	.2844	.2854	.2865	.2875	.2885	.2895	.2905	.2915
9.9	.2925	.2935	.2946	.2956	.2966	.2976	.2986	.2996	.3006	.3016

Use the properties of logarithms and $\ln 10 \approx 2.3026$ to find logarithms of numbers less than 1 or greater than 10.

Table D Powers and Roots

n	n^2	$\sqrt{n}$	n^3	$\sqrt[3]{n}$	n	n^2	$\sqrt{n}$	n^3	$\sqrt[3]{n}$
1	1	1.000	1	1.000	51	2,601	7.141	132,651	3.708
2	4	1.414	8	1.260	52	2,704	7.211	140,608	3.733
3	9	1.732	27	1.442	53	2,809	7.280	148,877	3.756
4	16	2.000	64	1.587	54	2,916	7.348	157,464	3.780
5	25	2.236	125	1.710	55	3,025	7.416	166,375	3.803
6	36	2.449	216	1.817	56	3,136	7.483	175,616	3.826
7	49	2.646	343	1.913	57	3,249	7.550	185,193	3.849
8	64	2.828	512	2.000	58	3,364	7.616	195,112	3.871
9	81	3.000	729	2.080	59	3,481	7.681	205,379	3.893
10	100	3.162	1,000	2.154	60	3,600	7.746	216,000	3.915
11	121	3.317	1,331	2.224	61	3,721	7.810	226,981	3.936
12	144	3.464	1,728	2.289	62	3,844	7.874	238,328	3.958
13	169	3.606	2,197	2.351	63	3,969	7.937	250,047	3.979
14	196	3.742	2,744	2.410	64	4,096	8.000	262,144	4.000
15	225	3.873	3,375	2.466	65	4,225	8.062	274,625	4.021
16	256	4.000	4,096	2.520	66	4,356	8.124	287,496	4.041
17	289	4.123	4,913	2.571	67	4,489	8.185	300,763	4.062
18	324	4.243	5,832	2.621	68	4,624	8.246	314,432	4.082
19	361	4.359	6,859	2.668	69	4,761	8.307	328,509	4.102
20	400	4.472	8,000	2.714	70	4,900	8.367	343,000	4.121
21	441	4.583	9,261	2.759	71	5,041	8.426	357,911	4.141
22	484	4.690	10,648	2.802	72	5,184	8.485	373,248	4.160
23	529	4.796	12,167	2.844	73	5,329	8.544	389,017	4.179
24	576	4.899	13,824	2.884	74	5,476	8.602	405,224	4.198
25	625	5.000	15,625	2.924	75	5,625	8.660	421,875	4.217
26	676	5.099	17,576	2.962	76	5,776	8.718	438,976	4.236
27	729	5.196	19,683	3.000	77	5,929	8.775	456,533	4.254
28	784	5.292	21,952	3.037	78	6,084	8.832	474,552	4.273
29	841	5.385	24,389	3.072	79	6,241	8.888	493,039	4.291
30	900	5.477	27,000	3.107	80	6,400	8.944	512,000	4.309
31	961	5.568	29,791	3.141	81	6,561	9.000	531,441	4.327
32	1,024	5.657	32,768	3.175	82	6,724	9.055	551,368	4.344
33	1,089	5.745	35,937	3.208	83	6,889	9.110	571,787	4.362
34	1,156	5.831	39,304	3.240	84	7,056	9.165	592,704	4.380
35	1,225	5.916	42,875	3.271	85	7,225	9.220	614,125	4.397
36	1,296	6.000	46,656	3.302	86	7,396	9.274	636,056	4.414
37	1,369	6.083	50,653	3.332	87	7,569	9.327	658,503	4.431
38	1,444	6.164	54,872	3.362	88	7,744	9.381	681,472	4.448
39	1,521	6.245	59,319	3.391	89	7,921	9.434	704,969	4.465
40	1,600	6.325	64,000	3.420	90	8,100	9.487	729,000	4.481
41	1,681	6.403	68,921	3.448	91	8,281	9.539	753,571	4.498
42	1,764	6.481	74,088	3.476	92	8,464	9.592	778,688	4.514
43	1,849	6.557	79,507	3.503	93	8,649	9.644	804,357	4.531
44	1,936	6.633	85,184	3.530	94	8,836	9.695	830,584	4.547
45	2,025	6.708	91,125	3.557	95	9,025	9.747	857,375	4.563
46	2,116	6.782	97,336	3.583	96	9,216	9.798	884,736	4.579
47	2,209	6.856	103,823	3.609	97	9,409	9.849	912,673	4.595
48	2,304	6.928	110,592	3.634	98	9,604	9.899	941,192	4.610
49	2,401	7.000	117,649	3.659	99	9,801	9.950	970,299	4.626
50	2,500	7.071	125,000	3.684	100	10,000	10.000	1,000,000	4.642

INDEX